Fundamentals of Communication Systems : Theory, Video Lectures, MATLAB and Mathcad Simulations

Janak Sodha

Janak Sodha received his B.Sc. (Hons.), M.Sc. and Ph.D. from the University of Manchester, England (UK). A telecommunications research engineer with ERA Technology (UK) for two years and currently a senior lecturer at the University of the West Indies, Cave Hill Campus, Barbados. This book is based on the experience of lecturing on communication theory for over eighteen years. Refer to the author's home page on http://janaksodha.com for details of his publications and search for the "Communications Systems" app for your smartphone or tablet using the author's surname.

Fundamentals of Communication Systems : Theory, Video Lectures, MATLAB and Mathcad Simulations
Copyright © 2015 AppBooke Publishing (UK)

ISBN 978-0-9928510-0-2

Contents

Preface

INTRODUCTION

The aim of this book is to provide a unique self-learning environment on the fundamentals of a communication system. The key features of this book are as follows:

Theory with mathematical steps: Traditional text books on the subject of communication systems take an encyclopedic approach, embed the essence of a given algorithm within a mathematical shell, present various performance graphs from other sources and nearly always leave out the inner-working details because these must be obvious. Without an excellent mathematical background, a student is left wondering on outer-edges of a *theory-tree* with *leaves* of equations, in the hope to identify the *branches* which should lead to the *fruit*. Ultimately, a student only makes significant progress with the help of a lecturer via physical lectures in a University. To combat this traditional difficulty, this book is written in the style of lecture notes **with all the mathematical steps**. The level of mathematics is developed throughout the book and for complex derivations, extra care has been taken to show all the steps. The appendix on probability theory assumes no prior knowledge. With simulations to illustrate each concept, you will gain a solid foundation in probability theory. Furthermore, **all the theory presented in this book** is verified via Mathcad and MATLAB simulations. By relating the simulation results to the theory (e.g. experimenting with the parameters of the simulation to observe the effect on the simulation results), one gains a much deeper understanding - together with a unique experience of the material - then via any other similar textbook. This experience will also help to develop your research skills.

Video Lectures: How many times have you read several paragraphs of a text book only to think to yourself, *"I wish the author would simply* *illustrate what this equation means. It took me over an hour to understand it!"*. And as if by magic, your wish has been granted! Just look for the video icon within a given chapter. Bullet points under the video icon indicate the items covered in the lecture. For example:

▶ **Fourier Transform**

- The effect of increasing the time period
  between each pulse

- Fourier transform fundamentals

- Theoretical analysis of a rectangular pulse

- Reconstruction of the pulse using a range of frequencies

Within each lecture, you will hear my voice and observe me sketch, derive and explain the theory. The experience will be similar to me sitting next to you as we explorer the wonders of a communication system on a "sketch-pad". To view the lectures, download the **app** "Communication Systems" from the Apple Store or Google Play Store for your smartphone or Tablet.

Mathcad & MATLAB Simulations: How many times have you fought your way through pages of mathematics wondering, *"wouldn't it be wonderful if the theory came to life, giving me the ability to see it action, experiment with parameters, underlying assumptions, etc. "*. You guessed it, your second wish has been granted! Nearly every aspect of all the theory presented in this entire book is simulated **both** in Mathcad and MATLAB. Just look for

the *simulation icon* ⭐ within any chapter or problems section. Next to this icon will be summary of the simulation contents. If you take the time to work through a given simulation, your understanding of the subject material will rapidly increase and you will gain the confidence to experiment with a given problem. The simulation code can be altered to investigate variations of a given algorithm or simulation conditions that would otherwise require a very high degree of mathematical ability to analyze.

- Mathcad-MATLAB simulations are embedded throughout a given section. At the end of a given section is a "Simulation" subsection, which combines the previous simulations in that section with added code to model the theory presented in that section. Therefore, its important that you work through the Mathcad/MATLAB simulations in order (as far as possible) unless you are an experienced programmer.

- Links to the Mathcad and MATLAB simulations are within the app "Communication Systems" available from the Apple Store or Google Play Store.

Interactive Problems and Solutions: How many times have you tried your level-best to work through a problem at the end of a chapter, wondering *"why doesn't my calculations match the numbers at the back of the book .....arrgh ... is it yet another typo! Wouldn't it be nice to see the full solution instead of a number and better still, what if I could watch the author solve the problem for me so that I may experience the thinking process"*. Well, your third and last wish has been granted! In addition to a full solutions manual, you can watch me solve selected problems. Once again, the experience will be similar to me sitting next to you as we work our way through a given problem. I have provided the corresponding Mathcad and MATLAB solutions to certain key problems so that you may experiment with the parameters of the problems, plot graphs, etc. to gain a further insight into the problem and its implications. You can make simple modifications to the existing code to create different problems. *Of course please do not cheat yourself from the learning process by viewing the full solution without first making a significant attempt.* Please note that I have chosen to highlight certain features of a given theory via the problems and solutions sections. The aim being to discover **new** and interesting features within the context of a problem. I hope the combination of video solutions and Mathcad/MATLAB simulations will encourage you to experiment, research and strive for excellence.

- Solutions manual available for purchase. Search online stores under the author's surname.

Updates: Finally, we have all come across text books which contain those annoying typos, etc. In particular, those moments when you wish you could strangle the author for not inserting extra lines of explanation or mathematical equations that took you hours to realize! Fortunately, I live in the Caribbean on a remote part of an island with two German shepherd dogs to protect me from such moments. However, my aim is to help you save those hours of pain by listening to your feedback and making updates available via the Communication Systems app which contains the form to contact me.

YOUR CONTRIBUTION

My fellow colleagues, please take the time to send me feedback on specific sentences which you feel may mislead a student or could be improved, or additional sentences, mathematics, diagrams that could be inserted to improve the overall quality. Of course, typos and simple updates (additional mathematics, sentences, restructuring of sentences, etc.,) will be given a very high priority. My aim is to make such updates available within a couple of weeks from the time it is brought to my attention.

There are several topics that I would still like to include (e.g. wireless communications), but its almost impossible to cover all the topics that would be a part of an ideal communication systems text book. Particularly because of the underlying methodology adopted which requires the theory to be verified by simulations, together with video lectures and solutions. Please, help me to help our students by contributing the material (in-depth analysis, subtopics, simulations, examples, problems with solutions) you wish I had included in this book. The material you present will be reviewed and if appropriate, be made available for download via the Communication Systems app and included in the next version of this textbook. Your name and brief biography will be stated at the beginning of your "subsection" or "section". *Together, we can create a unique book that will lay the foundations for a new generation of multimedia books which combine the efforts of professionals around the world.*

LEARNING METHODOLOGY

It is only by interacting with a given theory/problem/etc. via as many different means as possible can we understand its essence. As Einstein once said, "true knowledge is experience!".

1. View a given video lecture and read the corresponding section.

2. Run the Mathcad/MATLAB simulations to reinforce and clarify your understanding. Use the PDF file to understand the simulation code and to relate the graphs in the text book to the simulations. Advanced students, please take the time to modify the simulation code as suggested within a given simulation section. This task will give you an excellent introduction into the wonderful world of research, which I hope will encourage you to undertake a M.Sc./M.Phil./Ph.D. postgraduate degree in the field of Digital Communications.

3. Work through the problems at the end of the chapter. If you have any difficulty with a given problem, then watch my video solution if available. You can obviously opt to watch only the very beginning of the video to help you understand the problem and gain an insight as to how you may solve this problem. Please take the time to look at the corresponding Mathcad/MATLAB solution if available. Experiment with the variables to get a feel for what happens under different conditions.

4. Please send me feedback! You can help me to remove typos, ambiguities, emphasize key points, request for a derivation to be inserted, suggest hyperlinks to useful web sites, etc. Any such improvements will be made available via the Communication Systems app. Why should we all "fall-over" the same missing comma, mathematics, etc. when a quick update can help to relieve the stress of learning. Please help me make available the updates from which we all benefit.

A YOUNG TREE

We shall begin by climbing the *trunk* of our young communications *tree*. Within a given chapter or *limb*, we shall move along some of the *branches*. The overall goal is to make our way to one of the many outer edges of the tree (e.g. OFDM). Take the time to view the video lectures linked within this book. Experiment with the simulation code to fully grasp the concepts or *twigs*. I don't want you to fall! If while climbing, a sweet taste appears in your mouth, then its because you have fought your way through the details or *leaves* and tasted the fruit. As you make your way through the tree, I hope you will come to appreciate its unique nature and encourage me to add other branches accessible either via the app or a later edition. If you come across twigs which are weak, or if the fruit tastes bitter, then **please send me feedback**. My ongoing goal is to gradually mature this tree into full bloom with the incorporation of several other chapters and appendices.

MOTIVATION

Communication is part of our inherent nature. Limited to interact with the universe with our five senses, we use our ears, eyes, tongue, nose and skin, to gather information from our surroundings. With our senses, we communicate with other persons. For example, two persons talking together form a communication system in which the person speaking is the transmitter, the person listening is the receiver and the air in between is the channel through which information is conveyed via sound waves. From the early days of cave drawings, through to the use of paper and now together with electronic devices, communication has been and will always continue to be essential for the growth of mankind. Unfortunately, we have been successful in communicating information over an electronic communication system using only two of our senses (eyes and ears). Examples include, telephone, radio, television, computer network, cellular network, etc.

An ideal communication system would be fast, reliable, economical, efficient and mobile, with access to information from sources (humans, computers, etc.) around the world. Although mankind is far from achieving all these goals, we do live in a world of networks, created with the aid of devices which enable communication via wires, fiber optic cables, satellites in space, etc., to exchange information locally, or from another part of the world. Ongoing developments continue to concentrate on the goals of an ideal communication system, with a focus on the integration of different types of information to utilize our world of networks. For example, a cell phone can be used for the capture and communication of voice, photograph and video, radio (receiver and transmitter), music, internet access (browsing, email), and GPS tracking. More recently, applications running on various operating systems provide the everyday laptop features (e.g. games, banking, communication with hardware devices over the internet) with the added advantage of unique applications that take advantage of the GPS and motion sensing ability. We can expect the further integration of such smartphones with other hardware (e.g. robotics) in the near future.

BOOK OUTLINE

In chapter 1, the fundamentals of signals and Fourier analysis are presented. Armed with the tools necessary to analyze a signal in the time, frequency, energy and power domains, analog communication techniques are considered in chapter 2. Thereafter, we shift our focus to digital communication systems, with a first look at the fundamentals of a digital communication system (DCS) in chapter 3. In chapter 4, we rise above the details of a DCS to consider the minimum average number of binary digits per second required to fully represent a digital source and the maximum rate at which error-free communication can take place over a given channel. The conversion of an analog signal into a digital signal to enable its transfer over a DCS is considered in Chapter 5. Chapters 6 and 7 deal with the transfer of binary digits over a baseband and a bandpass channel via various types of modulation techniques, respectively. The remaining chapters concentrate on error-control coding, because of its key importance in the overall design of a DCS. A good error-control coding scheme can overcome the short comings of the modulator, channel and demodulator by embedding the information binary digits within a code structure that is exploited to ensure essentially error-free communication. Starting with block and convolutional codes in chapter 8, we then consider a combined error-control coding and modulation technique. In this chapter, we also pause to view the performance of all the communication systems considered in the previous chapters in the light of Shannon's channel capacity theorem.

BOOK APP

The **Communication Systems** App is available on the Apple and Google Play Stores. Search using the author's surname "Sodha". It provides access to book updates and the following:

	Multimedia	Number
	Video Lectures	+100
	Mathcad Simulations	+300
	MATLAB Simulations	+300

For my Mother and Father, Sita and Renuka

Chapter 1

Signals

1.1 Introduction

In a communication system, information is transmitted from the source to the destination via *signals* which represent the variation of measurable quantities (e.g. voltage, current, light intensity, pressure, etc.) with time t. Everything physical is simply energy in different forms. For example, what appears to be a blade of grass is a concentration of energy (collection of atomic particles e.g. electrons which are packets of energy) that is the same as the energy within the electric field of an electromagnetic wave. Given that energy is required to transfer information, it is critical that we understand the distribution of this energy within a signal. Thus, the aim of this chapter is to arm ourselves with the tools necessary to analyze signals in the time, frequency, energy and power domains.

1.1.1 Signals : Energy and Power in the Time Domain

▶ Periodic Signal

- A quick look at the features of a sinusoidal signal

- Energy and power signals

A signal $s(t)$ is *periodic* if $s(t + T) = s(t)$, where T is the *time period*. Its average value $\langle s(t) \rangle$ is given by

$$\langle s(t) \rangle = \frac{1}{T} \int_T s(t)dt \tag{1.1}$$

In general, the *energy* E and the *power* (time average of the energy) P of a signal are given by

$$E = \int_{-\infty}^{\infty} |s(t)|^2 \, dt \tag{1.2}$$

$$P = \lim_{T \to \infty} \frac{1}{T} \int_{\frac{-T}{2}}^{\frac{T}{2}} |s(t)|^2 \, dt \tag{1.3}$$

where for real signals, $|s(t)|^2$ is simply $s^2(t)$. If $0 < E < \infty$, then $P = 0$ and $s(t)$ is said to be an *energy signal*. However, if $E = \infty$, then $0 < P < \infty$ and $s(t)$ is said to be a *power signal*. Note that a signal cannot be both an energy and a power signal. Furthermore, there are signals which are neither energy or power signals. If the signal is periodic, then P is simply given by

$$P = \frac{1}{T} \int_T |s(t)|^2 \, dt \tag{1.4}$$

For example, for $s(t) = A \cos\left(\frac{2\pi t}{T}\right)$, Fig. 1.1 shows the variation of $s(t)$, $|s(t)|$ and $|s(t)|^2$ with $A = 2$ volts and $T = 2$ secs. Since E is the area under the $|s(t)|^2$ curve from $-\infty$ to $+\infty$, this is not an energy signal because $E = \infty$. By inspection, we expect the average value of $|s(t)|^2$ versus t from Fig. 1.1 to be $\frac{A^2}{2}$. Namely, the average power of this signal is given by

$$P = \lim_{T \to \infty} \frac{1}{T} \int_{\frac{-T}{2}}^{\frac{T}{2}} |s(t)|^2 \, dt = \frac{1}{T} \int_0^T A^2 \cos^2\left(\frac{2\pi t}{T}\right) dt = \frac{A^2}{2} \tag{1.5}$$

Since $0 < P < \infty$, this signal is a power signal.

- ⭐ SIMULATION **Signals:** Other examples of energy and power signals. In each case, the signal power and energy is evaluated using equations 1.2 and 1.3. Note that for a d.c. term, e.g. $s(t) = A$, the power of this signal is $\frac{1}{T}\int\limits_0^T A^2 dt = \frac{A^2}{T}\int\limits_0^T dt = A^2$.

1.1.2 Signal Power via Digital Samples and its Probability Histogram

Given that the average value $\langle f(x) \rangle$ of an integrable function $f(x)$ is given by

$$\langle f(x) \rangle = \frac{1}{b-a}\int\limits_a^b f(x)dx \tag{1.6}$$

a closer look the expression for the power of a periodic signal $P = \frac{1}{T}\int\limits_T |s(t)|^2 dt$ reveals not only that P is the average value of the function $|s(t)|^2$, but also if the signal $s(t)$ is sampled at a frequency of f_s samples per second for a total of N times, then to a good approximation

$$P \simeq \frac{1}{N}\sum_{n=0}^{N-1} |x_n|^2 \tag{1.7}$$

where the instantaneous digital samples $x_n = s\left(\frac{n}{f_s}\right)$ and $n = 0, 1, \cdots N - 1$. For example x_0, x_1, x_2, x_3 are given by $s\left(t = 0\right), s\left(t = \frac{1}{f_s}\right), s\left(t = \frac{2}{f_s}\right)$ and $s\left(t = \frac{3}{f_s}\right)$, respectively. Sampling is presented in greater detail in section

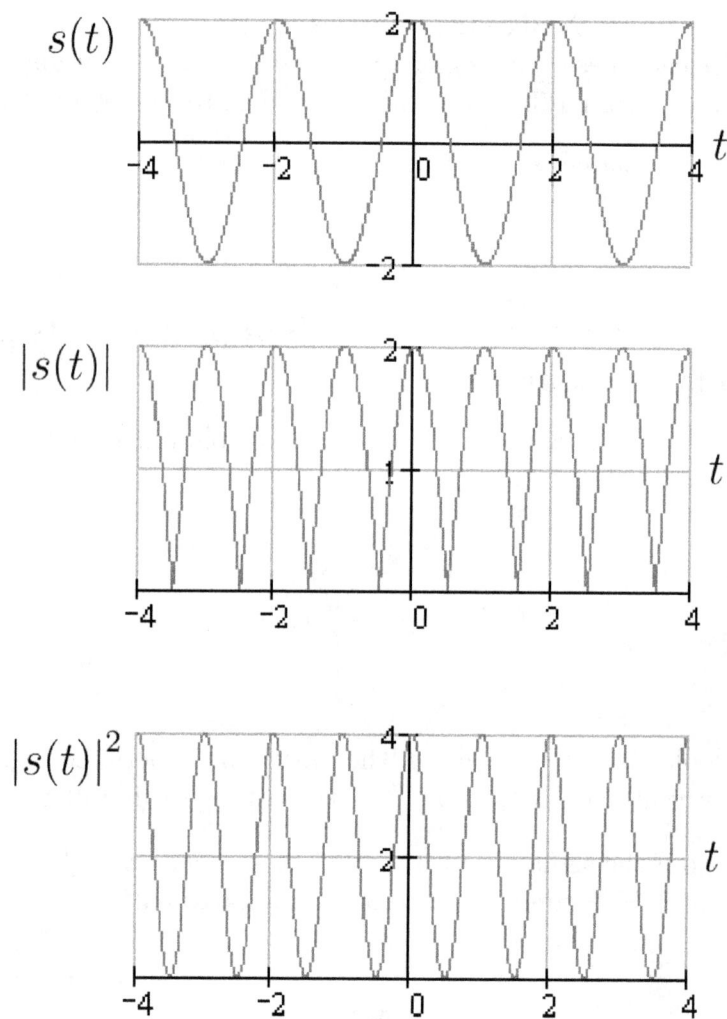

Figure 1.1: Power signal $s(t) = 2\cos(\pi t)$.

1.4.3 with supporting videos to determine the spectral components of an analog signal $s(t)$ using its digital samples x_n. If a value of x_n is equally likley, so that the probability $P(x_n)$ of x_n is $\frac{1}{N}$, then the average value of x_n is $\frac{1}{N}\sum_{n=0}^{N-1} x_n \equiv \sum_{n=0}^{N-1} P(x_n)x_n$ and from equation 1.7, $P \simeq \sum_{n=0}^{N-1} P(x_n)|x_n|^2$. Furthermore, the signal power P can be determined from its probability histogram (Appendix A.8) via

$$P \simeq \sum_{j=0}^{N_{bin}-1} P_j^{\text{bin}}\left(x_j^{\text{mid}}\right)^2 \tag{1.8}$$

where P_j^{bin} is the probability with which x_n falls within a given bin whose midpoint horizontal coordinate is x_j^{mid}.

- ⭐ SIMULATION **SinusoidalPDF**: The power of the signal $s(t) = A\cos\left(\frac{2\pi t}{T}\right)$ is determined via its digital samples and its probability histogram. Furthermore, you may select a signal with a uniform PDF to verify, as shown in section 5.2, that its average signal power is given by $\frac{A^2}{3}$.

1.1.3 Phasors

▶ **Phasors**

- Phasor representation of a sinusoidal signal

To develop an efficient communication system for a given physical channel, it is necessary to analyze a given signal in the frequency domain even though signals are inherently represented in the time domain. Let us begin by considering the sinusoidal signal $s(t) = A\cos\left(2\pi f t + \phi\right)$ as shown in Fig. 1.2, where the amplitude $A = 1$ volt, frequency $f = 0.5$ Hz, time period

$T = 1/f = 2$ seconds and the initial phase ϕ is 0 degrees. This signal can alternatively be represented by a *phasor* of length A, rotating *anti-clockwise* with an angular frequency $\omega = 2\pi f$ on a complex plane with a *imaginary* and *real* axis as illustrated in Fig. 1.3. Mathematically, this rotating phasor is represented as $Ae^{j(2\pi ft+\phi)}$.

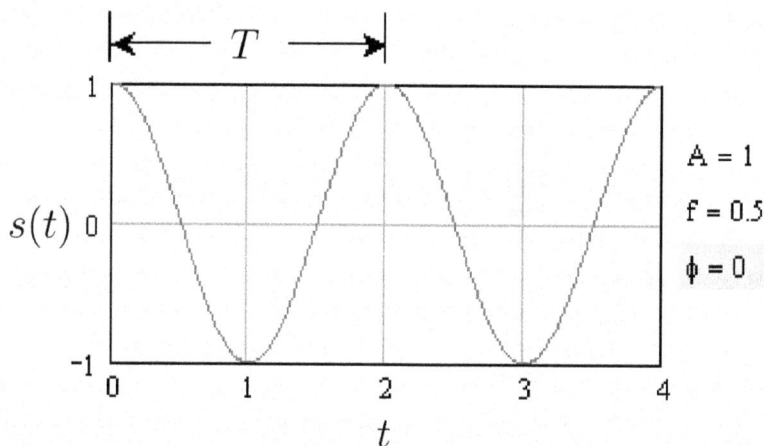

Figure 1.2: Sinusoidal signal.

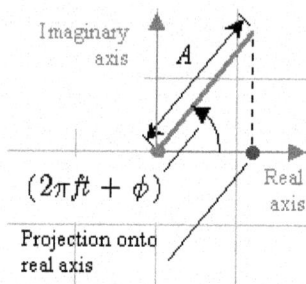

Figure 1.3: Phasor representation $Ae^{j(2\pi ft+\phi)}$ of the sinusoidal signal $s(t) = A\cos(2\pi ft + \phi)$.

Using *Euler's theorem* $e^{\pm j\theta} = \cos\theta \pm j\sin\theta$ where $j \triangleq \sqrt{-1}$, the signal $s(t)$ is the real part of the expression $Ae^{j(2\pi ft+\phi)}$, namely

$$s(t) = A \cos\left(2\pi f t + \phi\right) = \operatorname{Re}\left[A e^{j(2\pi f t + \phi)}\right] \tag{1.9}$$

Equivalently, the projection from the tip of the phasor onto the real axis is equivalent to the signal $s(t)$. Note that at time $t = 0$, the phasor makes an angle ϕ with respect to the real axis. To avoid having to extract the real part of a phasor, we can make use of two phasors which rotate in the opposite directions as shown in Fig. 1.4. The vector of length $\frac{A}{2}$ rotating in the anti-clockwise direction represents a *positive* frequency f and the vector of length $\frac{A}{2}$ rotating in the clockwise direction represents a *negative* frequency f. The resultant vector is given by

$$\frac{A}{2}e^{j(2\pi f t + \phi)} + \frac{A}{2}e^{j(2\pi(-f)t + (-\phi))} = A \cos\left(2\pi f t + \phi\right) \tag{1.10}$$

where we made use of the identity $\frac{e^{j\theta} + e^{-j\theta}}{2} = \cos\theta$.

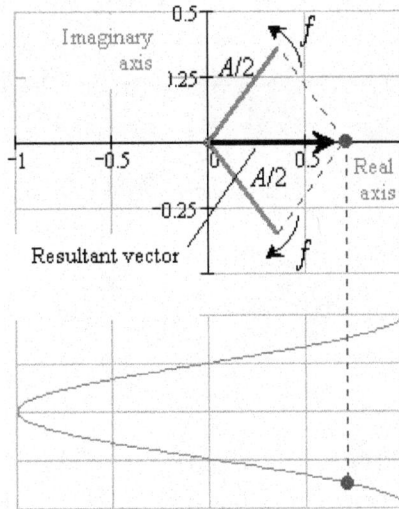

Figure 1.4: Two phasor representation of a sinusoidal signal.

1.1.4   Spectrum

▷ **Spectrum**

- The concept of an amplitude and phase spectrum

- Mathematical representation of spectral components

Based on the concept of creating a cosine waveform using two phasors rotating in opposite directions, Fig. 1.5 illustrates how the signal $s(t) = \cos\left(\pi t + \frac{\pi}{4}\right)$ can be represented in the frequency domain. The diagrams in part (a) and (b) are respectively referred to as the *amplitude* and *phase spectrum*. The two vertical lines in Fig. 1.5(a) represent two phasors, each of length $\frac{A}{2} = 0.5$ rotating in opposite directions as in Fig. 1.4 i.e. a pair of *conjugate* phasors. The phasors in a given spectrum are often referred to as the *spectral components* of the signal. From the spectrum in Fig. 1.5, we have

$$s(t) = 0.5e^{j\left(2\pi(-0.5)t - \frac{45\pi}{180}\right)} + 0.5e^{j\left(2\pi(0.5)t + \frac{45\pi}{180}\right)}$$

$$= 0.5\left[e^{-j\left(2\pi(0.5)t + \frac{45\pi}{180}\right)} + e^{j\left(2\pi(0.5)t + \frac{45\pi}{180}\right)}\right] \tag{1.11}$$

However since $e^{-j\theta} + e^{j\theta} = 2\cos\theta$, $s(t)$ may be simplified to

$$s(t) = \cos\left(2\pi(0.5)t + \frac{45\pi}{180}\right) = \cos\left(\pi t + \frac{\pi}{4}\right) \tag{1.12}$$

A standard amplitude spectrum is *double-sided* with both positive and negative frequencies as shown in Fig. 1.5. However, if the single phasor representation of Fig. 1.3 is used, then the amplitude spectrum in this case would be a *single-sided* spectrum in which only positive frequencies are represented. The single-sided spectrum of $s(t) = A\cos\left(2\pi ft + \phi\right)$ would be a vertical line of amplitude A at frequency f.

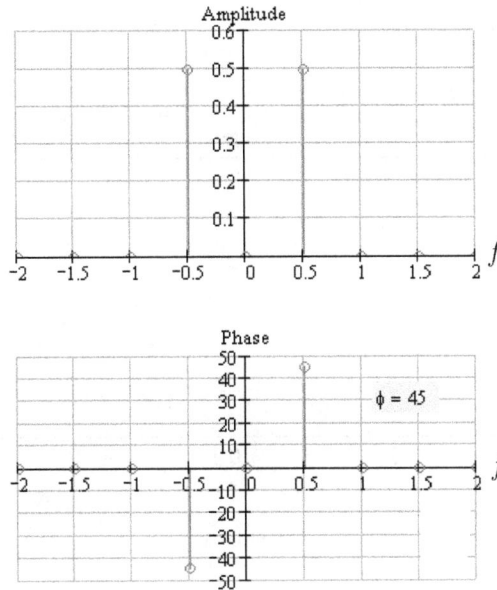

Figure 1.5: (a) Amplitude and (b) phase spectrum of the signal $\cos\left(\pi t + \frac{\pi}{4}\right)$.

Example: Reading a spectrum

Verify that the spectrum presented in Fig. 1.6 represents the signal $s(t) = 5 - 6\cos\left(20\pi t - \frac{\pi}{6}\right) + 3\sin\left(70\pi t\right)$ and draw the corresponding single-sided spectrum of this signal.

Solution

Since the phase angles are measured with respect to *cosine* waves, a sine wave is represented by a cosine wave using the trigonometry identity $\sin\theta = \cos(\theta - 90)$, where θ is in degrees. We shall switch between degrees and radians as required, where 2π rads $= 360$ degrees. Reading the amplitude and phase spectrum shown in Fig. 1.6 from left to right,

$$s(t) = 1.5e^{j\left(2\pi(-35)t + \frac{90\pi}{180}\right)} + 3e^{j\left(2\pi(-10)t - \frac{150\pi}{180}\right)}$$

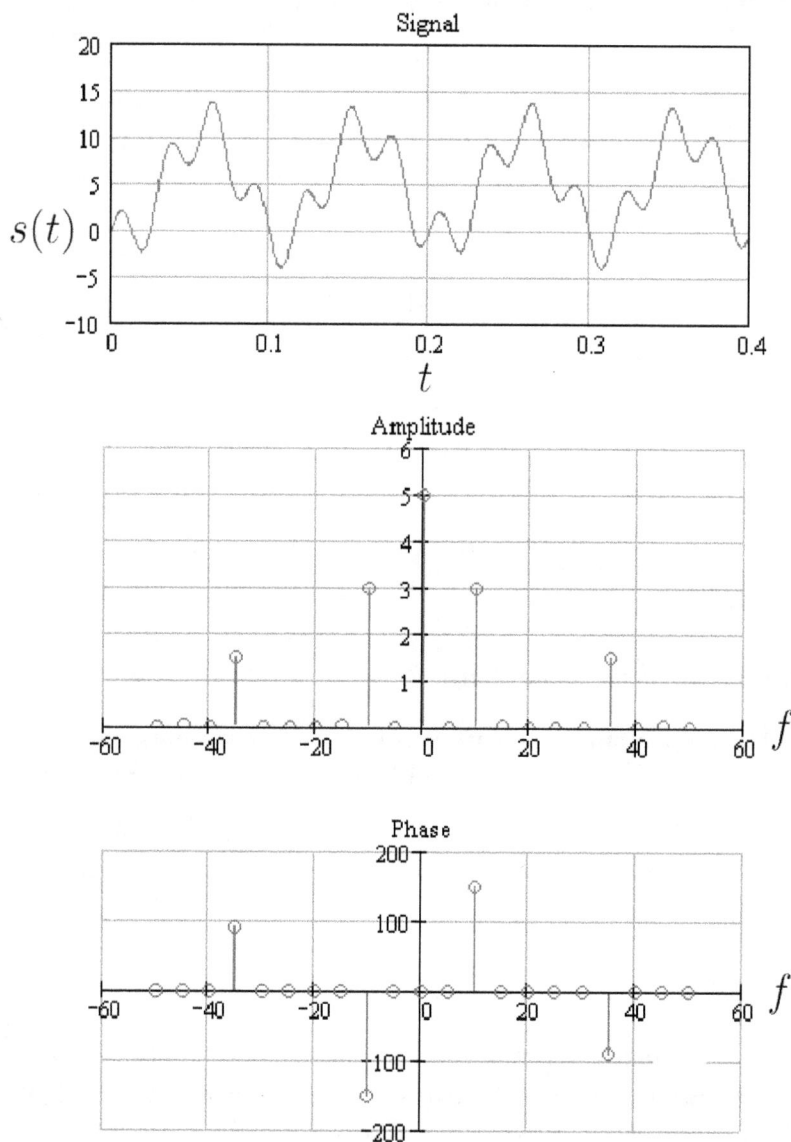

Figure 1.6: Signal and its amplitude and phase spectrum.

$$+5e^{j\left(2\pi(0)t+\frac{0\pi}{180}\right)} + 3e^{j\left(2\pi(10)t+\frac{150\pi}{180}\right)} + 1.5e^{j\left(2\pi(35)t-\frac{90\pi}{180}\right)}$$

$$= 5 + 3\left[e^{j\left(2\pi(10)t+\frac{150\pi}{180}\right)} + e^{-j\left(2\pi(10)t+\frac{150\pi}{180}\right)}\right] +$$

$$1.5\left[e^{j\left(2\pi(35)t-\frac{90\pi}{180}\right)} + e^{-j\left(2\pi(35)t-\frac{90\pi}{180}\right)}\right] \tag{1.13}$$

Simplifying this expression using the identities $e^{j\theta} = \cos\theta + j\sin\theta$, $e^{j\theta} + e^{-j\theta} = 2\cos\theta$, $\cos(\theta) = -\cos(\theta - \pi)$ and $\cos\left(\theta - \frac{\pi}{2}\right) = \sin\theta$, we find

$$s(t) = 5 + 6\cos\left(2\pi(10)t + \frac{150\pi}{180}\right) + 3\cos\left(2\pi(35)t - \frac{90\pi}{180}\right) \tag{1.14}$$

$$= 5 - 6\cos\left(20\pi t + \frac{150\pi}{180} - \pi\right) + 3\cos\left(70\pi t - \frac{\pi}{2}\right) \tag{1.15}$$

$$= 5 - 6\cos\left(20\pi t - \frac{\pi}{6}\right) + 3\sin(70\pi t) \tag{1.16}$$

The single-sided spectrum, which represents only positive frequencies, is found from the double-sided spectrum by folding over the negative frequency amplitudes and adding them to the positive frequency amplitudes, as shown in Fig. 1.7.

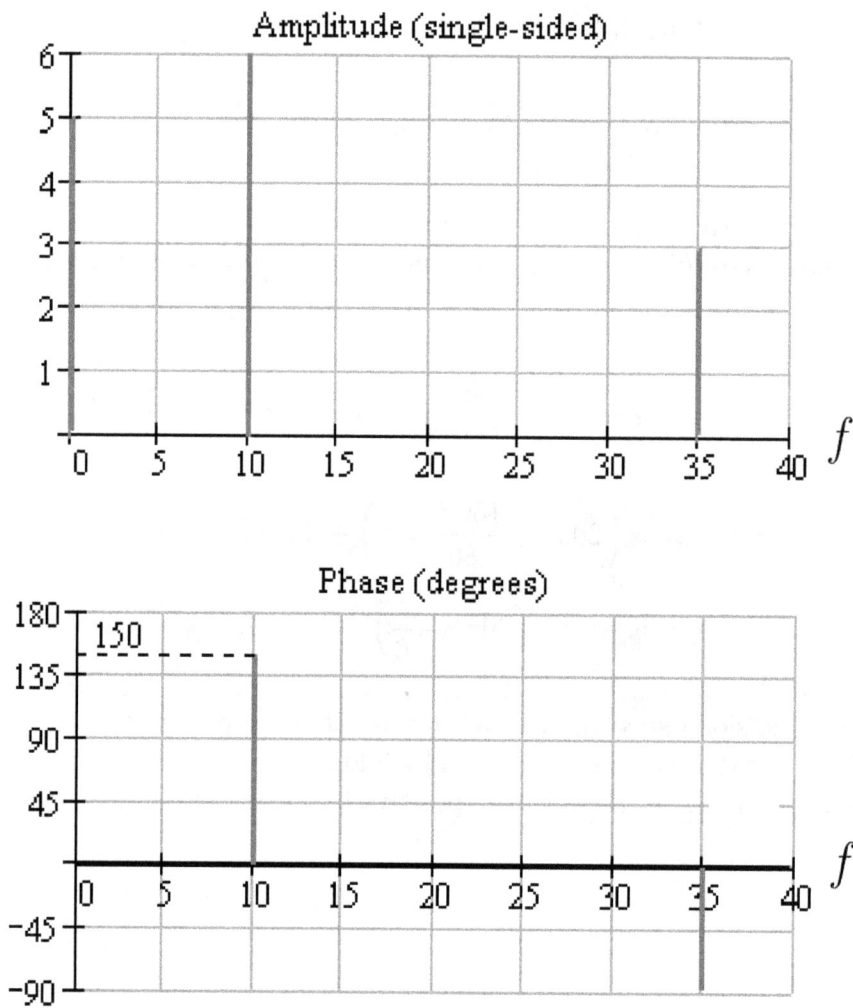

Figure 1.7: Single-sided spectrum.

1.2 Fourier Series

Fourier discovered that almost any *periodic* signal $s(t)$ of time period T can be reconstructed by adding together the appropriate sinusoidal waveforms. For example, the following series of cosines when added together will form a waveform that looks very similar to a square wave. As more terms are added, $s(t)$ will look more like a square wave.

$$s(t) = \cos\left(\frac{2\pi t}{T}\right) - \frac{1}{3}\cos\left(3\left(\frac{2\pi t}{T}\right)\right) + \frac{1}{5}\cos\left(5\left(\frac{2\pi t}{T}\right)\right) - \cdots \quad (1.17)$$

- ⭐ **SIMULATION** **Series**: Plot of $s(t)$ in which you may increase the number of terms in the series to observe how the shape of $s(t)$ changes. You will need an infinite number of terms to create a perfect square wave!

In general, any periodic signal of period T can be created using cosine and sine terms, referred to also as *components*, as follows

$$s(t) = A_0 + \sum_{n=1}^{\infty} A_n \cos\left(\frac{2\pi n t}{T}\right) + \sum_{n=1}^{\infty} B_n \sin\left(\frac{2\pi n t}{T}\right) \quad (1.18)$$

where $A_0 = \frac{1}{T}\int_T s(t)dt$ is the mean value or d.c. term of $s(t)$ and A_n and B_n are the amplitudes of the cosine and sine harmonics, respectively, referred to as the Fourier series *coefficients*. The integral $\int_T$ means that the integration is performed over any interval of T seconds e.g. $\int_{-T/2}^{T/2}$ or $\int_0^T$, etc.

If the signal $s(t)$ has *even* symmetry (signal to the left of origin is the mirror image of the signal to the right of the origin $t = 0$) then $B_n = 0$ and $A_n \neq 0$. Conversly, if the signal has *odd* symmetry (signal to the left of origin is the inverted mirror image of the signal to the right of the origin) then $B_n \neq 0$

and $A_n = 0$. The coefficients A_n and B_n, for $n = 1, 2, 3 \ldots$ are given by the *sifting* integrals

$$A_n = \frac{2}{T} \int_T s(t) \cos \left(\frac{2\pi nt}{T} \right) dt \qquad (1.19)$$

$$B_n = \frac{2}{T} \int_T s(t) \sin \left(\frac{2\pi nt}{T} \right) dt$$

where for example, $A_1 = \frac{2}{T} \int_T s(t) \cos \left(\frac{2\pi t}{T} \right) dt$ would be equal to 1 if $s(t) = \cos \left(\frac{2\pi t}{T} \right)$.

1.2.1   Exponential Fourier Series

In equation 1.18, given that the components at a single harmonic frequency are $A_n \cos \left(\frac{2\pi nt}{T} \right) + B_n \sin \left(\frac{2\pi nt}{T} \right)$, and making use of the relationships $\cos \theta = \frac{e^{j\theta} + e^{-j\theta}}{2}$ and $\sin \theta = \frac{e^{j\theta} - e^{-j\theta}}{2j}$, it is easily shown that

$$A_n \cos \left(\frac{2\pi nt}{T} \right) + B_n \sin \left(\frac{2\pi nt}{T} \right) = c_n e^{j\frac{2\pi nt}{T}} + c_{-n} e^{-j\frac{2\pi nt}{T}} \qquad (1.20)$$

where the *complex coefficient* $c_n = \frac{A_n - jB_n}{2}$ and c_{-n} is the complex conjugate $\frac{A_n + jB_n}{2}$, such that

$$c_n = \frac{A_n - jB_n}{2} \qquad (1.21)$$

$$= \frac{1}{2} \left[\begin{array}{c} \left(\frac{2}{T} \int_T s(t) \cos \left(\frac{2\pi nt}{T} \right) dt \right) \\ -j \left(\frac{2}{T} \int_T s(t) \sin \left(\frac{2\pi nt}{T} \right) dt \right) \end{array} \right] \qquad (1.22)$$

$$= \frac{1}{T} \int_T s(t) \left[\cos \left(\frac{2\pi nt}{T} \right) - j \sin \left(\frac{2\pi nt}{T} \right) \right] dt \qquad (1.23)$$

$$= \frac{1}{T} \int_T s(t) e^{\frac{-j2\pi nt}{T}} dt \qquad (1.24)$$

▶ **Fourier Series**

- Using the example of a rectangular
  periodic signal to introduce the Fourier Series

Hence, the exponential form of the Fourier series is given by

$$s(t) = \sum_{n=-\infty}^{\infty} c_n e^{\frac{j2\pi nt}{T}} \qquad (1.25)$$

where c_n is a complex coefficient given by

$$c_n = \frac{1}{T} \int_T s(t) e^{\frac{-j2\pi nt}{T}} dt \qquad (1.26)$$

Notice how the signal consists of phasors at frequencies $\frac{n}{T} = 0, \pm\frac{1}{T}, \pm\frac{2}{T}, \pm\frac{3}{T}, \cdots$ on a double-sided spectrum where the n_{th} harmonic phasor amplitude is $|c_n|$ and the phase angle is $\arg(c_n)$. This standard function $\arg(z)$ available within both mathcad and MATLAB is the angle θ in radians of the complex number $z = |z| e^{j\theta} = x + jy$ given by

$$\arg(z) = \begin{cases} \tan^{-1}\left(\frac{y}{x}\right) & if & x > 0 \\ \frac{\pi}{2} - \tan^{-1}\left(\frac{x}{y}\right) & if & y > 0 \\ -\frac{\pi}{2} - \tan^{-1}\left(\frac{x}{y}\right) & if & y < 0 \\ \pi + \tan^{-1}\left(\frac{y}{x}\right) & if & x < 0 \text{ and } y \geq 0 \\ -\pi + \tan^{-1}\left(\frac{y}{x}\right) & if & x < 0 \text{ and } y < 0 \\ \text{Undefined} & if & x = 0 \text{ and } y = 0 \end{cases} \qquad (1.27)$$

where $y = \text{Im}(z)$ and $x = \text{Re}(z)$. Equivalently, the signal $s(t)$ is given by

$$s(t) = \sum_n |c_n|\, e^{j\left(\frac{2\pi nt}{T} + \arg c_n\right)} \tag{1.28}$$

where $c_n = |c_n|\, e^{j\,\arg c_n}$. If $\arg c_n = 0$, then c_n will be a real positive value and if $\arg c_n = \pm\pi$, then c_n is a real negative value. Alternatively, in terms of a single-sided spectrum, since the amplitude of a given phasor will be equal to $2\,|c_n|$, the signal $s(t)$ is also given by

$$s(t) = c_o + \sum_{n=1}^{\infty} 2\,|c_n| \cos\left(\frac{2\pi nt}{T} + \arg c_n\right) \tag{1.29}$$

where c_o is the average value or d.c. (0 Hz) component of $s(t)$. Finally, on a double-sided spectrum, $|c_n| = |c_{-n}|$ and that the amplitude and phase spectrum will always have even and odd symmetry, respectively, because we are dealing conjugate phasors.

1.2.2   Example : Periodic Rectangular Pulse Waveform

Periodic Rectangular Signal

- Theoretical analysis using the exponential Fourier series

For example, consider the rectangular waveform presented in Fig. 1.8, in which each pulse is of height A and width τ. The pulses are separated by T seconds.

In this case, c_n is given by

$$c_n = \frac{1}{T} \int_T s(t) e^{\frac{-j2\pi nt}{T}}\, dt = \frac{1}{T} \int_{-\frac{\tau}{2}}^{\frac{\tau}{2}} A e^{\frac{-j2\pi nt}{T}}\, dt \tag{1.30}$$

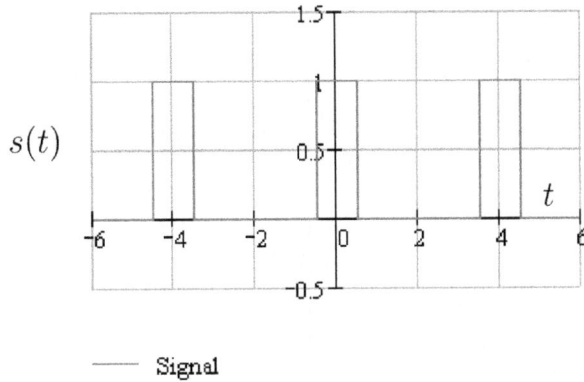

—— Signal

Figure 1.8: Periodic rectangular signal with $\tau = \frac{T}{4}$ and $T = 4$ seconds.

$$= \frac{A}{T} \left[\frac{-Te^{\frac{-j2\pi n t}{T}}}{j2\pi n} \right]_{-\frac{\tau}{2}}^{\frac{\tau}{2}} = \frac{A}{j2\pi n} \left[-e^{\frac{-j\pi n \tau}{T}} + e^{\frac{j\pi n \tau}{T}} \right] \qquad (1.31)$$

But since $\frac{e^{j\theta} - e^{-j\theta}}{2j} = \sin\theta$,

$$c_n = \frac{A}{\pi n} \left[\sin \frac{\pi n \tau}{T} \right] \qquad (1.32)$$

Alternatively, in terms of the sinc (x) function defined by

$$\text{sinc}(x) \triangleq \frac{\sin(\pi x)}{\pi x} \qquad (1.33)$$

and illustrated in Fig. 1.9, c_n may be expressed as

$$c_n = \frac{A\tau}{T} \left(\frac{T}{\pi n \tau} \right) \left[\sin \frac{\pi n \tau}{T} \right] = \frac{A\tau}{T} \frac{\sin\left(\frac{\pi n \tau}{T}\right)}{\frac{\pi n \tau}{T}} = \frac{A\tau}{T} \text{sinc}\left(\frac{n\tau}{T}\right) \qquad (1.34)$$

$$\text{sinc}(x)$$

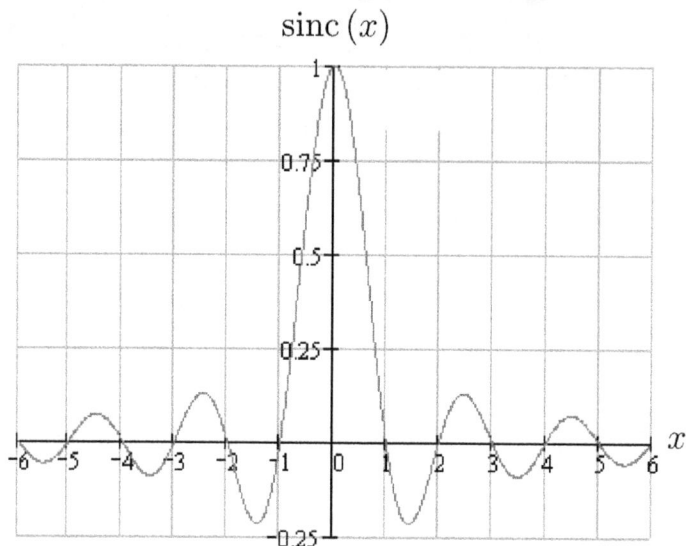

Figure 1.9: $\text{sinc}(x) \triangleq \frac{\sin(\pi x)}{\pi x}$ function.

For $\tau = \frac{T}{4}$, the first few coefficients are given by

$$c_0 = \frac{A\tau}{T} \text{sinc}(0) = \frac{A}{4} \tag{1.35}$$

$$c_1 = \frac{A\tau}{T} \text{sinc}\left(\frac{\tau}{T}\right) = \frac{A}{4} \text{sinc}\left(\frac{1}{4}\right) = \frac{A}{\pi\sqrt{2}} \tag{1.36}$$

$$c_{-1} = \frac{A\tau}{T} \text{sinc}\left(\frac{-\tau}{T}\right) = \frac{A}{4} \text{sinc}\left(\frac{-1}{4}\right) = \frac{A}{4} \frac{\sin(-\pi/4)}{-\pi/4} = \frac{A}{\pi\sqrt{2}} \tag{1.37}$$

Similarly, we find $c_2 = c_{-2} = \frac{A}{2\pi}$, $c_3 = c_{-3} = \frac{A}{3\pi\sqrt{2}}$, $c_4 = c_{-4} = 0$, $c_5 = c_{-5} = \frac{-A}{5\pi\sqrt{2}}$, $c_6 = c_{-6} = \frac{-A}{6\pi}$. The corresponding real and imaginary parts of c_n are presented in Fig. 1.10 and the amplitude and phase spectra in Fig. 1.11 for $A = 1$. Notice that even though the imaginary parts of c_5, c_6, and c_7 are zero, the corresponding phase angles are π because the real part of c_5, c_6, and c_7 are negative. Check this result using equation 1.27 taking into account that in practice, if $|c_n| < 0.01$, then $\arg(c_n)$ is taken to be zero because $|c_n|$ is insignificant.

Referring to the amplitude spectrum, if the tips of the vertical spectral lines were joined together as a continuous function, then this *envelope* function goes to zero at the frequencies $\pm\frac{1}{\tau}, \pm\frac{2}{\tau}, \pm\frac{3}{\tau}, \cdots$ The interval $\left(-\frac{1}{\tau}\text{ to }\frac{1}{\tau}\right)$ between the first two points where the envelope function goes to zero is called the *main lobe*. The intervals between successive zeros are called *sidelobes*. A comparison of the original and reconstructed signal is presented in Fig. 1.12, where the spectral coefficients in Fig. 1.11 are combined to reconstruct the signal $s(t)$ using either

$$s(t) = \sum_{n=-\infty}^{\infty} c_n e^{\frac{j2\pi nt}{T}} \tag{1.38}$$

or

$$s(t) = \sum_n |c_n| e^{j\left(\frac{2\pi nt}{T} + \arg c_n\right)} \tag{1.39}$$

or

$$s(t) = c_o + \sum_{n=1}^{\infty} 2|c_n| \cos\left(\frac{2\pi nt}{T} + \arg c_n\right) \tag{1.40}$$

- ⭐ SIMULATION **PeriodicRectPulses:** Implementation of the foregoing theory. Experiment with the height, width, time period and the number of spectral components used to reconstruct the signal. Observe how only the phase spectrum is effected by a time delay of the waveform.

To gain a better understanding of the reconstructed signal, consider the single-sided representation for which $s(t) = c_o + \sum_{n=1}^{\infty} 2|c_n| \cos\left(\frac{2\pi nt}{T} + \arg c_n\right)$. Using only the first six components, we find

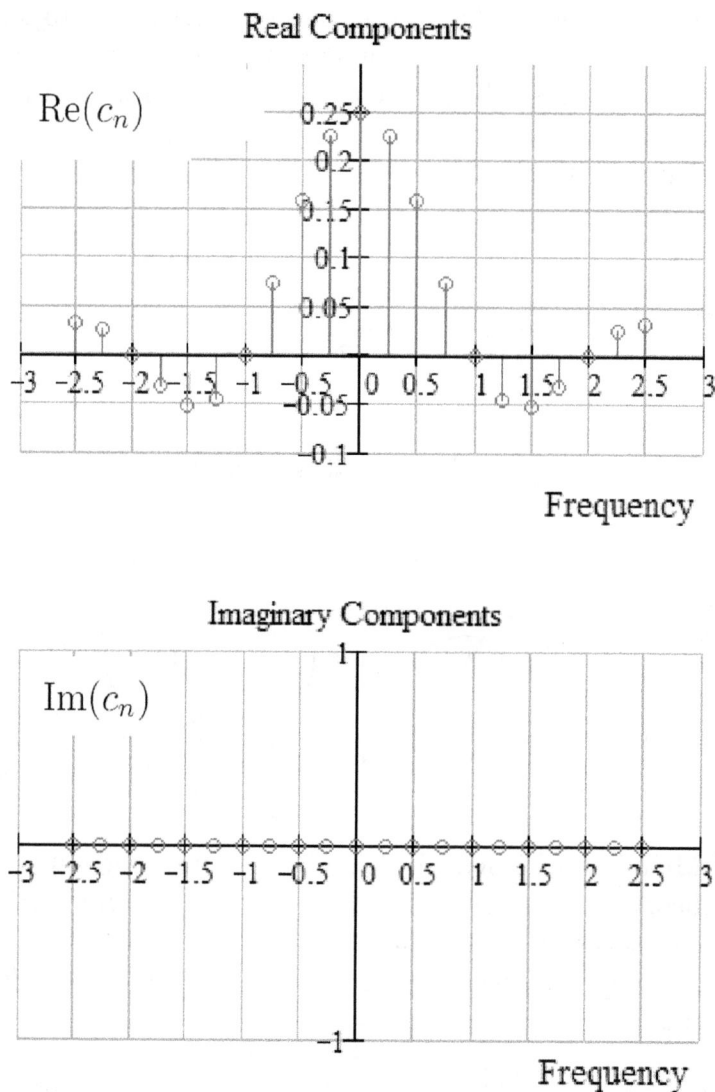

Figure 1.10: Real $\mathrm{Re}(c_n)$ and imaginary $\mathrm{Im}(c_n)$ parts c_n.

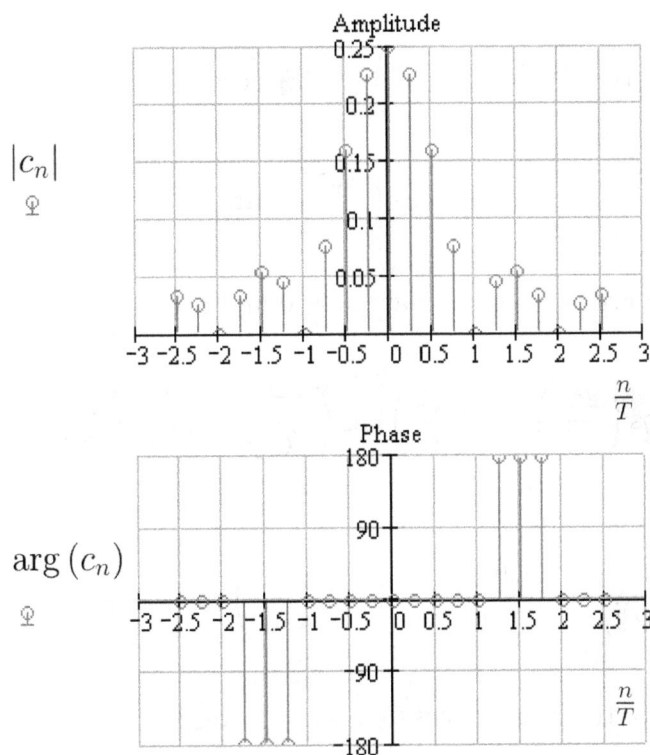

Figure 1.11: Amplitude and phase spectra of the periodic rectangular signal.

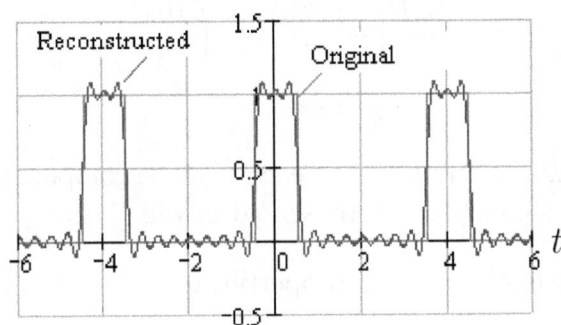

Figure 1.12: Original and reconstructed signal from the amplitude and phase spectra.

$$s(t) = c_o + 2\,|c_1|\cos\left(\frac{2\pi t}{T} + \arg c_1\right) + 2\,|c_2|\cos\left(\frac{4\pi t}{T} + \arg c_2\right) + \ldots \quad (1.41)$$

$$+ 2\,|c_6|\cos\left(\frac{12\pi t}{T} + \arg c_6\right) \qquad\qquad (1.42)$$

But $\arg c_1 = \arg c_2 = \arg c_3 = \arg c_4 = 0$ and $\arg c_5 = \arg c_6 = \pi$ so that

$$= \frac{A}{4} + \frac{2A}{\pi\sqrt{2}}\cos\left(\frac{2\pi t}{T}\right) + 2\left(\frac{A}{2\pi}\right)\cos\left(\frac{4\pi t}{T}\right) + 2\left(\frac{A}{3\pi\sqrt{2}}\right)\cos\left(\frac{6\pi t}{T}\right) + 0$$
$$(1.43)$$

$$+ 2\left(\frac{A}{5\pi\sqrt{2}}\right)\cos\left(\frac{10\pi t}{T} + \pi\right) + 2\left(\frac{A}{6\pi}\right)\cos\left(\frac{12\pi t}{T} + \pi\right) \qquad (1.44)$$

Making use of the identity $\cos(\theta + \pi) = -\cos(\theta)$,

$$
\begin{aligned}
s(t) \;=\; & \frac{A}{4} + \frac{\sqrt{2}A}{\pi}\cos\left(\frac{2\pi t}{T}\right) + \frac{A}{\pi}\cos\left(\frac{4\pi t}{T}\right) + \\
& \frac{\sqrt{2}A}{3\pi}\cos\left(\frac{6\pi t}{T}\right) - \frac{\sqrt{2}A}{5\pi}\cos\left(\frac{10\pi t}{T}\right) - \frac{A}{3\pi}\cos\left(\frac{12\pi t}{T}\right) (1.45)
\end{aligned}
$$

For $A = 1$ volt, an overlay plot of each of these components together with the reconstructed signal $s(t)$ are presented in Fig. 1.13.

Note the following interesting properties from Fig. 1.13.

- The time period of the *fundamental* spectral component ($n = 1$) is the same as the time period T of the signal.

- High frequency components of small amplitude contribute to improving the rectangular edges of the reconstructed signal.

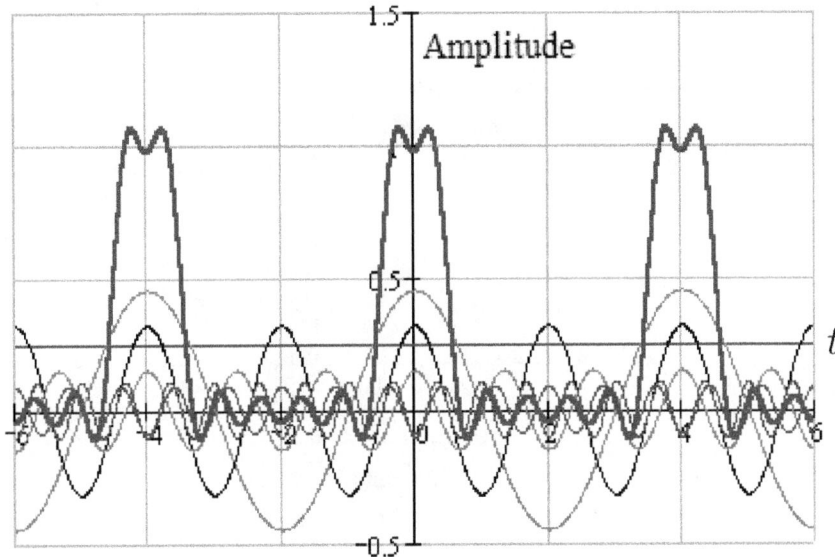

Figure 1.13: Single-sided spectral components and their summation.

- The fundamental component and a few other components are sufficient to give a reasonable representation of a signal. More specifically, most of the signal energy is contained within the main lobe of the amplitude spectrum. This will be illustrated by example in the next section.

- If a communication channel attenuates only high frequency components, then an input periodic rectangular signal to this channel will be output with a shape that still resembles the original signal.

If the signal in Fig. 1.8 is delayed by $\tau/2$, then the corresponding amplitude and phase spectra are presented in Fig. 1.14. Although both the amplitude and phase spectra are required to represent a given signal, it is the amplitude spectrum which is more useful, because it displays the signal's frequency content and provides an insight to the distribution of energy within a given signal.

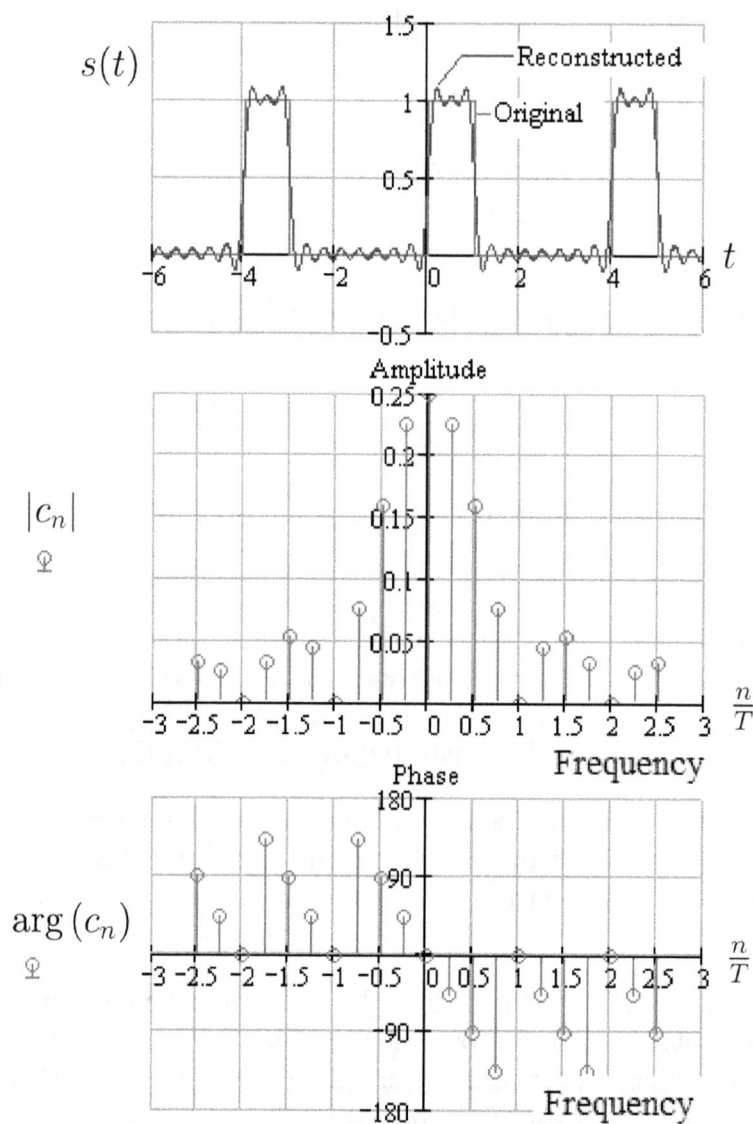

Figure 1.14: Shifted signal and corresponding spectra.

- SIMULATION **PulseShifted:** Simulation corresponding to Fig. 1.14, in which the spectral components of a time-shifted periodic rectangular waveform are determined. These are then used to reconstruct the signal and determine the signal power, which is confirmed via $P = \frac{1}{T} \int_T |s(t)|^2 \, dt$ i.e. the signal power is obtained from both the frequency and time domain representation of this signal. Modify the simulation to consider other types of signals.

1.2.3   Example : Half-Wave Sinusoid Signal

Consider the half-wave sinusoid signal, with a pulse width $\tau = \frac{T}{2}$ as shown in Fig. 1.15 together with its corresponding amplitude and phase spectra. Verify that the phasor on this double-sided spectrum at the fundamental positive frequency is given by $\frac{1}{4} e^{j\left(\frac{2\pi t}{T} - \frac{\pi}{2}\right)}$.

Solution

$$c_n = \frac{1}{T} \int_T s(t) e^{-j\frac{2\pi n t}{T}} \, dt = \frac{1}{T} \int_0^\tau \sin\left(2\pi f t\right) e^{-j\frac{n2\pi t}{T}} \, dt \qquad (1.46)$$

Since $\tau = \frac{T}{2}$ and $f = \frac{1}{T}$, we find

$$c_n = \frac{1}{T} \int_0^{T/2} \sin\left(\frac{2\pi t}{T}\right) e^{-j\frac{n2\pi t}{T}} \, dt \qquad (1.47)$$

$$= \frac{1}{T} \int_0^{T/2} \sin\left(\frac{2\pi t}{T}\right) \cos\left(\frac{n2\pi t}{T}\right) \, dt - j\frac{1}{T} \int_0^{T/2} \sin\left(\frac{2\pi t}{T}\right) \sin\left(\frac{n2\pi t}{T}\right) \, dt$$

$$(1.48)$$

For $n = 1$

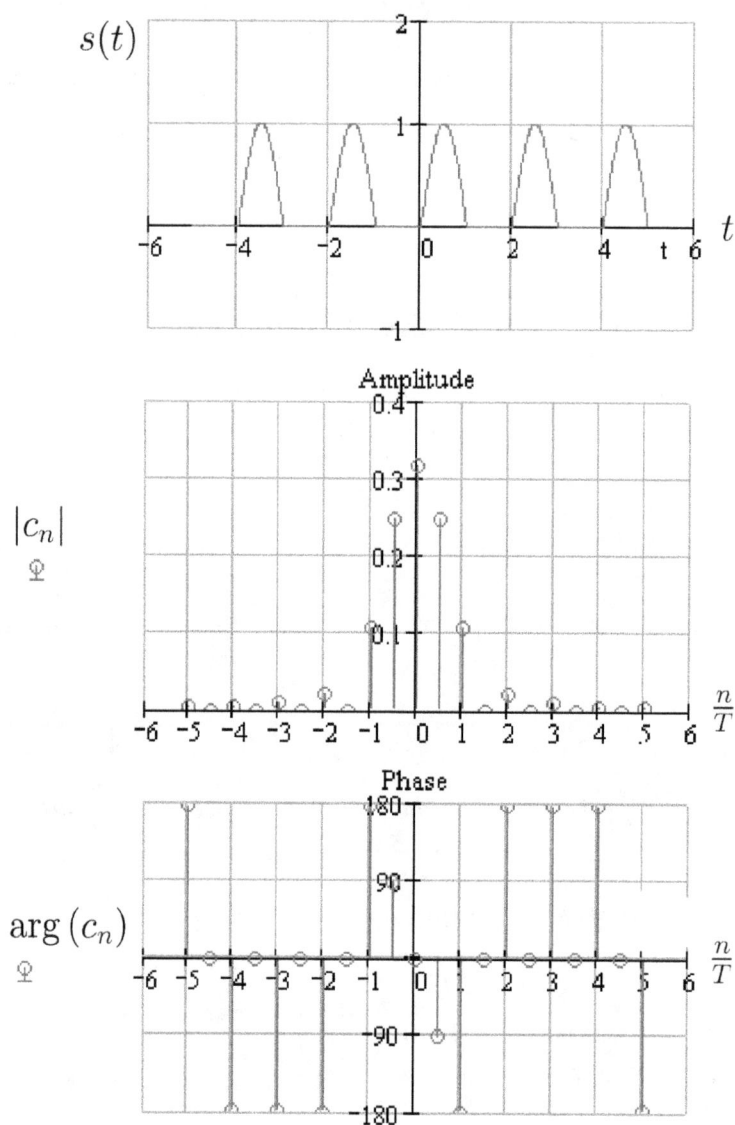

Figure 1.15: Have-wave rectified sinusoid signal and its spectrum.

$$c_1 = \frac{1}{T} \int_0^{T/2} \sin\left(\frac{2\pi t}{T}\right) \cos\left(\frac{2\pi t}{T}\right) dt - j\frac{1}{T} \int_0^{T/2} \sin^2\left(\frac{2\pi t}{T}\right) dt \qquad (1.49)$$

Making use of the identities $\sin\theta \cos\theta = \frac{\sin 2\theta}{2}$ and $\sin^2\theta = \frac{1-\cos 2\theta}{2}$, we find

$$c_1 = \frac{1}{2T} \int_0^{T/2} \sin\left(\frac{4\pi t}{T}\right) dt - j\frac{1}{2T} \int_0^{T/2} 1 - \cos\left(\frac{4\pi t}{T}\right) dt \qquad (1.50)$$

$$= \frac{1}{2T} \left[-\frac{T}{4\pi} \cos\left(\frac{4\pi t}{T}\right) \right]_0^{T/2} - j\frac{1}{2T} \left[t - \frac{T}{4\pi} \sin\left(\frac{4\pi t}{T}\right) \right]_0^{T/2} \qquad (1.51)$$

$$= \frac{1}{2T} \left[-\frac{T}{4\pi} \cos\left(2\pi\right) - \left(-\frac{T}{4\pi} \cos(0)\right) \right] - j\frac{1}{2T} \left[\frac{T}{2} - \frac{T}{4\pi} \sin\left(2\pi\right) - 0 \right] = -\frac{j}{4} \qquad (1.52)$$

Thus $|c_1| = \frac{1}{4}$ and $\arg c_1 = \tan^{-1}\left(\frac{-1/4}{0}\right) = -\frac{\pi}{2}$. Hence the coefficient $c_1 = |c_1| e^{j \arg c_1} = \frac{1}{4} e^{-j\frac{\pi}{2}}$ and the phasor on a double-sided spectrum at the positive fundamental frequency (rotating anti-clockwise) is given by $|c_1| e^{j\left(\frac{2\pi t}{T} + \arg c_1\right)} = \frac{1}{4} e^{j\left(\frac{2\pi t}{T} - \frac{\pi}{2}\right)}$.

- ⭐ SIMULATION **RectifiedSinusoid**: An implementation and verification of the foregoing analysis.

1.2.4   Simulation : Signal Distortion

SIMULATION **Channel:** The spectral components of a periodic rectangular signal are altered to investigate the effects of a channel which distorts the amplitude and phase of the incoming signal. Simulation results are presented in Fig. 1.16.

- **Experiment:**   Alter the amplitude and the phase variation functions to simulate different degrees of distortion.

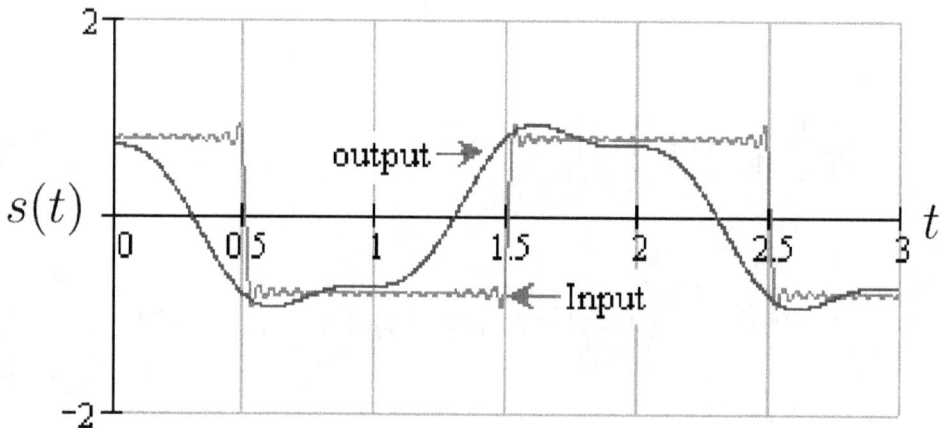

Figure 1.16: Channel Distortion.

1.3   Power and Bandwidth

1.3.1   Parseval Theorem for the Fourier Series

Parsevals Theorem

- Introduction to Parseval's theorem

- Its application to show that 90% of the signal power is contained within the main lobe of a periodic rectangular pulse signal

In the time-domain, recall that the average signal power is given by $P = \frac{1}{T} \int_T |s(t)|^2 \, dt$. An alternative method is to utilize the spectral components of the signal to determine the average signal power using *Parsevals theorem*

$$P = \sum_{n=-\infty}^{\infty} |c_n|^2 \qquad (1.53)$$

which is easily derived via

$$P = \frac{1}{T} \int_T |s(t)|^2 \, dt = \frac{1}{T} \int_T s(t) s^*(t) dt \qquad (1.54)$$

where $s^*(t)$ is the complex conjugate of $s(t)$. Using the exponential Fourier series spectral representation $s(t) = \sum_{n=-\infty}^{\infty} c_n e^{\frac{j2\pi nt}{T}}$, with $s^*(t) = \sum_{n=-\infty}^{\infty} c_n^* e^{-\frac{j2\pi nt}{T}}$, we find

$$P = \frac{1}{T} \int_T s(t) \left(\sum_{n=-\infty}^{\infty} c_n^* e^{-\frac{j2\pi nt}{T}} \right) dt = \sum_{n=-\infty}^{\infty} \left[\frac{1}{T} \int_T s(t) e^{-\frac{j2\pi nt}{T}} \right] c_n^* \qquad (1.55)$$

where $c_n = \frac{1}{T} \int_T s(t) e^{\frac{-j2\pi nt}{T}} dt$, thus

$$P = \sum_{n=-\infty}^{\infty} c_n c_n^* = \sum_{n=-\infty}^{\infty} |c_n|^2 \qquad (1.56)$$

or equivalently, given that $|c_n|^2 + |c_{-n}|^2 = \left| \frac{A_n - jB_n}{2} \right|^2 + \left| \frac{A_n + jB_n}{2} \right|^2 = \frac{(A_n)^2}{2} + \frac{(B_n)^2}{2}$

$$P = (A_0)^2 + \sum_{n=1}^{\infty} \frac{(A_n)^2}{2} + \frac{(B_n)^2}{2} \tag{1.57}$$

Recall that the amplitude of a given spectral component on a double-sided spectrum is denoted by $|c_n|$, where n ranges from $-\infty$ to $+\infty$. For a single-sided spectrum, the amplitude of a given spectral component, other than the d.c. component, is given by $2|c_n|$ and n ranges from 1 to $+\infty$ because the component c_0 at zero frequency is common to both the single and double-sided spectra. Now in section 1.1 it was shown that the average power of the signal $s(t) = A\cos(2\pi f t)$ of amplitude A is equal to $\frac{A^2}{2}$. Thus, the average power of a given cosine component $2|c_n|\cos\left(\frac{2\pi n t}{T} + \arg c_n\right)$ is equal to $\frac{(2|c_n|)^2}{2} = 2|c_n|^2$. Therefore, on a double-sided spectrum,

$$P = \sum_{n=-\infty}^{\infty} |c_n|^2 \tag{1.58}$$

Table 1.1 summarizes how to determine the average signal power from both a double and a single-sided spectrum.

To illustrate the use of Parsevals power theorem, consider once again the periodic rectangular signal shown in Fig. 1.8. Using only the spectral components within the main lobe of the amplitude spectrum presented in Fig. 1.11, the average signal power in the frequency-domain P_{freq} is given by

Double-Sided	Single-Sided
$P = \sum_{n=-\infty}^{\infty} |c_n|^2$	$P = |c_0|^2 + \sum_{n=1}^{\infty} \frac{|2c_n|^2}{2} = |c_0|^2 + 2\sum_{n=1}^{\infty} |c_n|^2$
$= |c_{-\infty}|^2 + \cdots + |c_0|^2 + \cdots + |c_\infty|^2$	$= |c_0|^2 + 2\left(|c_1|^2 + |c_2|^2 + \cdots + |c_\infty|^2\right)$

Table 1.1: Signal power from a spectrum.

$$P_{freq} = \sum_{n=-4}^{4} |c_n|^2 = |c_0|^2 + |c_1|^2 + |c_{-1}|^2 + |c_2|^2 + |c_{-2}|^2$$

$$+ |c_3|^2 + |c_{-3}|^2 + |c_4|^2 + |c_{-4}|^2 \qquad (1.59)$$

But since $|c_n| = |c_{-n}|$

$$P_{freq} = |c_0|^2 + 2\left(|c_1|^2 + |c_2|^2 + |c_3|^2 + |c_4|^2\right)$$

$$= \left(\frac{A}{4}\right)^2 + 2\left[\left(\frac{A}{\pi\sqrt{2}}\right)^2 + \left(\frac{A}{2\pi}\right)^2 + \left(\frac{A}{3\pi\sqrt{2}}\right)^2\right]$$

$$= \frac{A^2}{16} + \frac{2A^2}{\pi^2}\left[\frac{1}{2} + \frac{1}{4} + \frac{1}{18}\right] = A^2 0.226 \qquad (1.60)$$

However, the average signal power in the time-domain P_{time} is given by

$$P_{time} = \frac{A^2 \tau}{T} \qquad (1.61)$$

For $\tau = \frac{T}{4}$, we have $P_{time} = \frac{A^2}{4}$. Therefore, the ratio

$$\frac{P_{freq}}{P_{time}} = \frac{(A^2 0.226)\, 4}{A^2} = 0.9 \qquad (1.62)$$

or equivalently, $P_{freq} = 0.9 P_{time}$. The implication of this result is that most of the average signal power is contained within the main lobe of the amplitude spectrum. Of course, if additional spectral components beyond the main lobe are used to calculate P_{freq}, the ratio $\frac{P_{freq}}{P_{time}}$ would be even closer to one.

- ⭐ SIMULATION **Power:** The ratio $\frac{P_{freq}}{P_{time}}$ is analyzed over a range of spectral components. The signal is also reconstructed from the corresponding spectral components and compared with the original signal.

1.3.2   Bandwidth

▶ **Bandwidth**

- Bandwidth of a signal

- First-null bandwidth

- Bandwidth of a baseband or bandpass signal

- 3dB or half-power bandwidth

Fourier analysis reveals that a signal can be considered to be made up of sinusoidal components (cosines) of various amplitudes, phase and frequency. Thus, we may define the *bandwidth* of a signal as the frequency range, or the spectral width, which encompasses these components. Referring once again to our ongoing example of the periodic rectangular signal, it is clear that the absolute bandwidth of this signal is infinity, because an infinite number of components are required to reconstruct perfectly the original rectangular signal. However, recall from the previous subsection that most of signal power is contained within the main lobe of amplitude spectrum.   Specifically, if a filter is constructed to remove the frequency components higher than $\frac{1}{\tau}$ Hz, the signal would still contain 90% of original signal power.   Thus, from a practical viewpoint, the bandwidth of this signal can be approximated to $\frac{1}{\tau}$ Hz.   This bandwidth is commonly referred to as the *first null bandwidth*, because its the frequency range from 0 Hz to the first null, which in this case is at $\frac{1}{\tau}$ Hz.

In general, the bandwidth of a signal is usually taken to be the difference between two frequencies $f_{\max}$ and $f_{\min}$ above and below which the spectral components are assumed to be small. For example, consider the amplitude spectrum shown in Fig. 1.17(a). In this case, $f_{\min} = 0$ and the bandwidth B of this signal is simply equal to $f_{\max}$. A signal of this type is referred to as a *baseband* signal.   Note that on such a double-sided spectrum, B is the spectral width of either the positive or negative frequencies and **not** both.   A signal with an amplitude spectrum of the type presented in Fig. 1.17(b), which is a **single-sided** spectrum for convenience, is referred to as a *bandpass* signal.   In this case, the bandwidth B of the signal is equal to

$(f_{\max} - f_{\min})$. Since the selection of $f_{\max}$ is subjective, the bandwidth of a signal is usually taken to be the *null-to-null* bandwidth as illustrated in Fig. 1.18.

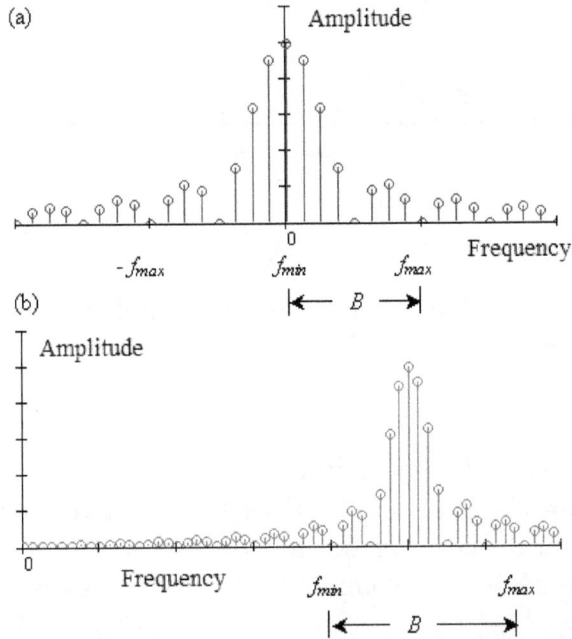

Figure 1.17: (a) Baseband signal bandwidth (b) Bandpass signal bandwidth.

There are several other popular measures of bandwidth. For example, the *half-power* bandwidth is the frequency interval within which the peak spectral component has reduced by a factor of $\frac{1}{\sqrt{2}}$. To understand this, recall that on a single-sided spectrum of $s(t)$, the average power of a given sinusoidal component $2\,|c_n| \cos\left(\frac{2\pi n t}{T} + \arg c_n\right)$ is equal to $2\,|c_n|^2$, or equivalently, $|c_n|^2$ on a double-sided spectrum. Thus, on a double-sided spectrum, let P_a represent the power of the peak spectral component of amplitude $|c_a|$ and let P_b represent the power of the spectral component with amplitude $|c_b| = \frac{|c_a|}{\sqrt{2}}$. The power ratio $\frac{P_b}{P_a} = \frac{|c_b|^2}{|c_a|^2} = \frac{|c_a|^2}{2|c_a|^2} = \frac{1}{2}$, or $P_b = \frac{P_a}{2}$ i.e. "half-power". On a decibel scale, $10\log_{10}\left(\frac{P_b}{P_a}\right) = 10\log_{10}(\frac{1}{2}) = -3$ dB. Since $10\log_{10}\left(\frac{P_b}{P_a}\right) =$

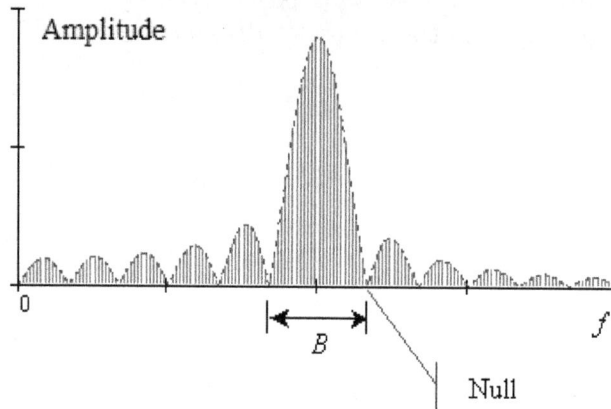

Figure 1.18: Null-to-null bandwidth B of a bandpass signal shown on a single-sided spectrum.

$10\log_{10}(P_b) - 10\log_{10}(P_a)$ is equivalent to $P_b(dB) - P_a(dB) = -3dB$, we find that $P_b(dB) = P_a(dB) - 3dB$. Thus a power reduction by a factor of 2 is equivalent to a 3 dB decrease on a decibel scale. Consequently, the *half-power* bandwidth is more commonly referred to as the *3 dB* bandwidth.

- ⭐ SIMULATION **Threebandwidth:** The power of the periodic rectangular signal shown in Fig. 1.8 is determined within the 3dB bandwidth. The spectral components within this bandwidth are used to reconstruct the signal. Experiment with the time period of this signal, etc. Also experiment with different types of periodic signals.

1.3.3   Simulation : Baseband versus Bandpass

▷ **Baseband versus Bandpass**

- Periodic rectangular pulse bandeband and bandpass signal components are analysed, filtered and used to reconstruct the signals

- Their signal powers are compared

⭐ SIMULATION **Bandwidth:** The average signal power of a filtered baseband and bandpass signal are determined from their amplitude spectra and compared with their original average power prior to filtering. The periodic rectangular signal in Fig. 1.8 is taken to be the baseband signal. The corresponding bandpass signal is created by multiplying the baseband signal with $A \cos(2\pi f_c t)$, with $A = 1$ volt and the carrier frequency $f_c = 5$ Hz. The result of eliminating all the spectral components outside the main lobe for both baseband and bandpass signals, together with the corresponding double and single-sided amplitude spectra are presented in Table 1.2.

- **Experiment:** Try a different carrier frequency f_c, and alter the width of the low-pass and high-pass filters. Finally, try different types of baseband signals.

Baseband	Bandpass
Average power = 0.25 W	Average power = 0.125 W
Double-Sided Spectrum	Double-Sided Spectrum
Filtered Double-Sided Spectrum	Filtered Double-Sided Spectrum
Filtered Single-Sided Spectrum	Filtered Single-Sided Spectrum
Reconstructed signal Average power = 0.226 W	Reconstructed signal Average power = 0.113 W

Table 1.2: Baseband and bandpass spectral analysis

1.4 Fourier Transform

1.4.1 Fundamentals

▶ Fourier Transform

- The effect of increasing the time period between each pulse

- Fourier transform fundamentals

- Theoretical analysis of a rectangular pulse

- Reconstruction of the pulse using a range of frequencies

Consider the rectangular periodic signal of time period T presented in Fig. 1.19. Using the exponential form of the Fourier series, recall that the complex coefficients are given by $c_n = \frac{A\tau}{T} \operatorname{sinc}\left(\frac{n\tau}{T}\right)$. For $A = 2$ volts, $T = 6$ secs and $\tau = 0.5$ secs, Fig. 1.20 shows the amplitude spectrum and the corresponding signal.

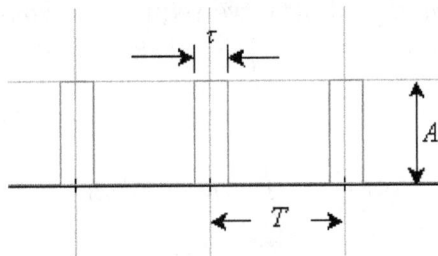

Figure 1.19: Rectangular periodic wave.

From Fig. 1.20, notice that as τ is decreased, the width of the main spectral lobe becomes wider. For a given τ, the spectral lines are separated by the fundamental frequency $\frac{1}{T}$. As the time period T is increased, the spectral lines come closer together. Indeed, as T tends to infinity, we end up with an

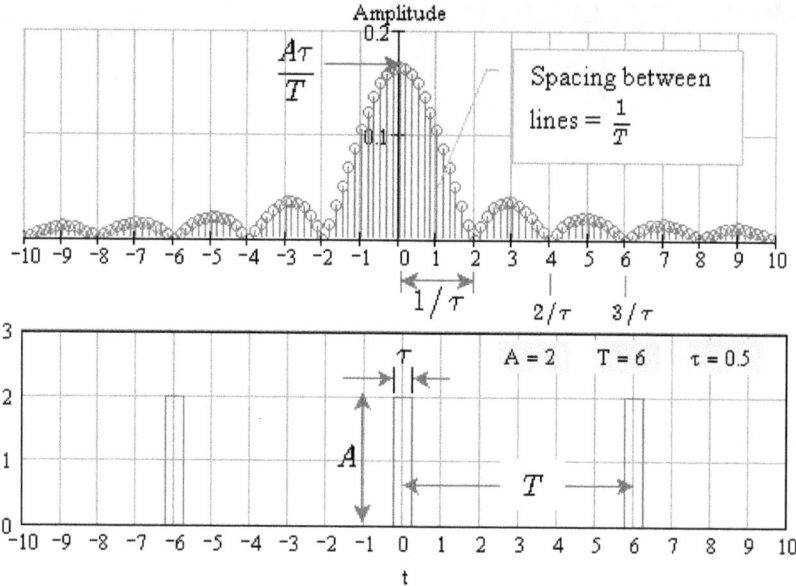

Figure 1.20: Periodic rectangular pulse signal and its amplitude spectrum.

isolated pulse and the frequency spectrum tends to a continuous spectrum rather than a discrete one.   To determine an expression for the continuos spectrum of an *aperiodic* signal $s(t)$, we require its *Fourier transform* (FT) $S(f)$ given by

$$S(f) = \int\limits_{-\infty}^{\infty} s(t)e^{-j2\pi ft}dt \qquad (1.63)$$

Notice that the signal $s(t)$ is multiplied with an exponential term $e^{-j2\pi ft} = \cos(2\pi ft) + j\sin(2\pi ft)$, at a frequency f and then integrated over all times. If the result of this integration is a large value, then the signal $s(t)$ has a dominant spectral component at frequency f. If $|S(f)| = 0$, then the signal does not contain a spectral component at the frequency f.   Notice how this information is independent of where in time this component appears. The key point being that the Fourier transform is **not** suitable for a *non-stationary* signal in which the signal contains time varying spectral components.   To recover the signal $s(t)$ from its FT $S(f)$, we apply the *inverse Fourier trans-form* given by

$$s(t) = \int_{-\infty}^{\infty} S(f)e^{j2\pi ft}df \tag{1.64}$$

Thus consider a rectangular pulse of width $\tau = 0.5$ centered at $t = 0$ with amplitude $A = 1$ volt as illustrated in Fig. 1.21.

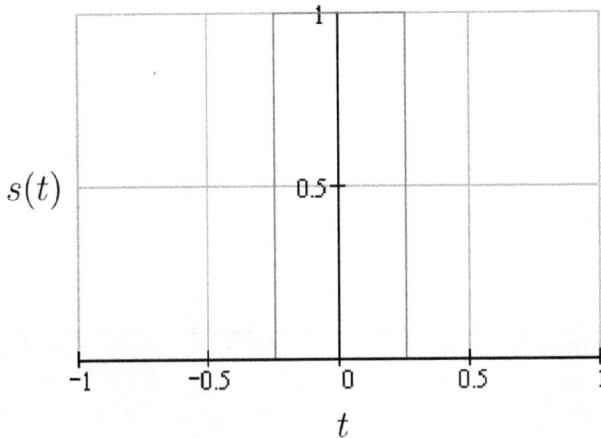

Figure 1.21: Rectangular pulse of width τ.

Its FT is given by

$$S(f) = \int_{-\tau/2}^{\tau/2} Ae^{-j2\pi ft}dt = A\left[\frac{e^{-j2\pi ft}}{-j2\pi f}\right]_{-\tau/2}^{\tau/2} = A\left(-\frac{e^{-j\pi f\tau}}{j2\pi f} + \frac{e^{j\pi f\tau}}{j2\pi f}\right)$$

$$= \frac{A}{\pi f}\left(\frac{e^{j\pi f\tau} - e^{-j\pi f\tau}}{2j}\right) \tag{1.65}$$

Making use of the identity $\frac{e^{j\theta} - e^{-j\theta}}{2j} = \sin\theta$ and the sinc(.) function (equation 1.33)

$$S(f) = A\tau \frac{\sin(\pi f \tau)}{\pi f \tau} = A\tau \operatorname{sinc}(f\tau) \tag{1.66}$$

The variation of $S(f)$ with f is presented in Fig. 1.22 and the corresponding amplitude $|S(f)|$ and phase $\operatorname{argc}(S(f))$ spectrum is presented in Fig. 1.23. Notice that for $f = 0$, $S(0) = A\tau e^{-0} \operatorname{sinc}(0) = A\tau$ and $S(\frac{1}{\tau}) = A\tau e^{-j\pi} \operatorname{sinc}(1) = 0$. Indeed, $S(\frac{1}{\tau}) = S(\frac{2}{\tau}) = S(\frac{3}{\tau}) = \cdots = 0$.

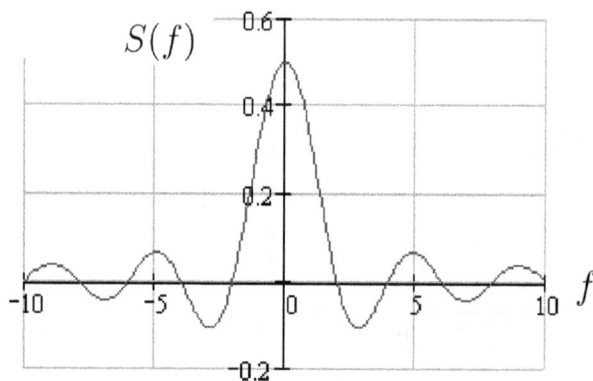

Figure 1.22: Fourier transform of a rectangular pulse.

• ⭐ SIMULATION **RectPulseFT:** For a verification of equation 1.66.

1.4.2   Example : Shifted Rectangular Pulse

▶ **Shifted Rectangular Pulse**

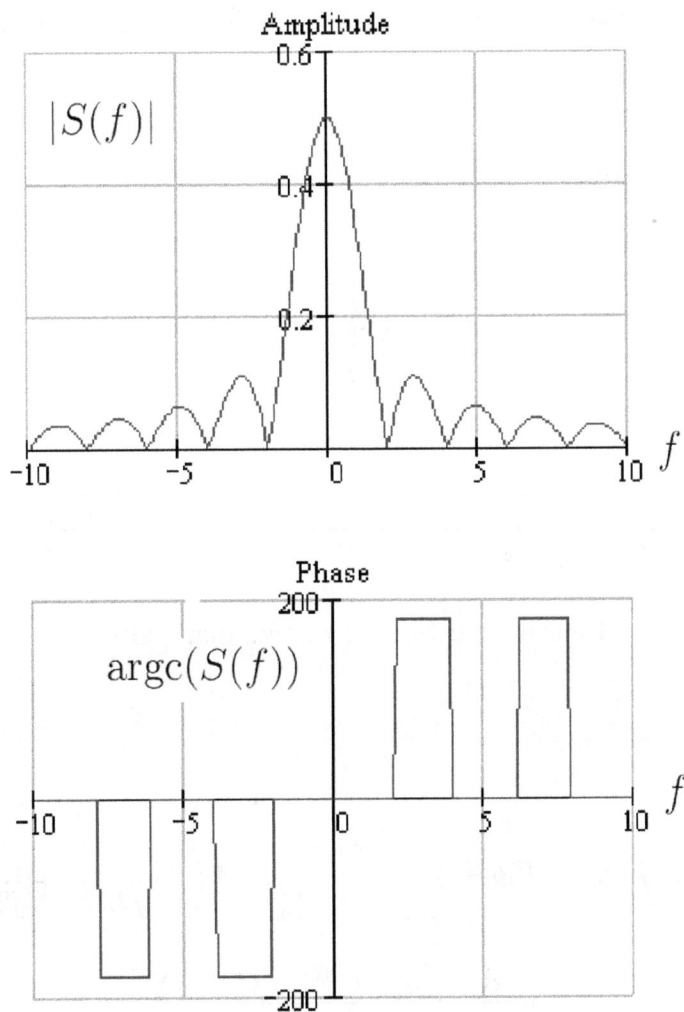

Figure 1.23: Rectangular pulse amplitude and phase spectrum envelope.

- Theoretical analysis to determine the
  Fourier transform of a shifted rectangular pulse

Determine the Fourier transform of the pulse presented in Fig. 1.24.

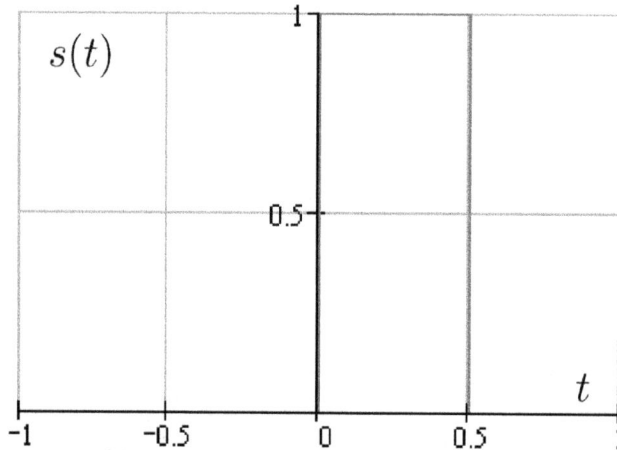

Figure 1.24: Shifted rectangular pulse.

Solution

$$S(f) = \int_0^\tau A e^{-j2\pi ft} dt = A \left[\frac{e^{-j2\pi ft}}{-j2\pi f} \right]_0^\tau = A \left(-\frac{e^{-j2\pi f\tau}}{j2\pi f} + \frac{1}{j2\pi f} \right)$$

$$= \frac{A}{\pi f} e^{-j\pi f\tau} \left(\frac{e^{j\pi f\tau} - e^{-j\pi f\tau}}{2j} \right) \tag{1.67}$$

Making use of the identity $\frac{e^{j\theta}-e^{-j\theta}}{2j} = \sin\theta$, we find

$$S(f) = A\tau e^{-j\pi f\tau} \frac{\sin(\pi f\tau)}{\pi f\tau} = A\tau e^{-j\pi f\tau} \operatorname{sinc}(f\tau) \tag{1.68}$$

1.4.3   Discrete Fourier Transform (DFT)

▶ **Sampling**

- Fundamentals of sampling

- Example of sampling a rectangular pulse

To determine the Fourier transform $S(f) = \int\limits_{-\infty}^{\infty} s(t)e^{-j2\pi ft}dt$ of an aperiodic signal $s(t)$ on a computer, which can only store and process digital signals, it is necessary to sample $s(t)$ at a frequency of f_s samples per second to create the digital signal $x[n]$ given by

$$x[n] = s\left(\frac{n}{f_s}\right) = s(nT_s) \tag{1.69}$$

for $n = 0, 1, \cdots, N - 1$ where $\frac{1}{f_s}$ is the time interval T_s between each sample.

▶ **DFT**

- DFT of a rectangular pulse

- Reconstruction of the pulse using the DFT spectral components

- Double-sided spectrum from the DFT and the FFT

The kth coefficient, denoted by X_k, of the N-point *discrete Fourier transform* (DFT) of a digital signal $x[n]$ (which we shall write as x_n for convenience) is given by

$$X_k = \sum_{n=0}^{N-1} x_n e^{-j\frac{2\pi kn}{N}} \tag{1.70}$$

where $k = 0, 1, \cdots N - 1$ and the frequency and amplitude of the kth component are given by $\frac{k f_s}{N}$ and $\frac{|X_k|}{N}$, respectively. Notice the DFT only searches for spectral components at the specific frequencies $0, \frac{f_s}{N}, \frac{2 f_s}{N}, \frac{3 f_s}{N}, \cdots, \frac{k f_s}{N}, \cdots, \frac{(N-1) f_s}{N}$ Hz. The separation between the DFT spectral components, referred to as the *frequency resolution*, is $\left(\frac{f_s}{N}\right)$ Hz. If a DFT spectral component is close to an actual spectral component within $s(t)$, there is a *smearing* effect within the DFT amplitude spectrum. The *Inverse Discrete Fourier Transform* (IDFT) x_n of the sequence $\{X_k\}$ is given by

$$x_n = \frac{1}{\sqrt{N}} \sum_{k=0}^{N-1} X_k e^{j \frac{(2 \pi n k)}{N}} \text{ for } n = 0, 1, \cdots N - 1 \qquad (1.71)$$

where $\frac{1}{\sqrt{N}}$ is a scale factor. Equation 1.71 is equivalent to $x = cfft(X)$ in Mathcad, where x is a vector of length N.

▶ DFT Example

- The signal $s(t) = \sin(\pi t)$ is sampled via $x_n = s\left(\frac{n}{f_s}\right)$

  with $f_s = 4$ Hz to determine its DFT as N is increased from 2 to 128.

- Notice how the smearing effect disappears when a DFT point lies

  exactly on the spectral component at 0.5 Hz. The positive and

  negative frequencies of a double-sided spectrum corresponds to points

  on the upper and lower half of the unit circle, respectively.

Taking the Nyquist sampling theorem (section 5.1.1) into account, we must ensure that the sampling frequency $f_s \geq 2 f_{max}$, where f_{max} is the highest frequency component within the signal $s(t)$. Once f_s is selected, the value N must be large enough to ensure the frequency resolution $\left(\frac{f_s}{N}\right)$ sufficiently small enough to minimize smearing effects and to extract all the significant spectral components within $s(t)$. If $f_s = 2 f_{max}$, then the highest frequency component corresponds to $\frac{f_s}{2}$, which in turn corresponds to $k = \frac{N}{2}$. Thus for example, if $f_s = 128$ Hz and $N = 1024$, then the frequency resolution is 0.125 Hz and the highest frequency component that can be extracted is

64 Hz. Given that the DFT is periodic (period of N values) with an even amplitude and odd phase spectrum, only the DFT components from $k = 0$ to $\frac{N}{2}$ are required to establish the double-sided spectrum.

A DFT spectral component has an associated *digital frequency*, defined by

$$\Omega = \frac{2\pi f}{f_s} \text{radians} \tag{1.72}$$

where the frequency f of the DFT spectral component is given by

$$f = \frac{k f_s}{N} \tag{1.73}$$

Hence, the digital frequency $\Omega = \frac{2\pi}{f_s}\left(\frac{k f_s}{N}\right) = \frac{2\pi k}{N}$ can be represented as an angle Ω rads from the horizontal of a unit length phasor $z = e^{j\Omega}$, where positive and negative frequencies correspond to angles ranging from 0 to $+\pi$ and 0 to $-\pi$, respectively. Thus on a circle of unit radius, the digital frequencies of the DFT spectral components correspond to N points on this circle, separated by an angle of $\frac{2\pi}{N}$ rads as illustrated in Fig. 1.25 for $N = 8$, because the first DFT component corresponds to $k = 0$, the second to $k = 1$, etc.

Other key features of the DFT are as follows:

- Both x_n and X_k have N values and these are said to lie in the DFT *window*. Samples of the signal that lie outside the window do not effect the analysis.

- The spectra of both nonperiodic and **periodic** signals may be obtained using the DFT.

- Finally, the DFT is only an approximation of the true spectrum of $s(t)$ because of windowing, aliasing and quantization errors.

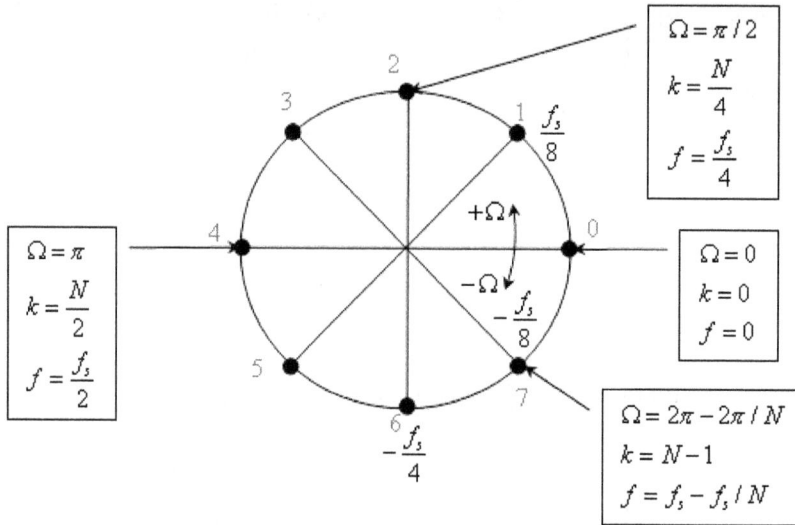

Figure 1.25: DFT points on a unit circle for $N = 8$.

1.4.4   Fast Fourier Transform (FFT)

▶ **Compare DFT with FT**

- Comparing the DFT with the FT of a rectangular pulse

- How to determine the FT envelope via the DFT

- Comparing the DFT spectral components with the FT envelope

- Summary of the samplings process, DFT, FFT and their relationships

A fast method to determine the DFT (which requires fewer calculations) is an algorithm known as the *fast Fourier transform* (FFT). In Mathcad, the FFT of a digital signal $x[n]$ (a vector x of length N) is given by

$$W = CFFT(x) \tag{1.74}$$

where $CFFT(.)$ is an in-built function within Mathcad such that $W_k = \frac{X_k}{N}$. In MATLAB, the equivalent function is $fft(x)$. Provided N (a power of 2) and f_s have been chosen appropriately, the DFT is related to the FT $S(f)$ as follows

$$S(f) \approx \frac{X_k}{f_s} = \left(\frac{N}{f_s}\right) W_k \tag{1.75}$$

where the scaling factor of $\frac{1}{f_s}$ is required to determine approximately the envelope of $S(f)$ using X_k. The analog signal $s(t)$ can be reconstructed from the DFT as follows

$$s_{DFT}(t) = \frac{|X_0|}{N} + \sum_{k=1}^{\frac{N}{2}} 2\frac{|X_k|}{N} \cos\left(2\pi\left(\frac{kf_s}{N}\right)t + \arg\left(X_k\right)\right) \tag{1.76}$$

or equivalently,

$$s_{CFFT}(t) = |W_0| + \sum_{k=1}^{\frac{N}{2}} 2|W_k| \cos\left(2\pi\left(\frac{kf_s}{N}\right)t + \arg\left(W_k\right)\right) \tag{1.77}$$

where $\frac{|X_0|}{N} = |W_0|$ is the d.c. component and $2\frac{|X_k|}{N} = 2|W_k|$ is the amplitude of a cosine of frequency $\left(\frac{kf_s}{N}\right)$ on a single-sided amplitude spectrum. To avoid numerical errors on a computer, if $|W_k| < 0.00001$ (i.e. an insignificant FFT component) then $\arg\left(W_k\right)$ is set zero in equation 1.77, otherwise $\arg\left(W_k\right)$ is calculated via equation 1.27.

8

1.4.5 Simulation I : Rectangular Pulse FT

SIMULATION **FSFT:** The FT of the rectangular pulse presented in Fig. 1.21 is determined and compared with the Fourier series components of the corresponding rectangular *periodic* signal. This simulation will also highlight the fact that evaluating the FT of a signal is computationally intensive.

1.4.6 Simulation II : Rectangular Pulse FFT and DFT

SIMULATION **ShiftRectPulse:** The FT, FFT and DFT together with its inverse functions are illustrated to solve the student exercise. In addition, the reconstruction of the signal from its FT, FFT and DFT are also illustrated. We shall make use of the FFT extensively throughout our study of signal spectra in this book. Therefore, please do take the time to experiment with this particular simulation.

- **Experiment:** Determine the FT of various types of pulses e.g. triangular, etc.

1.5 Fourier Transform Properties

Recall that the FT of a nonperiodic signal $s(t)$ is given by

$$S(f) = \int_{-\infty}^{\infty} s(t)e^{-j2\pi ft}dt \tag{1.78}$$

and its corresponding inverse is given by

$$s(t) = \int_{-\infty}^{\infty} S(f)e^{j2\pi ft}df \tag{1.79}$$

The above two equations are referred to as the *Fourier transform pair*. The standard notation used to denote the Fourier transform pair is $s(t) \longleftrightarrow S(f)$. Using this notation and $FT[s(t)]$ to represent the FT of $s(t)$, we will consider in this section, some of the key FT properties to be used later in this book.

1.5.1 Time Shift

If a signal $s(t)$ is delayed by t_d, then the FT of the delayed signal is given by $S(f)e^{-j2\pi f t_d}$, or more concisely, $s(t - t_d) \longleftrightarrow S(f)e^{-j2\pi f t_d}$. This is easily shown as follows.

$$FT[s(t - t_d)] = \int_{-\infty}^{\infty} s(t - t_d)e^{-j2\pi f t} dt \qquad (1.80)$$

Let $u = t - t_d$, then

$$FT[s(t - t_d)] = \int_{-\infty}^{\infty} s(u)e^{-j2\pi f(u+t_d)} du \qquad (1.81)$$

$$FT[s(t - t_d)] = e^{-j2\pi f t_d} \int_{-\infty}^{\infty} s(u)e^{-j2\pi f u} du = e^{-j2\pi f t_d} S(f) \qquad (1.82)$$

For example, consider the signal

$$s(t) = \begin{cases} \frac{A}{\pi} & \text{if} \quad t = 0 \\ \frac{\sin(At)}{\pi t} & \text{otherwise} \end{cases} \qquad (1.83)$$

The corresponding FT is easily shown (refer to problems section) to be given by

$$S(f) = \begin{cases} 1 & \text{if} \quad |f| \leq \frac{A}{2\pi} \\ 0 & \text{otherwise} \end{cases} \qquad (1.84)$$

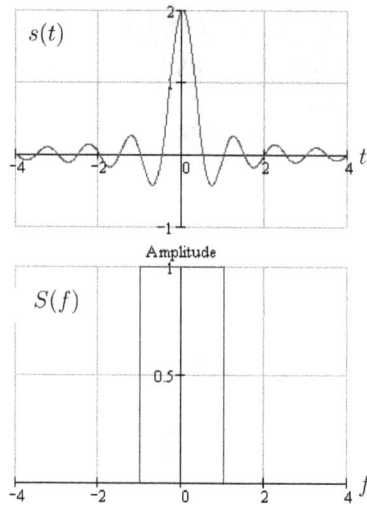

Figure 1.26: Sinc pulse and its Fourier transform.

The above transform pair are presented in Fig. 1.26 for $A = 2\pi$ volts.

If we now multiply $S(f)$ by $e^{-j2\pi f t_d}$ and reconstruct $s(t)$ from this altered spectrum via $s(t) = \int\limits_{-\infty}^{\infty} \left[S(f)e^{-j2\pi f t_d} \right] e^{j2\pi f t} df$, the corresponding signal is presented in Fig. 1.27. As expected, for $t_d = 2$, the signal is time shifted by 2 seconds.

- ⭐ SIMULATION **TimeShift:** Simulation corresponding to Fig. 1.27. Verify for yourself that the amplitude spectrum remains unchanged for different values of t_d. Discover what happens to the phase spectrum.

We shall on numerous occasions, shift or delay a signal to conveniently determine its FT using the discrete FFT functions within Mathcad and MAT-LAB.

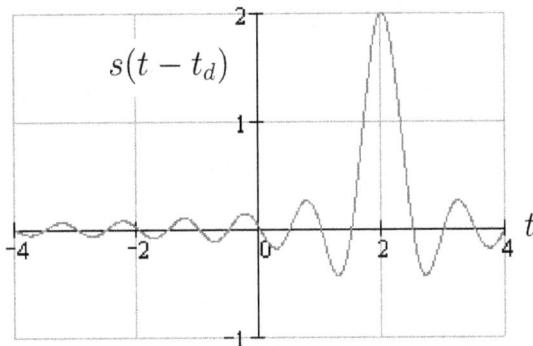

Figure 1.27: Since pulse shifted in time.

1.5.2   Frequency Shift

If a signal $s(t)$ is multiplied by $e^{j2\pi f_c t}$, then the FT is given by $S(f - f_c)$, or more concisely,

$$s(t)e^{j2\pi f_c t} \longleftrightarrow S(f - f_c) \tag{1.85}$$

This property is easily shown as follows.

$$FT\left[s(t)e^{j2\pi f_c t}\right] = \int_{-\infty}^{\infty} s(t)e^{j2\pi f_c t}e^{-j2\pi f t}dt \tag{1.86}$$

$$= \int_{-\infty}^{\infty} s(t)e^{-j2\pi(f - f_c)t}dt = S(f - f_c) \tag{1.87}$$

Using Eulers theorem,

$$\cos\left(2\pi f_c t\right) = \frac{1}{2}\left(e^{j2\pi f_c t} + e^{-j2\pi f_c t}\right) \tag{1.88}$$

and the frequency-shifting property, the FT of $s(t)\cos(2\pi f_c t)$ is given by the FT of $\frac{1}{2}\left(s(t)e^{j2\pi f_c t} + s(t)e^{-j2\pi f_c t}\right)$, which is equal to $\frac{1}{2}\left[S(f - f_c) + S(f + f_c)\right]$.

For example, consider the FT pair presented in Fig. 1.28. If this signal $s(t)$ is multiplied by $e^{j2\pi f_c t}$, where $f_c = 20$ Hz, the corresponding amplitude spectrum $|S(f - f_c)|$ is presented in Fig. 1.29. Notice how the amplitude spectrum of the original pulse is now centered on f_c. The frequency shifting property is also referred to as the *modulation property*.

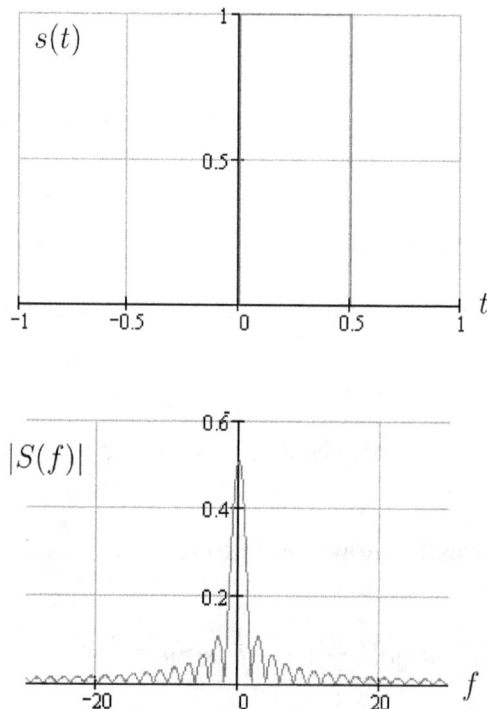

Figure 1.28: Rectangular pulse and its Fourier transform.

- ⭐ SIMULATION **FrequencyShiftExp:** Simulation corresponding to Fig. 1.29. Confirm for yourself that the first null baseband signal bandwidth is equal to $\frac{1}{\tau} = 2$ Hz and that the null-to-null bandpass signal bandwidth is equal to $\frac{2}{\tau} = 4$ Hz.

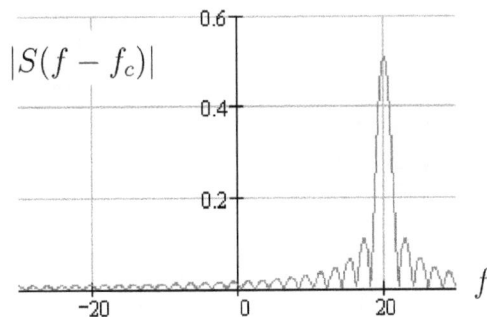

Figure 1.29: Fourier transform shifted to f_c Hz.

However, if the pulse $s(t)$ presented in Fig. 1.28 is multiplied by $\cos(2\pi f_c t)$, where $f_c = 20$ Hz, the new signal $s(t)\cos(2\pi f_c t)$ and its corresponding FT are presented in Fig. 1.30. As expected, we find the FT to be given by

$$S'(f) = \frac{1}{2}\left[S(f - f_c) + S(f + f_c)\right] \tag{1.89}$$

- ⭐ SIMULATION **FrequencyShiftCos:** Simulation corresponding to Fig. 1.30. Try different values for f_c including zero. Modify the code to consider what happens if $s(t)$ is multiplied by $\sin(2\pi f_c t)$.

1.5.3 Time Differentiation

The differential $\frac{ds(t)}{dt}$ is given by the inverse FT of $(j2\pi f)\,S(f)$ i.e. we have the FT pair

$$\frac{ds(t)}{dt} \longleftrightarrow (j2\pi f)\,S(f) \tag{1.90}$$

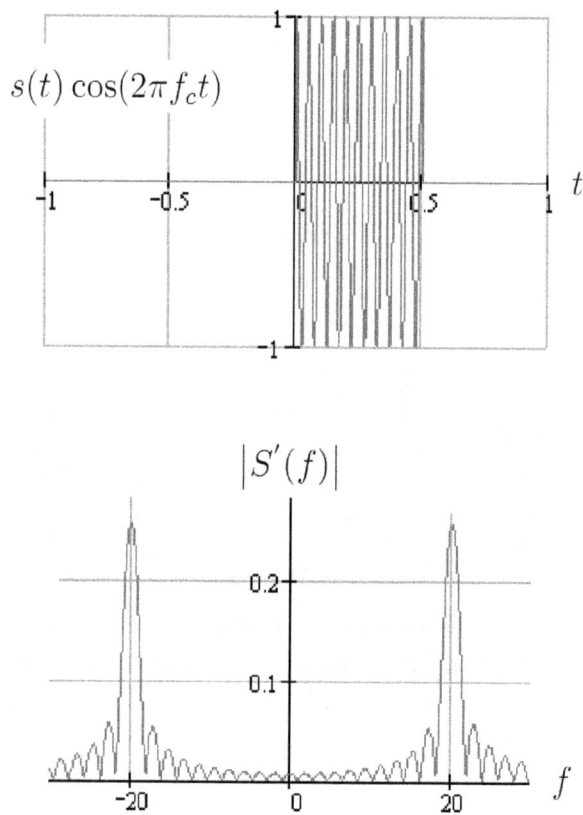

Figure 1.30: High frequency pulse and its Fourier transform $\left| S'(f) \right|$.

To prove this property,

$$\frac{ds(t)}{dt} = \frac{d}{dt}\left(\int_{-\infty}^{\infty} S(f)e^{j2\pi ft}df\right) = \int_{-\infty}^{\infty} j2\pi f S(f)e^{j2\pi ft}df \qquad (1.91)$$

which simply states that the inverse FT of $j2\pi f S(f)$ is given by

$$FT^{-1}\left[j2\pi f S(f)\right] = \frac{ds(t)}{dt} \qquad (1.92)$$

or equivalently

$$FT\left[\frac{ds(t)}{dt}\right] = (j2\pi f)\, S(f) \qquad (1.93)$$

⭐ SIMULATION **Differentiation:** Ilustrated by example using a sinc pulse. Note that in general,

$$\frac{d^n s(t)}{dt^n} \longleftrightarrow (j2\pi f)^n\, S(f) \qquad (1.94)$$

1.5.4   Duality

If $S(f) = FT\left[s(t)\right]$, then

$$s(f) = FT\left[S(-t)\right] \quad \text{and} \quad s(-f) = FT\left[S(-t)\right] \qquad (1.95)$$

To prove this property, consider

$$s(t) = \int_{-\infty}^{\infty} S(f)e^{j2\pi ft}df \qquad (1.96)$$

Let $u = -f$, then

$$s(t) = \int_{-\infty}^{\infty} S(-u)e^{-j2\pi ut}\,du \qquad (1.97)$$

Now let $t = f$,

$$s(f) = \int_{-\infty}^{\infty} S(-u)e^{-j2\pi uf}\,du \qquad (1.98)$$

comparing this with $FT\,[S(-t)] = \int_{-\infty}^{\infty} S(-t)e^{-j2\pi ft}dt$, we may replace u with t to get

$$s(f) = FT\,[S(-t)] \qquad (1.99)$$

For example, to show that $\frac{1}{\pi t} \longleftrightarrow -j\,\text{sgn}\,(f)$, where

$$\text{sgn}\,(t) = \begin{cases} 1 & if \quad t > 0 \\ -1 & if \quad t < 0 \\ 0 & it \quad t = 0 \end{cases} \qquad (1.100)$$

consider first the FT of

$$s\,(t) = \begin{cases} e^{-\frac{t}{n}} & if \quad t > 0 \\ -e^{\frac{t}{n}} & if \quad t < 0 \\ 0 & it \quad t = 0 \end{cases} \qquad (1.101)$$

given by

$$S(f) = \int_{-\infty}^{\infty} s(t)e^{-j2\pi ft}dt = \int_{-\infty}^{0} -e^{\frac{t}{n}}e^{-j2\pi ft}dt + \int_{0}^{\infty} e^{-\frac{t}{n}}e^{-j2\pi ft}dt \qquad (1.102)$$

$$= -\int_{-\infty}^{0} e^{t\left(\frac{1}{n}-j2\pi f\right)}dt + \int_{0}^{\infty} e^{-t\left(\frac{1}{n}+j2\pi f\right)}dt$$

$$= -\frac{1}{\left(\frac{1}{n} - j2\pi f\right)} + \frac{1}{\left(\frac{1}{n} + j2\pi f\right)} = \frac{-j4\pi f}{\frac{1}{n^2} + 4\pi^2 f^2} \qquad (1.103)$$

For $n \longrightarrow \infty$, $s(t) \longrightarrow \text{sgn}(t)$ and therefore

$$FT\left[\text{sgn}(t)\right] = \frac{-j4\pi f}{4\pi^2 f^2} = \frac{1}{j\pi f} \qquad (1.104)$$

Now making use of the duality theorem $s(f) = FT\left[S(-t)\right]$, we have

$$FT\left[\frac{1}{j\pi t}\right] = \text{sgn}(-f) = -\text{sgn}(f) \qquad (1.105)$$

or equivalently $\frac{1}{j\pi t} \longleftrightarrow -\text{sgn}(f)$, so that

$$\frac{1}{\pi t} \longleftrightarrow -j\,\text{sgn}(f) \qquad (1.106)$$

1.5.5 Scaling

The scaling property

$$s(at) \longleftrightarrow \frac{1}{|a|}S\left(\frac{f}{a}\right) \qquad (1.107)$$

is easily shown as follows. Let $u = at$ so that

$$FT\left[s(at)\right] = \int_{-\infty}^{\infty} s(at)e^{-j2\pi ft}dt = \frac{1}{|a|}\int_{-\infty}^{\infty} s(u)e^{-j2\pi f\frac{u}{a}}du \qquad (1.108)$$

but since $S\left(\frac{f}{a}\right) = \int_{-\infty}^{\infty} s(t)e^{-j2\pi \frac{f}{a}t}dt$, we may replace u with t to get

$$FT\left[s(at)\right] = \frac{1}{|a|}S\left(\frac{f}{a}\right) \qquad (1.109)$$

1.5.6   Other Properties

Some other FT properties which are easily proved or verified using Mathcad/MATLAB:

Linearity (or superposition)

$$a_1 s_1(t) + a_2 s_2(t) \longleftrightarrow a_1 S_1(f) + a_2 S_2(f) \tag{1.110}$$

where a_1 and a_2 are any constants.

Time Reversal

$$s(-t) \longleftrightarrow S(-f) \tag{1.111}$$

Spectral symmetry

$$S(-f) \longleftrightarrow S^*(f) \tag{1.112}$$

if $s(t)$ is a real signal.

1.5.7   Dirac Delta Function

The *Dirac delta function* $\delta(.)$ (or *impulse function*) is a mathematical abstraction such that

$$\delta(x) = \begin{cases} \infty & \text{if } x = 0 \\ 0 & x \neq 0 \end{cases} \tag{1.113}$$

such that

$$\int_{-\infty}^{\infty} \delta(x) dx = 1 \tag{1.114}$$

Notice that $\delta(.)$ sifts out the value $f(a)$ from the intergal

$$\int_{-\infty}^{\infty} f(x)\delta(x - a) dx = f(a) \tag{1.115}$$

in which $f(x)$ is any function that is continous at $x = 0$. Thus, the Fourier transform of the delta function $\delta(t)$ is given by

$$\int\limits_{-\infty}^{\infty} \delta(t)e^{-j2\pi ft}dt = e^0 = 1 \qquad (1.116)$$

The implication of this result that $FT[\delta(t)] = 1$ will become evident in section 1.8.6, when we consider the impulse response of a system. Conversly, $\delta(t) = FT^{-1}[1] = \int\limits_{-\infty}^{\infty} e^{+j2\pi ft}df$ and a very interesting feature is that for *even-sided* delta functions for which $\delta(-x) = \delta(x)$,

$$\delta(x) = \int\limits_{-\infty}^{\infty} e^{\pm j2\pi yx}dy \qquad (1.117)$$

where either the $+$ or the $-$ sign may used as required.

1.5.8   Hilbert Transform

The *Hilbert transform* phase shifts every frequency component of a signal $s(t)$ by $-\frac{\pi}{2}$ radians. For example, the Hilbert transform of $s(t) = A\cos(2\pi f + \theta)$ is $s(t) = A\cos(2\pi f + \theta - \frac{\pi}{2}) = A\sin(2\pi f + \theta)$. Thus, a phasor (positive frequency component) $Ae^{j2\pi ft}$ will transform to $Ae^{j\left(2\pi ft - \frac{\pi}{2}\right)} = Ae^{j(2\pi ft)}e^{j\left(-\frac{\pi}{2}\right)} = -jAe^{j(2\pi ft)}$ and $Ae^{-j2\pi ft}$ (negative frequency component) will transform to $jAe^{-j2\pi ft}$. This implies that in the frequency domain, the operation of the Hilbert transform is simply to multiply the spectrum of the signal by $-j$ and $+j$ over positive and negative frequencies, respectively. Thus for a real signal $s(t)$ with no d.c. component, let $s_h(t)$ denote the Hilbert transform of $s(t)$. Then the FT of $s_h(t)$ is given by

$$FT[s_h(t)] = -j\,\text{sgn}(f)S(f) \qquad (1.118)$$

$$\text{where sgn}\,(f) = \begin{cases} 1 & if \quad f > 0 \\ -1 & if \quad f < 0 \\ 0 & it \quad f = 0 \end{cases}. \quad \text{If we apply the Hilbert transform}$$

twice, then

$$[-j\,\text{sgn}(f)]^2 \, S(f) = -S(f) \tag{1.119}$$

i.e. a sign reversal of the signal. An example which demonstrates the use of the Hilbert transform is presented in section 2.3.2. It is shown in section 1.8.2 that the Hilbert transform of $s(t)$ is equivalent to its convolution with $\frac{1}{\pi t}$ in the time domain.

1.6   Spectral Density

In this section, we shall consider the signal processing tools necessary to analyze the distribution of energy and power within the spectral components of a signal $s(t)$.

1.6.1   Parseval's Theorem for the Fourier Transform

Consider two signals $x(t)$ and $y(t)$ with FTs $X(f)$ and $Y(f)$, then

$$\int_{-\infty}^{\infty} x(t)y^*(t)dt = \int_{-\infty}^{\infty} \left[\int_{-\infty}^{\infty} X(f)e^{j2\pi ft}df \right] y^*(t)dt \tag{1.120}$$

$$= \int_{-\infty}^{\infty}\int_{-\infty}^{\infty} X(f)y^*(t)e^{j2\pi ft}df\,dt \tag{1.121}$$

Assuming the order of integration can be changed

$$\int_{-\infty}^{\infty} x(t)y^*(t)dt = \int_{-\infty}^{\infty} X(f)\left[\int_{-\infty}^{\infty} y(t)e^{-j2\pi ft}dt \right]^* df = \int_{-\infty}^{\infty} X(f)Y^*(f)df$$

$$\tag{1.122}$$

Equation 1.122 is known as *Parseval's theorem or relation* for the Fourier transform.

1.6.2 Rayleigh's Energy Theorem

▶ **Energy Spectral Density**

- The ESD of a rectangular pulse is analyzed

Rayleigh's energy theorem follows from equation 1.122 for $x(t) = y(t)$ so that

$$\int\limits_{-\infty}^{\infty} x(t)x^*(t)dt = \int\limits_{-\infty}^{\infty} X(f)X^*(f)df \qquad (1.123)$$

or equivalently the energy E (from equation 1.2) of the signal $x(t)$

$$E = \int\limits_{-\infty}^{\infty} |x(t)|^2 \, dt = \int\limits_{-\infty}^{\infty} |X(f)|^2 \, df \qquad (1.124)$$

Thus, for a real-valued energy signal $s(t)$, the *energy spectral density* $ESD(f)$ (Joules/Hz) is given by

$$ESD(f) = |S(f)|^2 \qquad (1.125)$$

where $S(f)$ is the FT of $s(t)$. As the name suggests, the ESD is measure of how the energy of the signal $s(t)$ is distributed within its corresponding spectral components. The implication of equation 1.124 is that the area under a $ESD(f)$ versus f plot between two limits, say f_1 and f_2, is the summation of energy content of the spectral components between f_1 and f_2. In passing we note that a signal's energy is unaffected by the Hilbert transform because

$$E = \int_{-\infty}^{\infty} |x_h(t)|^2 \, dt = \int_{-\infty}^{\infty} |-j \, \text{sgn}(f) X(f)|^2 \, df = \int_{-\infty}^{\infty} |X(f)|^2 \, df \qquad (1.126)$$

where $x_h(t)$ is the Hilbert transform of a signal $x(t)$. Furthermore, $x(t)$ and $x_h(t)$ are orthogonal because given that $FT\left[x_h(t)\right] = X_h(f) = -j \, \text{sgn}(f) X(f)$, we have from Parsevals theorem

$$\int_{-\infty}^{\infty} x(t) x_h^*(t) dt = \int_{-\infty}^{\infty} X(f) X_h^*(f) df = \int_{-\infty}^{\infty} X(f) \left[-j \, \text{sgn}(f) X(f)\right]^* df \quad (1.127)$$

and provided $|X(f)|^2$ is even

$$\int_{-\infty}^{\infty} x(t) x_h^*(t) dt = -j \int_{-\infty}^{0} |X(f)|^2 \, df + j \int_{0}^{\infty} |X(f)|^2 \, df = 0 \qquad (1.128)$$

1.6.3   Time-Average Autocorrelation Function

▶ Time Average AutoCorrelation Function

- Fundamentals

- Autocorrelation function of a rectangular pulse

- Energy spectral density from the autocorrelation function

The energy E of a signal $s(t)$ may also be determined from the FT of the signal's *time-average autocorrelation function* $R(\tau)$ defined by

$$R(\tau) = \int\limits_{-\infty}^{\infty} s(t)s(t \pm \tau)dt \tag{1.129}$$

i.e. $R(\tau)$ is an even function of τ. For $\tau = 0$, $R(0) = \int\limits_{-\infty}^{\infty} [s(t)]^2\,dt = E$, where E is the energy content (normalized) of $s(t)$. The autocorrelation function $R(\tau)$ is a measure of how closely the original signal $s(t)$ and its copy $s(t-\tau)$ or $s(t+\tau)$ match at time τ. For example, consider the rectangular pulse shown in Fig. 1.31. In Fig. 1.32(a), a plot of $s(t)$ and $s(t-\tau)$ are shown for the specific example of $\tau = 0.4$. A plot of $s(t)s(t-\tau)$ is shown in Fig. 1.32(b). Notice that $s(t)s(t-\tau)$ is non-zero within their region of overlap. For $\tau = 0.4$,

$$R(0.4) = \int\limits_{-\infty}^{\infty} s(t)s(t-0.4)dt = 0.1 \tag{1.130}$$

is the shaded area in Fig. 1.32(b). This specific value of $R(0.4) = 0.1$ is shown as a large dot in Fig. 1.32(c). The solid lines correspond to calculating $R(\tau)$ for values of τ ranging from -1 to +1. For $\tau = 0$, we find $R(0) = 0.5$ which is the energy content of this signal. This is easily confirmed as follows:

$$E = \int\limits_{-\infty}^{\infty} [s(t)]^2\,dt = \int\limits_{0}^{0.5} 1^2 dt = [t]_0^{0.5} = 0.5 \text{ Joules} \tag{1.131}$$

• ⭐ SIMULATION **ESDAutocorrelation:** Simulation corresponding to Fig. 1.32. Try $R(\tau) = \int\limits_{-\infty}^{\infty} s(t)s(t+\tau)dt$ within the simulation. Does this make any difference ?

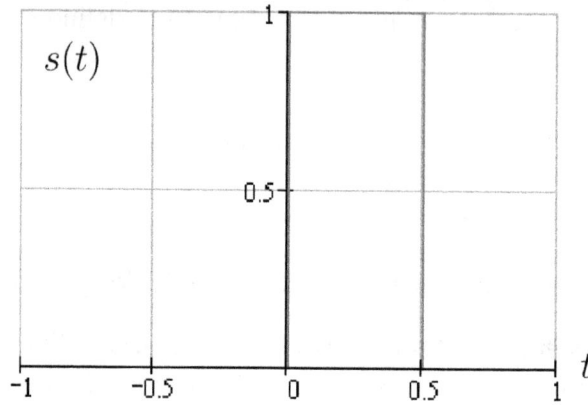

Figure 1.31: Rectangular pulse.

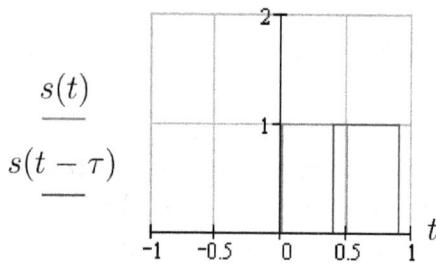

$$R(\tau, \text{Limit}) := \int_{-\text{Limit}}^{\text{Limit}} s(t) \cdot s(t - \tau)\, dt$$

$$\tau = 0.4 \qquad R(\tau, 1) = 0.1$$

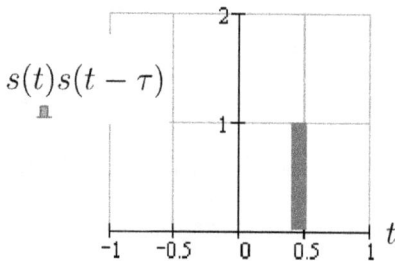

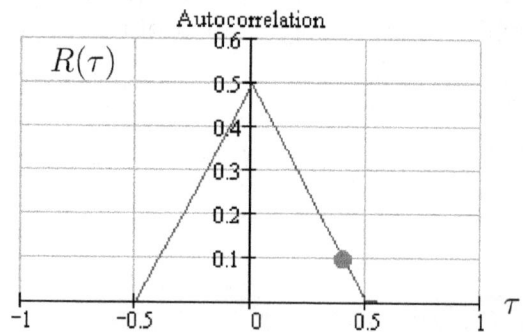

Figure 1.32: Autocorrelation example.

The autocorrelation function also provides an insight to the bandwidth of the signal. If $R(\tau)$ rapidly or gently decreases for values of τ just greater than or less than zero, then the bandwidth of the signal is large or small, respectively.

1.6.4   Power Spectral Density

▶ **Power Spectral Density**

- Autocorrelation function of a power signal

- Power spectral density of a sinusoidal signal

The *time-average autocorrelation function* $R(\tau)$ of a real **power signal** $s(t)$ is defined by

$$R(\tau) = \lim_{T\to\infty}\frac{1}{T}\int_{-T/2}^{T/2} s(t)s(t\pm\tau)dt \qquad (1.132)$$

Notice the similarity and the difference between the definitions of $R(\tau)$ for a energy and a power signal. For the special case of a *periodic* real-valued power signal $s(t)$ with time period T, the time-average autocorrelation function is given by

$$R(\tau) = \frac{1}{T}\int_{-T/2}^{T/2} s(t)s(t\pm\tau)dt \qquad (1.133)$$

where for $\tau = 0$,

$$R(0) = \frac{1}{T} \int\limits_{-T/2}^{T/2} [s(t)]^2 \, dt = P \tag{1.134}$$

where P is the average signal power. A key feature is that the *power spectral density* $PSD(f)$ (Watts/Hz) is the FT of $R(\tau)$, namely

$$PSD(f) = FT\,[R(\tau)] \tag{1.135}$$

The PSD is a measure of how the power of the signal $s(t)$ is distributed within its corresponding spectral components. Thus the signal power P from the PSD is given by

$$P = \int\limits_{-\infty}^{\infty} PSD(f) df \tag{1.136}$$

Power Spectral Density of a Sinusoid

Consider the sinusoidal power signal $s(t) = A\sin(2\pi f_o t + \phi)$ presented in Fig. 1.33 for $A = 1$, $\phi = 0$ and $f_o = 1$ Hz. The power of this signal (via the time-domian) is given by

$$P = \frac{1}{T} \int\limits_{-T/2}^{T/2} |s(t)|^2 \, dt = \frac{1}{T} \int\limits_{-T/2}^{T/2} A^2 \sin^2(2\pi f_o t) dt \tag{1.137}$$

Making use of the trigonometry identity $\sin^2\theta = \frac{1}{2}\left(1 - \cos 2\theta\right)$, $f_o = \frac{1}{T}$ and $\sin(2\pi f_o T) = 0$

$$P = \frac{A^2}{2} = 0.5 \text{ Watts} \tag{1.138}$$

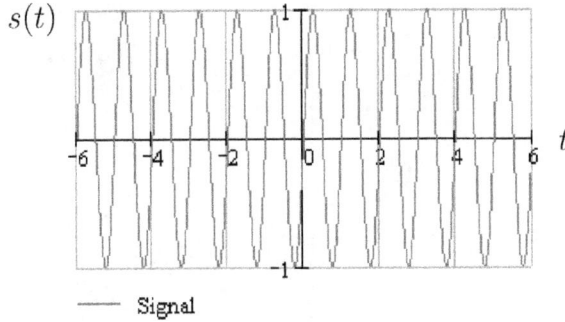

Figure 1.33: Sinusoidal power signal $s(t) = \sin(2\pi t)$.

The corresponding time-average autocorrelation function is given by

$$R(\tau) = \frac{1}{T} \int_{-T/2}^{T/2} s(t)s(t-\tau)dt \qquad (1.139)$$

$$= \frac{A^2}{T} \int_{-T/2}^{T/2} \sin(2\pi f_o t + \phi)\sin(2\pi f_o t - 2\pi f_o \tau + \phi)dt \qquad (1.140)$$

Using the trigonometry identity $\sin A \sin B = \frac{\cos(A-B)-\cos(A+B)}{2}$

$$R(\tau) = \frac{A^2}{2T} \int_{-T/2}^{T/2} \cos(2\pi f_o \tau)\, dt \qquad (1.141)$$

$$-\frac{A^2}{2T} \int_{-T/2}^{T/2} \cos(4\pi f_o t + 2\phi - 2\pi f_o \tau)\, dt \qquad (1.142)$$

$$= \frac{A^2 \cos(2\pi f_o \tau)}{2T}\left[\frac{T}{2} - \left(\frac{-T}{2}\right)\right] = \frac{A^2}{2}\cos(2\pi f_o \tau) \qquad (1.143)$$

Figure 1.34 illustrates the variation of autocorrelation function $R(\tau)$ with τ. The power spectral density, which is the FT of $R(\tau)$, is therefore given by

$$PSD(f) = FT\left[\frac{A^2}{2}\cos\left(2\pi f_o\tau\right)\right] \qquad (1.144)$$

$$= \frac{A^2}{4}\left[\delta(f - f_o) + \delta(f + f_o)\right] \qquad (1.145)$$

as represented in Fig. 1.35, where $\delta(.)$ is the Dirac delta function (section 1.5.7). The average power P using equation 1.136 is thus given by

$$P = \int_{-\infty}^{\infty} PSD(f)df = \int_{-\infty}^{\infty} \frac{A^2}{4}\left[\delta(f - f_o) + \delta(f + f_o)\right]df \qquad (1.146)$$

$$= \frac{A^2}{4} + \frac{A^2}{4} = \frac{A^2}{2} \qquad (1.147)$$

because $\int_{-\infty}^{\infty} \delta(f - f_o)df = 1$ only when $f = f_0$ and $\int_{-\infty}^{\infty} \delta(f + f_o)df = 1$ only when $f = -f_0$, otherwise its zero.

- ⭐ SIMULATION **PSDSinusoidal:** Simulation corresponding to Fig. 1.35.

Finally, note that in section 1.7.4, the random process $x(t, \theta) = A\cos\left(2\pi ft + \theta\right)$ is analyzed, where the amplitude A and frequency f are constants and the phase angle θ can be any value in the range 0 to 2π with equal probability.

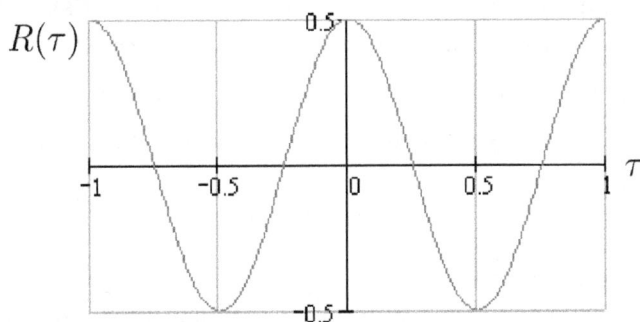

Figure 1.34: Autocorrelation of the sinusoidal signal.

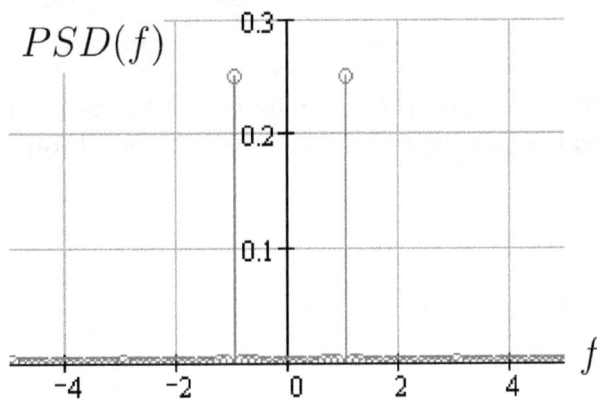

Figure 1.35: The FFT of $R(\tau) = \frac{A^2}{2} \cos{(2\pi f \tau)}$ or equivalently, the PSD of the sinusoidal signal.

1.6.5 Signal Power and Energy via Digital Samples and its FFT

In sections 1.4.3 and 1.4.4, we considered how an analog signal $s(t)$ may be instantaneously sampled N times (typically a power of 2) to create a digital signal $x_n = s\left(\frac{n}{f_s}\right)$, where $n = 0, 1, \ldots, N-1$, and that the amplitude and phase spectrum of either a periodic or non-periodic analog signal $s(t)$ can be determined via the FFT $W_k = \frac{X_k}{N}$, where $X_k = \sum_{n=0}^{N-1} x_n e^{-j\frac{2\pi kn}{N}}$ is the DFT of x_n. Recall the frequency resolution $\left(\frac{f_s}{N}\right)$ Hz is carefully selected to ensure the DFT points lie on the spectral components to avoid smearing effects. If $s(t)$ is a power signal, then from equation 1.7, the average power P of $s(t)$ can be estimated via

$$P \simeq \frac{1}{N} \sum_{n=0}^{N-1} |x_n|^2 = \sum_{k=0}^{N-1} |W_k|^2 \qquad (1.148)$$

because from equation 1.53, $|W_k|^2$ is the power of each spectral component within $s(t)$.

If $s(t)$ is an energy signal, then its energy can be estimated using the N-point FFT spectral components W_k of x_n as follows. From equation 1.125, the

$$ESD(f) = |S(f)|^2 \qquad (1.149)$$

where $S(f)$ (equation 1.75) is approximately given by $\left(\frac{N}{f_s}\right) W_k$. Thus, a good estimate of the ESD is $\left|\left(\frac{N}{f_s}\right) W_k\right|^2$. Given that the signal energy $E = \int_{-\infty}^{\infty} ESD(f)df$ is the area under the variation of $ESD(f)$ versus f, and that the separation between each FFT component is $\left(\frac{f_s}{N}\right)$ Hz, a good approximation of the signal $s(t)$ energy is

$$E \approx \sum_{k=0}^{N-1} \left| \left(\frac{N}{f_s} \right) W_k \right|^2 \left(\frac{f_s}{N} \right) = \left(\frac{N}{f_s} \right) \sum_{k=0}^{N-1} |W_k|^2 \qquad (1.150)$$

or equivalently, since $\sum_{k=0}^{N-1} |W_k|^2 = \frac{1}{N} \sum_{n=0}^{N-1} |x_n|^2$ from equation 1.148, the energy of the signal $s(t)$ can be estimated via

$$E \approx \frac{1}{f_s} \sum_{n=0}^{N-1} |x_n|^2 \qquad (1.151)$$

1.6.6 Digital Signal Energy

Unlike equation 1.151, which gives an estimate of the analog signal $s(t)$ energy using the digital samples x_n and knowledge of the sampling frequency, the energy E of a *digital* signal x_n is defined to be given by [Proakis and Manolakis, 1996]

$$E = \sum_{n=0}^{N-1} |x_n|^2 \qquad (1.152)$$

where E increases as more samples N are taken. It can be shown [Proakis and Manolakis, 1996] that $\sum_{n=0}^{N-1} |x_n|^2$ is also equal to

$$E = \frac{1}{2\pi} \int_{-\pi}^{\pi} |X(\Omega)|^2 \, d\Omega \qquad (1.153)$$

where $X(\Omega)$ is the discrete-time Fourier transform (DTFT).

1.6.7   Simulation I : PSD of Periodic Signals

⭐SIMULATION **PSDPeriodicSignal:** The power of the periodic signal $s(t) = 5 - 6\cos\left(20\pi t - \frac{\pi}{6}\right) + 3\sin\left(70\pi t\right)$, that was previously analyzed in section 1.1.4, is determined via its PSD and confirmed with equation 1.4. This analog signal is then sampled to a create a digital signal $x_n = s\left(\frac{n}{f_s}\right)$, where f_s is the sampling frequency and $n = 0, 1, \ldots, N-1$. The power P of $s(t)$ is determined using $P = \frac{1}{N}\sum_{n=0}^{N-1} |x_n|^2$ and also via the FFT spectral components using $P = \sum_{k=0}^{N-1} |W_k|^2$, where W_k is the FFT of x_n. For accurate results, the frequency resolution $\frac{f_s}{N}$ Hz is carefully selected to ensure the FFT points lie exactly on the spectral components within $s(t)$ i.e. no smearing effects.

1.6.8   Simulation II : Unipolar NRZ Line Code PSD

⭐SIMULATION **PSDLineCode:** A unipolar NRZ line code baseband signal is generated and analyzed to verify that approximately 91% of power is contained within the main lobe. Furthermore, the power of this signal is determined via the following methods:

1. $P = \frac{1}{T}\int_{\frac{-T}{2}}^{\frac{T}{2}} |s(t)|^2 \, dt$ because its a periodic signal.

2. $P = \frac{A^2\tau}{T}$ from equation 1.61 in section 1.3.   This is the theoretical result of applying method (1).

3. $P = \sum_{f=-\infty}^{\infty} PSD(f)$ where $PSD(f) = \sum_{n=-\infty}^{\infty} |c_n|^2 \delta(f - nf_o)$ using $c_n = \frac{A\tau}{T}\text{sinc}\left(\frac{n\tau}{T}\right)$ from equation 1.34 in section 1.2.   Equivalent to method (8) in which the PSD is used to determine the power content.

4. $P = \sum\limits_{-N}^{N} |c_n|^2$ where $c_n = \frac{A\tau}{T} \text{sinc}\left(\frac{n\tau}{T}\right)$. Equivalent to methods (3) and (6) in which the power contribution of each spectral component (cosine) is summed over a double-sided spectrum.

5. $P = \frac{1}{N}\sum\limits_{n=0}^{N-1} |x_n|^2$ where $x_n = s\left(\frac{n}{fs}\right)$ is the sampled version of the signal $s(t)$ and f_s is the sampling frequency. Here we are making of equation 1.7.

6. $P = \sum\limits_{k=0}^{N-1} |W_k|^2$ where W_k is the FFT of $x_n = s\left(\frac{n}{fs}\right)$. Equivalent to method (4). The frequency resolution $\frac{f_s}{N}$ is selected carefully to ensure the FFT (or DFT) points lie exactly on the spectral component frequencies.

7. $P = R(0)$ where $R(\tau) = \frac{1}{T} \int\limits_{-T/2}^{T/2} s(t)s(t+\tau)dt$ is the autocorrelation function of $s(t)$ from equation 1.133. Equivalent to method (1).

8. $P = \sum\limits_{k=0}^{N-1} |Wpsd_k|$ where $Wpsd$ is the FFT of $xAutoCor_n = R\left(\frac{n}{fs}\right)$, the sampled version of $R(\tau)$. Equivalent to method (3).

1.6.9   Simulation III : ESD of a Pulse

⭐ SIMULATION **PulseEnergy:** The rectangular pulse $s(t)$ in Fig. 1.31 is sampled at a frequency of f_s to create its digital representation x_n. The energy of $s(t)$ is estimated using equation 1.150 and the result compared with the expected value of $E = \int\limits_{-\infty}^{\infty} [s(t)]^2\, dt = \int\limits_{0}^{\tau} A^2 dt = A^2\tau$, where A is the height of the pulse and τ is the width.

1.7   Random Processes

All of the signals considered up to this point have been *deterministic* signals i.e. signals whose values are easily determined by a mathematical expression. In the real world, noise will transform a deterministic signal into a *random* signal whose future values cannot be predicted.   To analyze the behavior of a random signal, we must consider many instances of that signal or a collection of random signals as shown in Fig. 1.36.   Let these signals be the output from identical noise sources.   Please refer to Appendix A.11 for the fundamentals of discrete and continuous random variables as required throughout this section.

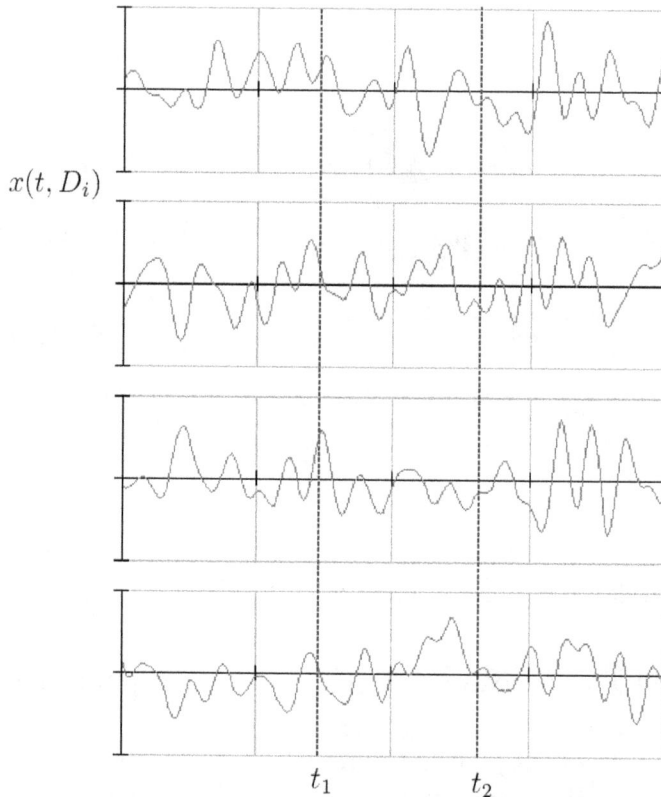

Figure 1.36: Sample functions of a random process $x(t, D)$.

The waveform output from any given noise source is referred to as a *sample function* and the set of all possible sample functions, obtained from an infinite number of noises sources, is called the *ensemble*. This ensemble of random signals defines the *random process*. For example, suppose we measure the temperature variation throughout the day and plot a graph of temperature versus time. This waveform would be a single sample function. If we repeat this experiment over N days, then we would have experimentally obtained N sample functions. Let $x(t, D_i)$ represent the temperature at time t on a given day D_i where $i = 1, 2, \cdots, N$. If we measure the temperature at the same time t_1 everyday, then we would obtain the temperatures $x(t_1, D_1)$, $x(t_1, D_2)$, $\cdots, x(t_1, D_N)$. Given that the value of $x(t_1, D)$ depends on D, then $x(t_1, D)$ is simply a random variable, denoted by $X(t_1)$, of the random process $x(t, D)$. If we sample $x(t, D)$ at a different time, say t_2, then the temperatures $x(t_2, D_1)$, $x(t_2, D_2)$, $\cdots, x(t_2, D_N)$ are the values of the random variable $X(t_2)$. Let x_i denote the value of the random variable $X(t_i)$. To completely describe the random process, we would need to determine the joint probability density function $f_X(x_1, x_2, \cdots, x_N)$, which is an impossible task. Fortunately, we typically require only partial information - formally referred to as the *second-order characterization* - about a random process to analyze random signals within a communication system as follows.

1.7.1 Ensemble Averages

The mean value of x_1, referred to as the *ensemble average* at a given time t_1, is defined by

$$\overline{x_1} = E\left[X(t_1)\right] = \int\limits_{-\infty}^{\infty} x_1 f_X(x_1) dx_1 \qquad (1.154)$$

where $f_X(x_1)$ is the PDF of the random variable $X(t_1)$. Refer to section A.11.2 for further details, in particular equation A.21. Notice the use of a specific time t_1 instead of simply time t to emphasize that this an **ensemble** average $\overline{x_1}$, which is also referred to as the *first* moment. Unlike other text books which utilize t and assume the reader appreciates that it is an ensemble average, the notation of t_1 will be maintained in the material to follow for clarity.

To determine $\overline{x_1}$ in a simulation, one would simply find all the average of the sample function amplitudes at time $t = t_1$. The *second* moment (equation A.24) is given by

$$\overline{x_1^2} = E\left[X^2(t_1)\right] = \int\limits_{-\infty}^{\infty} x_1^2 f_X(x_1)dx_1 \qquad (1.155)$$

Another important characterization of a (real) random process is the ensemble *autocorrelation function* $R_X(t_1, t_2)$ defined by

$$R_X(t_1, t_2) = \overline{x_1 x_2} = E[X(t_1)X(t_2)] = \int\limits_{-\infty}^{\infty}\int\limits_{-\infty}^{\infty} x_1 x_2 f_X(x_1, x_2)dx_1 dx_2 \quad (1.156)$$

i.e. multiply x_1 by x_2 over all the sample functions and find the average value. It will become evident later that the autocorrelation function provides a valuable insight to the spectral information of a random process.

Average Power of WSS Random Process

If $\overline{x_1} = \overline{x_2} = \cdots = \overline{x_N}$ and $R_X(t_1, t_2) = R_X(\tau)$, where $\tau = t_2 - t_1$, then the random process is referred to as a *wide-sense (or weakly) stationary* (WSS) process because $R_X(t_1, t_2)$ depends only on the time difference τ and not the actual value of t_1 and t_2. This typically the case for a communication system. Expressing $R_X(t_1, t_2)$ as $R_X(t_1, t_1 + \tau) = E[X(t_1)X(t_1 + \tau)]$, notice that for $\tau = 0$, the average power of a WSS process (as we shall see later in equation 1.176) is given by the second moment

$$R_X(0) = E[X^2(t_1)] = \overline{x_1^2} \qquad (1.157)$$

and is independent of time. Other properties are that $R_X(-\tau) = R_X(\tau)$ and $|R_X(\tau)| \leq R_X(0)$. Up to this point, we have only considered the ensemble averages i.e. the average of x_i at time t_i and $R_X(t_1, t_2) = \overline{x_1 x_2}$ which is the ensemble average of $x_1 x_2$. Next we shall consider time-averages.

1.7.2 Time Averages

For a given sample function, denoted by $x(t)$, the *time-averaged mean* $\langle x(t) \rangle$ is given by

$$\langle x(t) \rangle = \lim_{T \to \infty} \frac{1}{T} \int_{-T/2}^{T/2} x(t)dt \qquad (1.158)$$

Similarly, the *time-averaged autocorrelation* $R(\tau)$ (defined in section 1.6.4) of the sample function $x(t)$ is given by

$$R(\tau) = \lim_{T \to \infty} \frac{1}{T} \int_{-T/2}^{T/2} x(t)x(t + \tau)dt \qquad (1.159)$$

Of course the value of $\langle x(t) \rangle$ and $R(\tau)$ will depend on which sample function of the random process is used i.e. $\langle x(t) \rangle$ and $R(\tau)$ are also random variables.

1.7.3 Stationarity and Ergodicity

If a random process is *stationary*, then the expected value of the time-averaged mean (random variable) is equal to the ensemble mean so that

$$E\left[\langle x(t) \rangle\right] = \overline{x_1} = \overline{x_2} = \cdots = \overline{x_N} \qquad (1.160)$$

Furthermore, the expected value of the time-averaged autocorrelation $R(\tau)$ (random variable) is equal to the ensemble autocorrelation $R_X(\tau)$ i.e.

$$E\left[R(\tau)\right] = R_X(\tau) \qquad (1.161)$$

However, if $\langle x(t) \rangle$ is the **same** for all sample functions **and** $\langle x(t) \rangle = \overline{x_1} = \overline{x_2} = \cdots = \overline{x_N}$, then the random process is said to be *ergodic* in the mean. Similarly, if $R(\tau) = R_X(\tau)$, then the random process is said to be ergodic in the autocorrelation. Note that for an ergodic random process, all the statistics can be obtained from a single sample function i.e. we can use time averages instead of ensemble averages. This is equivalent to assuming that the ensemble of random signals consists of all possible time shifts of a single random signal. Since a time average cannot be a function of time, an ergodic process is inherently a stationary process but not vice versa. A reasonable assumption is that most communication signals are ergodic in the mean and in the autocorrelation. **Unless otherwise stated, we shall make this assumption throughout this book.** In summary, a random process is

- stationary if its statistical characteristics do not change with time and the ensemble autocorrelation function $R_X(t_1, t_2) = \overline{x_1 x_2} = R_X(\tau)$ where $\tau = t_2 - t_1$.

- wide-sense stationary (WSS) if the ensemble mean $\overline{x}$ at any time is a constant and $R_X(t_1, t_2) = \overline{x_1 x_2} = R_X(\tau)$.

- ergodic if the time average $\langle x(t) \rangle$ is the same for all sample functions, $\langle x(t) \rangle = \overline{x_1} = \overline{x_2} = \cdots = \overline{x_N}$ and the time-averaged autocorrelation is equal to the ensemble autocorrelation $R(\tau) = R_X(\tau)$.

1.7.4   Example : Random Phase Sinusoidal Signal

Consider the random process $x(t, \theta) = A \cos(2\pi f t + \theta)$, where the amplitude A and frequency f are constants and the phase angle θ can be any value in the range 0 to 2π with equal probability i.e. $f_\theta(\theta) = \frac{1}{2\pi}$. Some of the sample functions of $x(t, \theta)$ for $A = 1$ volt, $f = 1$ Hz are shown in Fig. 1.37.

Firstly, the ensemble average $\overline{x_1}$ at time t_1 is given by

$$\overline{x_1} = \int_{-\infty}^{\infty} x_1 f_X(x_1) dx_1 = \int_{-\infty}^{\infty} A \cos(2\pi f t_1 + \theta) f_\theta(\theta) d\theta \tag{1.162}$$

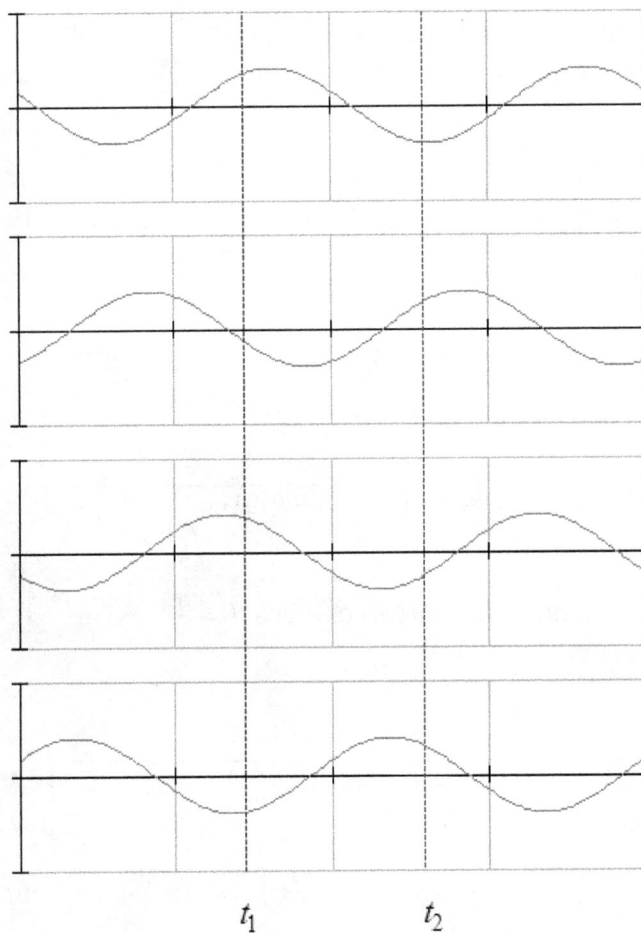

Figure 1.37: Sample functions of the random process $x(t, \theta)$.

$$\overline{x_1} = \frac{A}{2\pi} \int\limits_0^{2\pi} \cos\left(2\pi f t_1 + \theta\right) d\theta = 0 \qquad (1.163)$$

The same would be true for any time t i.e. $\overline{x} = \overline{x_1} = \overline{x_2} = \cdots = \overline{x_N} = 0$. Similarly, the second moment

$$\overline{x_1^2} = \int\limits_0^{2\pi} \left[A\cos\left(2\pi f t_1 + \theta\right)\right]^2 \frac{1}{2\pi} d\theta = \frac{A^2}{2} \qquad (1.164)$$

Secondly, the autocorrelation function $R_X(t_1, t_2)$ is given by

$$R_X(t_1, t_2) = \overline{x_1 x_2} = \overline{A\cos\left(2\pi f t_1 + \theta\right) A\cos\left(2\pi f t_2 + \theta\right)} \qquad (1.165)$$

Using the trigonometry identities $\cos\alpha\cos\beta = \frac{1}{2}\left[\cos\left(\alpha - \beta\right) + \cos\left(\alpha + \beta\right)\right]$ with $\cos(-\theta) = \cos\theta$,

$$= \frac{A^2}{2}\left\{\overline{\cos\left(2\pi f\left(t_2 - t_1\right)\right) + \cos\left(2\pi f\left(t_2 + t_1\right) + 2\theta\right)}\right\} \qquad (1.166)$$

$$= \frac{A^2}{2}\left\{\overline{\cos\left(2\pi f\left(t_2 - t_1\right)\right)} + \overline{\cos\left(2\pi f\left(t_2 + t_1\right) + 2\theta\right)}\right\} \qquad (1.167)$$

$$= \frac{A^2}{4\pi}\left\{\int\limits_0^{2\pi} \cos\left(2\pi f\left(t_2 - t_1\right)\right) d\theta + \int\limits_0^{2\pi} \cos\left[2\pi f\left(t_2 + t_1\right) + 2\theta\right] d\theta\right\} \qquad (1.168)$$

but the average value of a $\cos(.)$ function is zero and $\cos\left(2\pi f\left(t_2 - t_1\right)\right)$ does not depend on θ, thus

$$R_X(t_1, t_2) = \frac{A^2}{4\pi} \left\{ \cos\left(2\pi f\left(t_2 - t_1\right)\right) \int_0^{2\pi} d\theta + 0 \right\} = \frac{A^2}{2} \cos\left(2\pi f\left(t_2 - t_1\right)\right)$$

$$(1.169)$$

or equivalently with $\tau = (t_2 - t_1)$

$$R_X(\tau) = \frac{A^2}{2} \cos\left(2\pi f \tau\right) \tag{1.170}$$

Thus, the random process $x(t, \theta)$ is a wide-sense stationary (WSS) process because its autocorrelation function is only a function of τ and $\overline{x_i}$ is a constant (zero in this example). Furthermore as expected for WSS random process, $R_X(-\tau) = R_X(\tau)$, and $R_X(0) = \overline{x_1^2} = \frac{A^2}{2}$ and $|R_X(\tau)| \leq R_X(0)$. Notice that $R_X(\tau)$ is periodic (period of $\frac{1}{f}$) in which case the WSS random process is called *periodic*. Now consider a given sample function within the ensemble. The time-averaged autocorrelation is given by

$$R(\tau) = \lim_{T \to \infty} \frac{1}{T} \int_{-T/2}^{T/2} A\cos\left(2\pi f t + \theta\right) A\cos\left[2\pi f\left(t + \tau\right) + \theta\right] dt \tag{1.171}$$

$$= \lim_{T \to \infty} \frac{A^2}{T} \int_{-T/2}^{T/2} \frac{1}{2} \left\{\cos\left(2\pi f\tau\right) + \cos\left[4\pi f t + 2\theta + 2\pi f\tau\right]\right\} dt \tag{1.172}$$

$$= \lim_{T \to \infty} \frac{A^2 \cos\left(2\pi f\tau\right)}{2T} \int_{-T/2}^{T/2} dt = \frac{A^2}{2} \cos\left(2\pi f\tau\right) \tag{1.173}$$

Furthermore, by inspection, it is clear that the time-average $\langle x(t)\rangle$ of any given sample function is zero. Given that $\langle x(t)\rangle = \overline{x_i}$ for all i and $R(\tau) = R_X(\tau)$, the random process $x(t, \theta)$ is ergodic.

- ⭐ SIMULATION **SinusoidalRanProcess:** The autocorrelation function of the random process $x(t, \theta) = A \cos(2\pi ft + \theta)$ is determined and shown to agree with $R_X(\tau) = \frac{A^2}{2} \cos(2\pi f\tau)$. The average power P_X of the random process $x(t, \theta)$ is verified to be given by $P_X = R_X(0)$ and as will be proved in section 1.7.5 that $R_X(0)$ is also equal to the ensemble average of the amplitude squares $\overline{x^2}$ of the sample functions at a given time t.

1.7.5   Power Spectral Density of a Random Process

The *Wiener-Khintchine* theorem states that the power spectral density $PSD_X(f)$ of a WSS random process is given by the Fourier transform (FT) of the autocorrelation function $R_X(\tau)$. i.e.

$$PSD_X(f) = \int\limits_{-\infty}^{\infty} R_X(\tau)e^{-j2\pi f\tau}d\tau \qquad (1.174)$$

where $PSD_X(f)$ is real, $PSD_X(f) \geq 0$ and $PSD_X(-f) = PSD_X(f)$. Thus, $R_X(\tau)$ is the inverse FT of $PSD_X(f)$ (provided $R_X(\tau)$ becomes sufficiently small for large values of τ)

$$R_X(\tau) = FT^{-1}[PSD_X(f)] = \int\limits_{-\infty}^{\infty} PSD_X(f)e^{j2\pi f\tau}df \qquad (1.175)$$

Note that the PSD does **not** exist for a random process that is **not** a wide-sense stationary. Given that $R_X(\tau)$ is simply $R_X(t_1, t_2) = R_X(t_1, t_1 + \tau) = E[X(t_1)X(t_1+\tau)]$, notice that for $\tau = 0$, the *average* power P_X of the random process is given by

$$P_X = R_X(0) = E[X(t_1)X(t_1)] = \overline{x_1 x_1} = \overline{x_1^2} = \int\limits_{-\infty}^{\infty} PSD_X(f)df \qquad (1.176)$$

Finally if $E[X(t_1)] = 0$, then **variance** σ^2 of x_1 is equal to the second moment $\overline{x_1^2}$. Refer to Appendix A.11 for further details.

1.7.6 Example : Sum of Two Random Processes

Consider a random process $z(t) = x(t) + y(t)$, where $x(t)$ and $y(t)$ are random processes. Then the power P_Z (from equation 1.176) of $z(t)$ is given by

$$P_Z = \overline{z_1^2} = \overline{(x_1 + y_1)(x_1 + y_1)} \tag{1.177}$$

$$= \overline{x_1^2} + 2\overline{x_1 y_1} + \overline{y_1^2} \tag{1.178}$$

If $x(t)$ and $y(t)$ are **orthogonal**, then from equation A.40, $\overline{x_1 y_1} = 0$ and $P_Z = \overline{x_1^2} + \overline{y_1^2}$ is the sum of the power of $x(t)$ and $y(t)$. However, if $x(t)$ and $y(t)$ are **uncorrelated**, then from equation A.38, $\overline{x_1 y_1} = \overline{x_1}\,\overline{y_1}$ and $P_Z = \overline{x_1^2} + 2(\overline{x_1})(\overline{y_1}) + \overline{y_1^2}$. For ergodic processes, this is the same as $\langle x(t) \rangle \langle y(t) \rangle$ i.e. multiplication of the d.c. value of $x(t)$ and $y(t)$.

1.7.7 White Gaussian Noise

▶ White Noise PSD

- Single-sided noise power spectral density

- The power spectral density of band-limited white gaussian noise

- Noise power via its autocorrelation function

Natural noise (atmosphere, etc.) or man made noise (electromagnetic waves from various sources) can be reduced by taking appropriate measures. For example, shielding a wire cable. However, thermal noise, which is due to the random motion of electrons in the resistive components of electronic circuitry (resistors, wires, etc.) cannot be eliminated if the operating temperature is greater than zero degrees Kelvin. The random motion of electrons are currents which in turn produce random voltage fluctuations across for example, a resistor. The interesting feature of thermal noise is that its power spectral density is essentially flat as illustrated in Fig. 1.38 with a value of $\frac{N_o}{2}$ (Watts/Hz) over a double-sided spectrum or N_o over a single-sided spectrum, where N_o is referred to as the *single-sided noise power spectral density.*

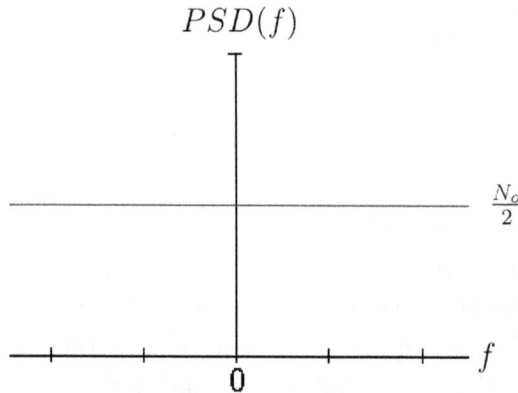

$$PSD(f)$$

$$\frac{N_o}{2}$$

$$0 \qquad f$$

Figure 1.38: White noise power spectral density.

Typically, N_o is constant over the frequency range d.c. (0 Hz) to approximately 10^{12} Hz. Hence the use of the term "white" to signify its similarity with white light which contains all the frequencies (or equivalently colors) within the visible range of electromagnetic radiation. Thus "white" in this context means "all frequencies". The term "gaussian" is used because thermal noise is a gaussian process. This is not surprising if we relate this process to the *central limit theorem* which essentially states that the sum of N independent random variables will have a probability density function (PDF) which tends to a gaussian function as N tends to infinity, irrespective of the original N random variable PDFs. Thus, the combined effect of the random voltage fluctuations due to free electrons is a gaussian noise signal. The autocorrelation function $R_n(\tau)$ for the theoretical white noise power spectral density in Fig. 1.38, given by

$$R_n(\tau) = FT^{-1}\left[\frac{N_o}{2}\right] = \frac{N_o}{2}\delta(\tau) \tag{1.179}$$

implies that any two samples at different times of the noise signal are uncorrelated because $R_n(\tau \neq 0) = 0$. In a practical communication system, the available bandwidth is within the frequency range of thermal noise and thus the thermal noise PSD will be flat over the communication bandwidth. For example, Fig. 1.39 illustrates the PSD of white gaussian noise filtered through an ideal low-pass filter of bandwidth B. In this example, the filter bandwidth $B = 15$ Hz and N_o has been set to ensure that the noise signal power, which is the area under the PSD given by $\int_{-\infty}^{\infty} PSD_n(f)df = \left(\frac{N_o}{2} * 2B\right)$, is equal to 30.5 W, from which we compute $N_o = 2.033$ W/Hz and the height $\frac{N_o}{2} = 1.017$ W/Hz.

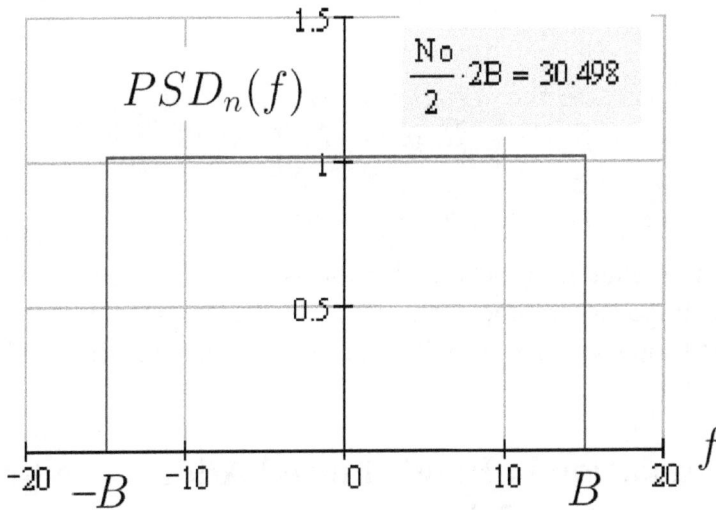

Figure 1.39: PSD of band-limited white Gaussian noise.

White gaussian noise is a wide-sense stationary random process. Thus, from the Wiener-Khintchine theorem (section 1.7.5), the power spectral density (PSD) is given by the Fourier transform (FT) of the ensemble autocorrelation function $R_X(\tau)$ i.e. $R_X(\tau)$ corresponding to the PSD shown in Fig. 1.39 is given by the inverse FT of the PSD as follows:

$$R_X(\tau) = \int_{-B}^{B} \frac{N_o}{2} e^{j2\pi f \tau} df = \frac{N_o}{2} \left[\frac{e^{j2\pi f \tau}}{j2\pi \tau} \right]_{-B}^{B}$$

$$= \frac{N_o}{2} \left[\frac{e^{j2\pi B \tau}}{j2\pi \tau} - \frac{e^{-j2\pi B \tau}}{j2\pi \tau} \right] = \frac{N_o}{2\pi \tau} \left[\frac{e^{j2\pi B \tau}}{2j} - \frac{e^{-j2\pi B \tau}}{2j} \right] \qquad (1.180)$$

Making use of the identity $\sin \theta = \frac{1}{2j} \left(e^{j\theta} - e^{-j\theta} \right)$ and $\operatorname{sinc}(x) = \frac{\sin(\pi x)}{\pi x}$,

$$R_X(\tau) = \frac{N_o}{2\pi \tau} \sin(2\pi B \tau) = \frac{N_o B}{2\pi B \tau} \sin(2\pi B \tau) = N_o B \operatorname{sinc}(2B\tau) \quad (1.181)$$

Fig. 1.40 illustrates the variation of $N_o B \operatorname{sinc}(2B\tau)$ with τ from which the noise signal power $P_X = R_X(0) = N_o B \operatorname{sinc}(0) = N_o B = 30.5$ W as expected.

Finally, note that if the channel noise is wide-sense stationary then its also very likely to be ergodic. An example of noise that is not ergodic or wide-sense stationary is the interference received in a moving vehicle.

1.7.8   Simulation I : Band-limited Additive White Gaussian Noise (AWGN)

▶ **Band-Limited Noise (Simulated)**

- A periodic noise signal is created using a large number if cosine signals of the same amplitude, each with a random phase

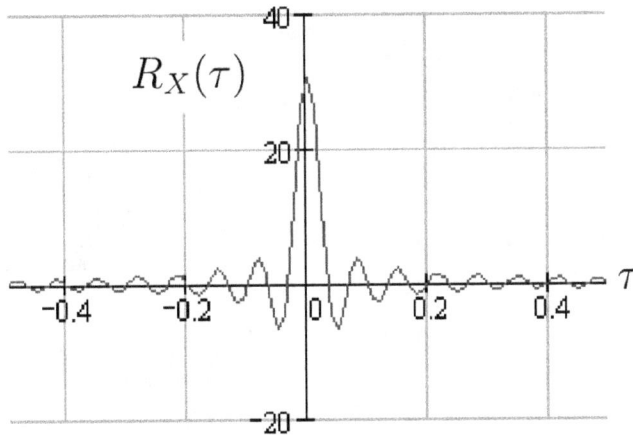

Figure 1.40: Autocorrelation function $R_X(\tau) = N_o B \, \text{sinc}\,(2B\tau)$.

- The power spectral density (PSD) of the noise signal is determined via the Fourier transform of the autocorrelation function

⭐ SIMULATION **WhiteGaussianNoisePSD:** A *rough* approximation to a band-limited gaussian noise signal is created by specifying the amplitude and phase spectra as follows. The amplitude of each component is fixed and the phase is randomly set between -180 to 180 degrees. Although this noise signal periodic, it is shown to approximately exhibit the required features of a AWGN noise signal. Fig. 1.41 shows a sample of the noise signal in the time domain and its corresponding amplitude and phase spectra. In this specific example, the ideal low-pass filter bandwidth $B = 10$ Hz. The noise signal power is determined both in the time and frequency domain (2 Watts in this case) and used to verify the single-sided noise power spectral density N_o (Watts/Hz) $= \frac{2}{10} = 0.2$ Watts/Hz.

- **Experiment:** Create your own noise signal and determines its PSD.

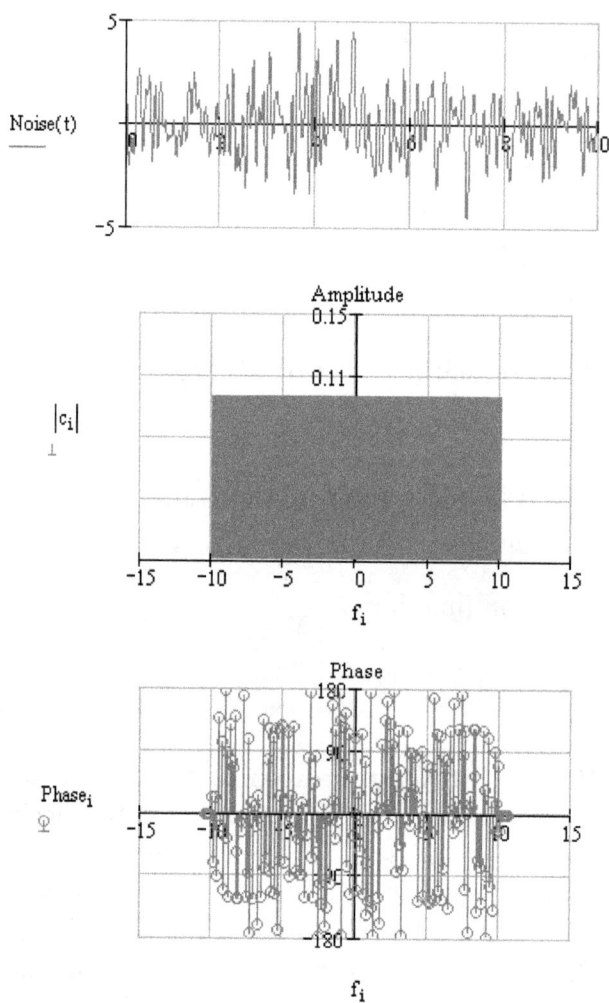

Figure 1.41: Simulated band-limited Gaussian noise signal and its spectrum.

The histogram of the noise signal compared with the expected histogram calculated from the theoretical gaussian distribution as shown in Fig. 1.42. Please refer to Appendix A.8 for the fundamentals of a probability histogram. Given that this noise signal is periodic, its exponential Fourier series is determined and used to reconstruct the noise signal (just for fun).

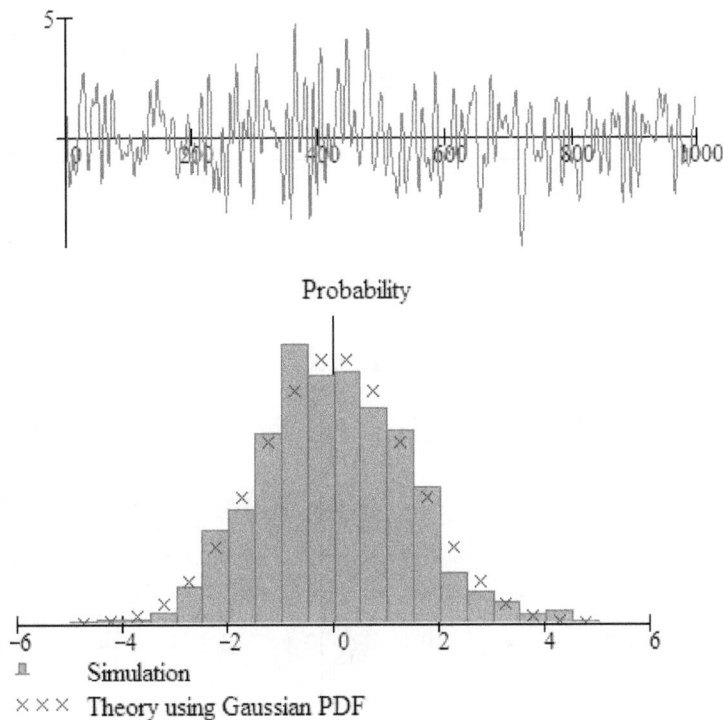

Figure 1.42: Simulated AWGN noise signal and its probability histogram.

The Fourier transform of $R_X(\tau)$ shown in Fig. 1.43 is then taken to determine the power spectral density of the noise signal as shown in Fig. 1.44, from which the value of N_o is calculated and shown to agree with the expected value of 0.2 Watts/Hz. Finally, the inverse Fourier transform of the theoretical power spectral density is shown to agree with $R_X(\tau) = N_o B \operatorname{sinc}(2B\tau)$. The time-average autocorrelation function $R(\tau)$ of the noise signal is shown to agree with $R_X(\tau)$ as shown in Fig. 1.43 i.e. this random process is ergodic in the autocorrelation.

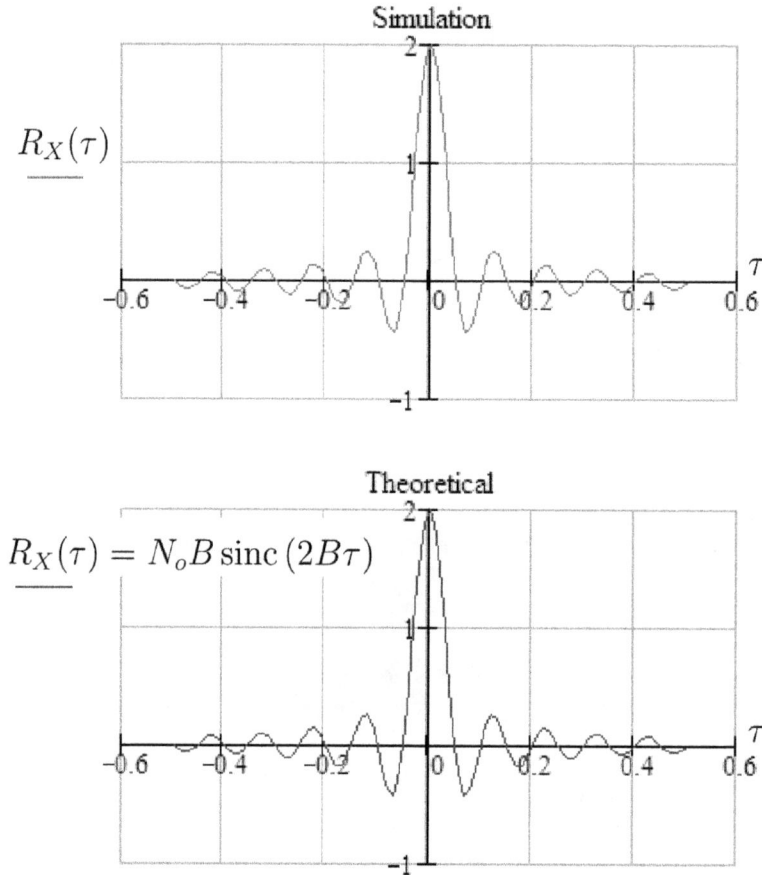

Figure 1.43: Autocorrelation function of the AWGN signal.

1.7.9   Simulation II : AWGN Random Process

⭐ SIMULATION **AWGNRandomProcess:** An ensemble of random noise signals is created using the noise signal analyzed in section 1.7.8. The av-

$$PSD_n(f)$$

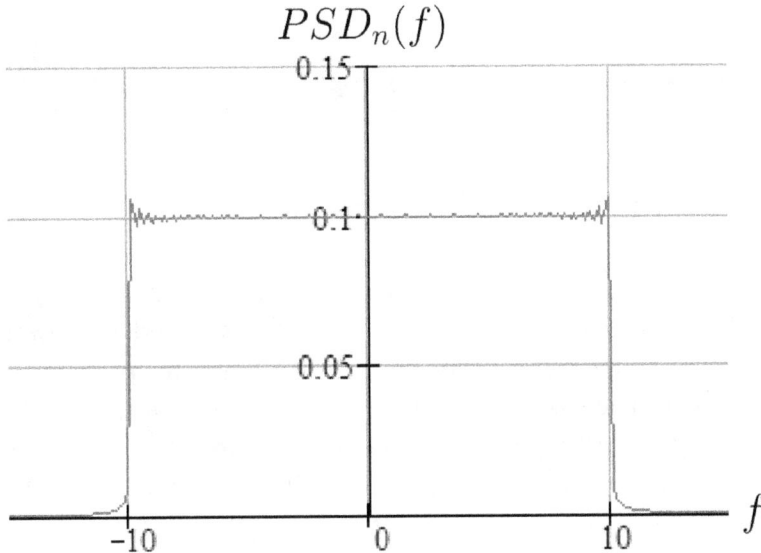

Figure 1.44: AWGN PSD determined from the FT of the autocorrelation function.

erage power P_X of this random process is determined via its ensemble autocorrelation function $R_X(\tau) = E[X(t)X(t+\tau)]$ and shown to be equal to $P_X = \int_{-B}^{B} \frac{N_o}{2} df = N_o B$. Recall that P_X of a wide-sense stationary random process is given by $P_X = R_X(\tau = 0) = \overline{x^2}$, where $\overline{x^2}$ is the ensemble average of the amplitude squares of noise signals (i.e. sample functions) at a given time t.

1.8 Linear Time-Invariant (LTI) System

If the input signal to a system is $s(t)$ and the corresponding output signal is $y(t)$ such that $y(t) = F[s(t)]$, where $F[.]$ (**not** to be confused with the Fourier transform $FT[.]$) represents the operation of the system on the input signal, then for a *linear system*, for any two input signals $s_1(t)$ and $s_2(t)$, we find

$$F\left[s_1(t) + s_2(t)\right] = F\left[s_1(t)\right] + F\left[s_2(t)\right] = y_1(t) + y_2(t). \tag{1.182}$$

$$F\left[\alpha s(t)\right] = \alpha F\left[s(t)\right] \tag{1.183}$$

where α is a scalar. A system that satisfies equation 1.182 is called *additive* and that which satisfies equation 1.183 is called *homogeneous*. A nonlinear system is one which does not satisfy these two properties. Notice that if the input to the system is $s_1(t) + s_2(t)$, then the output is the same as if each signal, $s_1(t)$ and $s_2(t)$, is sent separately through the system and the corresponding outputs $y_1(t)$ and $y_2(t)$ are added together. From equation 1.183, we note that $F\left[0\right] = 0$ i.e. $y(t) = 0$ to a zero input.

The RC filter to be considered next is not only a linear system, but is in fact a *linear time-invariant* (LTI) system for which

$$F\left[s(t - t_d)\right] = y(t - t_d) \tag{1.184}$$

where t_d is a constant time delay i.e. a delayed input signal gives a delayed output signal. A random process through a linear system is considered later in section 1.8.11.

1.8.1   RC Filter

It is necessary to analyze the RC filter in detail because it will be used to illustrate the properties of a LTI system, convolution and spectral density relationships (input and output) in the sections to follow. Consider the circuit shown in Fig. 1.45 which consists of a resistor R in series with a capacitor C. If the capacitor is initially uncharged and the input to the circuit is the d.c. voltage $V_{in} = V_o$, where V_o is a constant, then it can be shown (refer to the problems section) that the voltage $V_{out}(t)$ across the capacitor after time t seconds is given by $V_{out}(t) = V_o\left(1 - e^{\frac{-t}{RC}}\right)$. As shown in Fig. 1.46 for $V_o = 2$ volts, the capacitor exponentially charges from zero

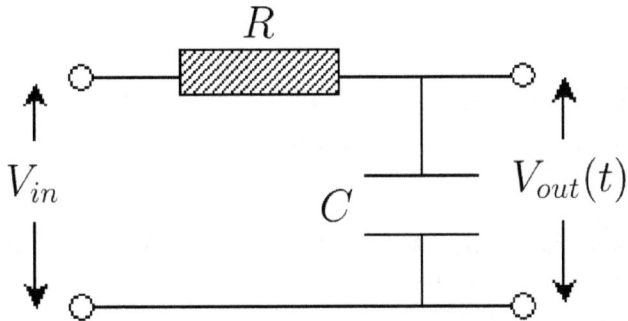

Figure 1.45: RC circuit.

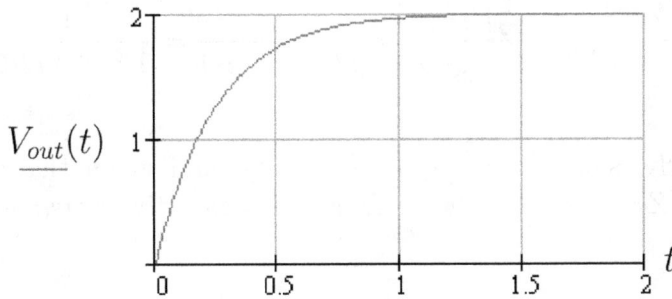

Figure 1.46: Charging a capacitor via a resistor.

to V_o. The rate at which the capacitor is charged is dictated by the *time constant* $t_c = RC$ seconds.

- ⭐ SIMULATION **Charging:** Plot of the equation $V_{out}(t) = V_o \left(1 - e^{\frac{-t}{RC}}\right)$. Experiment with different values for the time constant t_c.

If the input to the circuit is a sinusoidal variation such that $V_{in} = V_o \sin(2\pi ft)$, then the *impedance* of the resistor $Z_1 = R$.   However, the impedance of the capacitor $Z_2 = \frac{1}{j2\pi fC}$, where we have used the standard complex notation of $j = \sqrt{-1}$ which corresponds to a 90 degree phase difference. Using Kirchoff's voltage law,

$$V_{in} = IZ_1 + IZ_2 = I\left(Z_1 + Z_2\right) \tag{1.185}$$

$$V_{out} = IZ_2 = \frac{V_{in}Z_2}{(Z_1 + Z_2)} \tag{1.186}$$

$$\frac{V_{out}}{V_{in}} = \frac{Z_2}{(Z_1 + Z_2)} = \frac{\frac{1}{j2\pi fC}}{R + \frac{1}{j2\pi fC}} = \frac{1}{j2\pi fCR + 1} = \frac{1}{1 + j2\pi fRC} \tag{1.187}$$

Equivalently, since $\frac{1}{j} = \frac{j}{j*j} = \frac{j}{-1} = -j$, and given that Z_2 may be expressed as $Z_2 = -jX_c$ where X_c is the *capacitive reactance* given by $X_c = \frac{1}{2\pi fC}$,

$$\frac{V_{out}}{V_{in}} = \frac{Z_2}{(Z_1 + Z_2)} = \frac{-jX_c}{R - jX_c} \tag{1.188}$$

For a given R and C, notice that the ratio $\frac{V_{out}}{V_{in}}$ is a function of the input signal frequency f since $X_c = \frac{1}{2\pi fC}$.   Accordingly, we take the *frequency response* $H(f)$ of the RC circuit in Fig. 1.45 to be given by

$$H(f) = \frac{V_{out}}{V_{in}} = \frac{1}{1 + j2\pi f RC} \tag{1.189}$$

where the magnitude $|H(f)|$ is determined by multiplying by its complex conjugate to give

$$|H(f)|^2 = \frac{-jX_c}{(R - jX_c)} \frac{jX_c}{(R + jX_c)} = \frac{X_c^2}{R^2 + X_c^2} \tag{1.190}$$

$$|H(f)| = \frac{X_c}{\sqrt{R^2 + X_c^2}} \tag{1.191}$$

Since $H(f) = \frac{1}{1+j2\pi f RC} \equiv \frac{1+j0}{1+j2\pi f RC}$, the corresponding phase variation with frequency $\theta(f)$ is given by

$$\theta(f) = \tan^{-1}(\frac{0}{1}) - \tan^{-1}(\frac{2\pi f RC}{1}) = -\tan^{-1}(2\pi f RC) \tag{1.192}$$

Taking for example $R = 10^6$ Ω and $C = 0.25$ μF, the time constant $t_c = 0.25$ seconds. The corresponding frequency response curves of $|H(f)|$ and $\theta(f)$ are shown in Fig. 1.47. For example, for $f = \frac{2}{\pi}$ Hz, we find $|H(f = \frac{2}{\pi})| = \frac{1}{\sqrt{2}}$, or equivalently, $|V_{out}| = \frac{|V_{in}|}{\sqrt{2}}$ i.e. the amplitude of the output signal is reduced by a factor of $\frac{1}{\sqrt{2}}$. Furthermore, the phase shift of the output signal $\theta(f = \frac{2}{\pi})) = -45$ degrees. However, for an input signal with $f \gg \frac{2}{\pi}$ Hz, the amplitude of the output signal will be close to zero and phase shift essentially -90 degrees. Given that the frequency response of the "RC circuit" is such that it attenuates high input frequency signals, it is referred to as a *low-pass filter* (LPF).

The frequency at which the output amplitude is reduced by a factor of $\frac{1}{\sqrt{2}}$ is known as the *cutoff* frequency f_c of the filter, or equivalently as the 3

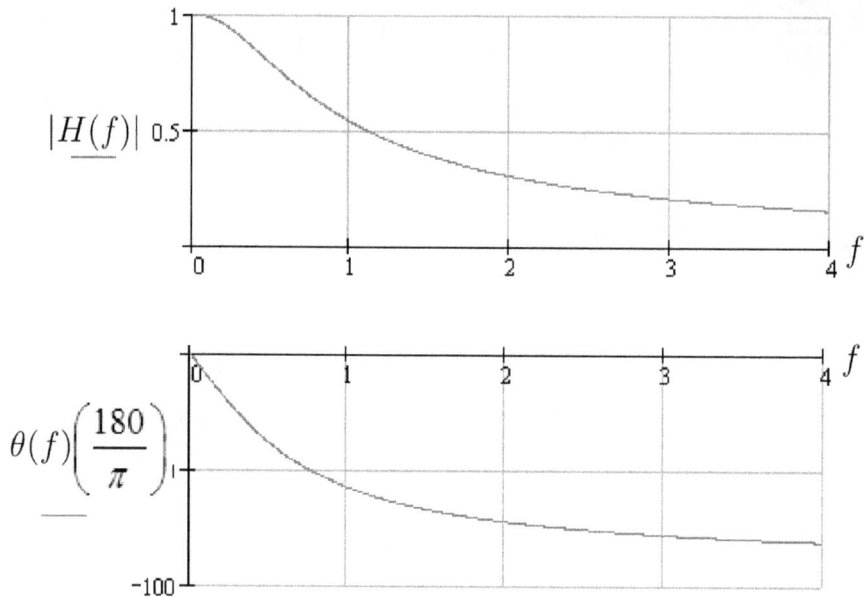

Figure 1.47: Frequency response of the RC circuit.

dB bandwidth of the filter. In this example, we found $f_c = \frac{2}{\pi}$. In general, it is easily shown that for a *RC* filter, $f_c = \frac{1}{2\pi RC} = \frac{1}{2\pi t_c}$ Hz.

- ⭐ SIMULATION **Complexz:** Simulation corresponding to Fig. 1.47.

We are now faced with a very interesting question! What if the input signal is not sinusoidal, but for example, a rectangular pulse which we recall from Fourier analysis can be reconstructed with sinusoidal waveforms of various amplitudes and phase. Can we predict the output signal by passing each spectral component of the rectangular pulse through the filter and adding the corresponding outputs ? The answer is yes, provided the RC filter is a *linear system*. Analyzing a given LTI system via the spectral components of the incoming signal turns out to be elegantly related to its corresponding analysis in the time domain as we shall see next.

1.8.2 Convolution

For an LTI system, let the input signal be $s(t)$ and let the corresponding output signal be $y(t)$. If $S(f)$ denotes the FT of $s(t)$ and $Y(f)$ denotes the FT of $y(t)$, then

$$Y(f) = S(f)H(f) \tag{1.193}$$

where $H(f)$ is the frequency response of the LTI system. Alternatively, in the time domain, if the *inverse* FT of $H(f)$ is denoted by $h(t)$, then

$$y(t) = s(t) * h(t) \tag{1.194}$$

where the operator $*$ denotes *convolution*, which is defined for any two signals $s_1(t)$ and $s_2(t)$ to be given by

$$s_1(t) * s_2(t) = \int_{-\infty}^{\infty} s_1(\tau)s_2(t-\tau)d\tau \tag{1.195}$$

Thus

$$y(t) = s(t) * h(t) = \int_{-\infty}^{\infty} s(\tau)h(t-\tau)d\tau = \int_{-\infty}^{+\infty} s(t-\tau)h(\tau)d\tau \tag{1.196}$$

To prove equation 1.193, we have

$$Y(f) = \int_{-\infty}^{+\infty} y(t)e^{-j2\pi ft}dt \tag{1.197}$$

$$= \int_{-\infty}^{+\infty}\int_{-\infty}^{+\infty} s(t-\tau)h(\tau)d\tau e^{-j2\pi ft}dt \tag{1.198}$$

$$= \int\limits_{-\infty}^{+\infty}\int\limits_{-\infty}^{+\infty} s(t-\tau)h(\tau)d\tau e^{-j2\pi f(t-\tau)}e^{-j2\pi f\tau}\,dt \qquad (1.199)$$

$$= \int\limits_{-\infty}^{+\infty} s(t-\tau)e^{-j2\pi f(t-\tau)}\,dt \int\limits_{-\infty}^{+\infty} h(\tau)e^{-j2\pi f\tau}\,d\tau \qquad (1.200)$$

Let $u = t - \tau$, so that

$$Y(f) = \int\limits_{-\infty}^{+\infty} s(u)e^{-j2\pi fu}\,du \int\limits_{-\infty}^{+\infty} h(\tau)e^{-j2\pi f\tau}\,d\tau \qquad (1.201)$$

$$Y(f) = S(f)H(f) \qquad (1.202)$$

1.8.3   Hilbert Transform in Light of Convolution

The Hilbert transform of $s(t)$ is equivalent to its convolution with $\frac{1}{\pi t}$ in the time domain. To verify this, recall from equation 1.106 that $FT\left[\frac{1}{\pi t}\right] = -j\,\mathrm{sgn}\,(f)$. Thus

$$\frac{1}{\pi t} * s(t) = \int\limits_{-\infty}^{\infty} \frac{1}{\pi(t-\tau)} s(\tau)d\tau \qquad (1.203)$$

via equation 1.202 corresponds to $-j\,\mathrm{sgn}\,(f)\,S(f)$ in the frequency domain, which in turn from equation 1.118, is the $FT\left[s_h(t)\right]$, where $s_h(t)$ denotes the Hilbert transform of $s(t)$. Thus

$$s_h(t) = \frac{1}{\pi t} * s(t) \qquad (1.204)$$

1.8.4 Input and Output Spectral Densities

If the input signal $s(t)$ and output signal $y(t)$ are **energy signals**, then from equation 1.125, the input and output energy spectral densities, given by $ESD_{in}(f) = |S(f)|^2$ and $ESD_{out}(f) = |Y(f)|^2$, are related by

$$ESD_{out}(f) = ESD_{in}(f)\,|H(f)|^2 \qquad (1.205)$$

Similarly, if $s(t)$ and $y(t)$ are **power signals**, then

$$PSD_{out}(f) = PSD_{in}(f)\,|H(f)|^2 \qquad (1.206)$$

1.8.5 Example: Signals through a RC Filter

Consider an input signal $s(t) = \begin{cases} e^{-at} & if \ \ t \geq 0 \\ 0 & if \ \ t < 0 \end{cases}$, where $a > 0$ is a constant, that is passed through a filter $H(f) = \frac{1}{1+j2\pi f}$. Then the FT of $s(t)$ is given by

$$S(f) = \int_{-\infty}^{\infty} s(t)e^{-j2\pi ft}dt = \int_{0}^{\infty} e^{-t(a+j2\pi f)}dt = \frac{1}{(a+j2\pi f)} \qquad (1.207)$$

and therefore

$$ESD_{in}(f) = |S(f)|^2 = \frac{1}{(a+j2\pi f)}\frac{1}{(a-j2\pi f)} = \frac{1}{a^2+(2\pi f)^2} \qquad (1.208)$$

and

$$ESD_{out}(f) = ESD_{in}(f)\,|H(f)|^2 = \left(\frac{1}{a^2+(2\pi f)^2}\right)\left(\frac{1}{1+(2\pi f)^2}\right) \qquad (1.209)$$

From Parseval's theorem, the input signal energy

$$E_{in} = \int_{-\infty}^{\infty} ESD_{in}(f)df = \int_{-\infty}^{\infty} \frac{1}{a^2 + (2\pi f)^2} df = \int_{-\infty}^{\infty} |s(t)|^2\, dt = \int_{0}^{\infty} e^{-2at} dt = \frac{1}{2a}$$

$$(1.210)$$

Taking $a > 1$, we have $\frac{1}{1+(2\pi f)^2} - \frac{1}{a^2+(2\pi f)^2} = \frac{a^2-1}{\left(1+(2\pi f)^2\right)\left(a^2+(2\pi f)^2\right)}$. Thus we may write

$$ESD_{out}(f) = \frac{1}{a^2 - 1} \left[\frac{1}{1 + (2\pi f)^2} - \frac{1}{a^2 + (2\pi f)^2} \right]$$

and thus by inspection,

$$E_{out} = \frac{1}{a^2 - 1} \left(\frac{1}{2} - \frac{1}{2a} \right) = \frac{a - 1}{2a\,(a^2 - 1)} = \frac{a - 1}{(a^2 - 1)} E_{in} \qquad (1.211)$$

where for example if $a = 3$, then $E_{out} = \frac{1}{4} E_{in}$. If $s(t)$ is a power signal with a constant spectral density $PSD_{in}(f) = K$, then

$$PSD_{out}(f) = PSD_{in}(f)\, |H(f)|^2 = K \left(\frac{1}{1 + (2\pi f)^2} \right) = \frac{K}{2} \left(\frac{2}{1 + (2\pi f)^2} \right)$$

$$(1.212)$$

From problem 10 in chapter 1, the FT of $e^{-a|t|}$ is given by $\frac{2a}{a^2+(2\pi f)^2}$. Thus, the time-average autocorrelation function $R(\tau)$, given by the inverse FT of $PSD_{out}(f)$ from equation 1.135, is $\frac{K}{2} e^{-|t|}$. Hence, from equation 1.134, the output power is $R(0) = \frac{K}{2}$ watts.

1.8.6 Impulse Response

Figure 1.48 summaries the foregoing results using the FT pair $\longleftrightarrow$ notation. In the FT pair $h(t) \longleftrightarrow H(f)$, the output $h(t)$ is referred to as the *impulse response* of the system, because $h(t)$ is the output of the system when the input is $\delta(t)$ i.e. $h(t) = F[\delta(t)]$, where $\delta(t)$ is the Dirac delta (or impulse) function. Thus, analyzing a LTI system with $\delta(t)$ as the input is equivalent to analyzing the frequency response of the system using sinusoidal signals over all frequencies.

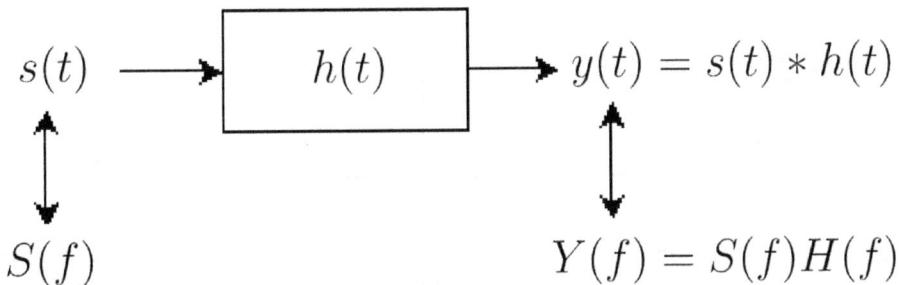

Figure 1.48: LTI system input and output Fourier transform pairs.

To determine the impulse response $h(t)$ of a system in practice, we could input a rectangular pulse and analyze the output as the width of the pulse tends to zero, whilst maintaining the area under the pulse equal to 1. To illustrate this, Fig. 1.49 shows a rectangular pulse of width τ with height $A = \frac{1}{\tau}$ and its corresponding FT. As τ is reduced, the height increases to ensure the area under the pulse is equal to 1. The Dirac or impulse function is more closely approximated as τ tends to zero.

- ⭐ SIMULATION **Dirac:** Simulation corresponding to Fig. 1.49. What happens as τ is increased? The Dirac function can also be approximated as a zero mean Gaussian function with a standard deviation that tends to zero, namely $\delta(t) = \lim_{\sigma \to \infty} \left(\frac{1}{\sigma\sqrt{2\pi}} e^{-\frac{t^2}{2\sigma^2}} \right)$. Verify

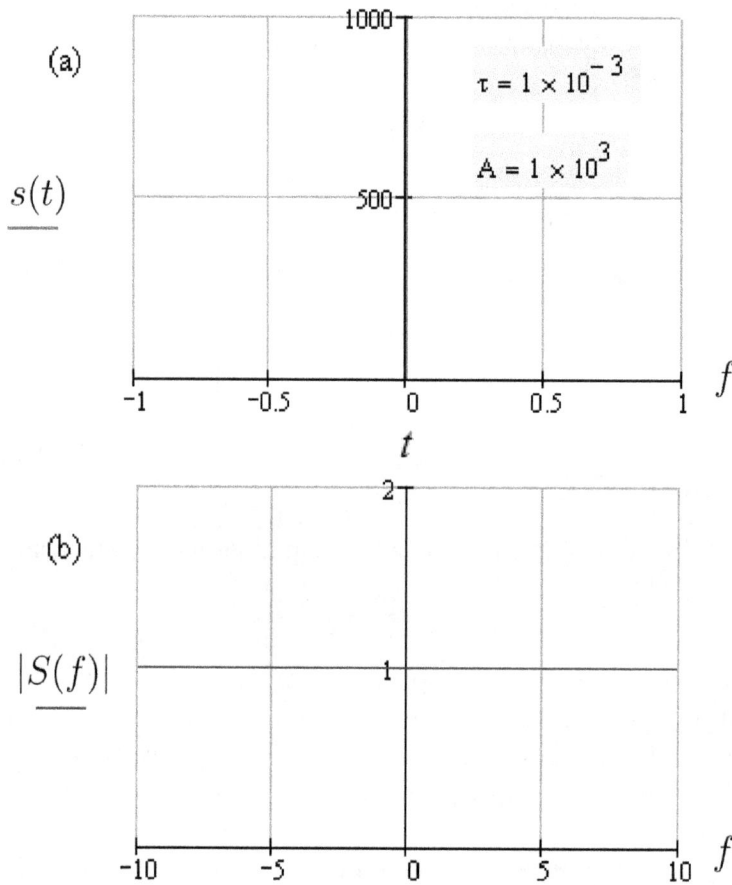

Figure 1.49: Dirac delta function approximation and its Fourier transform.

that $\int\limits_{-\infty}^{\infty} \delta(t) = 1$ and show that its Fourier transform gradually tends toward $S(f) = 1$ over all frequencies.

1.8.7 Complex Exponential through a LTI System

Given that a signal can be viewed as the summation of cosines with varying amplitude and phase (i.e. spectral components), the analysis of a single spectral component (cosine signal) through an LTI system is critical to understanding how signals are processed by a linear system. Thus consider the complex signal $s(t) = Ae^{j(2\pi ft+\theta)}$ passing through a LTI system with an impulse response $h(t)$. Then from equation 1.196, the output signal $y(t)$ is given by

$$y(t) = s(t) * h(t) = \int\limits_{-\infty}^{+\infty} s(t-\tau)h(\tau)d\tau \tag{1.213}$$

$$= \int\limits_{-\infty}^{+\infty} Ae^{j(2\pi f(t-\tau)+\theta)}h(\tau)d\tau = \int\limits_{-\infty}^{+\infty} Ae^{j(2\pi ft+\theta)}e^{j(-2\pi f\tau)}h(\tau)d\tau \tag{1.214}$$

$$= Ae^{j(2\pi ft+\theta)} \int\limits_{-\infty}^{+\infty} e^{-j2\pi f\tau}h(\tau)d\tau \tag{1.215}$$

Since the FT of $h(t)$ is $H(f)$ given by

$$H(f) = \int\limits_{-\infty}^{+\infty} h(\tau)e^{-j2\pi f\tau}d\tau \equiv |H(f)|\,e^{j\,\arg(H(f))} \tag{1.216}$$

then

$$
\begin{aligned}
y(t) &= Ae^{j(2\pi ft+\theta)}H(f) \tag{1.217}\\
&= Ae^{j(2\pi ft+\theta)}|H(f)|\,e^{j\,\arg(H(f))} = A\,|H(f)|\,e^{j(2\pi ft+\theta+\arg(H(f)))} \tag{1.218}
\end{aligned}
$$

The implication of equation 1.217 is that a real spectral component $A\cos(2\pi ft+\theta)$ passing through a system with frequency response $H(f)$ will have its amplitude altered by a factor of $|H(f)|$ to become $A\,|H(f)|$ and its phase θ will be shifted by $\arg(H(f))$. Notice how the frequency of the input sinusoidal signal does not change.

⭐ SIMULATION **CosineThroughRCFilter:** The input signal $s(t) = A\cos(2\pi ft+\theta)$ is passed through a RC filter with frequency response $H(f) = \frac{1}{1+j2\pi fRC}$ and corresponding impulse response $h(t) = \begin{cases} \frac{1}{t_c}e^{\frac{-t}{t_c}} & for & t>0 \\ 0 & otherwise \end{cases}$.

Equation 1.216 is verified first before showing that the output signal $y(t) = A\,|H(f)|\cos(2\pi ft+\theta+\arg(H(f))) = \int_{-\infty}^{+\infty} s(t-\tau)h(\tau)d\tau$. Experiment with different values for f to verify that this RC filter is a low-pass filter. A similar approach is implemented in the simulation of section 1.8.14.

1.8.8   Example: RC Low-Pass Filter

Recall from equation 1.189 that for an RC circuit, the frequency response $H(f)$ is given by $H(f) = \frac{1}{1+j2\pi fRC}$. The inverse FT $h(t)$ is given by

$$h(t) = \begin{cases} \frac{1}{t_c}e^{\frac{-t}{t_c}} & for & t>0 \\ 0 & otherwise \end{cases} \tag{1.219}$$

For a time constant $t_c = 0.25$ seconds, the impulse response $h(t)$ is shown in Fig. 1.50.

Let the input to the RC circuit be the rectangular pulse shown in Fig. 1.51. To determine the output of the RC circuit, the following operation is undertaken.

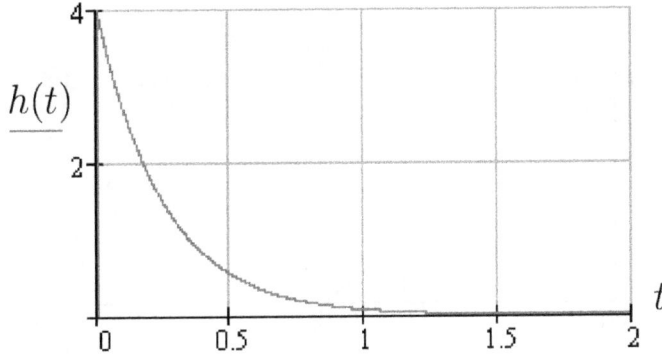

Figure 1.50: Impulse response of an RC circuit.

1. Determine the FT of $h(t)$ to give $H(f)$

2. Determine the FT of $s(t)$ to give $S(f)$

3. Multiply the two FTs to give $Y(f) = S(f)H(f)$

4. Take the inverse FT of $Y(f)$ to determine $y(t)$

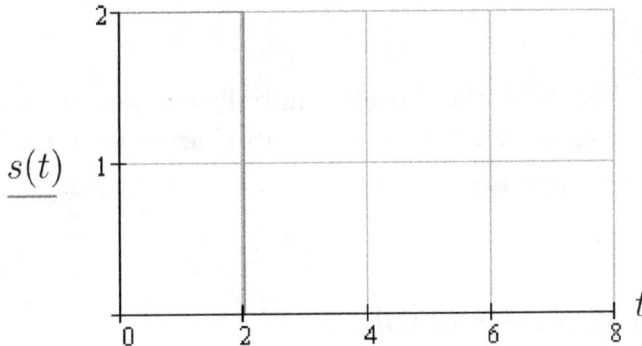

Figure 1.51: Input pulse to the RC circuit.

The result of undertaking the above four steps is presented in Fig. 1.52 and compared with the result of convolving $s(t)$ with $h(t)$. As expected, both methods are equivalent.

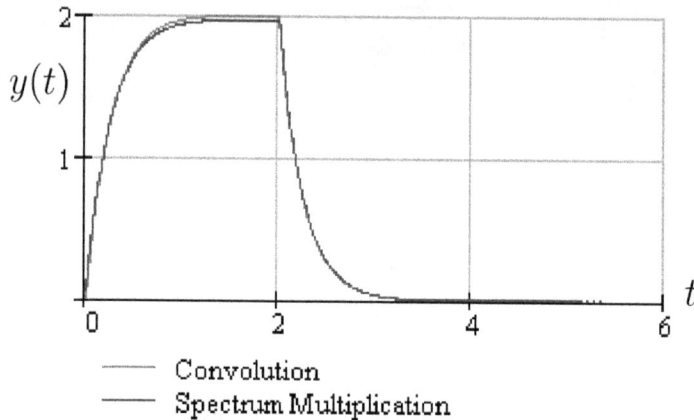

Figure 1.52: RC circuit output signal.

- ⭐ SIMULATION **RC:** Implementation of the foregoing analysis. Experiment with different input signals and verify that both the time and frequency domain methods are equivalent.

Referring to Fig. 1.52, the capacitor initially charges to 2 volts in exactly the same manner as in Fig. 1.46 and then discharges to zero. The equation of the discharging curve is easily shown to be given by $V_{out}(t) = V_o \left(e^{\frac{-t}{RC}} \right)$.

1.8.9   Ideal Low-Pass Filter

The ideal frequency response $H(f)$ of a *low-pass* filter is shown in Fig. 1.53 for which

$$H(f) = \begin{cases} 1 & for \quad |f| \leq B \\ \\ 0 & otherwise \end{cases} \tag{1.220}$$

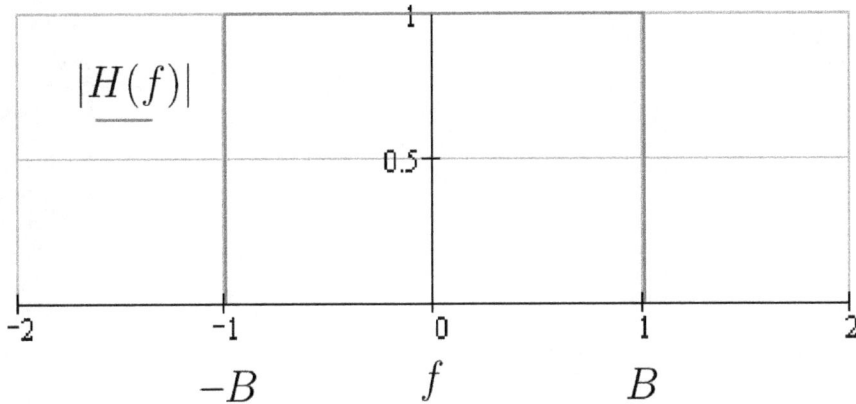

Figure 1.53: Ideal low-pass filter frequency response.

where B is the bandwidth of the filter. The impulse response $h(t) = 2B \operatorname{sinc}(2Bt)$ is given by the inverse FT of $H(f)$. For example, Fig. 1.54 shows the impulse response variation for $B = 1$ Hz.

Let the input signal $s(t)$ to this ideal low-pass filter be a rectangular pulse as shown in Fig. 1.55. Its corresponding amplitude spectrum $|S(f)|$ is shown in Fig. 1.56. Recall that to determine the output signal in the frequency domain, we multiply $S(f)$ with $H(f)$ to give $Y(f) = S(f)H(f)$, and then find the inverse FT of $Y(f)$ to determine $y(t)$. Since $H(f)$ is equal to 1 for all frequencies within B and zero otherwise, the process of multiplying by $H(f)$ simply eliminates the frequency components outside B. To demonstrate this feature, Fig. 1.57 shows the result of eliminating all the frequency components outside $B = 1$ Hz. If the remaining spectral components are used to reconstruct the signal, then we can expect the signal to be the same as if the pulse was transmitted through the ideal low-pass filter frequency response of Fig. 1.53. This is confirmed in Fig. 1.58, which also shows an overlay of the curve found by convolving $s(t)$ with $h(t)$.

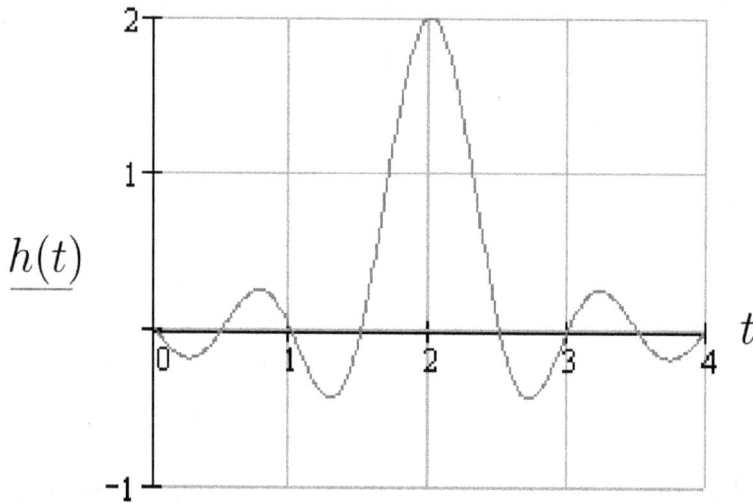

Figure 1.54: Impulse response of an ideal low-pass filter.

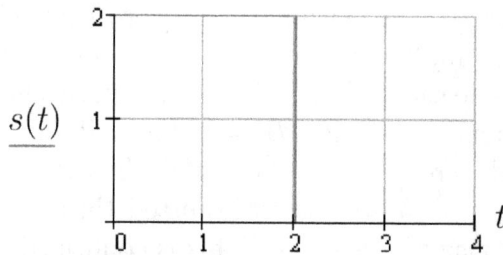

Figure 1.55: Input signal to the LPF.

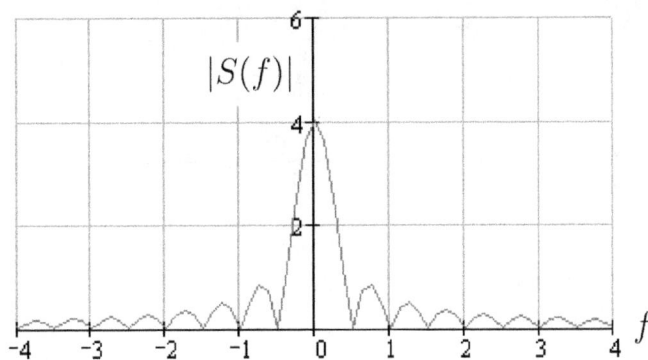

Figure 1.56: Amplitude spectrum of the input pulse.

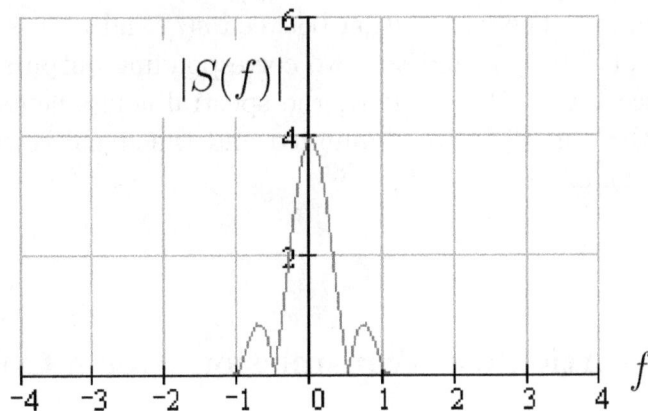

Figure 1.57: Amplitude spectrum filtered through an ideal LPF.

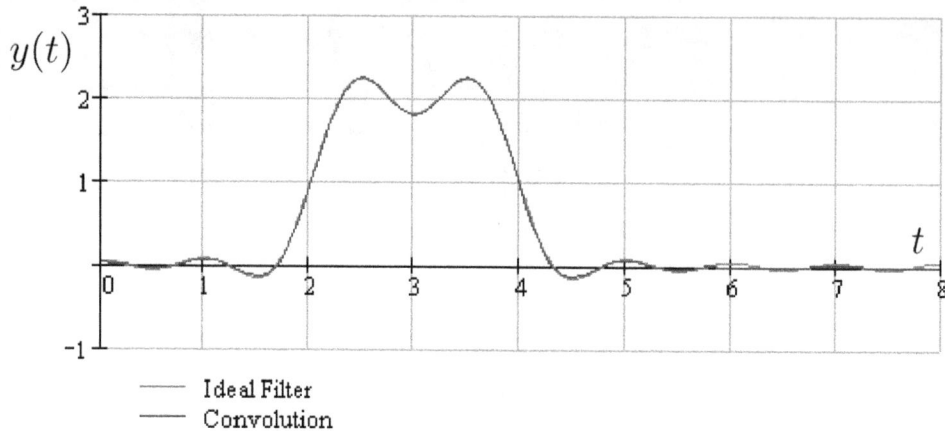

Figure 1.58: Reconstructed signal compared with convolving $s(t)$ with $h(t)$.

- ⭐ SIMULATION **IdealLowPass:** The FT of the impulse response presented in Fig. 1.54 is determined and overlaid with the ideal frequency response shown in Fig. 1.53. The FT $S(f)$ of the input rectangular pulse is then determined and multiplied by the FT $H(f)$ of the impulse response to determine $Y(f)$, from which the output signal $y(t)$ is found. The convolution between $s(t)$ and $h(t)$ is also used to determine the output and the two corresponding outputs are overlaid to produce Fig. 1.58. Finally, the spectral components outside the bandwidth B in $S(f)$ are eliminated and the signal reconstructed in the time domain to give Fig. 1.58.

1.8.10   Distortionless Transmission over a Channel

Distortionless transmission over a channel with a frequency response $H_c(f)$ occurs when the input signal $s(t)$ and the output signal $y(t)$ satisfy the condition

$$y(t) = ks(t - t_d) \qquad (1.221)$$

where k is a constant and t_d is the time-delay. Taking the FT of both sides and making use of the property $s(t - t_d) \longleftrightarrow S(f)e^{-j2\pi f t_d}$

$$Y(f) = kS(f)e^{-j2\pi f t_d} \qquad (1.222)$$

but $Y(f) = S(f)H_c(f)$, and therefore

$$H_c(f) = |H_c(f)| \, e^{j \, \arg(H_c(f))} = |H_c(f)| \, e^{j\theta_c(f)} = ke^{-j2\pi f t_d} \qquad (1.223)$$

Thus for distortionless transmission, we require $|H_c(f)|$ equal to a constant k (e.g. if $|H_c(f)| = 1$, then the *gain* is 1) and $\theta_c(f) = -2\pi f t_d$. Notice that $\theta_c(f)$ is a linear function of frequency f to ensure that the delay is the same for all the spectral components of the signal. If $\theta_c(f) \neq -2\pi f t_d$, then the signal would be phase distorted.

1.8.11 Random Process Through a Linear System

As in section 1.8.1 for the transmission of deterministic signals through a linear time-invariant (LTI) system, the following two relationships are still valid if the input $s(t)$ and the corresponding output $y(t)$ are random processes

$$Y(f) = S(f)H(f) \qquad (1.224)$$

$$PSD_{out}(f) = PSD_{in}(f) |H(f)|^2 \qquad (1.225)$$

The autocorrelation function $R_Y(\tau)$ at the output is given by the inverse FT of $PSD_{out}(f)$ can be shown [Couch, 2001] to be given by

$$R_Y(\tau) = h(-\tau) * h(\tau) * R_X(\tau) \qquad (1.226)$$

where $R_X(\tau)$ is the autocorrelation function of the input random process and $h(.)$ is the impulse response of the LTI system. Note that the average power in the output random process is given by $R_Y(0)$.

1.8.12   Example : Random Process Through a RC Filter

Consider once again the RC filter for which the frequency response $H(f) = \frac{1}{1+j2\pi fRC}$ (equation 1.189). If the input is white noise with $PSD_{in}(f) = \frac{N_o}{2}$, then

$$PSD_{out}(f) = PSD_{in}(f)\,|H(f)|^2 = \frac{N_o}{2}\left(\frac{1}{1+(2\pi fRC)^2}\right) \qquad (1.227)$$

$$= \frac{N_o}{4}\left(\frac{1}{RC}\right)\left(\frac{2\left(\frac{1}{RC}\right)}{\left(\frac{1}{RC}\right)^2+(2\pi f)^2}\right) \qquad (1.228)$$

From problem 10 in chapter 1, the FT of $e^{-a|t|}$ is given by $\frac{2a}{a^2+(2\pi f)^2}$. Comparing this with equation 1.228, the inverse FT of $PSD_{out}(f)$ is given by

$$R_Y(\tau) = \frac{N_o}{4RC}e^{-\frac{|t|}{RC}} \qquad (1.229)$$

and hence the output noise power is given by

$$P_Y = R_Y(0) = \frac{N_o}{4RC} \qquad (1.230)$$

Equivalently, in terms of the 3 dB bandwidth of the filter $f_c = \frac{1}{2\pi RC}$ Hz,

$$P_Y = \frac{N_o(2\pi f_c)}{4} = \frac{N_o\pi f_c}{2} \text{ Watts} \qquad (1.231)$$

1.8.13 Sum of Random Processes

Consider the addition of two WSS processes $x(t)$ and $y(t)$ to form $z(t) = x(t) + y(t)$. Then

$$R_Z(\tau) = \overline{z(t)z(t+\tau)} = \overline{[x(t) + y(t)]\,[x(t+\tau) + y(t+\tau)]}$$

$$= R_X(\tau) + R_Y(\tau) + R_{XY}(\tau) + R_{YX}(\tau) \qquad (1.232)$$

If $x(t)$ and $y(t)$ are uncorrelated, then $R_{XY}(\tau) = R_{YX}(\tau) = \overline{x}\overline{y}$. Furthermore, if either mean value $\overline{x}$ or $\overline{y}$ is zero, which is typically the case in a communication system, then $R_{XY}(\tau) = R_{YX}(\tau) = 0$ and $R_Z(\tau) = R_X(\tau) + R_Y(\tau)$, in which case the PSD

$$PSD_Z(f) = PSD_X(f) + PSD_Y(f) \qquad (1.233)$$

1.8.14 Simulation : Frequency Response of an RC Filter

SIMULATION **RCFreqResponse:** The frequency response is determined by convolving the input signal $\sin(2\pi f t)$ with the impulse response $h(t)$ of the RC filter over a range of frequencies. The amplitude and phase variation with frequency are shown to overlay the curves in Fig. 1.47.

- **Experiment:** Change the input signal to a periodic square wave signal and pass it through the RC filter. Exchange the positions of the resistor and capacitor and modify the simulation accordingly. Is it still a low-pass filter ?

1.9   Problems

1. (a) Let the instantaneous voltage $v(t)$ across a resistor R be given by $v(t) = V_o \cos\left(2\pi f t\right)$, where V_o is the maximum amplitude and the frequency $f = \frac{1}{T}$. Determine the average voltage $\langle v(t) \rangle$ and the average power $\langle p(t) \rangle$ delivered to the resistor.

(b) Determine whether the signal shown below is a power or an energy signal and find its value.

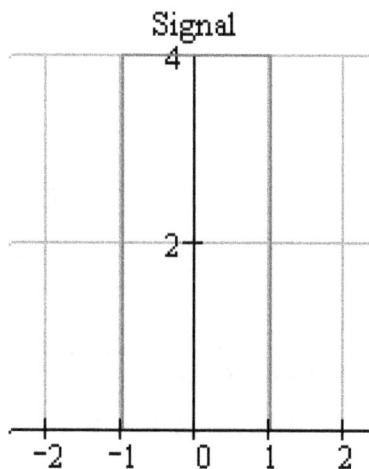

Signal

2. Sketch the double sided amplitude and phase spectrum of the following signal $s(t) = 2 \cos\left(5\pi t - \dfrac{2\pi}{3}\right) + 3 \cos\left(10\pi t - \dfrac{\pi}{2}\right)$.

3. (a) Determine the exponential Fourier series for the periodic signal shown below. (b) Verify your solution via Mathcad/MATLAB.

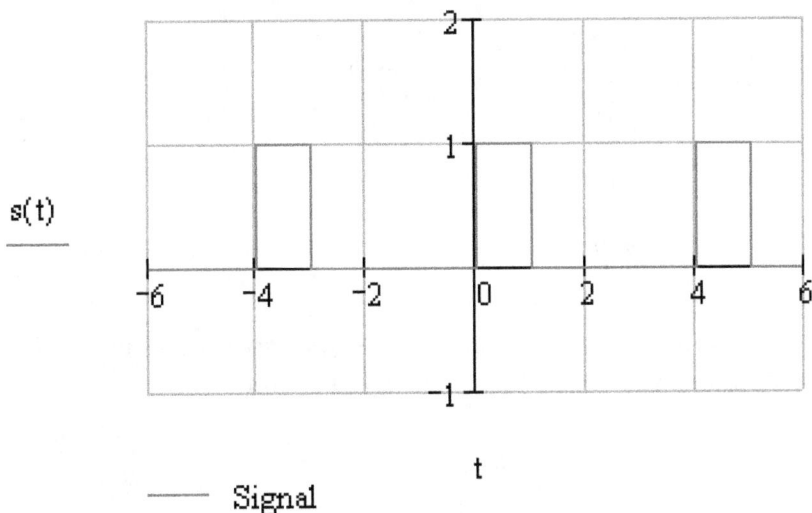

Signal

4. (a) Determine the exponential Fourier series of the signal $s(t) = A\cos\left(\pi t + \dfrac{\pi}{4}\right)$ via Mathcad/MATLAB.

(b) Determine the signal power using the amplitude spectrum and compare it with the expected result.

(c) Also, reconstruct the signal using the amplitude and phase spectrum.

(d) Experiment the amplitude, phase and frequency of this signal and observe the effect on the amplitude and phase spectrum.

5. (a) Repeat problem 4 but now for the signal $s(t) = 5-6\cos\left(20\pi t - \dfrac{\pi}{6}\right) + 3\sin\left(70\pi t\right)$ which was used as an example in section 1.1.

(b) Determine the amplitude and phase spectrum.

(c) Experiment with the signal expression $s(t)$ to gain a better understanding of the Mathcad/MATLAB simulation and the effect on the amplitude and phase spectrum.

6. Using Mathcad or MATLAB, determine the Fourier series for the Binary phase shift keying (BPSK) signal shown below which represents an alternating binary digit zero and one sequence baseband signal. This baseband signal, which is a square waveform with $T = 8$ and $A = 1$, is overlaid on this diagram for convenience. Multiplying the square wave with a high frequency sinusoidal waveform generates the BPSK signal.

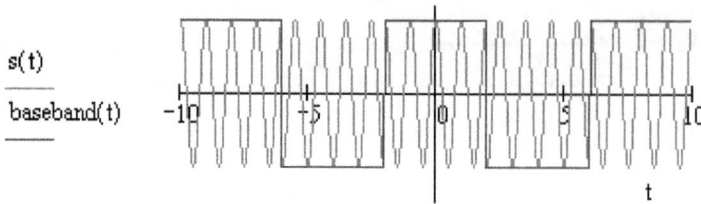

7. Verify that the exponential Fourier series for the well known triangular periodic signal shown below is given by $c_n = \dfrac{1}{2}\operatorname{sinc}^2\left(\dfrac{n}{2}\right)$. Calculate the average signal power, plot the corresponding power spectral density and use this to verify your answer for the average signal power.

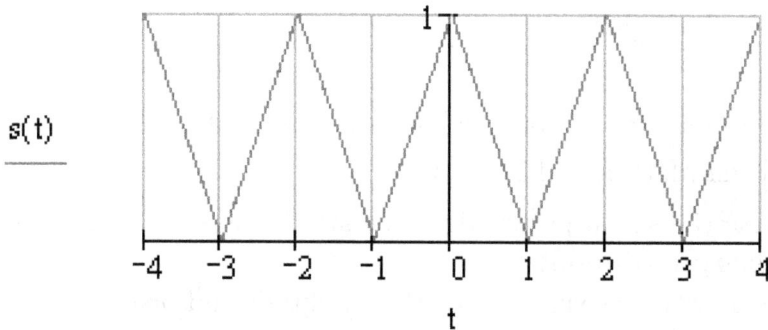

8. From section 1.4, recall that the FT of a rectangular pulse of amplitude A and width τ is given by $S(f) = A\tau\operatorname{sinc}(f\tau)$.

(a) Determine the Fourier Transform (FT) of the sinc(.) pulse $s(t) = 2AB\operatorname{sinc}(2Bt)$ by making use of the FT duality property, that if $s(t) \leftrightarrow S(f)$, then $S(t) \leftrightarrow s(-f)$.

(b) Verify your solution via Mathcad/MATLAB.

(c) Determine the output of passing this pulse through an ideal low-pass filter, which eliminates all frequencies greater than B Hz.

(d) From its energy spectral density, determine the signal energy.

9. Consider a sinusoidal signal given by $s(t) = A\sin(2\pi f_c t)$, where f_c is the carrier frequency and A is the amplitude. Determine the Fourier transform this pulse. Make use of the Dirac property $\delta(f) = \displaystyle\int_{-\infty}^{\infty} e^{-j2\pi ft}dt.$

10. Consider an exponential pulse given by

$$s(t) = \begin{cases} e^{-\frac{t}{T}} & t > 0 \\ 0 & otherwise \end{cases}$$

(a) Determine the Fourier transform this pulse.

(b) If the pulse with $T = 1$ second is passed through an ideal low-pass filter with a cut-off frequency $f_c = 2$ Hz, determine the corresponding filtered pulse.

(c) From its energy spectral density, determine the percentage amount of energy passed through the low-pass filter in part (b).

(d) If $s(t) = e^{-a|t|}$, where a is a constant, what is the corresponding Fourier transform ?

11. Consider a damped sinusoid pulse given by

$$x(t) = \begin{cases} e^{-t/T} \sin\left(\frac{2\pi t}{T}\right) & t > 0 \\ 0 & otherwise \end{cases}$$

(a) Determine the Fourier transform of this pulse using the real signal frequency translation theorem that the FT of the signal $s(t) \cos(2\pi f_c t + \theta)$ is given by $\frac{1}{2} \left[e^{j\theta} S(f - f_c) + e^{-j\theta} S(f + f_c) \right]$, where $s(t) \longleftrightarrow S(f)$.

(b) Estimate its bandwidth for $T = 0.5$ seconds.

(c) From the pulse energy spectral density, determine the pulse energy.

12. Consider a triangular pulse $s(t)$ given by

$$s(t) = \begin{cases} A^2 \left(1 - \frac{|t|}{T_b}\right) & for \quad |t| \le T_b \\ 0 & otherwise \end{cases}$$

(a) Determine the Fourier transform of this pulse.

(b) From its energy spectral density, determine the pulse energy.

13. Consider the sinc function shown below for which $s(t) = \frac{\sin(At)}{\pi t}$ where $A = 2\pi$. Verify the FT scaling property $s(at) \longleftrightarrow \frac{1}{|a|} S\left(\frac{f}{a}\right)$.

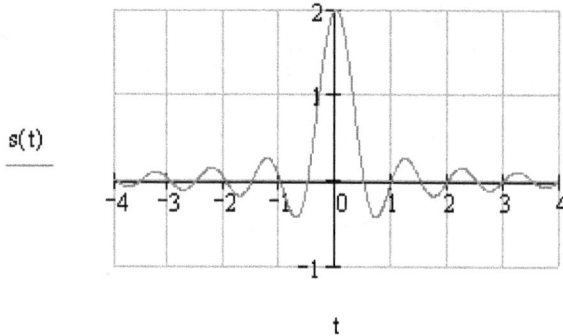

14. Consider the pulse shown below.

(a) Determine the FT using Mathcad/MATLAB and reconstruct the signal using only first side-lobe of the amplitude spectrum.

(b) What is the pulse bandwidth ?

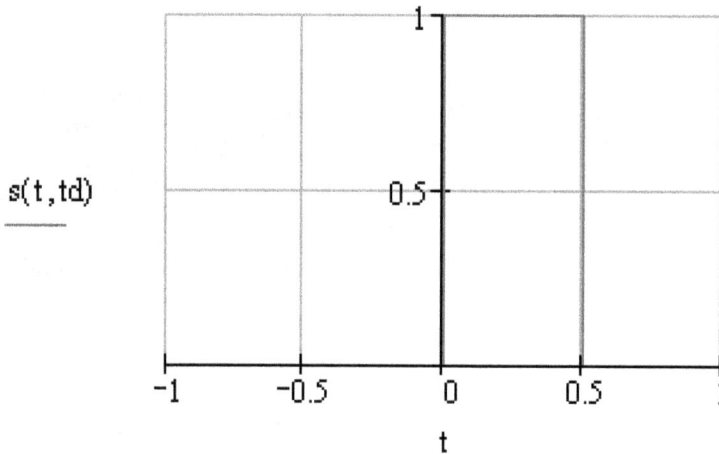

15. Consider the rectangular pulse shown below.

(a) Determine the signal energy by evaluating $\int_{-\infty}^{\infty} |s(t)|^2 \, dt$. Confirm your answer using the following two methods.

(b) Use the Fourier Transform (FT) of its autocorrelation function.

(c) Use the FT of the signal. Confirm that 90% of the energy is contained within the main spectral lobe.

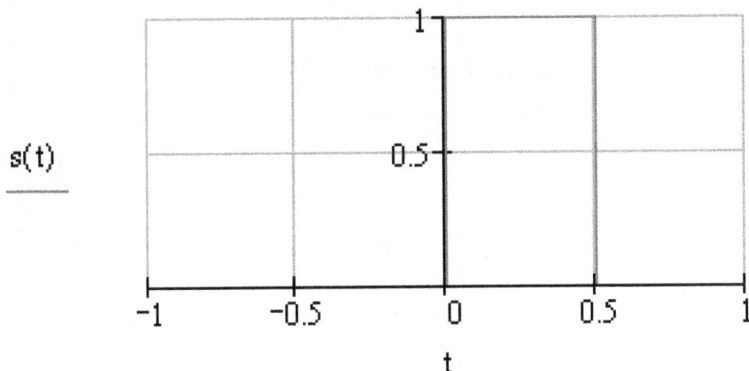

16. Determine the power of the sinusoidal signal $s(t) = A\sin\left(\dfrac{2\pi t}{T}\right)$ with $A = 3$ volts and $T = 0.5$ secs. Verify your answer by finding the power spectral density of the sinusoidal signal via its autocorrelation function.

17. Consider the RC circuit shown below. If the capacitor is initially uncharged and the input to the circuit is the constant d.c. voltage V_o, prove that the voltage across the capacitor after time t seconds is given by $V_{out}(t) = V_o\left(1 - e^{-t/RC}\right)$.

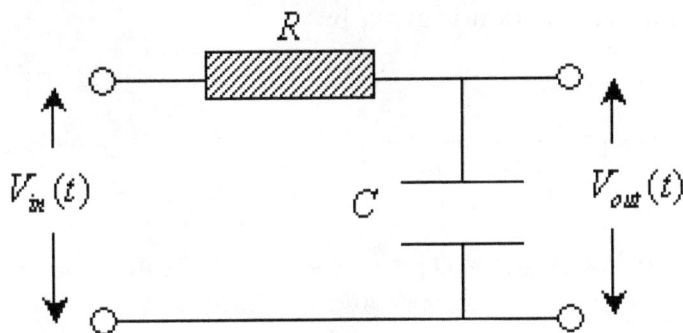

18. (a) Show that the Fourier transform of the signal $s(t) = u(t)e^{-at}$ is given by $S(f) = \dfrac{1}{a + j2\pi f}$, where $u(t) = \begin{cases} 1 & t > 0 \\ 0 & t < 0 \end{cases}$ is a unit step function and $a > 0$ is a constant.

(b) Hence determine the impulse response of the RC filter for which the frequency response is given by $H(f) = \dfrac{1}{1 + j2\pi f RC}$.

19. The frequency response of a RC filter is given by $H(f) = \dfrac{1}{1 + j2\pi f}$.
The Fourier Transform (FT) of the input signal is given by

$$S(f) = \begin{cases} 1 & if & f \leq \dfrac{A}{2\pi} \\ 0 & otherwise \end{cases}.$$

(a) Determine the input and the output signal via the inverse FT of $H(f)S(f)$.

(b) Verify your answer by finding the convolution of the input signal with the impulse response of the RC filter.

20. The periodic signal $s(t) = 1 + 2\cos(2\pi f_o t)$ with $f_o = 1$ Hz is input to an RC low-pass filter with a cut-off constant $f_c = 1$ Hz.

(a) Determine the input power spectral density $PSD_{in}(f)$ and hence the input signal power.

(b) Determine the output power spectral density $PSD_{out}(f)$ of the filter making use of the fact that $PSD_{out}(f) = |H(f)|^2 PSD_{in}(f)$ and hence the output signal power.

21. The Fourier Transform of a rectangular pulse amplitude A and width τ is given by $S(f) = A\tau \operatorname{sinc}(f\tau)$. This pulse is input to a raised cosine Nyquist filter whose transfer function is given by

$$H(f) = \begin{cases} 1 & if & |f| < (2f_o - B) \\ 0.5\left[1 + \cos\left[\pi\dfrac{|f| - (2f_o - B)}{2(B - f_o)}\right]\right] & if & (2f_o - B) \leq |f| < B \\ 0 & if & |f| \geq B \end{cases}$$

where $B = f_o(r + 1)$. For A = 1 volt, $\tau = 1$ s, $f_o = 1$ Hz and $r = 0$, 0.5 and 1, determine the filter impulse response and the filter output.

22. (a) Determine the power of the signal $s(t) = 3\sin\left(\frac{2\pi t}{T}\right)$ where $T = 0.5$ seconds. You may make use of the fact that $\frac{1}{T}\int_T \sin^2\left(\frac{2\pi t}{T}\right) dt = \frac{1}{2}$.

(b) Given that the corresponding autocorrelation function is $R(\tau) = \frac{A^2}{2}\cos\left(\frac{2\pi\tau}{T}\right)$, determine the power spectral density (PSD) of $s(t)$.

(c) Use your PSD to verify your answer to part (a).

23. The FFT spectrum of an analog signal is shown below.

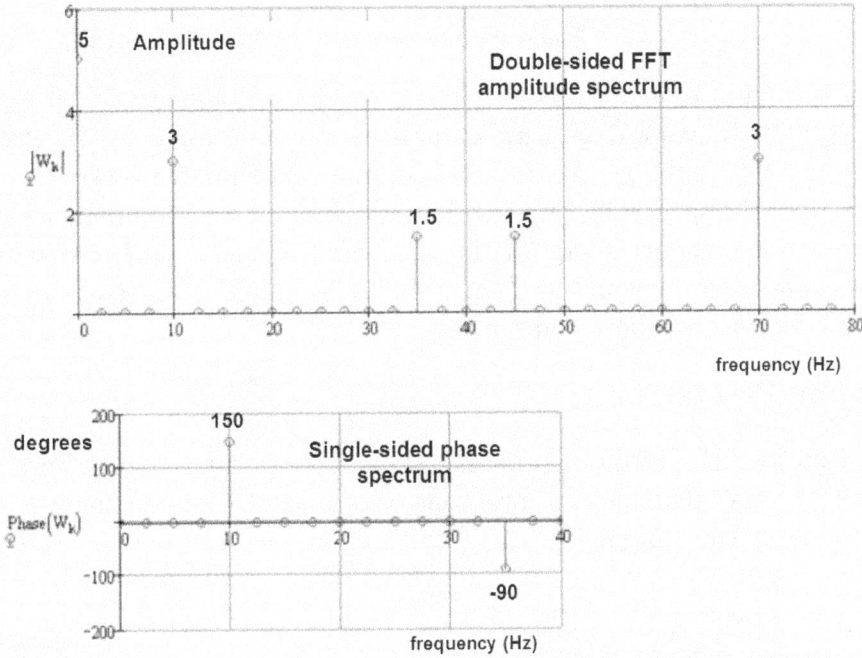

Double-sided FFT amplitude spectrum

Single-sided phase spectrum

(a) Write down an expression for this signal in terms of cosines.

(b) How many FFT points were used and what is the sampling frequency?

(c) What is the highest frequency this N-point FFT is available to extract?

(d) What would happen to the amplitude spectrum if the frequency resolution was not correct?

24. Consider the following expression designed to simulate band-limited AWGN.

$$Noise\,(t) = \sum_{n=1}^{100} 0.2 \cos\left(\frac{2\pi n t}{10} + \varphi\right)$$

where the phase φ is set randomly between $-\pi$ and $+\pi$.

(a) Without any derivation, sketch its double-sided amplitude spectrum and determine the power of this noise signal using your amplitude spectrum.

(b) What should the resolution of the FFT be to ensure the spectral components of this noise signal are accurately determined?

(c) The N-point Discrete Fourier Transform (DFT) of a digital signal $x[n]$

is given by $X_k = \sum\limits_{n=0}^{N-1} x_n e^{\frac{-j2\pi kn}{N}}$. Without using this expression for X_k, sketch the FFT spectrum of $\left|\frac{X_k}{N}\right|$ versus k if the sampling frequency $f_s = 30Hz$ and $N = 300$. In your diagram (as large as possible for clarity), clearly label the first two and last two spectral components that correspond to the positive half of the spectrum, and the first two and last two components that correspond to the negative side of the spectrum. Thus, you should have 8 components labeled on your diagram. State the amplitude, frequency and the k value for each of these components.

Random Processes

25. Consider the random process $x(t) = A\cos(2\pi f_o t + \theta)$, where the amplitude A and frequency f_o are constants and the phase angle θ can be any value in the range $-\frac{\pi}{4}$ to $+\frac{\pi}{4}$ with equal probability i.e. $f_\theta(\theta) = \begin{cases} \frac{2}{\pi} & \text{if } -\frac{\pi}{4} \le \theta \le \frac{\pi}{4} \\ 0 & \text{otherwise} \end{cases}$.

(a) Determine ensemble average $\overline{x_1}$ at time t, the second moment $\overline{x_1^2}$ and the time average $\langle x^2(t) \rangle$. What do you conclude?

(b) The power spectral density (PSD) of this random process via the Fourier transform of its autocorrelation function.

26. Consider the random process $r(t) = s(t) + n(t)$, where $s(t) = A\cos(2\pi f_o t + \theta)$ is not a random process and $n(t)$ is a zero mean WSS random process with autocorrelation function $R_n(t_1, t_2)$.

(a) Is $r(t)$ a WSS random process?

(b) Determine the autocorrelation function $R_r(t_1, t_2)$.

27. Consider the random process $n(t) = n_1(t) + n_2(t)$, where $n_1(t)$ and $n_2(t)$ represent ergodic random noise both of power 4 W with d.c. values of 1 V and -1 V, respectively. What is the power of $n(t)$ if $n_1(t)$ and $n_2(t)$ are orthogonal or if they are uncorrelated?

28. Determine an expression for the power spectral density of a random process for which its autocorrelation function $R_X(\tau) = A + Be^{-a|\tau|}$, where A, B and a are integers. Make use of the Dirac property $\delta(f) = \int\limits_{-\infty}^{\infty} e^{-j2\pi ft} dt$

and the Fourier Transform pair (from question 10) $e^{-a|t|} \longleftrightarrow \frac{2a}{a^2 + (2\pi f)^2}$.

29. The power spectral density of a random process is given by $PSD(f) =$
$$\begin{cases} \left(1 - \frac{|f|}{B}\right) & \text{for} & |f| \leq B \\ 0 & \text{otherwise} \end{cases}.$$

(a) Determine the average power of this random process.

(b) The corresponding autocorrelation function $R_X(\tau)$.

30. If the input to a linear filter is $r(t) = A \sin(2\pi f_o t) + n(t)$, where $n(t)$ is an AWGN signal with a power spectral density (PSD) of $\frac{N_0}{2}$ Watts/Hz over a doubel-sided spectrum, determine the

(a) PSD of the input sinusoid.

(b) PSD of the signal at the output of the filter and hence the power of the signal component.

(c) PSD of the noise at the output of the filter and hence the power of the noise component.

True knowledge is experience.

— *Albert Einstein*

Chapter 2

Analog Communications

2.1   Introduction

A communication system consists of an *information source, transmitter, channel* and a *receiver*. For an *analog communication system* (ACS), a transducer is typically required to convert the information source (speech, video, etc.) output into an electrical signal that is transmitted over the communication channel via carrier modulation and demodulation. This received electrical signal is then passed through another transducer, which performs the inverse operation of the source transducer, to complete the transfer of information from the source to the destination. Specifically, the analog message (or information bearing) signal $m(t)$ is a baseband signal (spectrum centered on 0 Hz) that is typically not suitable for direct transmission over a given channel. Examples include telephone speech (300 Hz - 3.5 kHz), high quality audio (up to 15 kHz), video (0-6 MHz), etc. The solution is to shift its spectrum to be centered over a high frequency f_c via the *modulation* of a sinusoidal carrier i.e. the amplitude, phase or frequency of a carrier is varied in accordance with $m(t)$. A key advantage of modulation is that a given channel can be used to transfer several signals simultaneously via *frequency-division multiplexing* (FDM) in which the spectra of the message signals are separated in frequency. Furthermore, the use of a high frequency sinusoidal carrier is essential for transmission via an antenna. For example, the propagation of electromagnetic waves of wavelength λ via an antenna of length L requires that $L \approx \frac{\lambda}{2}$. e.g. an audio signal of 4 kHz requires $L \approx \frac{1}{2}\frac{3(10^8)}{4(10^3)} = 37.5$ km! However, the transmission of a high frequency carrier of say 900 MHz requires $L \approx 0.17$ m which is practically feasible.

2.1.1   Power Signal-to-Noise Ratio (SNR) in Baseband Systems

Before analyzing modulation techniques, it is necessary to establish the **power** *signal-to-noise ratio* (SNR) in a baseband communication system that will serve as a reference point for comparison. Consider the simple analog baseband communication shown in Fig. 2.1, in which the signal $m(t)$ is a zero-mean ergodic random process bandlimited to B Hz transmitted directly over the channel **without** any modulation. For example, over short pair of wires, optical fibre or coaxial cable.

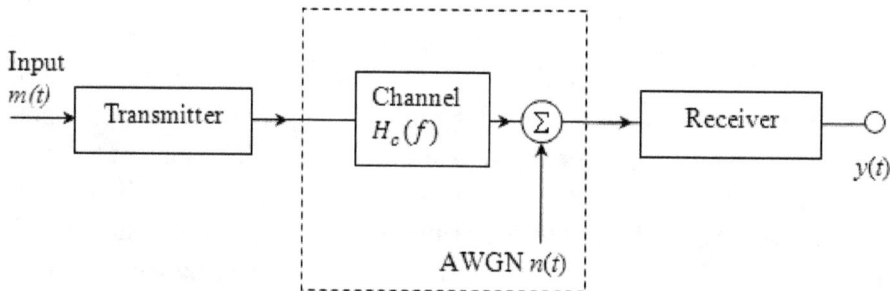

Figure 2.1: A baseband communication system.

The analysis to follow will also establish a foundation for the baseband communication system to be covered later in sections 6.2 and 6.3. Let $S_m(f)$ represent the Fourier transform (FT) of $m(t)$ so that

$$m(t) = FT^{-1}\left[S_m(f)\right] = \int_{-\infty}^{\infty} S_m(f)e^{j2\pi ft}df \qquad (2.1)$$

and the average signal power P_m of $m(t)$ is given by

$$P_m = \int_{-\infty}^{\infty} PSD_m(f)df \qquad (2.2)$$

where $PSD_m(f)$ represent the power spectral density of $m(t)$. Given that $m(t)$ is bandlimited to B Hz, $P_m = \int_{-B}^{B} PSD_m(f)df$. Furthermore, let the frequency response of the transmitter $H_t(f)$ and the receiver $H_r(f)$, referred to respectively as the *preemphasis* and *deemphasis* filters, to be simply that of an ideal low-pass filter (LPF) (section 1.8.9) so that

$$H_t(f) = H_r(f) = \begin{cases} 1 & if \quad -B \le f \le B \\ 0 & \text{otherwise} \end{cases} \qquad (2.3)$$

In section 2.8, we shall take a look at how these filters can be used to increase the output power signal-to-noise ratio by slightly distorting the input signal $m(t)$. For now however, the *overall* communication system transfer function $H(f)$ required for **distortionless** transmission (section 1.8.10) is given by

$$H(f) = H_t(f)H_c(f)H_r(f) = \begin{cases} ke^{-j2\pi f t_d} & if \quad -B \le f \le B \\ 0 & \text{otherwise} \end{cases} \qquad (2.4)$$

where k is a constant, t_d is the time-delay of the system and the channel frequency response $H_c(f) = ke^{-j2\pi f t_d}$. Taking $k = 1$ for simplicity, the output of the receiver

$$y(t) = s_o(t) + n_o(t) \qquad (2.5)$$

where $n_o(t)$ is the output noise component and $s_o(t)$ is the output signal component given by (via equations 1.193 and 2.4)

$$s_o(t) = \int_{-\infty}^{\infty} S_m(f)H(f)e^{j2\pi ft}df = \int_{-\infty}^{\infty} S_m(f)e^{j2\pi f(t-td)}df = m(t - t_d) \quad (2.6)$$

i.e. its simply a delayed version of the input. Using the relation $PSD_{out}(f) = PSD_{in}(f)|H(f)|^2$ from equation 1.206, the average signal **power** P_{in} at the receiver input is

$$P_{in} = \int_{-\infty}^{\infty} PSD_m(f)\,|H_t(f)H_c(f)|^2\,df = \int_{-B}^{B} PSD_m(f)df = P_m \quad (2.7)$$

because $H_t(f) = \begin{cases} 1 & if \quad -B \le f \le B \\ 0 & \text{otherwise} \end{cases}$ and $|H_c(f)|^2 = 1$. Notice that the average signal power P_t at the output of the transmitter is equal to

$$P_t = \int_{-\infty}^{\infty} PSD_m(f)\,|H_t(f)|^2\,df = \int_{-B}^{B} PSD_m(f)df = P_m \quad (2.8)$$

and thus we could have expressed P_{in} as simply

$$P_{in} = \int_{-\infty}^{\infty} PSD_m(f)\,|H_c(f)|^2\,df \quad (2.9)$$

The average signal power P_{out} at the receiver output is

$$
\begin{aligned}
P_{out} &= E\left[s_o^2(t)\right] = E\left[m^2(t - t_d)\right] = E\left[m^2(t)\right] & (2.10) \\
&= \int_{-\infty}^{\infty} PSD_m(f)\,|H(f)|^2\,df \\
&= \int_{-B}^{B} PSD_m(f) = P_m = P_{in} & (2.11)
\end{aligned}
$$

- Conventionally, for an **analog** communication system, the **power** signal-to-noise is written as S/N or $\frac{S}{N}$. However, to avoid confusion with the FT notation $S(f)$, the symbol P is retained to emphasize power. For a **digital** communication system (chapter 3 onwards), the signal-to-noise ratio (SNR) is defined to be $\left(\frac{E_b}{N_o}\right)$ in decibels.

For additive noise $n(t)$ that is a zero-mean white gaussian with a power spectral density $\frac{N_o}{2}$W/Hz, the receiver output noise power spectral density is $\frac{N_o}{2}\left|H_r(f)\right|^2$ and hence, the average noise power N_{out} at the output of the receiver is given by

$$N_{out} = \int_{-B}^{B} \frac{N_o}{2} \left|H_r(f)\right|^2 df = \frac{N_o}{2} \int_{-B}^{B} df = N_o B \qquad (2.12)$$

Hence, the *signal-to-noise* **power** ratio (SNR) at the output of the low-pass filter

$$\frac{P_{out}}{N_{out}} = \frac{P_{in}}{N_o B} \qquad (2.13)$$

where if we let $\gamma = \frac{P_{in}}{N_o B}$, then we may express the output power SNR as $\left(\frac{P}{N}\right)_{out} = \gamma$. The output power SNR $\left(\frac{P}{N}\right)_{out}$ is the figure of merit for an analog communication system. Although γ is directly proportional to P_{in}, constraints such as the transmitter power, type of channel, international regulations, etc. inherently limit the output power SNR.

Example : Baseband Communication over a Distorting Channel

In the foregoing analysis, if the channel frequency response $H_c(f) = \frac{B}{B+jf}$ and $H_r(f) = \begin{cases} \frac{1}{H_c(f)} & if \quad -B \leq f \leq B \\ 0 & \text{otherwise} \end{cases}$, then the average signal power P_{out} at the receiver output is given by

$$P_{out} = \int_{-B}^{B} PSD_m(f) \left|H_c(f)\right|^2 \left|H_r(f)\right|^2 df \qquad (2.14)$$

where $H(f) = H_t(f)H_c(f)H_r(f)$ with $H_t(f) = \begin{cases} 1 & if \ -B \leq f \leq B \\ 0 & \text{otherwise} \end{cases}$.
Given that $H_c(f)H_r(f) = 1$,

$$P_{out} = \int_{-B}^{B} PSD_m(f)df = P_m \qquad (2.15)$$

The average noise power N_{out} at the output of the receiver is given by

$$N_{out} = \int_{-B}^{B} \frac{N_o}{2} |H_r(f)|^2 \, df \qquad (2.16)$$

where $H_r(f) = \frac{1}{H_c(f)} = 1 + j\frac{f}{B}$ so that $|H_r(f)|^2 = \left(1 + j\frac{f}{B}\right)\left(1 - j\frac{f}{B}\right) = 1 + \left(\frac{f}{B}\right)^2$. Thus

$$N_{out} = \frac{N_o}{2} \int_{-B}^{B} 1 + \left(\frac{f}{B}\right)^2 df = \frac{N_o}{2} \left[f + \frac{f^3}{3B^2} \right]_{-B}^{B} = \frac{4N_o B}{3} \qquad (2.17)$$

which yields an output power SNR of

$$\frac{P_{out}}{N_{out}} = \frac{3P_m}{4N_o B} \qquad (2.18)$$

2.1.2   Narrowband Noise

In the sections to follow, we shall consider the modulation (various techniques) of a carrier of frequency f_c and utilize a band-pass filter (BPF) with a very narrow bandwidth prior to the demodulator as shown in Fig. 2.2. This *narrowband* band-pass filter is referred to as the *predetection* filter. Within the demodulator, the low-pass filter (LPF) is referred to as the *postdetection* filter. Details of the detector will be revealed later.

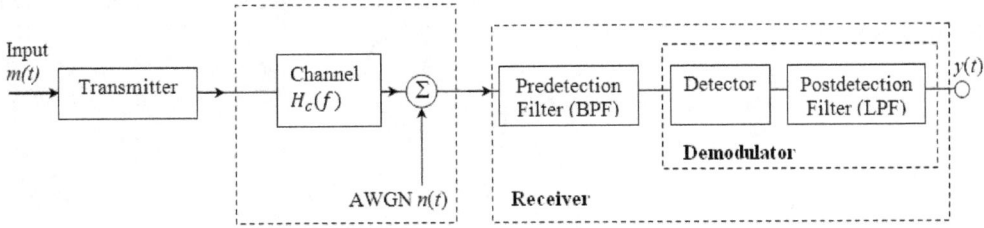

Figure 2.2: Analog communication system block diagram.

2.1.3 Predetection Band-pass Filter

Suppose the predetection filter in Fig. 2.2 is an ideal band-pass filter (BPF) of bandwidth $2B$ with the transfer function

$$H_r(f) = \begin{cases} 1 & if \quad f_c - B \leq |f| \leq f_c + B \\ 0 & \text{otherwise} \end{cases} \tag{2.19}$$

where $B \ll f_c$, centered at the frequency f_c. Narrowband noise is produced when white noise with a PSD of $\frac{N_o}{2}$ W/Hz is passed through such a narrowband filter. A sample $n_{band}(t)$ of the narrowband noise at the output of the predetection filter can be expressed as a sinusoid of random amplitude $E_{noise}(t)$ and phase $\theta_{noise}(t)$ such that

$$n_{band}(t) = E_{noise}(t) \cos\left(2\pi f_c t + \theta_{noise}(t)\right) \tag{2.20}$$

where $E_{noise}(t)$ is the *envelope function* and $\theta_{noise}(t)$ is the *phase function* of $n_{band}(t)$. Equivalently,

$$n_{band}(t) = E_{noise}(t) \cos\left(2\pi f_c t\right) \cos\left(\theta_{noise}(t)\right) - E_{noise}(t) \sin\left(2\pi f_c t\right) \sin\left(\theta_{noise}(t)\right)$$

$$= n_c(t) \cos\left(2\pi f_c t\right) - n_s(t) \sin\left(2\pi f_c t\right) \tag{2.21}$$

where equation 2.21 is referred to the *quadrature representation* of $n_{band}(t)$ in which the *in-phase* component is

$$n_c(t) = E_{noise}(t) \cos \left(\theta_{noise}(t) \right) \qquad (2.22)$$

and the *quadrature* component is

$$n_s(t) = E_{noise}(t) \sin \left(\theta_{noise}(t) \right) \qquad (2.23)$$

so that $n_c^2(t) + n_s^2(t) = E_{noise}^2(t) \left[\cos^2 \left(\theta_{noise}(t) \right) + \sin^2 \left(\theta_{noise}(t) \right) \right] = E_{noise}^2(t)$ and $\frac{n_s(t)}{n_c(t)} = \tan \left(\theta_{noise}(t) \right)$, to give

$$E_{noise}(t) = \sqrt{n_c^2(t) + n_s^2(t)} \qquad (2.24)$$

and

$$\theta_{noise}(t) = \tan^{-1} \left(\frac{n_s(t)}{n_c(t)} \right) \qquad (2.25)$$

Using equation 1.206 with $PSD_{in}(f) = \frac{N_o}{2}$ W/Hz, the output noise power spectral density is $\frac{N_o}{2} |H_r(f)|^2$ and the average noise power at the output of the band-pass predetection narrowband filter N_{band} is given by

$$N_{band} = E \left[n_{band}^2(t) \right] = \int_{-\infty}^{\infty} \frac{N_o}{2} |H_r(f)|^2 \, df \qquad (2.26)$$

$$= 2 \int_{f_c - B}^{f_c + B} \frac{N_o}{2} df = N_o \left[f_c + B - (f_c - B) \right] = 2 N_o B \qquad (2.27)$$

Notice how N_{band} does not depend on the narrowband center frequency f_c. To extract the in-phase component, we multiply $n_{band}(t)$ by $2 \cos(2\pi f_c t)$

and pass the result through a low-pass filter because $n_{band}(t)2\cos(2\pi f_c t) = n_c(t) + n_c(t)\cos(4\pi f_c t) - n_s(t)\sin(4\pi f_c t)$. An ideal low-pass filter removes the $2f_c$ frequency components and allows $n_c(t)$ to pass through. Similarly, $n_{band}(t)$ multiplied by $-2\sin(2\pi f_c t)$ and filtered by a low-pass filter extracts the quadrature component $n_s(t)$. It can be shown that the noise components $n_c(t)$ and $n_s(t)$ have the following properties:

- The PSD of $n_c(t)$ and $n_s(t)$,

$$PSD_{n_c}(f) = PSD_{n_s}(f) = \begin{cases} PSD_{n_{band}}(f - f_c) + PSD_{n_{band}}(f + f_c) & |f| \leq B \\ 0 & \text{otherwise} \end{cases}$$

(2.28)

from which we notice that $PSD_{n_c}(f) = PSD_{n_s}(f) = N_o$ W/Hz over $-B$ to $+B$ Hz and zero otherwise.

- $n_c(t)$ and $n_s(t)$ both have zero mean and the same variance as $n_{band}(t)$.

- $n_c(t)$ and $n_s(t)$ are uncorrelated with each other i.e. $E\left[n_c(t)n_s(t)\right] = 0$.

- If $n_{band}(t)$ is gaussian, then $n_c(t)$ and $n_s(t)$ are also gaussian.

- $E_{noise}(t)$ is a random variable with a rayleigh distribution and $\theta_{noise}(t)$ is a random variable with a uniform distribution.

2.1.4 Postdetection Low-pass Filter

If the in-phase or quadrature component is then passed through an ideal LPF of bandwidth B, the noise power at the output of the receiver in Fig. 2.2 is given by

$$E\left[n_s^2(t)\right] = E\left[n_c^2(t)\right] = \int_{-B}^{B} N_o df = 2N_o B \tag{2.29}$$

Notice how $E\left[n_{band}^2(t)\right] = E\left[n_s^2(t)\right] = E\left[n_c^2(t)\right] = 2N_o B$. This will be verified by simulation in the section 2.1.5.

2.1.5 Simulation : Narrowband Noise

⭐ SIMULATION **NarrowBandNoise** (Mathcad, MATLAB): An approximation to a bandlimited AWGN signal $Noise(t)$ is generated using cosine spectral components of the same amplitude but with random phase, as done previously in section 1.7.8. This signal $Noise(t)$ is then sampled to generate the digital signal $x_n = Noise\left(\frac{n}{f_s}\right)$ in which f_s is the sampling frequency given by $f_s = N f_{resolution}$, where N is the number of spectral components within the DFT, n is sample index $n = 0, 1, 2, \cdots N - 1$ and $f_{resolution}$ is the frequency resolution that is set equal to the separation between the spectral components used to generate $Noise(t)$. The amplitude spectrum of x_n, determined via its fast Fourier transform (FFT), is used to calculate the average power of this noise signal and shown to be equal to the power of $Noise(t)$ in the time-domain. Working in the frequency domain, the noise signal x_n is passed through an ideal narrowband filter of bandwidth B centered over a carrier frequency f_c (equation 2.19). The power of this filtered signal N_{band} is shown to be equal to $2N_oB$ using its amplitude spectrum to verify equation 2.27.

The narrowband filtered spectrum is then used to reconstruct (via the inverse FFT) the filtered noise signal $xNoiseFiltered_n$ from which the sampled in-phase and quadrature phase components n_{c_n} and n_{s_n} are extracted by multiplying $xNoiseFiltered_n$ with $2\cos(2\pi f_c \frac{n}{f_s})$ and $-2\sin(2\pi f_c \frac{n}{f_s})$ and using a low-pass filter, respectively. These are then used to determine the envelope variation $E_{noise_n} = \sqrt{n_{c_n}^2 + n_{s_n}^2}$ and $\theta_{noise_n} = \tan^{-1}\left(\frac{n_{s_n}}{n_{c_n}}\right)$ to generate $n_{band_n} = E_{noise_n} \cos\left(2\pi f_c \frac{n}{f_s} + \theta_{noise_n}\right)$, which is shown to correspond to narrowband filtered noise signal $xNoiseFiltered_n$ with an average noise power $N_{band} = \overline{n_{band}^2} = \overline{n_c^2} = \overline{n_s^2} = \sigma^2 = 2N_oB$ (refer to section 1.7.5 as required). Graphs of $xNoiseFiltered_n$ and n_{band_n} versus time $\frac{n}{f_s}$ are overlaid with E_{noise_n} to verify it does accurately represent the envelope variation. We shall make use of this feature when we consider an envelope detector in section 2.5. Finally, the probability histogram of E_{noise_n} is overlaid with the expected values using the Rayleigh probability density function (PDF)

$$f_{noise}(r, \sigma) = \begin{cases} \frac{r}{\sigma^2} \exp\left(\frac{-r^2}{2\sigma^2}\right) & if \qquad r > 0 \\ 0 & otherwise \end{cases}$$, where σ is the standard

deviation of n_c and n_s given by $\sigma = \sqrt{2N_oB}$. Also, the probability his-

togram of n_{band_n} is overlaid with the expected values using the Gaussian
PDF $f(x) = \frac{1}{\sigma\sqrt{2\pi}} \exp\left[\frac{-x^2}{2\sigma^2}\right]$.

- **Experiment:** Verify that n_c and n_s are uncorrelated and confirm that
  the PSD of the noise components $PSD_{n_c}(f) = PSD_{n_s}(f) = N_o$ W/Hz.

2.2 Amplitude Modulation

In *amplitude* (or equivalently *linear*) *modulation* (AM), the amplitude A_c of
a high frequency sinusoidal carrier $s(t)$ with frequency f_c is varied in accordance with the message signal. The frequency f_c is constant and the phase
of the carrier may be taken to be zero without a loss of generality. There are
four basic types of AM. Namely, *double-sideband suppressed-carrier* (DSB-
SC), *double-sideband large-carrier* (DSB-LC) which is also referred to as or-
dinary/conventional/full amplitude modulation, *single-sideband* (SSB) and
vestigial-sideband (VSB) modulation. We shall consider each of these variations in the sections to follow before considering angle modulation in section
2.7.

2.2.1 Double-Sideband Suppressed-Carrier (DSB-SC) Modulation

In DSB-SC modulation, the carrier signal is given by

$$s_{DSB}(t) = m(t) \cos\left(2\pi f_c t\right) \tag{2.30}$$

where f_c is the carrier frequency. From section 1.5, the FT $S_{DSB}(f)$ of
the modulated signal is given by

$$S_{DSB}(f) = \frac{1}{2}\left[S_m(f + f_c) + S_m(f - f_c)\right] \tag{2.31}$$

where typically f_c is very much greater than the bandwidth B of $m(t)$.
Notice how the spectrum $S_m(f)$ of the baseband signal $m(t)$ has been shifted

by f_c and that the bandwidth of the modulated signal is now $2B$.   If the carrier is *coherently* (or *synchronously*) demodulated by multiplying the carrier by $\cos(2\pi f_c t)$ under a noiseless environment, then

$$s_{DSB}(t)\cos(2\pi f_c t) = m(t)\cos^2(2\pi f_c t) = \left[\frac{m(t)}{2} + \frac{m(t)}{2}\cos(4\pi f_c t)\right].$$

$$(2.32)$$

Using a low-pass filter to eliminate the term $\frac{m(t)}{2}\cos(4\pi f_c t)$, the recovered signal is $\frac{m(t)}{2}$. To avoid this factor of $1/2$, $s_{DSB}(t)$ is multiplied by $2\cos(2\pi f_c t)$ as in the next section, where we shall take the channel noise into account to determine the power SNR at the demodulator output. A key feature of this modulation technique is that there are no discrete components of the carrier frequency at $\pm f_c$ i.e. the carrier is suppressed.

2.2.2   SNR in DSB-SC Modulation

The receiver in Fig. 2.2 for a DSB system is shown in Fig. 2.3 in which the predetection filter is a narrowband ideal bandpass filter with bandwidth $2B$ used to limit the out-of-band noise. The output of the predetection narrowband filter enters the demodulator, which multiplies the incoming signal by $2\cos(2\pi f_c t)$ and passes it through a postdection low-pass filter.

The input to the narrowband filter in Fig. 2.3 is $s_{DSB}(t) + n(t)$, where $s_{DSB}(t) = m(t)\cos(2\pi f_c t)$ is the transmitter output assumed to be transferred over the channel without distortion. Given that the narrowband filter is an ideal bandpass filter which allows the signal component $s_{DSB}(t)$ through unperturbed, the signal at the output of the narrowband filter $s_{band}(t)$ is

$$s_{band}(t) = s_{DSB}(t) + n_{band}(t) \qquad (2.33)$$

$$= [m(t) + n_c(t)]\cos(2\pi f_c t) - n_s(t)\sin(2\pi f_c t) \qquad (2.34)$$

(making use of equation 2.21) and the input to the low-pass filter (LPF) is given by

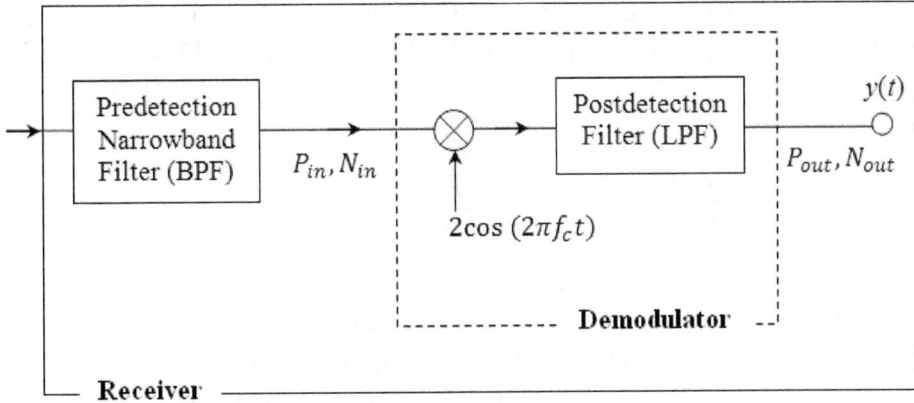

Figure 2.3: DSB receiver.

$$s_{band}(t)2\cos(2\pi f_c t) = 2\left[m(t) + n_c(t)\right]\cos^2(2\pi f_c t) - 2n_s(t)\sin(2\pi f_c t)\cos(2\pi f_c t) \tag{2.35}$$

$$= m(t) + n_c(t) + \left[m(t) + n_c(t)\right]\cos(4\pi f_c t) - n_s(t)\sin(4\pi f_c t) \tag{2.36}$$

hence the LPF filter output simplifies to

$$y(t) = m(t) + n_c(t) \tag{2.37}$$

because the high frequency terms at $2f_c$ have been eliminated by the LPF filter. Notice that the quadrature component $n_s(t)$ has been eliminated. Having analyzed the input and output of the DSB receiver, we may now determine the input and output power SNRs. The signal power P_{in} at the input of the **demodulator** is given by

$$\begin{aligned}
P_{in} &= E\left[s_{DSB}^2(t)\right] = \overline{\left[m(t)\cos(2\pi f_c t)\right]^2} \\
&= \overline{m^2(t)}\left(\overline{\cos^2(2\pi f_c t)}\right) = \frac{E\left[m^2(t)\right]}{2} = \frac{P_m}{2} \tag{2.38}
\end{aligned}$$

and the noise power at the demodulator input is given by $N_{in} = N_{band} = 2N_oB$ (section 2.1.3). Thus, the power SNR at the input of the demodulator is

$$\left(\frac{P}{N}\right)_{in} = \frac{P_{in}}{N_{in}} = \frac{P_{in}}{2N_oB} = \frac{P_m}{4N_oB} = \frac{\gamma}{2} \tag{2.39}$$

because (from section 2.1.1) $\gamma = \frac{P_{in}}{N_oB}$. At the output of the low-pass filter in Fig. 2.3, since $y(t) = m(t) + n_c(t)$, the signal power $P_{out} = E\left[m^2(t)\right] = P_m$ and the noise power $N_{out} = E[n_c^2(t)] = 2N_oB$ (from equation 2.29). Thus, the power SNR at the output of the low-pass filter

$$\left(\frac{P}{N}\right)_{out} = \frac{P_{out}}{N_{out}} = \frac{P_m}{2N_oB} = \frac{P_{in}}{N_oB} = \gamma \tag{2.40}$$

which we note is same as the power SNR at the output of the baseband system in section 2.1.1, equation 2.13. A figure of merit for the demodulator is the *detector gain*, which for the DSB system is given by $\frac{(P/N)_{out}}{(P/N)_{in}} = \frac{\gamma}{\gamma/2} = 2.$

2.2.3   Simulation : DSB-SC Communication System

⭐ SIMULATION **DSBSC:** The communication system presented in Fig. 2.4 is modeled.

The signal and its spectrum at various points throughout the communication system in Fig. 2.4 are presented in Table 2.1. A brief description of the parts A-H is as follows.

A Message signal $m(t)$ to be transmitted.

B Amplitude spectrum of the message signal $m(t)$. The spectral range of the message signal is referred as the *baseband* and accordingly, $m(t)$ is referred to as the *baseband signal*. We shall constraint the bandwidth B of $m(t)$ to 3 Hz.

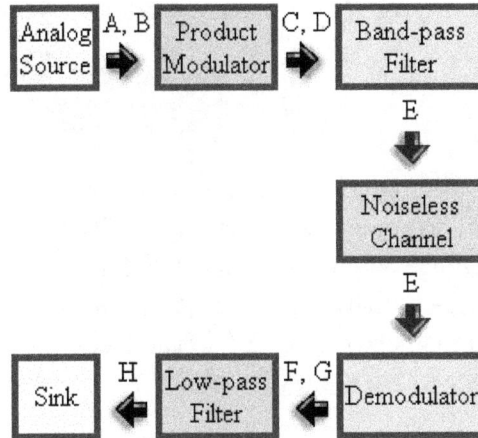

Figure 2.4: Noiseless DSB-SC communication system.

C The amplitude modulated signal overlaid with $m(t)$. This amplitude modulated signal is created by multiplying the baseband signal with $\cos(2\pi f_c t)$, where in this example, $f_c = 6$ Hz. Notice the change in phase of the carrier signal each time $m(t)$ goes to zero.

D Amplitude spectrum of the modulated signal. Notice how the spectral components are centered on either side of the carrier frequency $f_c = 6$ Hz and that the bandwidth of the modulated signal is $2B$. The portion of the spectrum which lies above f_c is referred to the *upper sideband* and the portion below f_c as the *lower sideband* as labeled in Fig. 2.5.

E The spectral components on either side of $(f_c \pm 3)$ Hz are eliminated using a bandpass filter.

F This is the result of multiplying the filtered signal from part E with $2\cos(2\pi f_c t)$ at the demodulator. We have assumed the ideal conditions of a noiseless channel and no phase or frequency difference between these two signals.

G The amplitude spectrum of the signal in part F. Notice how the sidebands are now centered on $2f_c$. To recover the original base band signal, a low-pass filter is required to eliminate the spectral components on either side of ± 3 Hz.

H A comparison between the message and the demodulated baseband signal.

- **Experiment:** Change the carrier frequency. Be sure to set the band-pass filter cut-off frequencies accordingly. Try other types of message signals. Investigate what happens if the received modulated carrier from the channel is multiplied by $\cos(2\pi f_c t + \phi)$ or $\cos(2\pi(f_c + \Delta f)t)$, where ϕ is the phase error and Δf is the frequency error. Can we tolerate small errors in frequency or phase ?

- **Noise:** Unlike the results presented in Table 2.1, the signal amplitude is increased by a factor of 10 and noise is taken into account within the simulation **DSBSC**.

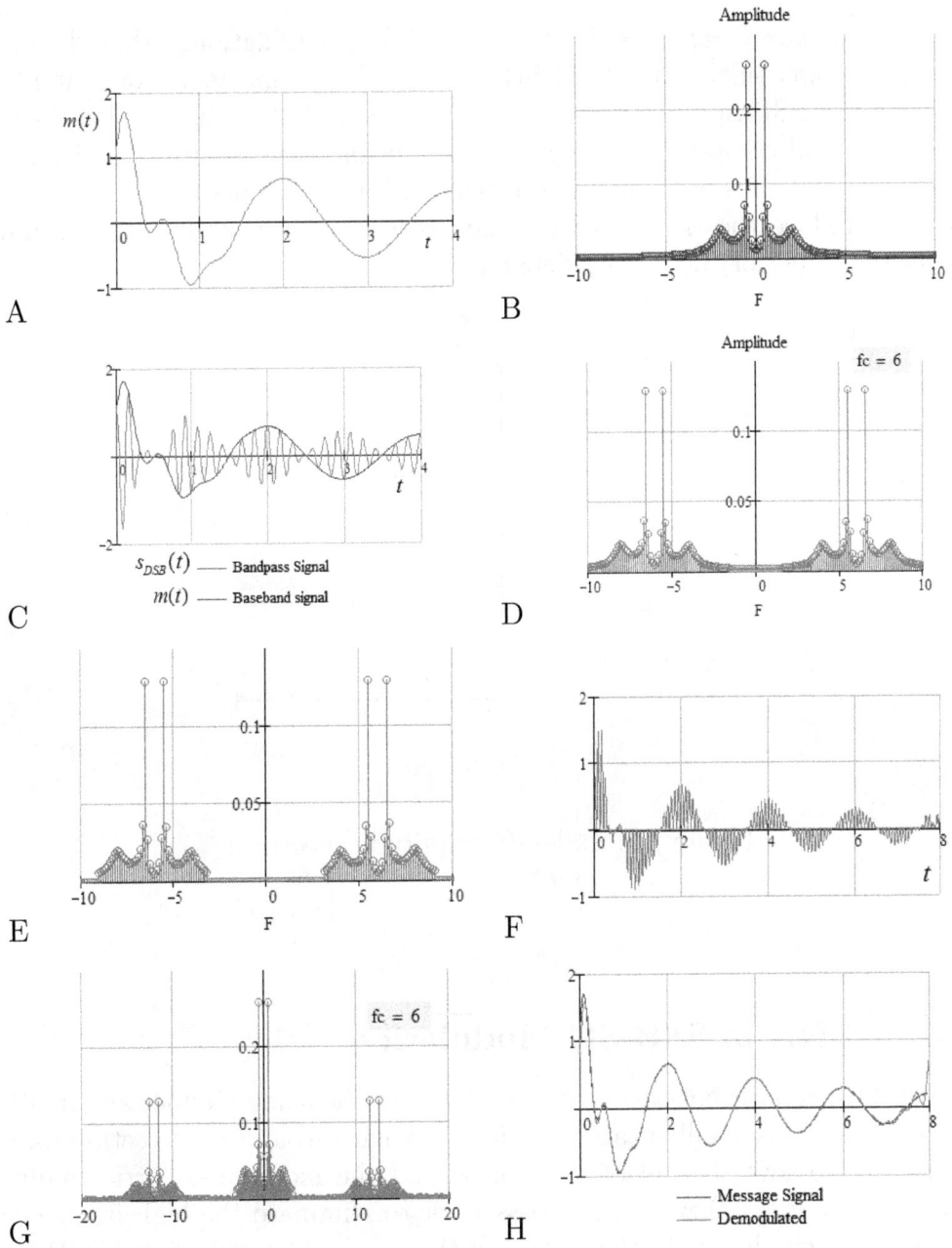

Table 2.1: DSB-SC communication system signals and spectra

2.3   SSB-SC Modulation

In *single-sideband Suppressed-Carrier* (SSB-SC) modulation, either the upper or the lower sideband of the DSB-SC signal is transmitted. For example, consider the DSB-SC spectrum shown in Fig. 2.5. Notice that the baseband signal spectral components are contained within both the upper and lower sidebands and thus accordingly, the bandwidth of the modulated carrier is $2B$. To reduce the transmission bandwidth to B, we need only transmit either the upper or the lower sideband.

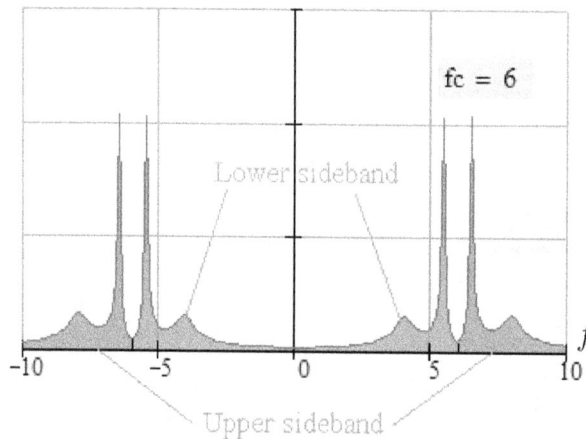

Figure 2.5: DSB-SC amplitude spectrum.

2.3.1   SNR in SSB-SC Modulation

A SSB-SC signal is coherently demodulated in the same manner as a DSB-SC signal. Thus, as illustrated in Fig 2.3, a narrowband predetection filter is used to remove the out-of-band noise and the modulated carrier multiplied by $2\cos\left(2\pi f_c t\right)$ with a low-pass filter to eliminate the high frequency terms. Given that only the upper or the lower sideband of the DSB-SC signal is transmitted, the narrowband filter bandwidth required is only B Hz. Accordingly, its output average noise power is reduced in half to

$$N_{band} = N_o B \qquad (2.41)$$

A SSB signal $s_{SSB}(t)$ can be expressed as

$$s_{USB}(t) = m(t)\cos{(2\pi f_c t)} - m_p(t)\sin{(2\pi f_c t)} \qquad (2.42)$$

$$s_{LSB}(t) = m(t)\cos{(2\pi f_c t)} + m_p(t)\sin{(2\pi f_c t)} \qquad (2.43)$$

where $s_{USB}(t)$ and $s_{LSB}(t)$ are the SSB-SC signals corresponding to the upper and the lower sideband, respectively and $m_p(t)$ is the message signal after every frequency component of $m(t)$ is phase shifted by $-\frac{\pi}{2}$ radians, given by the Hilbert transform of $m(t)$ (section 1.5.8), so that

$$S_{m_p}(f) = -j\,\mathrm{sgn}(f)S_m(f) \qquad (2.44)$$

where $S_{m_p}(f)$ and $S_m(f)$ are the Fourier transforms of $m_p(t)$ and $m(t)$, respectively. Thus, $m_p(t)$ is given by the inverse Fourier transform

$$m_p(t) = \int\limits_{-\infty}^{\infty} S_{m_p}(f)e^{j2\pi ft}df = \int\limits_{-\infty}^{\infty} -j\,\mathrm{sgn}(f)S_m(f)e^{j2\pi ft}df \qquad (2.45)$$

Alternatively, recall from section 1.5 that $\frac{1}{\pi t} \longleftrightarrow -j\,\mathrm{sgn}\,(f)$, we have

$$m_p(t) = \frac{1}{\pi t} * m(t) \qquad (2.46)$$

or equivalently by defining $h(t) = \frac{1}{\pi t}$,

$$m_p(t) = \int_{-\infty}^{\infty} h(\tau)m\,(t - \tau)\,d\tau \tag{2.47}$$

Without any loss of generality, for the transmission of the SSB signal $s_{LSB}(t)$, the signal power P_{in} at the input of the demodulator is given by

$$
\begin{aligned}
P_{in} &= E\left[s_{LSB}^2(t)\right] = \overline{[m(t)\cos(2\pi f_c t)]^2} + \overline{[m_p(t)\sin(2\pi f_c t)]^2} \\
&= \frac{E\left[m^2(t)\right]}{2} + \frac{E\left[m_p^2(t)\right]}{2} = E\left[m^2(t)\right]
\end{aligned} \tag{2.48}
$$

The noise power at the demodulator input from equation 2.41 is given by $N_{in} = N_o B$ so that the power SNR at the input of the demodulator is

$$\left(\frac{P}{N}\right)_{in} = \frac{P_{in}}{N_{in}} = \frac{P_{in}}{N_o B} = \gamma \tag{2.49}$$

Similarly, at the output of the low-pass filter in Fig. 2.3, the signal power $P_{out} = E\left[m^2(t)\right]$ and the noise power $N_{out} = E[n_c^2(t)] = N_o B$. Thus, the power SNR at the receiver output is

$$\left(\frac{P}{N}\right)_{out} = \frac{P_{out}}{N_{out}} = \frac{E\left[m^2(t)\right]}{N_o B} = \frac{P_{in}}{N_o B} = \gamma \tag{2.50}$$

which indicates that a SSB provides the same $\left(\frac{P}{N}\right)_{out}$ as both the DSB and baseband systems, assuming the use of ideal filters and coherent demodulation, with a detector gain of $\frac{(P/N)_{out}}{(P/N)_{in}} = 1$.

2.3.2 Simulation : SSB-SC via Phase Shifting

SIMULATION **Hilbert:** Both the time and frequency domain methods are illustrated to generate $m_p(t)$ using the sinc pulse $m(t) = \dfrac{\sin(At)}{\pi t}$. In this case, $S_m(f) = \begin{cases} 1 & if & |f| \leq \dfrac{A}{2\pi} \\ 0 & otherwise \end{cases}$. The $m(t)$ signal and its corresponding $m_p(t)$ signal are shown in Fig. 2.6 for $A = 2\pi$. The corresponding $s_{USB}(t)$ is shown in Fig. 2.7 and the amplitude spectrum $|S_{USB}(f)|$ and $|S_{LSB}(f)|$ of $s_{USB}(t)$ and $s_{LSB}(t)$ for $f_c = 6$ Hz in Fig. 2.8. The result of multiplying $s_{USB}(t)$ by $2\cos(2\pi f_c t)$ is shown in Fig. 2.9 from which its clear that a low-pass filter will extract the baseband signal $m(t)$.

- **Experiment**: Try a different input pulse $m(t)$. Be sure to set the filter bandwidths accordingly to recover $m(t)$.

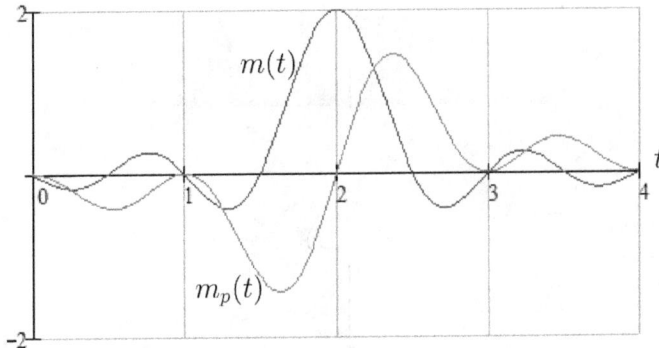

Figure 2.6: $m(t)$ and $m_p(t)$ signals.

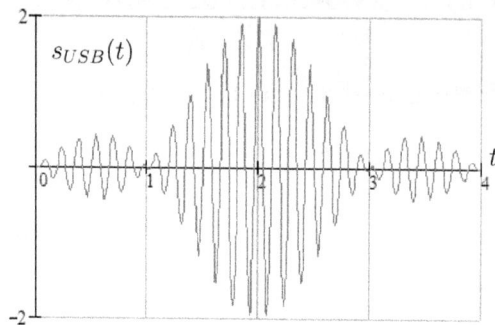

Figure 2.7: SSB-SC $s_{USB}(t)$ signal.

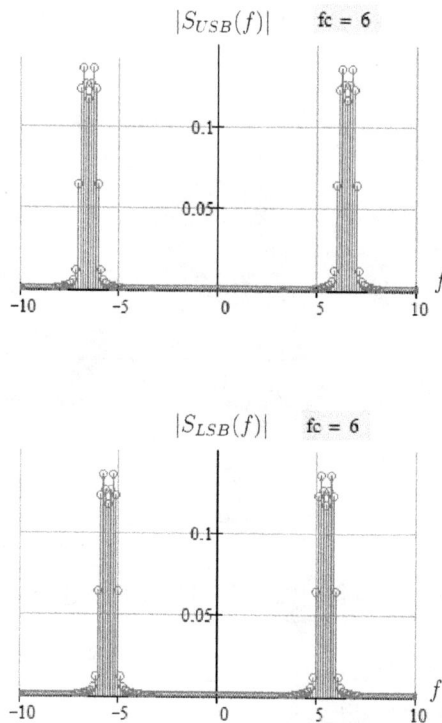

Figure 2.8: FFT amplitude spectrum of the upper and lower sideband SSB-SC signals.

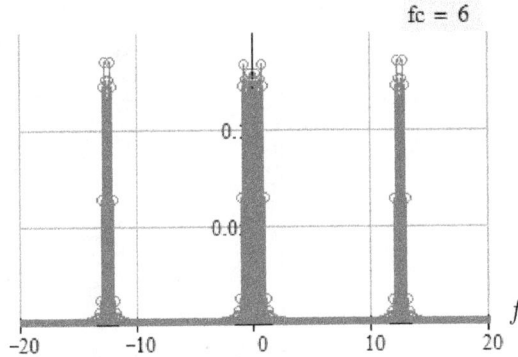

Figure 2.9: FFT amplitude spectrum of $s_{USB}(t)2\cos(2\pi f_c t)$.

2.4 Vestigial-Sideband (VSB) Modulation

An ideal filter or phase shifter is unrealizable in practice i.e. it is not possible to generate an ideal SSB-SC signal. A practical solution is to employ *vestigial-sideband* (VSB) modulation in which one sideband and a trace (i.e. vestige) of the other sideband is also transmitted as illustrated in Fig. 2.10 using a shaping (or vestigial) filter with frequency response $H_{vsb}(f)$.

VSB modulation is a compromise between DSB and SSB modulation. The bandwidth requirement is now less than DSB and more than SSB i.e. VSB is a simple method to reduce the transmission bandwidth (typically used in an analog TV system). The Fourier transform $S_{VSB}(f)$ of the modulated signal is given by

$$S_{VSB}(f) = \frac{1}{2}\left[S_m(f + f_c) + S_m(f - f_c)\right]H_{vsb}(f) \qquad (2.51)$$

The baseband signal $m(t)$ can be recovered via coherent demodulation by multiplying the VSB signal by $2\cos(2\pi f_c t)$ and using a low-pass filter. To ensure that $m(t)$ is recovered without distortion, the frequency response of the shaping filter used to create the VSB signal from the DSB signal must be antisymmetrical about the carrier frequency f_c such that

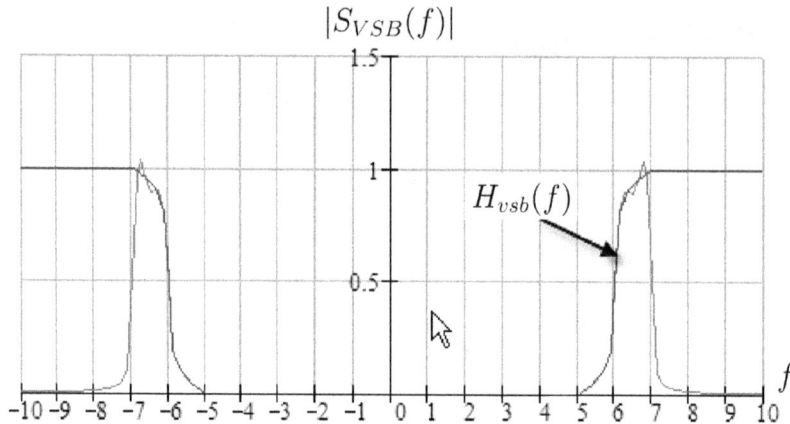

Figure 2.10: Amplitude spectrum $|S_{VSB}(f)|$ of the VSB signal and the frequency response $H_{vsb}(f)$.

$$H_{vsb}(f + f_c) + H_{vsb}(f - f_c) = 1 \; for \; |f| \leq B_m \qquad (2.52)$$

where B_m is the bandwidth of $m(t)$.

2.4.1   Simulation : VSB

SIMULATION **VSB:** The communication system presented in Fig. 2.4 is modeled once again and simulation results presented for both DSB and VSB.

- **Experiment:** Try various types of shaping filters (e.g. refer to problems section). Convince yourself that we require $H_{vsb}(f+f_c)+H_{vsb}(f-f_c)$ equal to a constant for the distortionless recovery of $m(t)$.

2.5   Conventional Amplitude Modulation

In conventional amplitude modulation, which is also referred to as *ordinary* AM or DSB-LC, a large carrier signal with amplitude A_c is added to the

DSB-SC signal so that

$$s_{AM}(t) = m(t)\cos\left(2\pi f_c t\right) + A_c \cos\left(2\pi f_c t\right) = \left[m(t) + A_c\right]\cos\left(2\pi f_c t\right) \quad (2.53)$$

In this case, the spectrum $S_{AM}(f)$ of the modulated signal is given by

$$S_{AM}(f) = \frac{1}{2}\left[S_m(f + f_c) + S_m(f - f_c) + A_c\delta(f + f_c) + A_c\delta(f - f_c)\right] \quad (2.54)$$

Unlike in DSB-SC, there is now a discrete component of the carrier frequency at $\pm f_c$. Although the modulated carrier can be coherently demodulated as in DSB-SC, the reason for adding $A_c \cos\left(2\pi f_c t\right)$ to the DSB-SC signal is to avoid coherent demodulation, which is an expensive and complex option. By ensuring that $(A_c + m(t)) \geq 0$, a simple *envelope detector* together with a low-pass filter as shown in Fig. 2.11 can be used to extract the message signal. Let $v_{cap}(t)$ denote the voltage across the capacitor at time t. The capacitor is charged during a positive half-cycle of $s_{AM}(t)$ if $s_{AM}(t) > v_{cap}(t)$, otherwise it discharges slowly through the resistor R. The variation of $v_{cap}(t)$ is illustrated in Fig. 2.12 together with the modulated carrier $s_{AM}(t)$ and $m(t) + A_c$. To follow the envelope of $s_{AM}(t)$, it is necessary to ensure that the discharge of the capacitor through the resistor is not too slow, or equivalently, we must ensure that the time constant RC seconds is not too large. Otherwise, as evident in Fig. 2.13, $v_{cap}(t)$ is not able to accurately follow the envelope. Finally, a low-pass filter is used to smooth out the ragged edges of $v_{cap}(t)$ and the d.c. level in the filtered output is removed by a.c. coupling. We can expect a degradation in performance in comparison to coherent detection (section 2.5.2). This is acceptable because the AM envelope receiver is not complex and inexpensive to build i.e. its ideally suited for the consumer market of broadcast AM radio.

- ⭐ SIMULATION **EnvelopeDemodulator:** An implementation of the envelope detector. Verify that envelope detection is only possible if $(A_c + m(t)) \geq 0$. Experiment with other types of $m(t)$ signals. Increase the value of f_c to improve performance.

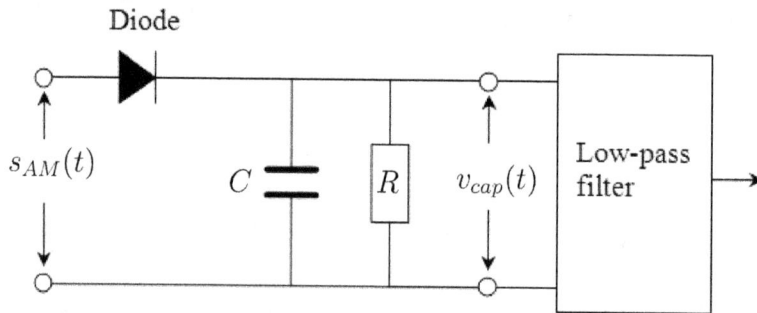

Figure 2.11: Envelope demodulator.

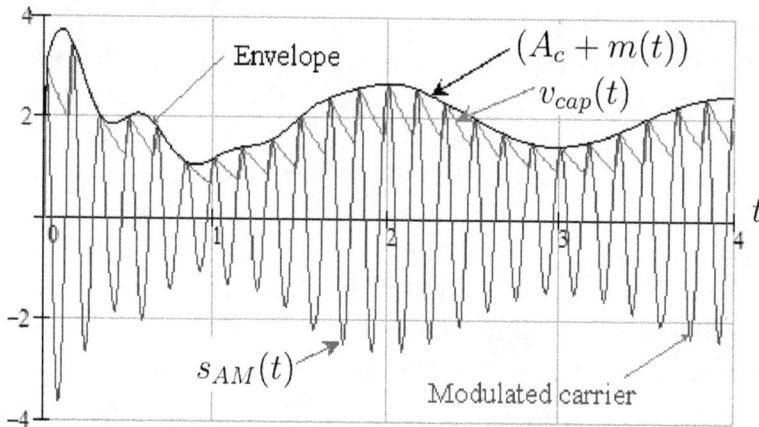

Figure 2.12: Envelope detection.

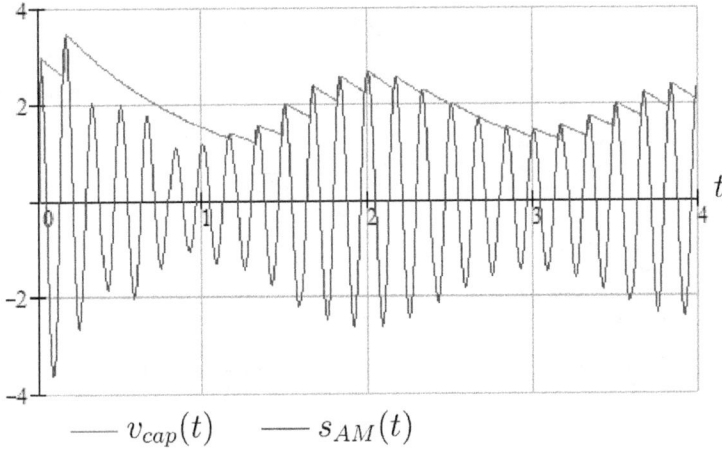

$$\underline{\quad\quad} \; v_{cap}(t) \qquad \underline{\quad\quad} \; s_{AM}(t)$$

Figure 2.13: Large time constant.

If $(A_c + m(t)) \geq 0$ for all t, then $A_c \geq |m(t)_{\min}|$, where $|m(t)_{\min}|$ is the minimum value of $m(t)$. Equivalently, in terms of the *modulation index* μ defined as $\mu = \frac{|m(t)_{\min}|}{A_c}$, we require $\mu \leq 1$ for the envelope demodulation of an AM signal.   If this condition cannot be satisfied, then we are forced to use coherent demodulation as in DSB-SC.

2.5.1   Efficiency

The *power efficiency* η of conventional AM is defined by $\eta = \frac{P_s}{P_t}$, where P_s is power carried by the sidebands and P_t is the total power of the AM signal i.e. η is the percentage of the total power being used to convey the message. Let P_c represent the carrier power.   Then $P_t = P_s + P_c$, and hence

$$\eta = \frac{P_s}{P_t} = \frac{P_s}{P_s + P_c} \tag{2.55}$$

where $\eta \leq 1$. For conventional amplitude modulation, $s_{AM}(t) = A_c \cos\left(2\pi f_c t\right) + m(t) \cos\left(2\pi f_c t\right)$, where the first and second terms of this expression represent the carrier and sidebands, respectively. Thus $P_c = \frac{A_c^2}{2}$ and $P_s = \frac{\overline{m^2(t)}}{2}$ so that

$$\eta = \frac{P_s}{P_s + P_c} = \frac{\overline{m^2(t)}}{\overline{m^2(t)} + A_c^2} \tag{2.56}$$

Now consider the simple case of tone modulation where $m(t) = A_m \cos(2\pi f_m t)$ for which $\mu = \frac{|m(t)_{min}|}{A_c} = \frac{A_m}{A_c}$. In this case, $\overline{m^2(t)} = \frac{A_m^2}{2} = \frac{(\mu A_c)^2}{2}$ and therefore

$$\eta = \frac{\frac{(\mu A_c)^2}{2}}{\frac{(\mu A_c)^2}{2} + A_c^2} = \frac{\mu^2}{\mu^2 + 2} \tag{2.57}$$

Equivalently, using $m(t) = \mu A_c \cos(2\pi f_m t)$ we have

$$s_{AM}(t) = \mu A_c \cos(2\pi f_m t) \cos(2\pi f_c t) + A_c \cos(2\pi f_c t) \tag{2.58}$$

But given that

$$\cos\left[2\pi\left(f_c + f_m\right)t\right] + \cos\left[2\pi\left(f_c - f_m\right)t\right] = 2\cos(2\pi f_c t)\cos(2\pi f_m t) \tag{2.59}$$

we may express $s_{AM}(t)$ as

$$s_{AM}(t) = A_c \cos(2\pi f_c t) + \frac{\mu A_c}{2} \cos\left[2\pi\left(f_c + f_m\right)t\right] + \frac{\mu A_c}{2} \cos\left[2\pi\left(f_c - f_m\right)t\right] \tag{2.60}$$

From equation 1.56 in section 1.3, the power contribution of each spectral component on a double-sided spectrum is $|c_n|^2$, where $|c_n|$ is the amplitude of the component on a double-sided spectrum. Thus, the sideband terms $\frac{\mu A_c}{2} \cos\left[2\pi\left(f_c + f_m\right)t\right]$ and $\frac{\mu A_c}{2} \cos\left[2\pi\left(f_c - f_m\right)t\right]$ **each** contribute $\left(\frac{\mu A_c}{4}\right)^2 + \left(\frac{\mu A_c}{4}\right)^2$ power to yield $P_s = 2\left(\frac{\mu A_c}{4}\right)^2 + 2\left(\frac{\mu A_c}{4}\right)^2 = \frac{(\mu A_c)^2}{4}$ and therefore

$$\eta = \frac{P_s}{P_s + P_c} = \frac{\frac{(\mu A_c)^2}{4}}{\frac{(\mu A_c)^2}{4} + \frac{A_c^2}{2}} = \frac{\mu^2}{\mu^2 + 2} \tag{2.61}$$

From the variation of η versus μ shown in Fig. 2.14, the maximum efficiency $\eta_{\max}$ occurs for $\mu = 1$, for which $\eta_{\max} = 0.333$ or equivalently 33.3%. This means that $\frac{1}{3}$ of the transmitted power P_t is used to carry the message i.e. ordinary AM is not a power efficient modulation scheme.

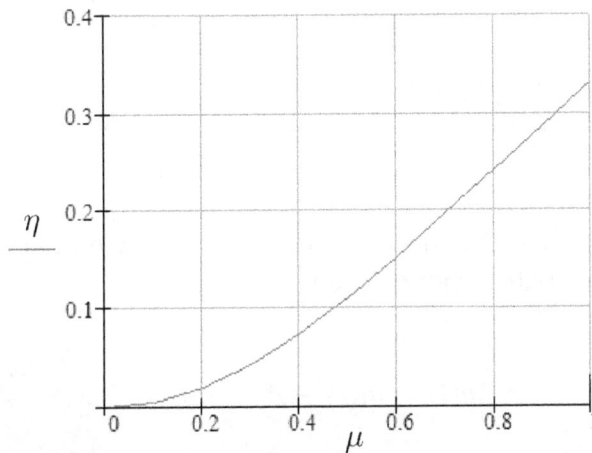

Figure 2.14: Efficiency of ordinary AM for a tone message signal.

2.5.2 SNR in Coherent AM

Using once again the receiver in Fig. 2.3, the input to the predetection narrowband filter is $s_{AM}(t) + n(t)$, where from equation 2.53, $s_{AM}(t) = [m(t) + A_c] \cos(2\pi f_c t)$ is the transmitter output assumed to be transferred over the channel without distortion. The signal at the output of the narrowband filter $s_{band}(t)$ is given by

$$s_{band}(t) = s_{AM}(t) + n_{band}(t) \tag{2.62}$$

$$= [m(t) + A_c + n_c(t)] \cos (2\pi f_c t) - n_s(t) \sin (2\pi f_c t) \qquad (2.63)$$

and the input to the postdetection low-pass filter (LPF) is given by

$$
\begin{aligned}
s_{band}(t)2\cos(2\pi f_c t) \;=\; & 2\left[m(t) + A_c + n_c(t)\right]\cos^2(2\pi f_c t) \qquad (2.64)\\
& -2n_s(t)\sin(2\pi f_c t)\cos(2\pi f_c t)
\end{aligned}
$$

$$
\begin{aligned}
=\; & m(t) + A_c + n_c(t) + \qquad\qquad\qquad (2.65)\\
& [m(t) + A_c + n_c(t)]\cos(4\pi f_c t) - n_s(t)\sin(4\pi f_c t)
\end{aligned}
$$

from which the high frequency terms at $2f_c$ are eliminated using a LPF filter of bandwidth B to produce the receiver output

$$y(t) = m(t) + A_c + n_c(t) \qquad (2.66)$$

which indicates the need for a d.c. suppressor to eliminate A_c from $y(t)$. The signal power with the d.c. component suppressed is given by

$$P_{out} = E\left[m^2(t)\right] \qquad (2.67)$$

assuming $E\left[m(t)\right] = 0$. The noise power N_{out} at the receiver output is given by

$$N_{out} = E\left[n_c^2(t)\right] = \int_{-B}^{B} N_o df = 2N_o B \qquad (2.68)$$

and the signal power P_{in} at the input of the demodulator is given by

$$P_{in} = E\left[s^2_{AM}(t)\right] = E[[m(t) + A_c]^2 \cos^2(2\pi f_c t)]$$

$$= \frac{1}{2}E\left[m^2(t)\right] + A_c E\left[m(t)\right] + \frac{A_c^2}{2} \tag{2.69}$$

where if $E\left[m(t)\right] = 0$, then

$$P_{in} = \frac{1}{2}\left(A_c^2 + E\left[m^2(t)\right]\right) \tag{2.70}$$

The noise power at the demodulator input is given by $N_{in} = N_{band} = 2N_oB$ (section 2.1.2) so that the power SNR at the input of the demodulator in terms of $\gamma = \frac{P_{in}}{N_oB}$ is given by

$$\left(\frac{P}{N}\right)_{in} = \frac{P_{in}}{N_{in}} = \frac{P_{in}}{2N_oB} = \frac{\gamma}{2} \tag{2.71}$$

Now expressing P_{out} in terms of P_{in},

$$P_{out} = \frac{E\left[m^2(t)\right] P_{in}}{P_{in}} = \frac{2E\left[m^2(t)\right] P_{in}}{(A_c^2 + E\left[m^2(t)\right])} \tag{2.72}$$

Finally, the power SNR at the output of the low-pass filter

$$\left(\frac{P}{N}\right)_{out} = \frac{P_{out}}{N_{out}} = \frac{E\left[m^2(t)\right]}{(A_c^2 + E\left[m^2(t)\right])}\gamma \tag{2.73}$$

Comparing this with the baseband system, the ratio $\frac{\left(\frac{P}{N}\right)_{out}}{\left(\frac{P}{N}\right)_{baseband}} = \frac{E\left[m^2(t)\right]}{(A_c^2 + E[m^2(t)])}$. For example, for tone modulation with $m(t) = A_m \cos(2\pi f_m t)$, $A_m = A_c$ so that $\mu = 1$ and $E[m^2(t)] = \frac{A_c^2}{2}$, $\frac{\left(\frac{P}{N}\right)_{out}}{\left(\frac{P}{N}\right)_{baseband}} = \frac{1}{3}$ or equivalently $\left(\frac{P}{N}\right)_{out}(dB) = \left(\frac{P}{N}\right)_{baseband}(dB) - 10\log_{10}(3)$, which indicates that the amplitude modulation output power SNR is at least 3dB worse than the previous DSB and SSB systems. Finally, the detector gain

$$\frac{(P/N)_{out}}{(P/N)_{in}} = \frac{2E\left[m^2(t)\right]}{\left(A_c^2 + E\left[m^2(t)\right]\right)} \tag{2.74}$$

is $\frac{2}{3}$ for $\mu = 1$ tone modulation.

2.5.3   SNR in Envelope Detection

In practice, an ordinary AM signal $s_{band}(t)$ is not multiplied by $2\cos\left(2\pi f_c t\right)$ as in the section 2.5.2 but sent through a envelope demodulator so that

$$
\begin{aligned}
s_{band}(t) &= s_{AM}(t) + n_{band}(t) \\
&= \left[m(t) + A_c + n_c(t)\right]\cos\left(2\pi f_c t\right) - n_s(t)\sin\left(2\pi f_c t\right) \\
&= E(t)\cos\left(2\pi f_c t + \theta(t)\right)
\end{aligned} \tag{2.75}
$$

where the envelope function $E(t)$ and phase functions $\theta(t)$ are given by

$$
\begin{aligned}
E(t) &= \sqrt{\left[m(t) + A_c + n_c(t)\right]^2 + n_s^2(t)} \\
\theta(t) &= \tan^{-1}\left(\frac{n_s(t)}{\left[m(t) + A_c + n_c(t)\right]}\right)
\end{aligned} \tag{2.76}
$$

Alternatively, using a phasor representation,

$$s_{band}(t) = \operatorname{Re}\left(E(t)e^{j2\pi f_c t}\right) \tag{2.77}$$

where $E(t) = m(t) + A_c + n_c(t) + jn_s(t)$ as shown in Fig. 2.15.

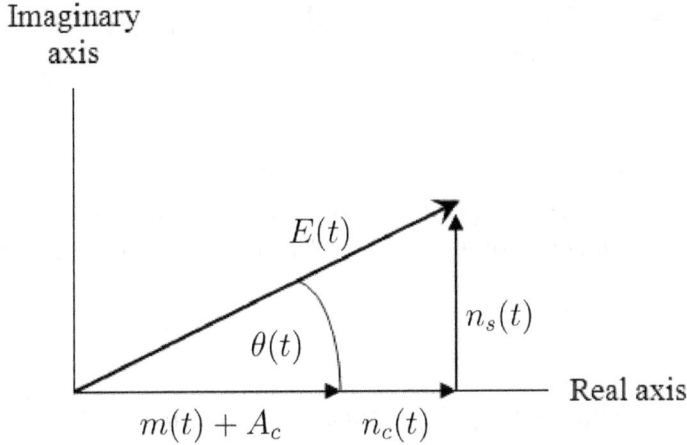

Figure 2.15: Low noise phasor diagram for $E(t) = m(t) + A_c + n_c(t) + jn_s(t)$.

Low-Noise If with a high probability, $m(t) + A_c \gg n_c(t)$ and $n_s(t)$, then $E(t) \simeq m(t) + A_c + n_c(t)$ and a capacitor is used within the envelope detector to eliminate the d.c. component A_c, then

$$E(t) \simeq m(t) + n_c(t) \tag{2.78}$$

Thus as before, $P_{out} = E\left[m^2(t)\right]$, $N_{out} = E\left[n_c^2(t)\right] = 2N_oB$ (section 2.1.4) and given that $P_{in} = \frac{1}{2}\left(A_c^2 + E\left[m^2(t)\right]\right)$ with $\gamma = \frac{P_{in}}{N_oB}$, we have

$$\left(\frac{P}{N}\right)_{out} = \frac{E\left[m^2(t)\right]}{2N_oB} = \frac{E\left[m^2(t)\right]}{\left(A_c^2 + E\left[m^2(t)\right]\right)}\frac{P_{in}}{N_oB} = \frac{E\left[m^2(t)\right]}{\left(A_c^2 + E\left[m^2(t)\right]\right)}\gamma \tag{2.79}$$

which is the same as the output power SNR of AM with coherent detection i.e. the performance of the envelope detector is equivalent to that of coherent (or synchronous) detector if the input power SNR is large. For example, for tone modulation in which $m(t) = A_m \cos(2\pi f_m t)$, $\mu = \frac{|m(t)_{min}|}{A_c} = \frac{A_m}{A_c}$, and

$E[m^2(t)] = \frac{A_m^2}{2}$, we have $\left(\frac{P}{N}\right)_{out} = \frac{A_m^2}{(2A_c^2+A_m^2)}\gamma = \frac{\mu^2}{(2+\mu^2)}\gamma$. Notice that for 100% modulation $(\mu = 1)$, $\left(\frac{P}{N}\right)_{out} = \frac{\gamma}{3}$.

High-Noise If the input power SNR is very small, then the envelope is dominated by the noise as shown in Fig. 2.16 using the phasor representation. In this case,

$$E(t) \simeq E_{noise}(t) + [m(t) + A_c] \cos \theta_{noise}(t) \qquad (2.80)$$

where $E_{noise}(t) = \sqrt{n_c^2(t) + n_s^2(t)}$ and $\theta_{noise}(t) = \tan^{-1}\left(\frac{n_s(t)}{n_c(t)}\right)$ are the envelope and phase functions of the noise $n_{band}(t)$ as in section 2.1.2. To extract $m(t)$ from $E(t)$ that contains the term $m(t) \cos \theta_{noise}(t)$ is not possible.

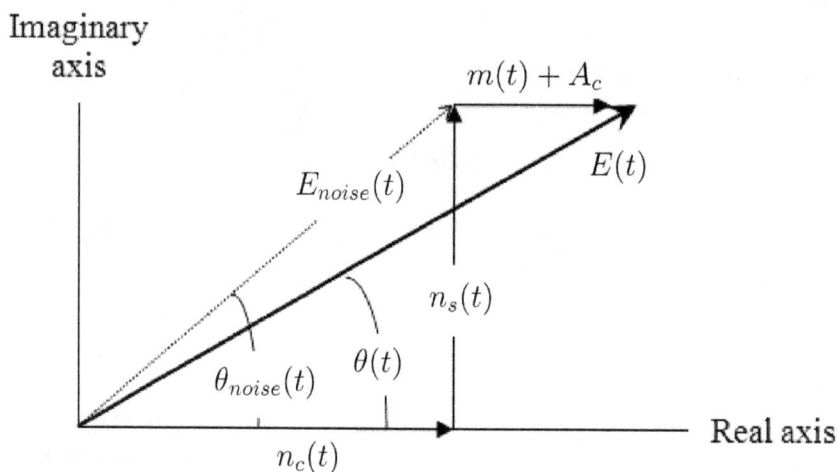

Figure 2.16: High noise phasor diagram for $E(t) \simeq E_{noise}(t) + [m(t) + A_c] \cos \theta_{noise}(t)$.

Threshold Effect Comparing the high and low noise cases, clearly there is input power SNR below which envelope detection is not possible. This is referred to as the *threshold effect*. Specifically, it is defined as the value of $\left(\frac{P}{N}\right)_{in}$ for which $E_{noise}(t) \ll A_c$ with probability 0.99. Recall the PDF of $E_{noise}(t)$ from section 2.1.5 is given by

$$f_{noise}(r, \sigma) = \begin{cases} \frac{r}{\sigma^2} \exp\left(\frac{-r^2}{2\sigma^2}\right) & if \quad r > 0 \\ 0 & otherwise \end{cases} \qquad (2.81)$$

where σ is the standard deviation of n_c and n_s given by $\sigma = \sqrt{2N_oB}$. Consider the probability $P\left(E_{noise}(t) < A_c\right) = 0.99$, which is equivalent to

$$P\left(E_{noise}(t) > A_c\right) = 0.01 = \int_{A_c}^{\infty} \frac{r}{\sigma^2} \exp\left(\frac{-r^2}{2\sigma^2}\right) dr.$$ Let $u = \frac{r^2}{2\sigma^2}$, so that $\frac{du}{dr} = \frac{r}{\sigma^2}$.

Changing the limits, for $r = A_c$ we have $u = \frac{A_c^2}{2\sigma^2}$ and for $r = \infty$, $u = \infty$.

Thus $\int_{\frac{A_c^2}{2\sigma^2}}^{\infty} \exp\left(-u\right) du = \exp\left(\frac{-A_c^2}{2\sigma^2}\right) = 0.01$, from which $A_c = \sqrt{2\sigma^2 \log_e(100)}$.

Given that $\sigma = \sqrt{2N_oB}$, the value of A_c at the threshold effect is given by

$$A_c = \sqrt{4N_oB \log_e(100)} \qquad (2.82)$$

If $m(t)$ is normalized such that $|m(t)_{min}| = 1$, then $\mu = \frac{1}{A_c}$ and

$$P_{in} = \frac{A_c^2}{2}\left(1 + \mu^2 E\left[m^2(t)\right]\right) \qquad (2.83)$$

For example, if $E\left[m^2(t)\right] = 1$ and $\mu = 1$, then $P_{in} = A_c^2 = 4N_oB \log_e(100)$, $N_{in} = 2N_oB$ and therefore, $\left(\frac{P}{N}\right)_{in} = 2 \log_e(100) = 9.21$ or 9.64 dB. Now for tone modulation in which $m(t) = A_m \cos(2\pi f_m t)$, $\mu = \frac{|m(t)_{min}|}{A_c} = \frac{A_m}{A_c}$,

$$E[m^2(t)] = \frac{A_m^2}{2} \qquad (2.84)$$

$$P_{in} = \frac{1}{2}\left(A_c^2 + E\left[m^2(t)\right]\right) = \frac{A_c^2}{2}\left(1 + \frac{\mu^2}{2}\right) = 2N_oB\log_e(100)\left(1 + \frac{\mu^2}{2}\right)$$

$$(2.85)$$

and therefore $\left(\frac{P}{N}\right)_{in} = \log_e(100)\left(1 + \frac{\mu^2}{2}\right)$. If $\mu = 1$, then $\left(\frac{P}{N}\right)_{in} = 6.908$ or equivalently 8.4 dB i.e. the input power SNR should be greater than 8.4 dB to make use of envelope detection for 100% ($\mu = 1$) tone modulation.

2.5.4 Simulation I: Threshold Effect

⭐SIMULATION **ThresholdEffect:** AWGN noise is simulated and filtered through a narrowband filter to verify that $N_{in} = 2N_oB$. The noise components $n_c(t)$ and $n_s(t)$ are extracted to determine the envelope variation

$E(t) = \sqrt{[m(t) + A_c + n_c(t)]^2 + n_s^2(t)}$ for $m(t) = A_m\cos(2\pi f_m t)$ with $A_m = A_c$. Under low noise, this is shown to correspond well with the approximation $E(t) \simeq m(t) + A_c + n_c(t)$ and the detector gain $\frac{(P/N)_{out}}{(P/N)_{in}} = \frac{2E[m^2(t)]}{(A_c^2 + E[m^2(t)])}$ is verified to be $\frac{2}{3}$. Similarly, under high noise, $E(t)$ is shown to correspond well with the approximation $E(t) \simeq E_{noise}(t) + [m(t) + A_c]\cos\theta_{noise}(t)$. At the threshold of $A_c = \sqrt{4N_oB\log_e(100)}$, $\left(\frac{P}{N}\right)_{in} = 6.908$ is verified for tone modulation. Also, $P(E_{noise}(t) > A_c) = \int\limits_{A_c}^{\infty} \frac{r}{\sigma^2}\exp\left(\frac{-r^2}{2\sigma^2}\right)dr = 0.01$ is verified and determined for the simulated narrowband noise.

2.5.5 Simulation II: Coherent Demodulator

⭐SIMULATION **CoherentAM:** The analysis presented in section 2.5.2 is verified for $m(t) = A_m\cos(2\pi f_m t)$ tone modulation. The amplitude spectrum of the signal and simulated AWGN are determined at every stage of the receiver. The simulated values for the $P_{in}, N_{in}, P_{out}, N_{out}$ and the detector gain are compared with theory and the power efficiency η verified to be $\frac{1}{3}$ for $\mu = 1$.

- **Experiment**: Use the other function for $m(t)$ provided within this simulation.

2.5.6   Simulation III: Square Law Detector

⭐SIMULATION **SquareLaw**: An alternative to envelope detection is illustrated in which the modulated signal $s_{AM}(t)$ is squared and then passed through a low-pass filter. In this case,

$$[s_{AM}(t)]^2 = [m(t) + A_c]^2 \cos^2(2\pi f_c t) = [m(t) + A_c]^2 \frac{[1 + \cos(4\pi f_c t)]}{2}$$

$$= \frac{[m(t) + A_c]^2}{2} + \frac{[m(t) + A_c]^2}{2} \cos(4\pi f_c t) \qquad (2.86)$$

The output $y(t)$ of the low-pass filter is therefore given by

$$y(t) = \frac{[m(t) + A_c]^2}{2} = \frac{A_c^2}{2}\left[\left(\frac{m(t)}{A_c}\right)^2 + 2\frac{m(t)}{A_c} + 1\right] \qquad (2.87)$$

If we ensure that $A_c \gg m(t)$, so that the term $\left(\frac{m(t)}{A_c}\right)^2$ is very small, then

$$y(t) \simeq \left[A_c m(t) + \frac{A_c^2}{2}\right] \qquad (2.88)$$

from which the d.c. component $\frac{A_c^2}{2}$ has to be removed and the amplitude reduced by a factor of A_c to extract $m(t)$. A comparison between the demodulated signal and the $m(t)$ is shown in Fig. 2.17. The performance can be improved by ensuring that the carrier component is very much larger than the other components i.e. increase A_c to a large value. Of course, the penalty incurred is an inefficient use of transmitter power.

- **Experiment:** Increase the amplitude A_c. Does the performance improve? Verify that $y(t) \simeq \left[A_c m(t) + \frac{A_c^2}{2} \right]$. Try different expressions for $m(t)$.

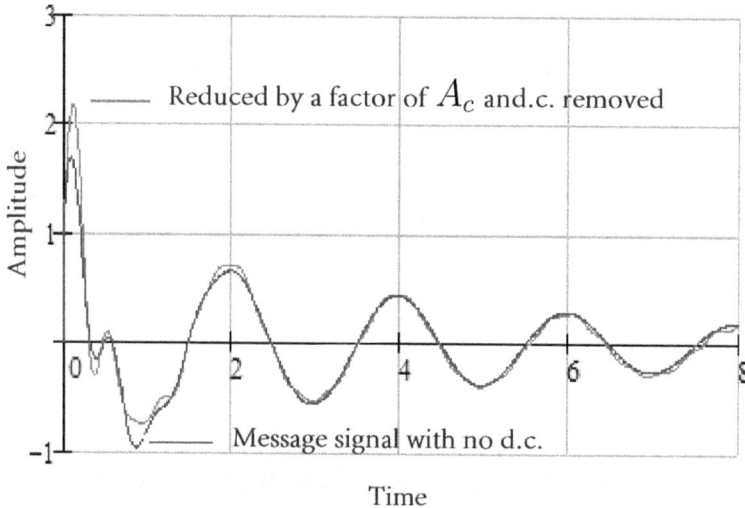

Figure 2.17: Comparison between the square law demodulator output and the message signal.

2.6 Heterodyning

If the modulated signal is shifted to a fixed known frequency band, then the filtering, amplification and demodulation processing stages can be pre-designed accurately. For example, in commercial AM radio receivers, the modulated carrier is typically shifted down to the *intermediate-frequency* f_I of 455 kHz for processing. This operation, which is referred to as *heterodyning* or frequency *mixing,* is illustrated in shown in Fig. 2.18, where f_c is the frequency of the incoming carrier and f_L is the frequency of the local oscillator.

The output $y(t)$ of the mixer is given by

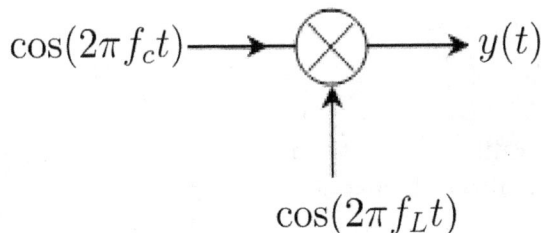

Figure 2.18: Mixer.

$$y(t) = \cos(2\pi f_c t)\cos(2\pi f_L t) = \frac{1}{2}\cos\left[2\pi\left(f_c + f_L\right)t\right] + \frac{1}{2}\cos\left[2\pi\left(|f_c - f_L|\right)t\right]$$
$$(2.89)$$

- ⭐ SIMULATION **Mixing:** The mixer shown in Fig. 2.18 is implemented. You may view the spectrum of $y(t)$. Try $(f_c > f_L)$ and $(f_c < f_L)$.

If we now use a filter to remove the high frequency component, the filtered output signal $\frac{1}{2}\cos\left[2\pi\left(|f_c - f_L|\right)t\right]$ consists of the input signal spectrum shifted down to the intermediate frequency $f_I = |f_c - f_L|$. Suppose f_L is generated such that $f_L = f_c + f_I$, and let the incoming signal $s(t) = m(t)\cos\left(2\pi f_c t\right)$, where $m(t)$ is the message signal of bandwidth B_m. If $s(t)$ is multiplied by $2\cos\left(2\pi f_L t\right)$, then the spectrum of $y(t) = s(t)2\cos\left(2\pi f_L t\right)$ will consist of side-bands centered on

$$f_L + f_c = f_L + f_L - f_I = (2f_L - f_I) \qquad (2.90)$$

or equivalently

$$f_L + f_c = (2f_c + f_I) \qquad (2.91)$$

and

$$f_L - f_c = f_L - (f_L - f_I) = f_I \qquad (2.92)$$

Using a band-pass filter centered on f_I with cut-off frequencies at $(f_I + B_m)$ and $(f_I - B_m)$, the filtered output will consist of side-bands (upper and lower) centered on f_I only as required. Now suppose

$$s(t) = m(t) \cos \left[2\pi \left(f_c + 2f_I \right) t \right] \qquad (2.93)$$

then the spectrum of $y(t)$ will consist of side-bands centered on

$$f_L + (f_c + 2f_I) = f_L + (f_L - f_I) + 2f_I = 2f_L + f_I \qquad (2.94)$$

$$(f_c + 2f_I) - f_L = (f_L - f_I) + 2f_I - f_L = f_I \qquad (2.95)$$

i.e. there is another frequency, referred to as the *image* frequency $f_{image} = f_c + 2f_I$, which will also pass through the band-pass filter centered on f_I and interfere with the desired signal $s(t) = m(t) \cos(2\pi f_c t)$. If $f_L = f_c - f_I$, then it is easily shown (refer to problems section) that in this case, $f_{image} = f_c - 2f_I$. The solution is to simply filter out the image frequency prior to mixing. If $f_L = f_c + f_I$, then the mixing operation is referred to as *up-conversion*. Conversely, mixing with $f_L = f_c - f_I$ is referred to as *down-conversion*. In both cases, notice that the image frequency is separated from f_c by $2f_I$.

2.6.1   Superhetrodyne Receiver

The *superhetrodyne* (or *superhet*) receiver shown in Fig. 2.19 is used in a AM radio. The RF (radio-frequency) block consists of an amplifier and a filter that is linked with the local oscillator frequency $f_L = f_c + f_I$ to ensure that the image frequency of the desired radio station is eliminated. The up-conversion mixer shifts the spectrum of the desired radio station signal to be centered over f_I. The IF (intermediate-frequency) block consists of an amplifier and a very sharp band-pass filter that cuts-out all other radio frequencies. Finally, an envelope detector is used to extract the audio signal as explained in section 2.5.

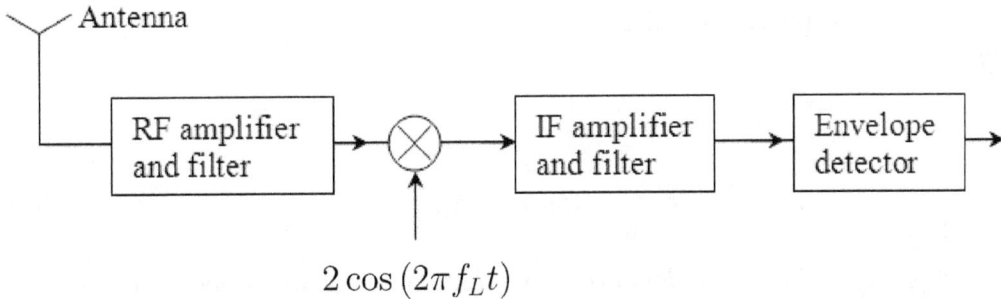

Figure 2.19: Superhetrodyne receiver.

2.6.2 Simulation : Mixer

SIMULATION **UpConversion** The incoming signal $s(t) = m(t) \cos (2\pi f_c t)$ is mixed with $2 \cos (2\pi f_L t)$, where

$$f_L = f_c + f_I \text{ and } m(t) = \begin{cases} \text{sinc}(10t) & \text{if} \quad |t| \leq 0.5 \\ 0 & \text{otherwise} \end{cases} \qquad (2.96)$$

The spectra of $m(t)$, $s(t)$ and $y(t) = s(t)2 \cos (2\pi f_L t)$ is determined. Using an ideal band-pass filter centered on f_I, the spectrum of $y(t)$ is filtered. You may then select the option to see what happens if the incoming signal is given by $m(t) \cos (2\pi f_{image} t)$, where $f_{image} = f_c + 2f_I$.

- **Student Exercise**: DownConversion: Repeat the simulation for $f_L = f_c - f_I$.

- **Experiment**: What are consequences of $f_L \neq f_c + f_I$ and $f_L \neq f_c - f_I$?

2.7   Angle Modulation

In *angle modulation or exponential modulation* (EM), the *generalized angle* $(2\pi f_c t + \phi(t))$ of a high frequency sinusoidal carrier $s_{EM}(t) = A_c \cos(2\pi f_c t + \phi(t))$ is varied in accordance with the message signal $m(t)$, where A_c and f_c are constants. In particular, if the *instantaneous phase deviation* $\phi(t)$ is given by

$$\phi(t) = k_p m(t) \tag{2.97}$$

where k_p is the *phase deviation* constant, then the angle modulation is referred to as *phase modulation* (PM). If the *instantaneous frequency deviation* $\frac{d\phi(t)}{dt}$ is given by

$$\frac{d\phi(t)}{dt} = k_f m(t) \tag{2.98}$$

where k_f is the *frequency deviation* constant, then its referred to as *frequency modulation* (FM). We shall only consider the angle-modulated signals $s_{PM}(t) = A_c \cos(2\pi f_c t + k_p m(t))$ and $s_{FM}(t) = A_c \cos\left(2\pi f_c t + k_f \int\limits_{-\infty}^{t} m(\tau)d\tau\right)$, although the expression for $\phi(t)$ need not be restricted to only two expressions $(k_p m(t)$ or $k_f \int\limits_{-\infty}^{t} m(\tau)d\tau)$ i.e. PM and FM are only two of the many possible angle modulation schemes. The key features being a high bandwidth requirement and good performance in the presence of noise. Examples of a PM and an FM signal are shown in Figs. 2.20 and 2.21 for a square wave message signal.

- ⭐ SIMULATION **AngleModulation**: Simulation corresponding to Figs. 2.20 and 2.21. Experiment with other types of $m(t)$ signals and analyze the influence of k_p and k_f.

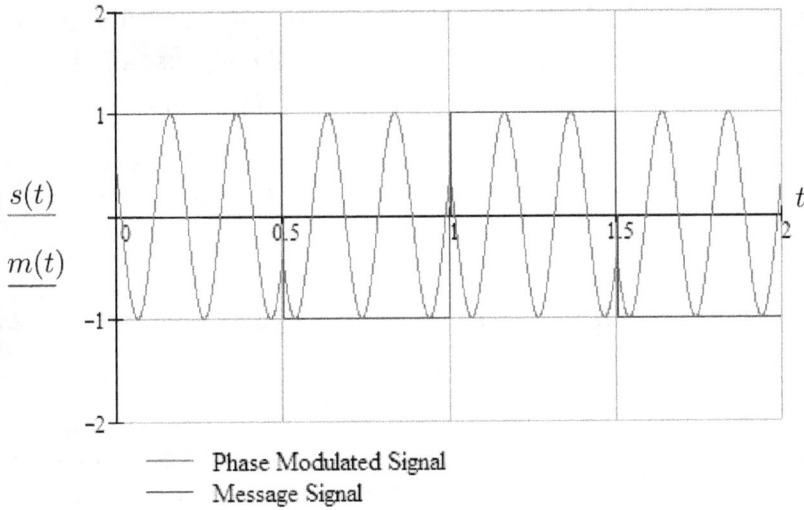

Figure 2.20: Phase modulated carrier.

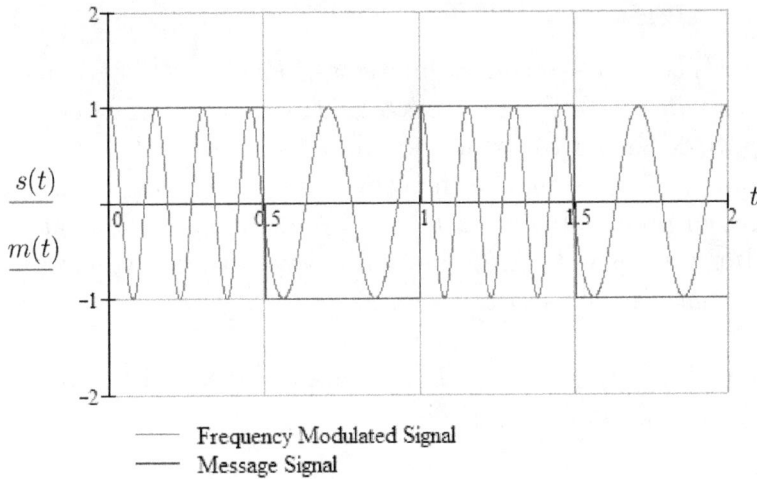

Figure 2.21: Frequency modulated carrier.

2.7.1　Fourier Spectra of Angle-Modulated Signals

The carrier $s_{EM}(t) = A_c \cos\left(2\pi f_c t + \phi(t)\right)$ can be expressed as $s_{EM}(t) =$ $\mathrm{Re}\left(A_c e^{j(2\pi f_c t + \phi(t))}\right) = \mathrm{Re}\left(A_c e^{j(2\pi f_c t)} e^{j(\phi(t))}\right)$. Making use of the series expansion $e^x = 1 + x + \frac{x^2}{2!} + \frac{x^3}{3!} + \cdots$, with $x = j\phi(t)$, yields

$$s_{EM}(t) = \mathrm{Re}\left(A_c \left(\cos(2\pi f_c t) + j\sin(2\pi f_c t)\right)\left[1 + j\phi(t) - \frac{\phi^2(t)}{2!} - \frac{j\phi^3(t)}{3!} + \cdots\right]\right)$$

$$= A_c\left[\cos\left(2\pi f_c t\right) - \phi(t)\sin(2\pi f_c t) - \frac{\phi^2(t)}{2!}\cos\left(2\pi f_c t\right) + \frac{\phi^3(t)}{3!}\sin(2\pi f_c t)\right]$$

$$\text{(2.99)}$$

Unlike in amplitude modulation (AM), notice how the message signal spectrum is not easily extracted with the use of filters from the spectrum of $s_{EM}(t)$, which consists of the unmodulated carrier plus other terms. For *narrowband* angle modulation, $|\phi(t)|_{\max} \ll 1$ (i.e. $\phi^2(t) \simeq 0$ making higher terms essentially zero), the narrowband signal $s_{NB}(t)$ is given by

$$s_{NB}(t) \approx A_c\left[\cos\left(2\pi f_c t\right) - \phi(t)\sin(2\pi f_c t)\right] \qquad \text{(2.100)}$$

Now $s_{NB}(t)$ contains the unmodulated carrier $A_c \cos\left(2\pi f_c t\right)$ and the modulated carrier $\phi(t)\cos(2\pi f_c t + \frac{\pi}{2})$ which generates a pair of sidebands. Thus, the bandwidth of the narrowband modulated signal is $2B_m$, where B_m is the bandwidth of $m(t)$. Note that in the previous sections, $m(t)$ was a zero-mean ergodic random process bandlimited to B and the bandwidth at the output low-pass filter was also B. A distinction between these two bandwidths will become necessary in this section.

The bandwidth B_{EM} which contains approx. 98% of the power is given by *Carson's rule* as

$$B_{EM} \simeq 2\left(D + 1\right)B_m \qquad \text{(2.101)}$$

where D is the *deviation ratio* defined as

$$D = \frac{\Delta f}{B_m} \qquad (2.102)$$

where

$$\Delta f = \frac{1}{2\pi} \left[\frac{d\phi(t)}{dt} \right]_{\max} \qquad (2.103)$$

is the maximum frequency deviation. If $D \ll 1$ (typically $D < 0.2$) then the modulated signal is referred to as a *narrowband* (NB) angle-modulated signal. In this case, $B_{EM} \simeq 2\,(D+1)\,B_m \approx 2B_m$ as before. Conversely if $D \gg 1$, then its referred to as *wideband* (WB) angle-modulated signal and its bandwidth $B_{EM} \simeq 2\left(\frac{\Delta f}{B_m}+1\right)B_m \approx 2\Delta f$. For broadcast FM, D is typically 5.

- ⭐ SIMULATION **FMBandwidth:** The narrowband bandwidth of an FM modulated signal is determined to verify that $B_{EM} \simeq 2\,(D+1)\,B_m$ contains approximately 98% of the power in $s_{EM}(t)$. Experiment with the parameters to ensure that $D \simeq 5$ and verify that in this case, $B_{EM} \approx 2\Delta f$.

- **Student Exercise**: Repeat simulation for PM.

2.7.2 Ideal Demodulator

Let $s_{EM}(t) = A_c \cos \alpha(t)$, where $\alpha(t) = 2\pi f_c t + \phi(t)$. Then the *instantaneous* frequency f of $s_{EM}(t)$ is given by $f = \frac{1}{2\pi} \frac{d\alpha(t)}{dt}$. For both PM and FM, we find

$$f_{PM} = \frac{1}{2\pi} \frac{d}{dt}\left(2\pi f_c t + k_p m(t)\right) = \qquad (2.104)$$

$$\frac{1}{2\pi}\left(2\pi f_c + k_p \frac{dm(t)}{dt}\right) = f_c + \frac{k_p}{2\pi}\frac{dm(t)}{dt} \tag{2.105}$$

and

$$f_{FM} = \frac{1}{2\pi}\frac{d}{dt}\left(2\pi f_c t + k_f \int\limits_{-\infty}^{t} m(\tau)d\tau\right) \tag{2.106}$$

$$= \frac{1}{2\pi}\left(2\pi f_c + k_f m(t)\right) = f_c + \frac{k_f}{2\pi}m(t) \tag{2.107}$$

i.e. f_{PM} varies linearly with $\frac{dm(t)}{dt}$ and f_{FM} varies linearly with $m(t)$. For example, for $m(t) = \text{sinc}(10t)$ over a time duration of 0.5 seconds only, as shown in Fig. 2.22, the corresponding instantaneous frequency variation for FM and PM are presented in Fig. 2.23 for $f_c = 100$ Hz.

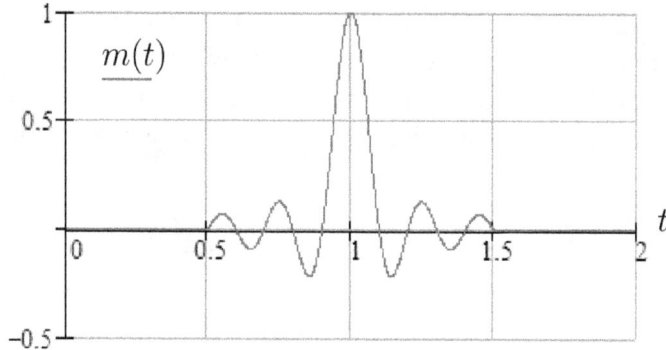

Figure 2.22: Message signal $m(t)$.

- ⭐ SIMULATION **InstantaneousFrequency:** Simulation corresponding to Fig. 2.23. Experiment with other types of $m(t)$ signals.

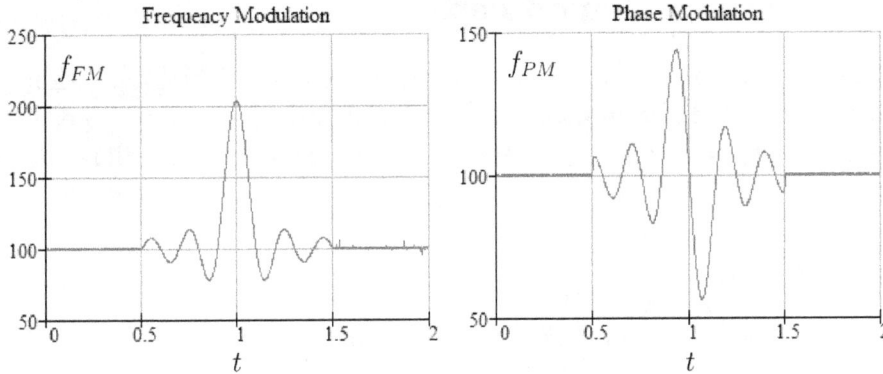

Figure 2.23: Instantaneous frequency.

From Fig. 2.23, it is clear that to demodulate the carrier, we require a device known as a *discriminator* that outputs a value proportional to the instantaneous frequency of the input signal. Ideally, the output $y_d(t)$ of a discriminator is given by

$$y_d(t) = k_d \frac{d\phi(t)}{dt} \tag{2.108}$$

where k_d is the *discriminator* constant, then

$$y_d(t)_{PM} = k_d \frac{d\phi(t)}{dt} = k_d k_p \frac{dm(t)}{dt} \tag{2.109}$$

and

$$y_d(t)_{FM} = k_d \frac{d\phi(t)}{dt} = k_d k_f m(t) \tag{2.110}$$

With the scaling factors taken into account appropriately, $m(t)$ with FM is recovered directly, whereas with PM, we require also the use of an integrator to extract $m(t)$.

2.7.3   Practical Demodulator

A simple approximation to the ideal discriminator considred in section 2.7.2 is to make use of a differentiator followed by an envelope detector i.e. for the input signal $s(t) = A_c \cos\left(2\pi f_c t + \phi(t)\right)$, the output of the differentiator is given by

$$\frac{ds(t)}{dt} = -A_c \left[2\pi f_c + \frac{d\phi(t)}{dt}\right] \sin\left(2\pi f_c t + \phi(t)\right) \qquad (2.111)$$

for which the envelope detector output is $A_c \left[2\pi f_c + \frac{d\phi(t)}{dt}\right]$. In this case,

$$y_d(t)_{PM} = A_c \left[2\pi f_c + k_p \frac{dm(t)}{dt}\right] \qquad (2.112)$$

and

$$y_d(t)_{FM} = A_c \left[2\pi f_c + k_f m(t)\right] \qquad (2.113)$$

as illustrated in Fig. 2.24 and 2.25 for FM and PM, respectively for $m(t) = \text{sinc}(10t)$. In both cases, the d.c. component $A_c 2\pi f_c$ is easily removed and as before an integrator is required to recover $m(t)$ for PM. Comparing equations 2.112 and 2.113 with 2.109 and 2.110, the discriminator constant $k_d = A_c$, asumming an ideal differentiator and envelope detector.

- ⭐ SIMULATION **Discriminator:** Simulation corresponding to Figs. 2.24 and 2.25.

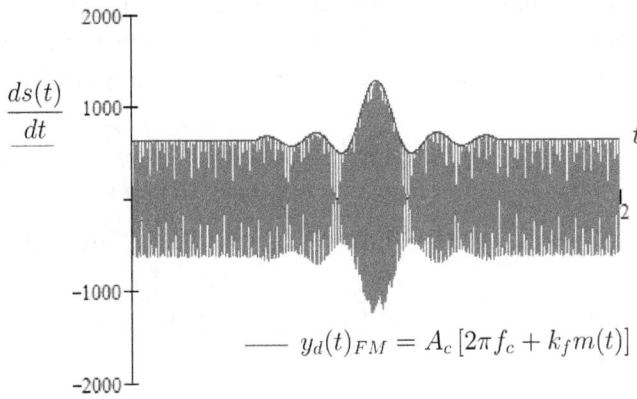

Figure 2.24: Differential of $s(t)$ and its envelope for FM.

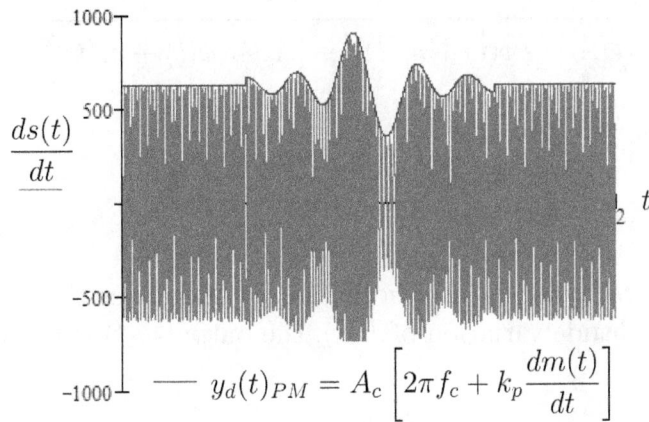

Figure 2.25: Differential of $s(t)$ and its envelope for PM.

2.7.4   SNR in Angle Modulation

Unlike the previous linear modulation schemes which allow the output signal and noise powers to be considered separately, angle modulation is non-linear. Fortunately, narrowband angle modulation is approximately linear. In this case, the input to the discriminator within the angle demodulation reciever of Fig. 2.26 is

$$s_{band}(t) = s_{EM}(t) + n_{band}(t) = A_c \cos\left[2\pi f_c t + \phi(t)\right] + n_{band}(t) \qquad (2.114)$$

where we recall from equation 2.21 that $n_{band}(t) = n_c(t) \cos\left(2\pi f_c t\right) - n_s(t) \sin\left(2\pi f_c t\right)$. Therefore

$$s_{band}(t) = \cos\left(2\pi f_c t\right)\left[A_c \cos\phi(t) + n_c(t)\right] - \sin\left(2\pi f_c t\right)\left[A_c \sin\phi(t) + n_s(t)\right]$$

$$= E(t) \cos\left(2\pi f_c t + \theta(t)\right) \qquad (2.115)$$

where the envelope function $E(t)$ and phase functions $\theta(t)$ are given by

$$E(t) = \sqrt{\left[A_c \cos\phi(t) + n_c(t)\right]^2 + \left[A_c \sin\phi(t) + n_s(t)\right]^2} \qquad (2.116)$$

$$\theta(t) = \tan^{-1}\left(\frac{A_c \sin\phi(t) + n_s(t)}{A_c \cos\phi(t) + n_c(t)}\right) \qquad (2.117)$$

Using a *limiter* prior to the dicriminator, as indiacted in Fig. 2.26, to supress any amplitude variation of $E(t)$, the output $y(t)$ is ideally given by

$$y(t) = \begin{cases} \theta(t) & \text{for } \text{PM} \\ \frac{d\theta(t)}{dt} & \text{for } \text{FM} \end{cases} \qquad (2.118)$$

Notice how a PM demodulator followed by a differentiator can be used to demodulate a FM signal.

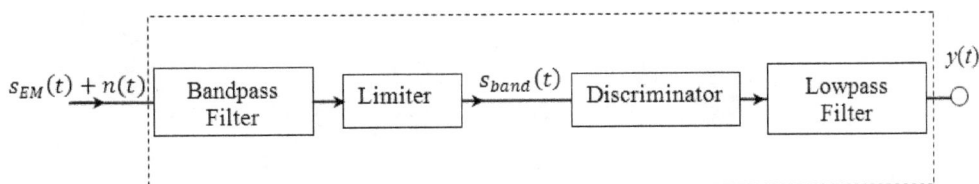

Figure 2.26: Angle demodulation.

Low Noise

Recall from section 2.1.2 that $E_{noise}(t) = \sqrt{n_c^2(t) + n_s^2(t)}$ and $\theta_{noise}(t) = \tan^{-1}\left(\frac{n_s(t)}{n_c(t)}\right)$. If $A_c \gg n_c(t)$ and $n_s(t)$, then as evident from Fig. 2.27, the length L of the arc AB is

$$L \approx E(t)\,[\theta(t) - \phi(t)] \approx E_{noise}(t)\sin\left[\theta_{noise}(t) - \phi(t)\right] \qquad (2.119)$$

and

$$E(t) \approx A_c + E_{noise}(t)\cos\left[\theta_{noise}(t) - \phi(t)\right] \approx A_c \qquad (2.120)$$

because $[\theta(t) - \phi(t)]$ is a small angle.

Thus

$$\theta(t) - \phi(t) \approx \frac{E_{noise}(t)}{A_c}\sin\left[\theta_{noise}(t) - \phi(t)\right] \qquad (2.121)$$

$$\theta(t) \approx \phi(t) + \Delta\phi(t) \qquad (2.122)$$

where

$$\Delta\phi(t) = \frac{E_{noise}(t)}{A_c}\sin\left[\theta_{noise}(t) - \phi(t)\right] \qquad (2.123)$$

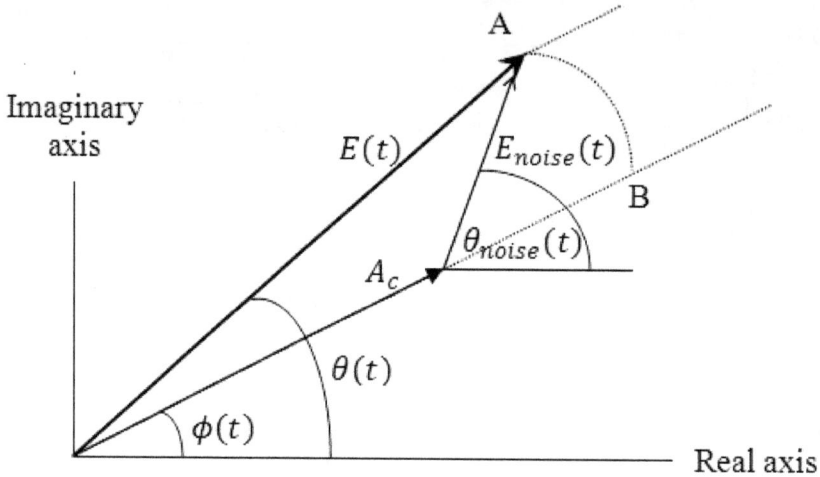

Figure 2.27: Phasor diagram for angle modulation.

$$= \frac{n_s(t)}{A_c} \cos \phi(t) - \frac{n_c(t)}{A_c} \sin \phi(t) \qquad (2.124)$$

Thus

$$y(t) = \begin{cases} \theta(t) \approx \phi(t) + \Delta\phi(t) = k_p m(t) + \Delta\phi(t) & \text{for} \quad \text{PM} \\ \frac{d\theta(t)}{dt} \approx \frac{d\phi(t)}{dt} + \frac{d(\Delta\phi(t))}{dt} = k_f m(t) + \frac{1}{A_c} \frac{d(\Delta\phi(t))}{dt} & \text{for} \quad \text{FM} \end{cases} \qquad (2.125)$$

Given that the power spectral density of $n_c(t)$ and $n_s(t)$ are equal i.e. $PSD_{n_c}(f) = PSD_{n_s}(f)$, and taking $\phi(t)$ to be a constant ϕ relative to $\theta_{noise}(t)$, because it varies much slowly than $\theta_{noise}(t)$, then from equation 2.124 (also refer to section 1.8.13), the PSD of $\Delta\phi(t)$ is given by

$$PSD_{\Delta\phi}(f) = \frac{\cos^2 \phi}{A_c^2} PSD_{n_s}(f) + \frac{\sin^2 \phi}{A_c^2} PSD_{n_c}(f) \qquad (2.126)$$

Recall from section 2.1.5 that the PSD of white noise is $\frac{N_o}{2}$ with $PSD_{n_c}(f) = N_o$ W/Hz, and given that $\cos^2 \phi + \sin^2 \phi = 1$,

$$PSD_{\Delta\phi}(f) = \begin{cases} \frac{N_o}{A_c^2} & \text{for} \quad |f| \le \Delta f + B_m \\ 0 & \text{otherwise} \end{cases} \qquad (2.127)$$

The demodulated noise bandwidth is $\Delta f + B_m$, which when passed through a low-pass filter of bandwidth $B = B_m$ removes the out-of-band noise. Thus, the PSD of the noise at the output of the low-pass filter is given by

$$PSD_{n_o}(f) = \begin{cases} \frac{N_o}{A_c^2} & \text{for} \quad |f| \le B \\ 0 & \text{otherwise} \end{cases} \qquad (2.128)$$

Threshold Effect If $A_c \gg n_c(t)$ and $n_s(t)$, then the tip of the phasor $E(t)$ will move around the tip of the carrier phasor because $\theta_{noise}(t)$ is random with uniform distribution in the range 0 to 2π. However, if the noise becomes large enough such that $E_{noise}(t)$ is near the value of A_c, then the tip of the phasor $E(t)$ may encircle the origin. This would be perceived as a crackling sound because of a spike in the variation of $\frac{d\theta}{dt}$ as $E(t)$ rapidly changes by 2π.

Phase Modulation SNR At the receiver input, since the carrier amplitude remains constant, $P_{in} = \frac{A_c^2}{2}$ and $N_{in} = N_o B_{EM}$ where $B_{EM} \simeq 2(D+1)B_m$ is the predetection filter bandwidth. Thus

$$\left(\frac{P}{N}\right)_{in} = \frac{A_c^2}{2N_o B_{EM}} \qquad (2.129)$$

which is independent of $m(t)$. The power SNR $\left(\frac{P}{N}\right)_{in}$ in equation 2.129 is also referred to as the *carrier-to-noise* power ratio in other texts. At the output, using equation 2.128, the noise power

$$N_{out} = 2B\left(\frac{N_o}{A_c^2}\right) = \frac{2N_o B}{A_c^2} \qquad (2.130)$$

Alternatively, from equation 2.125,

$$N_{out} = E\left[\Delta\phi^2(t)\right] = E\left[\left(\frac{n_s(t)}{A_c}\cos\phi(t) - \frac{n_c(t)}{A_c}\sin\phi(t)\right)^2\right] \qquad (2.131)$$

Expanding this expression and taking into account that $E\left[n_c(t)n_s(t)\right] = 0$, $E\left[n_c^2(t)\right] = E\left[n_s^2(t)\right] = 2N_0B$ and $\cos^2\phi + \sin^2\phi = 1$, the output noise power

$$N_{out} = E\left[n_s^2(t)\right]\frac{\cos^2\phi}{A_c^2} + E\left[n_c^2(t)\right]\frac{\sin^2\phi}{A_c^2} = \frac{2N_0B}{A_c^2} \qquad (2.132)$$

The output signal power from equation 2.125 is given by

$$P_{out} = k_p^2 E\left[m^2(t)\right] \qquad (2.133)$$

so that the output power SNR

$$\left(\frac{P}{N}\right)_{out} = \frac{(A_c k_p)^2\, E\left[m^2(t)\right]}{2N_0B} \qquad (2.134)$$

Given that $\gamma = \frac{P_{in}}{N_0B} = \frac{A_c^2}{2N_0B}$, the output power SNR

$$\left(\frac{P}{N}\right)_{out} = k_p^2 E\left[m^2(t)\right]\gamma \qquad (2.135)$$

Since $\phi(t) = k_p m(t)$, the maximum frequency deviation

$$\Delta f = \frac{1}{2\pi}\left[\frac{d\phi(t)}{dt}\right]_{\max} = \frac{k_p}{2\pi}\left[\frac{dm(t)}{dt}\right]_{\max} \tag{2.136}$$

and therefore

$$\left(\frac{P}{N}\right)_{out} = (2\pi\Delta f)^2 \left(\frac{E\left[m^2(t)\right]}{\left[\frac{dm(t)}{dt}\right]_{\max}^2}\right)\gamma \tag{2.137}$$

from which for example, if the wide-band bandwidth $2\Delta f$ is doubled, $\left(\frac{P}{N}\right)_{out}$ will increase by a factor of 4 or equivalently $10\log_{10}4$ dB $\simeq 6$ dB provided there is no threshold effect i.e. carrier power is much larger than the noise power. For tone modulation with $m(t) = A_m\cos(2\pi f_m t)$, $E[m^2(t)] = \frac{A_m^2}{2}$, the output power SNR is given by $\left(\frac{P}{N}\right)_{out} = k_p^2\frac{A_m^2}{2}\gamma$ or equivalently, since $\left[\frac{dm(t)}{dt}\right]_{\max} = 2\pi f_m A_m$,

$$\left(\frac{P}{N}\right)_{out} = \frac{1}{2}\left(\frac{\Delta f}{f_m}\right)^2\gamma \tag{2.138}$$

Frequency Modulation SNR   As evident from equation 2.118, we may utilize a PM demodulator followed by a differentiator to demodulate an FM signal. The input power SNR remains as $\left(\frac{P}{N}\right)_{in} = \frac{A_c^2}{2N_o B_{EM}}$. At the output, from equation 2.125, $P_{out} = k_f^2 E\left[m^2(t)\right]$. Given that the noise passes through a differentiator whose transfer function is $H(f) = j2\pi f$, and from section 1.8.11 that $PSD_{out}(f) = PSD_{in}(f)\left|H(f)\right|^2$, the PSD of the noise at the output of the differentiator is given by

$$PSD_{n_o}(f) = \begin{cases} \frac{N_o(2\pi f)^2}{A_c^2} & \text{for} \quad |f| \le B \\ 0 & \text{otherwise} \end{cases} \tag{2.139}$$

Thus, the noise power at the output

$$N_{out} = \int_{-B}^{B} PSD_{n_o}(f)df = \int_{-B}^{B} \frac{N_o(2\pi f)^2}{A_c^2}df = \frac{8N_o\pi^2 B^3}{3A_c^2} \qquad (2.140)$$

Notice how N_{out} decreases as the carrier power $\frac{A_c^2}{2}$ increases. This effect is called *noise quieting*. The output power SNR is given by

$$\left(\frac{P}{N}\right)_{out} = \frac{3A_c^2 k_f^2 E[m^2(t)]}{8N_o\pi^2 B^3} = 3\left(\frac{k_f^2 E[m^2(t)]}{(2\pi B)^2}\right)\gamma \qquad (2.141)$$

where $\gamma = \frac{P_{in}}{N_o B} = \frac{A_c^2}{2N_o B}$. Since $\frac{d\phi(t)}{dt} = k_f m(t)$, the maximum frequency deviation $\Delta f = \frac{1}{2\pi}\left[\frac{d\phi(t)}{dt}\right]_{max} = \frac{k_f [m(t)]_{max}}{2\pi}$ and therefore

$$\left(\frac{P}{N}\right)_{out} = 3\left(\frac{\Delta f}{B}\right)^2\left(\frac{E[m^2(t)]}{[m(t)]_{max}^2}\right)\gamma = 3D^2\left(\frac{E[m^2(t)]}{[m(t)]_{max}^2}\right)\gamma \qquad (2.142)$$

where $D = \frac{\Delta f}{B}$ is determined by the bandwidth B of the postdection low-pass filter and not B_m. Ideally, for $m(t)$ bandlimited to B_m, the LPF bandwidth $B = B_m$. If message signal $m(t)$ is normalized such that $[m(t)]_{max} = 1$, then $\left(\frac{P}{N}\right)_{out} = 3D^2 E[m^2(t)]\gamma$. Finally, for tone modulation with $m(t) = A_m\cos(2\pi f_m t)$, $E[m^2(t)] = \frac{A_m^2}{2}$, $[m(t)]_{max} = A_m$ the output power SNR $\left(\frac{P}{N}\right)_{out} = \frac{3D^2\gamma}{2}$.

Comparison of PM and FM   Given $\left(\frac{P_o}{N_o}\right)_{PM} = (2\pi\Delta f)^2\left(\frac{E[m^2(t)]}{\left[\frac{dm(t)}{dt}\right]_{max}^2}\right)\gamma$

and $\left(\frac{P_o}{N_o}\right)_{FM} = 3\left(\frac{\Delta f}{B}\right)^2\left(\frac{E[m^2(t)]}{[m(t)]_{max}^2}\right)\gamma$, the ratio

$$\frac{\left(\frac{P_o}{N_o}\right)_{PM}}{\left(\frac{P_o}{N_o}\right)_{FM}} = \frac{(2\pi B)^2}{3}\left(\frac{[m(t)]_{max}^2}{\left[\frac{dm(t)}{dt}\right]_{max}^2}\right) \qquad (2.143)$$

which indicates that PM is better than FM if $(2\pi B)^2 \left[m(t)\right]^2_{\max} > 3 \left[\frac{dm(t)}{dt}\right]^2_{\max}$.
Typically, $m(t)$ consists of mainly low-frequency components, in which case $\left[\frac{dm(t)}{dt}\right]_{\max}$ is small and PM is better than FM. However, if $m(t)$ consists of mainly high-frequency components, then $\left[\frac{dm(t)}{dt}\right]_{\max}$ is large and FM is better than PM. For example, for tone modulation with $m(t) = A_m \cos(2\pi f_m t)$, $\left[m(t)\right]_{\max} = A_m$ and $\left[\frac{dm(t)}{dt}\right]_{\max} = 2\pi f_m A_m$, $\frac{\left(\frac{S_o}{N_o}\right)_{PM}}{\left(\frac{S_o}{N_o}\right)_{FM}} = \frac{1}{3}$ (for $B = f_m$).

2.7.5   Simulation I : FM Demodulation using a Limiter

⭐SIMULATION **FMLimiter:** We take advantage of the fact that the information is contained within the frequency and not the amplitude of the carrier. The modulated carrier is first sent through a limiter which sets all positive values to +1 and all negative values to -1. This signal is then sent through a band-pass filter of bandwidth B_{EM} centered at the carrier frequency f_c before passing it through a differentiator and then an envelope detector. Notice that the differentiation is undertaken within the frequency domain by making use of the Fourier transform property that $\frac{ds(t)}{dt} \longleftrightarrow (j2\pi f)\, S(f)$. Finally, a low-pass filter of bandwidth B_m is used to recover the message signal $m(t)$.

- **Experiment:** Modify the code to perform the differentiation in the time-domain.

2.7.6   Simulation II : PM Demodulation

⭐SIMULATION **PM:** The phase modulation technique is implemented in which the differentiation is now performed within the time-domain. Once again an envlope detector with a low-pass filter of bandwidth B_m is implemented as in FM demodulation. The only addition is a simple integrator to finally recover the message signal $m(t)$.

- **Experiment**: Determine the influence of k_p. Try other types of $m(t)$ signals. Modify the code to undertake the differentation in frequency domain.

2.7.7   Simulation III : FM SNR

⭐ SIMULATION **FMsnr**: AWGN is first simulated and its spectra determined. Then the FM modulated signal $s_{EM}(t)$ is shown to be a narrowband signal $s_{NB}(t)$ with bandwidth $B_{EM} \simeq 2(D+1)B_m$. The AWGN noise is then passed through a narrowband filter of bandwidth $2B$, where $B = \frac{B_{EM}}{2}$, centered over the carrier frequency f_c to extract the components n_c and n_s. These in turn are used to construct the narrowband noise signal $n_{band}(t)$ via $E_{noise}(t) = \sqrt{n_c^2(t) + n_s^2(t)}$ and $\theta_{noise}(t) = \tan^{-1}\left(\frac{n_s(t)}{n_c(t)}\right)$. The expected filtered noise power of $2N_0B$ is also verified. Thereafter, the signal $s_{band}(t) = s_{EM}(t) + n_{band}(t)$ is shown to correpond to $s_{band}(t) = E(t)\cos(2\pi f_c t + \theta(t))$.

For low noise ($A_c \gg n_c$ and n_s), the following are determined: firstly, the ideal output $\theta_{ideal}(t) = \phi(t)$, secondly, the lownoise output $\theta_{lownoise}(t) = \phi(t) + \frac{n_s(t)}{A_c}\cos\phi(t) - \frac{n_c(t)}{A_c}\sin\phi(t)$ and thirdly, the noise only output $\theta_{noiseonly}(t) = \frac{n_s(t)}{A_c}\cos\phi(t) - \frac{n_c(t)}{A_c}\sin\phi(t)$. Thereafter, the expected input signal and noise powers $P_{in} = \frac{A_c^2}{2}$ and $N_{in} = N_o B_{EM}$ are shown to correpond well with simulation values. To determine the output signal and noise powers, $\theta_{ideal}(t)$ is passed through a differentiator and a low-pass filter of bandwith B_m used to recover the message signal under ideal conditions. This is ovelayed with $m(t)$ and P_{out} is verified with the expected value of $P_{out} = k_f^2 E[m^2(t)]$. Secondly, $\theta_{noiseonly}(t)$ is passed through a differentiator and a low-pass filter of bandwith B_m used to confirm that $N_{out} = \frac{8N_o\pi^2 B^3}{3A_c^2}$. Then, the simulated ratio $\left(\frac{P_{out}}{N_{out}}\right) dB$ in decibels is verified with the expected value. Finally, $\theta(t) = \tan^{-1}\left(\frac{A_c\sin\phi(t) + n_s(t)}{A_c\cos\phi(t) + n_c(t)}\right)$ is passed through a differentiator followed by a low-pass filter of bandwidth B_m and overlayed with $m(t)$ to demonstrate the performance under noise.

- **Experiment**: Determine the influence of k_f. Try other types of $m(t)$ signals.

2.7.8   Simulation IV : PM SNR

SIMULATION **PMsnr:** Essentially identical to simulation **FMsnr** with the key diifference being that a differentiator is not used. The other slight differences are that $\theta_{ideal}(t) = \phi(t) = k_p m(t)$, $P_{out} = k_p^2 E\left[m^2(t)\right]$ and $N_{out} = \frac{2N_o B}{A_c^2}$ where $B = B_m$ in this case.

- **Experiment**: Determine the influence of k_p. Try other types of $m(t)$ signals.

2.8   Preemphasis and Deemphasis Filters

The preemphasis filter $H_t(f)$ and the deemphasis filter $H_r(f)$ in Fig. 2.1, section 2.1.1, can be cleverly designed to maximize the output power SNR $\left(\frac{P_{out}}{N_{out}}\right)$ as follows. Let the overall system transfer function $H(f)$ be given by

$$H(f) = H_t(f)H_c(f)H_r(f) = ke^{-j2\pi f t_d} \tag{2.144}$$

where k is a constant, t_d is the time-delay and $H_c(f)$ is the frequency response (also referred to as the channel transfer function) of the channel. The average signal power P_{out} at the receiver output (following equation 2.10) is given by

$$P_{out} = \int_{-\infty}^{\infty} PSD_m(f)\left|H(f)\right|^2 df = k^2 \int_{-\infty}^{\infty} PSD_m(f)df \tag{2.145}$$

and the average noise power N_{out} at the receiver output (following equation 2.12) is

$$N_{out} = \int_{-\infty}^{\infty} PSD_{noise}(f)\left|H_r(f)\right|^2 df \tag{2.146}$$

where $PSD_{noise}(f)$ is the noise power spectral density. Therefore, the output power SNR that we wish to maximize for a given transmitted power P_T is

$$\left(\frac{P_{out}}{N_{out}}\right) = \frac{k^2 \int_{-\infty}^{\infty} PSD_m(f)df}{\int_{-\infty}^{\infty} PSD_{noise}(f) \left|H_r(f)\right|^2 df} \qquad (2.147)$$

where

$$P_T = \int_{-\infty}^{\infty} PSD_m(f) \left|H_t(f)\right|^2 df \qquad (2.148)$$

From equation 2.147,

$$\left(\frac{P_{out}}{N_{out}}\right) = \left(\frac{P_T}{P_T}\right) \frac{k^2 \int_{-\infty}^{\infty} PSD_m(f)df}{\int_{-\infty}^{\infty} PSD_{noise}(f) \left|H_r(f)\right|^2 df} \qquad (2.149)$$

$$= \frac{k^2 P_T \int_{-\infty}^{\infty} PSD_m(f)df}{\int_{-\infty}^{\infty} PSD_m(f) \left|H_t(f)\right|^2 df \int_{-\infty}^{\infty} PSD_{noise}(f) \left|H_r(f)\right|^2 df} \qquad (2.150)$$

Given that the numerator of equation 2.150 is fixed for a given P_T and $m(t)$, to maximize $\left(\frac{P_{out}}{N_{out}}\right)$ we need to minimize the denominator using the *Cauchy-Schwarz's inequality* (Appendix A.12.1)

$$\left|\int_{-\infty}^{\infty} A(f)B(f)df\right|^2 \leq \int_{-\infty}^{\infty} \left|A(f)\right|^2 df \int_{-\infty}^{\infty} \left|B(f)\right|^2 df \qquad (2.151)$$

where in this case,

$$\left|A(f)\right|^2 = PSD_m(f) \left|H_t(f)\right|^2 \qquad (2.152)$$

and

$$|B(f)|^2 = PSD_{noise}(f)\,|H_r(f)|^2 \tag{2.153}$$

i.e. the minimum value of the denominator in equation 2.150 is

$$\left|\int_{-\infty}^{\infty} A(f)B(f)df\right|^2 \tag{2.154}$$

or equivalently,

$$\left|\int_{-\infty}^{\infty} \sqrt{PSD_m(f)PSD_{noise}(f)}\,|H_t(f)|\,|H_r(f)|\,df\right|^2 \tag{2.155}$$

The equality in equation 2.151 holds if $A(f) = cB^*(f)$, where c is an arbitary constant. Since $|A(f)|^2 = c^2\,|B(f)|^2$, we have

$$PSD_m(f)\,|H_t(f)|^2 = c^2 PSD_{noise}(f)\,|H_r(f)|^2 \tag{2.156}$$

Now from equation 2.144, $|H(f)|^2 = |H_t(f)H_c(f)H_r(f)|^2 = k^2$ so that $|H_r(f)| = \frac{k}{|H_c(f)||H_t(f)|}$ substituted into equation 2.156 gives

$$|H_t(f)|^2 = c\frac{k}{|H_c(f)|}\sqrt{\frac{PSD_{noise}(f)}{PSD_m(f)}} \tag{2.157}$$

Similary, given that $|H_t(f)| = \frac{k}{|H_c(f)||H_r(f)|}$, we get from equation 2.156

$$|H_r(f)|^2 = \frac{1}{c}\frac{k}{|H_c(f)|}\sqrt{\frac{PSD_m(f)}{PSD_{noise}(f)}} \tag{2.158}$$

Substituing equation 2.157 into equation 2.148,

$$P_T = c\int_{-\infty}^{\infty} \frac{k\sqrt{PSD_m(f)PSD_{noise}(f)}}{|H_c(f)|}df \tag{2.159}$$

from which the constant c is given by

$$c = \frac{P_T}{k \int_{-\infty}^{\infty} \frac{\sqrt{PSD_m(f)PSD_{noise}(f)}}{|H_c(f)|} df} \tag{2.160}$$

Finally substituting equation 2.160 into equations 2.157 and 2.158 yields the optimum transfer functions

$$|H_t(f)|^2_{opt} = \frac{P_T}{|H_c(f)|} \sqrt{\frac{PSD_{noise}(f)}{PSD_m(f)}} \left(\frac{1}{\int_{-\infty}^{\infty} \frac{\sqrt{PSD_m(f)PSD_{noise}(f)}}{|H_c(f)|} df} \right) \tag{2.161}$$

$$|H_r(f)|^2_{opt} = \frac{k^2}{P_T |H_c(f)|} \sqrt{\frac{PSD_m(f)}{PSD_{noise}(f)}} \left(\int_{-\infty}^{\infty} \frac{\sqrt{PSD_m(f)PSD_{noise}(f)}}{|H_c(f)|} df \right) \tag{2.162}$$

where $|H_t(f)|_{opt} |H_r(f)|_{opt} = \frac{k}{|H_c(f)|}$. The preemphasis filter $|H_t(f)|_{opt}$ magnifies the weaker frequency components within $m(t)$ and reduces the stronger frequency components within $m(t)$. The deemphasis filter $|H_r(f)|_{opt}$ performs the inverse operation of $|H_t(f)|_{opt}$. Since the noise signal passes through only the deemphasis filter, which reduces the strength of the noise components, the output power SNR is effectively increased in comparison to simply using $H_t(f) = H_r(f) = \begin{cases} 1 & if \quad -B \leq f \leq B \\ 0 & otherwise \end{cases}$ as in section 2.1.1. Given that equation 2.150 is maximized when its denominator is given by equation 2.155, the optimum output power SNR is

$$\left(\frac{P_{out}}{N_{out}} \right)_{opt} = \frac{k^2 P_T \int_{-\infty}^{\infty} PSD_m(f) df}{\left| \int_{-\infty}^{\infty} \sqrt{PSD_m(f)PSD_{noise}(f)} |H_t(f)| |H_r(f)| df \right|^2} \tag{2.163}$$

$$= \frac{P_T \int_{-\infty}^{\infty} PSD_m(f)df}{\left| \int_{-\infty}^{\infty} \frac{\sqrt{PSD_m(f)PSD_{noise}(f)}}{|H_c(f)|} df \right|^2} \qquad (2.164)$$

2.9   Phase-Locked Loop (PLL)

For the coherent demodulation of an amplitude or angle-modulated signal, we assumed in the previous sections that the local oscillator frequency was exactly the same as the incoming carrier frequency f_c. In practice, a *phase-locked loop* (PLL) is used to adjust the local oscillator frequency until it is equal to the frequency of the incoming carrier signal. At that point, the phase difference between the two will either be a constant or zero. Note that a PLL can be used for FM demodulation, filtering and frequency synthesis.

A block diagram of a PPL is shown in Fig. 2.28, in which the *voltage-controlled oscillator* (VCO) is simply an oscillator that outputs a sinusoidal waveform $s_{vco}(t)$. The frequency of $s_{vco}(t)$ is linearly dependent on the input voltage to the VCO. Let the incoming signal $s(t)$ be given by $s(t) = A\sin\left(2\pi\left(f_c + \Delta f\right)t + \theta\right)$, where Δf is a small increase in the carrier frequency and θ is a constant.

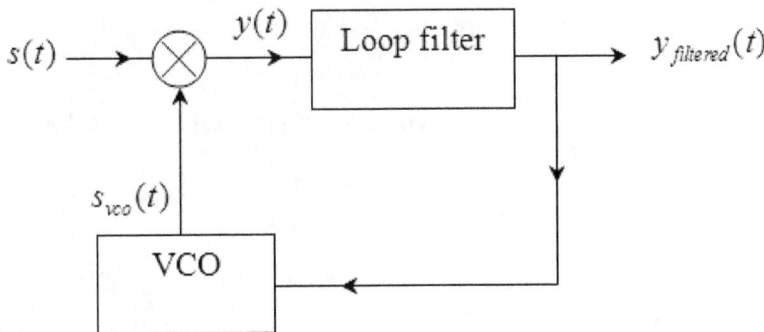

Figure 2.28: Phase-locked loop.

We may express $s(t)$ as

$$s(t) = A\sin\left(2\pi f_c t + 2\pi\Delta f t + \theta\right) = A\sin\left(2\pi f_c t + \beta\right) \tag{2.165}$$

where $\beta = 2\pi\Delta f t + \theta$. Let the VCO output

$$s_{vco}(t) = B\cos\left(2\pi f_c t + \alpha\right) \tag{2.166}$$

where α is a constant. Then

$$y(t) = s(t)s_{vco}(t) = AB\sin\left(2\pi f_c t + \beta\right)\cos\left(2\pi f_c t + \alpha\right) \tag{2.167}$$

Using the trigonometric identity $\sin A\cos B = \frac{1}{2}\sin\left(A - B\right)+\frac{1}{2}\sin\left(A + B\right)$

$$y(t) = \frac{AB}{2}\sin\left(2\pi f_c t + \beta - 2\pi f_c t - \alpha\right) + \frac{AB}{2}\sin\left(2\pi f_c t + \beta + 2\pi f_c t + \alpha\right) \tag{2.168}$$

$$= \frac{AB}{2}\sin\left(\beta - \alpha\right) + \frac{AB}{2}\sin\left(4\pi f_c t + \beta + \alpha\right) \tag{2.169}$$

Hence the output $y_{filtered}(t)$ of the loop filter, which is a low-pass filter (narrow-band), is given by

$$y_{filtered}(t) = \frac{AB}{2}\sin\left(\beta - \alpha\right). \tag{2.170}$$

If $\Delta f = 0$, then $\beta = \theta$ and $y_{filtered}(t) = \frac{AB}{2}\sin\left(\theta - \alpha\right)$ i.e. a constant value independent of time. If $\Delta f \neq 0$, the input voltage $y_{filtered}(t)$ to the VCO will increase with time. Consequently, the frequency of $s_{vco}(t)$ will increase until it is equal to $(f_c + \Delta f)$. When the frequency difference between $s_{vco}(t)$ and $s(t)$ becomes zero, the input to the VCO will once again be a constant voltage and the frequency of $s_{vco}(t)$ will no longer change. Of course if Δf is negative, then the input voltage to the VCO will reduce with time and the frequency of $s_{vco}(t)$ will reduce to match the decrease in the input carrier frequency. If θ is a function of time, then the PPL will track the phase of the incoming carrier to ensure $(\theta - \alpha)$ is a constant.

2.9.1 Simulation : PLL

SIMULATION **PLL:** The operation of the PLL in Fig. 2.28 is implemented. You may view the spectrum of $s(t)$, $y(t)$ and $y_{filtered}(t)$. The filtered spectrum of $y(t)$ is used to reconstruct the signal $y_{filtered}(t)$ and this is shown to be the same as $y_{filtered}(t) = \frac{AB}{2} \sin(2\pi \Delta f t + \theta - \alpha)$.

- **Experiment:** Observe what happens for $\Delta f = 0$ and non-zero values of Δf.

2.10 Problems

1. In DSB-SC modulation, the carrier signal is given by $s_{DSB}(t) = m(t) \cos(2\pi f_c t)$. (a) If this modulated carrier is multiplied by $\cos(2\pi f_c t + \phi)$, where ϕ is the phase error and then sent through a low-pass filter, (i) determine the recovered signal. (ii) What would happen if $\phi = \pm\frac{\pi}{2}$? (b) Consider what happen if the modulated carrier is multiplied by $\cos[2\pi(f_c + f)t]$, where f is the frequency error. (c) Which of the two effects considered in parts(a) and (b) is worse ? You may assume a noiseless channel.

2. Show that for a periodic DSB-SC signal given by $s_{DSB}(t) = m(t)A_c \cos(2\pi f_c t)$, the power of the modulated signal $P_{DSB} = \frac{A_c^2 P_m}{2}$, where P_m is the message signal power.

3. Consider a periodic message signal $m(t)$ given by

$$m(t) = \begin{cases} 2 & \text{if} & 0 \leq t < \frac{T}{4} \\ -1 & \text{if} & \frac{T}{4} \leq t < \frac{3T}{4} \\ 0 & \text{otherwise} \end{cases}$$

Using the DSB-SC modulation scheme with $s_{DSB}(t) = m(t)A_c \cos(2\pi f_c t)$, where A_c is the amplitude of the carrier, with $T = 0.2$ s, $A_c = 2$ volts, $f_c = 200$ Hz. Determine,

(a) the message signal power P_m and the modulated signal power P_{DSB}. Verify that $P_{DSB} = \frac{A_c^2 P_m}{2}$.

(b) the amplitude spectrum of $m(t)$ and $s_{DSB}(t)$.

(c) the message signal power after passing $m(t)$ through an ideal low-pass filter with a cut-off frequency $f_{cut} = 20$ Hz.

(d) the modulated signal power after passing $s_{DSB}(t)$ through a band-pass centered on f_c that cuts of frequencies above and below $(f_c + f_{cut})$ and $(f_c - f_{cut})$, respectively. Determine the percentage of the original power retained after the filter.

(e) Finally, reconstruct both filtered signals and overlay them on a single graph.

4. Using the message signal in the previous problem with ordinary amplitude modulation such that $s_{AM}(t) = A_c[1 + \mu m_n(t)] \cos(2\pi f_c t)$, where $m_n(t)$ is the normalized message signal, determine the following for $\mu = 0.9$, $A_c = 1$ volt and $f_c = 200$ Hz:

(a) normalized message signal power

(b) modulated signal power

(c) carrier power P_c and the power P_s carried by the sidebands

(d) power efficiency η

5. Consider a conventional amplitude modulation scheme in which the message signal $m(t) = 2\sin(2\pi t)$. If the modulation index $\mu = 0.2$, what is the amplitude A_c of the added carrier signal ? Sketch the modulated carrier and overlay this with a plot of $m(t) + A_c$.

6.(a) If the average value of the message signal $m(t)$ is zero, (a) show that for a periodic AM signal given by $s_{AM}(t) = [m(t) + A_c] \cos(2\pi f_c t)$, the power of the modulated signal $P_{AM} = \frac{(P_m + A_c^2)}{2}$, where P_m is the message signal power. (b) Verify the formula of part(a) for the periodic message signal $m(t)$ shown below. Take $f_c = 200$ Hz, $T = 0.2$ s and $A_c = 2$ volts.

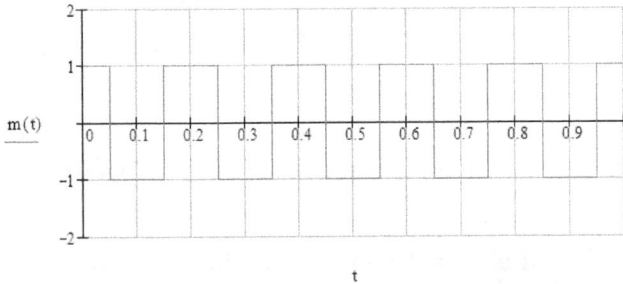

7. (a) If conventional amplitude modulation is coherently demodulated as in DSB-SC modulation, verify that the output of the low-pass filter is given by $\frac{m(t)+A_c}{2}$. (b) Modify the simulation **DSBSC** of section 2.1 to implement the coherent demodulation of $s_{AM}(t)$. Confirm the result of part(a) via simulation.

8. (a) Show that $s_{AM}(t) = [1 + \mu\cos(2\pi f_m t)]\, A_c \cos(2\pi f_c t)$ for $m(t) = A_m \cos(2\pi f_m t)$. (b) Verify by simulation that for the envelope detection in this case, we require that the time constant $t_c \leq \frac{\sqrt{1-\mu^2}}{2\pi f_m \mu}$. Select appropriate values for all the variables.

9. Consider conventional amplitude modulation for which
$$s_{AM}(t) = [m(t) + A_c]\cos(2\pi f_c t).$$ Let the message signal $m(t)$ be given by

$$m(t) = \begin{cases} e^{-t/T}\sin(\frac{2\pi t}{T}) + \cos(\pi t)\, e^{-0.2t} & \text{if} \quad t > 0 \\ 0 & \text{otherwise} \end{cases}$$

where $T = 0.5$ seconds, $A_c = 2$ volts and $f_c = 6$ Hz.

(a) Plot a graph of $m(t)$ for over a time duration of 8 seconds.
(b) Determine the maximum value $V_{\max}$ of the envelope signal $e(t) = [m(t) + A_c]$.
(c) Using the envelope detector in section 2.5, consider the voltage across the capacitor on the discharge cycle given by $v_{cap}(t) = V_{\max} e^{-\frac{t}{t_c}}$, where t_c

is time constant.   (i) Show that the first two terms of the Taylor series approximation of $v_{cap}(t)$ is given by $v_{approx}(t) = V_{\max}\left(1 - \frac{t}{t_c}\right)$. (ii) Plot a graph of $v_{cap}(t)$ and overlay on this graph a plot of $v_{approx}(t)$.

(d) Plot a graph of $\left|\frac{de(t)}{dt}\right|$ and overlay this graph with $\left|\frac{dv_{approx}(t)}{dt}\right|$. Choose t_c such that $\left|\frac{dv_{approx}(t)}{dt}\right| > \left|\frac{de(t)}{dt}\right|$.

(e) Using your value of t_c from part(d), confirm that the capacitor is able to follow the envelope $e(t)$.

(f) Repeat parts (d) and (e) for $t_c = 1$ second. Confirm for yourself that the range for which $\left|\frac{dv_{approx}(t)}{dt}\right| \geq \left|\frac{de(t)}{dt}\right|$, the capacitor is able to follow the envelope.

10.   Verify that $s_{USB}(t) = m(t)\cos(2\pi f_c t) - m_p(t)\sin(2\pi f_c t)$ and $s_{LSB}(t) = m(t)\cos(2\pi f_c t) + m_p(t)\sin(2\pi f_c t)$ for $m(t) = A_m\cos(2\pi f_m t)$, where $m_p(t)$ is the message signal after every frequency component of $m(t)$ is phase shifted by $-\frac{\pi}{2}$ radians.

11. (a) If a SSB signal is coherently demodulated, verify that the output of the low-pass filter is given by $\frac{m(t)}{2}$. (b) Implement in either Mathcad/MATLAB, the coherent demodulation of a USB and LSB signal.

12. (a) If $y_1(t)$ represents the result of phase shifting every frequency component of $s_{USB}(t) = m(t)\cos(2\pi f_c t) - m_p(t)\sin(2\pi f_c t)$ by $-\frac{\pi}{2}$ radians, determine $y_1(t)$.

(b) Show that $m(t)$ can be recovered from $s_{USB}(t)$ via a demodulator whose output $y_2(t)$ is given by $y_2(t) = s_{USB}(t)\cos(2\pi f_c t) + y_1(t)\sin(2\pi f_c t)$.

13. Modify the simulation VSB in section 2.4 to implement the frequency response $H_{vsb}(f)$ as shown below to recover $m(t)$.

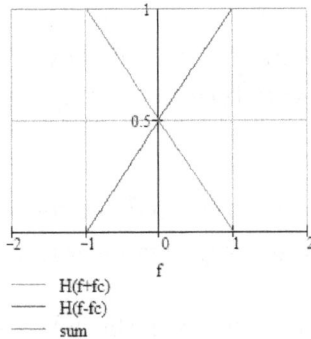

H(f+fc)
H(f-fc)
sum

14. In *quadrature amplitude modulation* (QAM), two message signals are sent simultaneously via $s_{QAM}(t) = m_1(t)\cos(2\pi f_c t) + m_2(t)\sin(2\pi f_c t)$.

(a) Show that $m_1(t)$ can be recovered by multiplying the received signal with $\cos(2\pi f_c t)$ and using a low-pass filter to remove high frequency components.

(b) Show that $m_2(t)$ can be recovered by multiplying the received signal with $\sin(2\pi f_c t)$ and using a low-pass filter to remove high frequency components.

(c) Implement QAM in either Mathcad or MATLAB to verify the results of parts (a) and (b). Choose any two message signals for $m_1(t)$ and $m_2(t)$.

15. Referring to the mixing operation presented in section 2.6, verify that the image frequency $f_{image} = f_c - 2f_I$ if $f_L = f_c - f_I$.

16. Determine local oscillator frequencies f_L within a superheterodyne receiver that can be used to tune the receiver to a signal at 900 kHz ? Assume the IF frequency to be 455 kHz.

17. In a superhetrodyne AM receiver, up-conversion is used instead of down-conversion because it is easier to design a local oscillator with a frequency f_L that is tunable over a smaller frequency ratio $\frac{f_{L\max}}{f_{L\min}}$, where $f_{L\max}$ and $f_{L\min}$ are the maximum and minimum values of f_L, respectively. (a) Verify this statement for the AM broadcast range of 550 to 1600 kHz with f_I set to 455 kHz. (b) How does the ratio $\frac{f_{L\max}}{f_{L\min}}$ depend on f_I ?

18. In problem 17, determine the image frequency range (i.e. image frequency band). (a) Do the desired and image frequency band overlap for $f_I = 455$ kHz. (b) What is the minimum value of f_I above which the two bands will not overlap.

19. Consider the angle modulated signal $s(t) = A_c \cos\left(2\pi f_c t + \phi(t)\right)$ with $A_c = 2$ volt and $f_c = 2$ Hz. Determine the instantaneous frequency for (a) $\phi(t) = \frac{\pi t^2}{2}$ (b) $\phi(t) = \frac{1}{2}\sin\left(4\pi t\right)$. In part(b), also determine the maximum frequency and phase deviation.

20. Verify for that the signal power of an angle modulated carrier signal $s(t) = A_c \cos\left(2\pi f_c t + \phi(t)\right)$ is given by $\frac{A_c^2}{2}$ for any message signal $m(t)$.

21. If the tone signal $m(t) = A_m \cos\left(2\pi f_m t\right)$ is sent via the FM signal $s(t) = A_c \cos\left(2\pi f_c t + \phi(t)\right)$, determine and expression for the

(a) instantaneous frequency
(b) maximum frequency deviation
(c) deviation ratio
(d) bandwidth of the FM signal.
(e) If $m(t) = A_m \sin\left(2\pi f_m t\right)$ with $A_m = 1$ volt, $f_m = 2$ Hz, $A_c = 10$ volts, $f_c = 50$ Hz and $k_f = 30$, determine the bandwidth and the power of the FM signal.

22. If the tone signal $m(t) = A_m \sin\left(2\pi f_m t\right)$ is sent via the PM signal $s(t) = A_c \cos\left(2\pi f_c t + \phi(t)\right)$, determine and expression for the

(a) instantaneous frequency
(b) maximum frequency deviation
(c) deviation ratio
(d) bandwidth of the PM signal.
(e) If $m(t) = A_m \cos\left(2\pi f_m t\right)$ with $A_m = 1$ volt, $f_m = 1$ Hz, $A_c = 10$ volts, $f_c = 100$ Hz and $k_p = 30$, determine the bandwidth and the power of the PM signal.

23. If the tone signal $m(t) = A_m \sin\left(2\pi f_m t\right)$ is sent via the PM signal $s(t) = A_c \cos\left(2\pi f_c t + \phi(t)\right)$ (a) determine the spectrum of the angle-modulated signal in problem 22 assuming that $D \ll 1$. (b) Verify your answer via Mathcad/MATLAB for $A_m = A_c = 1$ volt, $f_c = 100$ Hz, $f_m = 10$ Hz, $k_p = 0.1$. If the modulated signal is passed through an ideal band-pass filter of bandwidth $2\left(D + 1\right) f_m$, determine the power of the filtered signal as a percentage of the original signal power for $k_p = 0.1$ and $k_p = 5$. Under what condition is Carson's rule satisfied ?

24.   Consider the angle-modulated signal $s(t) = A_c \cos(2\pi f_c t + \phi(t))$ with $\phi(t) = 100 \cos(2\pi t)$ with $f_c = 100$ Hz, $A_c = 1$ volt.   (a) Determine the deviation ratio (b) and the bandwidth of signal. Is $s(t)$ a narrowband or wideband angle-modulated signal ? (c) Confirm Carson's rule.

25.   If the message signal $m(t) = \begin{cases} \text{sinc}(10t) & \text{if} & |t| \leq T \\ 0 & \text{otherwise} \end{cases}$ for $T = 0.5$ s is sent via the FM signal $s(t) = A_c \cos(2\pi f_c t + \phi(t))$ with $f_c = 200$ Hz, $A_c = 1$ volt and $k_f = 300$, determine the
   (a) maximum phase and frequency deviation
   (b) bandwidth B_m from the spectrum of $m(t)$
   (c) deviation ratio
   (d) spectrum of $s(t)$
   (e) power and the bandwidth of $s(t)$
   (f) If the modulated signal $s(t)$ is passed through an ideal band-pass filter of bandwidth $2(D+1)B_m$, determine the power of the filtered signal as a percentage of the original signal power.

26. (a) Show that if a FM signal is filtered by a high pass filter as shown below with a time constant $t_c \ll 1$, then a envelope detector can be used to extract the message signal $m(t)$ from the filter output. (b) Modify the simulation **FM2** in section 2.7 to verify part(a) via simulation.

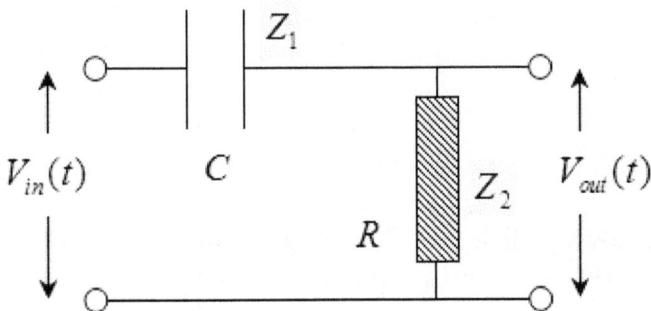

High pass filter.

There are only two ways to live your life. One is as though nothing is a miracle. The other is as though everything is a miracle.

— *Albert Einstein*

Chapter 3

Digital Communications

3.1 Introduction

In this chapter we begin with a first look at the key processing blocks of a *digital communication system* (DCS). The chapters to follow will consider each processing block or combinations of blocks in further detail. Please refer to Appendix A.1 for an introduction to probability theory as required.

3.1.1 Digital Communication System (DCS)

▶ DCS

- Fundamentals of the key processing blocks

- Example of a popular modulation scheme

In a DCS, the information (e.g. voice, data, etc.) to be transferred from the source to the destination is first represented by a finite set of digital symbols (e.g. binary digits 0 or 1). Any given digital symbol is sent to the destination by mapping a symbol to a signal (e.g. high or low frequency waveform) that is transmitted over the channel (e.g. wires, space, optical fibers, etc.). Unlike an analog communication system (ACS), the signal transmitted over the channel is from a finite set of possible signals (or waveforms).

197

The objective of the receiver is not to reproduce the transmitted signal as in an analog communication system, but to simply determine, which signal from the known set of possible signals was transmitted and thereby identify the corresponding symbol. The inherent advantages of a DCS are as follows:

- Unlike an ACS in which noise will always distort the received signal, a digital communication system can withstand channel noise and distortion. This is because the task of the receiver is not to recover the transmitted signal with precision, but to simply decide which digital symbol corresponds to the received signal. For example, the transmission of binary digits over a given channel requires the use of only two different signals. Suppose signal $s_1(t)$ is mapped to a binary digit 1 and signal $s_2(t)$ is mapped to a binary digit 0 i.e. to send a binary digit 1, signal $s_1(t)$ is transmitted for a fixed finite time interval. At the receiver, it is only necessary to decide which of the two possible signals, $s_1(t)$ or $s_2(t)$, was transmitted!

- Information from various different types of sources can be represented by a given set of symbols, then multiplexed (i.e. essentially grouped together) and transmitted over a single communication channel. For example, telephone conversations, computer data, etc. can be represented by binary digits which are then multiplexed and sent via an undersea fiber optic cable from one continent to another.

- Several repeaters can be inserted in-between the source and destination, where each repeater recovers the digital symbols and retransmits the corresponding signal into the next segment of the communication link. Since the information is embedded within the sequence of digital symbols, information is not lost at each repeater provided the symbols are recovered without error. Compare this with an ACS in which the signal progressively becomes weaker through the channel. Amplification of this signal in between the source and destination will only serve to amplify both the noise and the signal. Consequently for a ACS, the transmitter power essentially dictates the distance between the source and the sink.

- A sequence of digital symbols can be segmented into packets, where each packet contains a predefined number of symbols. These packets can be sent via different routes to the destination, as for example in a computer network. Can you imagine the nightmare of trying to segment an analog signal and multiplexing it with other analog signals in a manner that will allow the receiver to demultiplex the received signal!

- Digital circuits are less susceptible to interference from external noise sources e.g. a washing machine. They are more reliable, flexible and can be produced at a low cost e.g. it is cheaper to replace a digital circuit than to repair it.

- Extra digital symbols can be added to the information bearing symbols to combat channel noise.

- Digital symbols are easily encrypted for secure communication.

- Information (voice, video, data, etc.) stored digitally has several advantages i.e. it can be searched, accessed from a different part of the world (e.g. via the internet), reproduced without deterioration and relatively inexpensive!

A block diagram of a DCS is shown in Fig. 3.1. Other blocks will be added to this diagram as we continue our study of a DCS. We shall begin with a brief overview of each processing block before considering them in greater detail in the chapters to follow.

INFORMATION SOURCE The information source outputs a message which is either *analog* or a *digital* in nature. For example, a human voice converted to an electrical signal via a transducer, or text, computer data, sensor readings, etc.

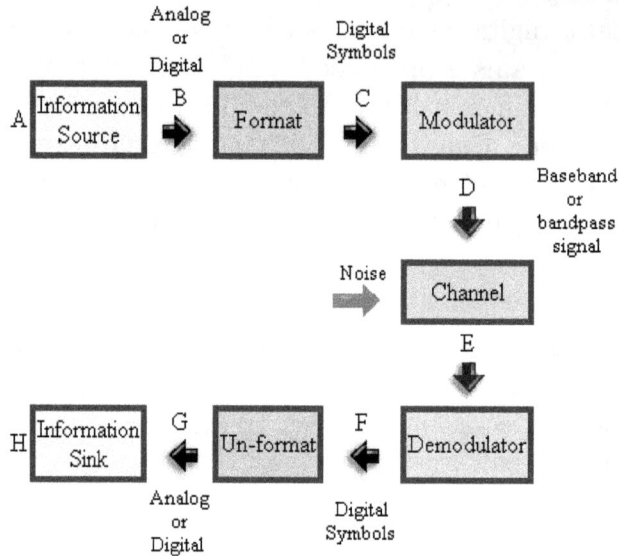

Figure 3.1: Block diagram of a digital communication system.

FORMAT   To gain the benefits of a DCS, the format block transforms the information source output to digital symbols. Thus the combination of the information source and the format block can be thought of as a *digital source*.   Examples of digital symbols are listed in Table 3.1.   A digital message is constructed using a finite number of digital symbols e.g. a Morse code telegraph message is constructed using only two symbols.

Type	Digital Symbols
Binary	0, 1
Morse code	dot (.), dash (-)
Group n binary digits to form a symbol. In this case, $n = 2$.	00, 01, 10, 11
Integers	0, 1, 2, 3, etc.
Alphabet	A, B, C, D, etc.

Table 3.1: Examples of digital symbols.

MODULATOR A modulator transmits the signal which corresponds to the input symbol. For example, suppose the signal $s_1(t)$ is mapped to a binary digit 1 and signal $s_2(t)$ is mapped to a binary digit 0. If the input symbol to the modulator is the binary digit 0, the modulator will transmit the signal $s_2(t)$ into the channel.

The transmitted signal is either a *baseband* or a *bandpass* signal depending on the type of channel. The spectrum of a baseband signal starts from (or close to) d.c. whereas the spectrum of a bandpass signal is centered on a high frequency. Figure 3.2 presents an overlay of these two types of signals for the input binary sequence 1100101. Notice how in this specific example, the bandpass signal is created by switching the phase of a high frequency sinusoidal waveform by a 180 degrees depending on the binary digit type.

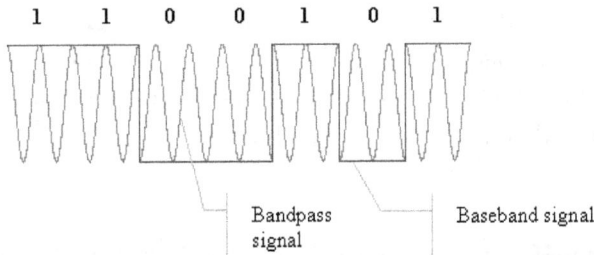

Figure 3.2: Overlay of a baseband and a bandpass signal.

CHANNEL The modulator and demodulator are separated by a channel. Key examples are as follows:

- A pair of wires that carry an electrical signal e.g. the telephone wire which has a usable bandwidth of a few hundred kHz. A signal through this type of a channel will be distorted both in amplitude and phase.

- A fiber optical cable over which light is transported via internal reflections e.g. transatlantic optical fiber cables which carry voice, data, video, etc. The light source intensity is modulated in accordance with the binary digits to be transferred. Repeaters are used between segments of the cable to amplify the light and a photodiode is used at the receiver to extract an electrical representation of the received binary digits. A usable bandwidth of a few hundred GHz.

- Free space through which electromagnetic waves are transmitted. A usable bandwidth from approximately 1 MHz (e.g. AM radio) up to 100 GHz (e.g. satellite systems).

- Data storage devices such as a magnetic tape (cassette), compact disc (CD-RW), etc.

The type of channel will dictate the available bandwidth and the signal power required for communication over this physical medium. Typically, a channel behaves as a filter i.e. it distorts the signal. The effects of distortion are reduced with the use of an equalizer. Atmospheric noise, man-made noise, thermal noise within electronic devices, etc. contribute to randomly corrupting the signal sent through the channel.

DEMODULATOR The demodulator performs the inverse operation of the modulator. Specifically, the demodulator identifies which of the finite set of possible transmitted signals has been received and outputs the corresponding symbol.

Note that with the use of error-control coding (refer to section 3.3), to improve the overall performance of the DCS, the demodulator and the channel decoder blocks are typically combined into a single processing block. In this case, the demodulation part would not make a decision on which symbol corresponds to the received signal.

UN-FORMAT The un-format block performs the inverse operation of the format block. For example, if an analog signal was output by the information source, the un-format block would use the input digital symbols to reconstruct the analog signal. The combination of the un-format block and the information sink can be thought of as a *digital sink*.

INFORMATION SINK The information sink is the final destination of the information transmitted by the information source.

3.1.2 Disadvantages of a DCS

- Synchronization is essential. Without proper synchronization at every stage within a DCS, the output of the un-format block would be "garbage", even if the communication channel is noiseless.

- Requires a greater channel bandwidth than a ACS to communicate the same information in a digital format.

3.1.3 Performance Criteria

Performance Criteria

- Probability of an error in a binary digit (P_e)

- What is meant by a probability of $P_e = 0.1$?

For an analog communication system, the performance criteria is the quality (fidelity criterion) of the analog signal reproduced at the information sink. This is because the information is embedded within the variation of the signal. However, for a DCS, it is not the shape of the received signal that is important. The demodulator need only determine which digital symbol is represented by the received signal. Thus for a DCS, a key performance criteria is the *probability* of incorrectly receiving a digital symbol at the input of the un-format block. Please refer to Appendix A.1 for an introduction to probability theory. In most cases, the symbols are binary digits, in which case, the performance of the DCS would be judged by the probability of an error in a binary digit P_e at point F in Fig. 3.1. The channel bandwidth required to achieve a given P_e is an important aspect of the overall performance of a given DCS. We shall focus on this aspect in section 8.11.

3.1.4   Simulation : Binary Source

⭐ SIMULATION **Probability:** A binary source is modeled. The probability with which a binary digit zero is output from this source may be controlled to observe the output binary stream. The total number of binary digit zeros and ones are counted to calculate the probability with which a binary digit 0 and 1 are output from the binary source.

- **Experiment:** Verify the advantage of specifying the required number of events as oppose to the total number of trials.

3.2   Discrete Memoryless Channel (DMC)

▷ **DMC**

- Model of a digital communication system

  as a binary source, a BSC and a binary sink

- Demonstrate that P_e at the sink is equal to

  the BSC crossover probability

Referring back to Fig. 3.1, notice that digital symbols are input at point C and digital symbols exit at point F. Hence, this block diagram may be simplified by replacing the blocks between points C and F by a single block, referred to a *discrete memoryless channel* (DMC) and the combination of the information source and format block by a digital source with M symbols, as shown in Fig. 3.3. For $M = 2$, the digital source is referred to as a binary source. Simplification of a DCS in this manner is extremely useful, because as we shall see in chapter 4, it is possible to analyze a DCS to a great extent without even considering the mechanism of the hidden (information source, format, modulator, physical channel, demodulator, un-format) sub-blocks! In the chapters to follow, additional processing blocks will be inserted within the basic DCS block diagram of Fig. 3.3.

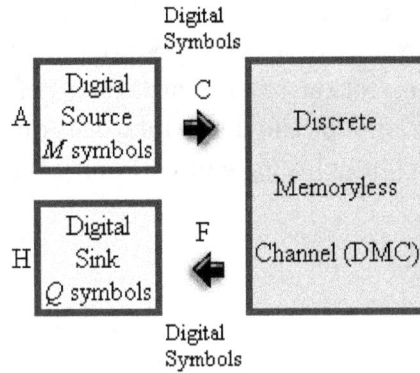

Figure 3.3: Simplified model of a DCS.

Given that the DMC encompasses a modulator, the physical channel and a demodulator, only digital symbols enter and exit the DMC. In general, M symbols enter the DMC and Q symbols exit, where $Q \geqslant M$. For example, suppose the signals transmitted by the modulator represent binary digits i.e. $M = 2$. At the receiver, if the demodulator is unable to confidently establish whether the received waveform represents a binary digit 1 or 0, it could be designed to output a third possible symbol, say 'U', to indicate the symbol is unknown, in which case $Q = 3$.

Let x_i represent a digital symbol that enters the DMC, where $i = 1, 2, \cdots, M$ and let y_j represent the corresponding output digital symbol from the DMC, where $j = 1, 2, \cdots, Q$. Under a noiseless channel, $y_j = x_i$ with $j = i$. However, for a real channel, there is a probability with which say symbol x_1 is transmitted and due to noise within the channel, the demodulator misinterprets the received symbol as y_2. Mathematically, this is simply the conditional probability $P(y_2|x_1)$, which is referred to as a *forward transition probability* i.e. $P(y_2|x_1)$ is the probability that y_2 is received **given that** x_1 was transmitted. Please refer to Appendix A.4 for an introduction to conditional probabilities.

To visually represent all possible forward transition probabilities between the M input symbols and the Q output symbols, a transition probability diagram is used as shown in Fig. 3.4(a), which illustrates the simple case where $M = Q = 2$. Notice how a straight line between an input symbol and an output symbol is labeled with the corresponding transition probability.

Fig. 3.4(b) shows another example, in which $M = 2$ and $Q = 8$. In this case, for each input binary digit, one of eight possible symbols (0, 1, 2, .., 7) is output from the DMC. In section 8.6, the advantage of $Q \geqslant M$ to gain extra information about how confident the demodulator is on deciding whether the signal received from the channel represents a binary digit one or zero will be illustrated.

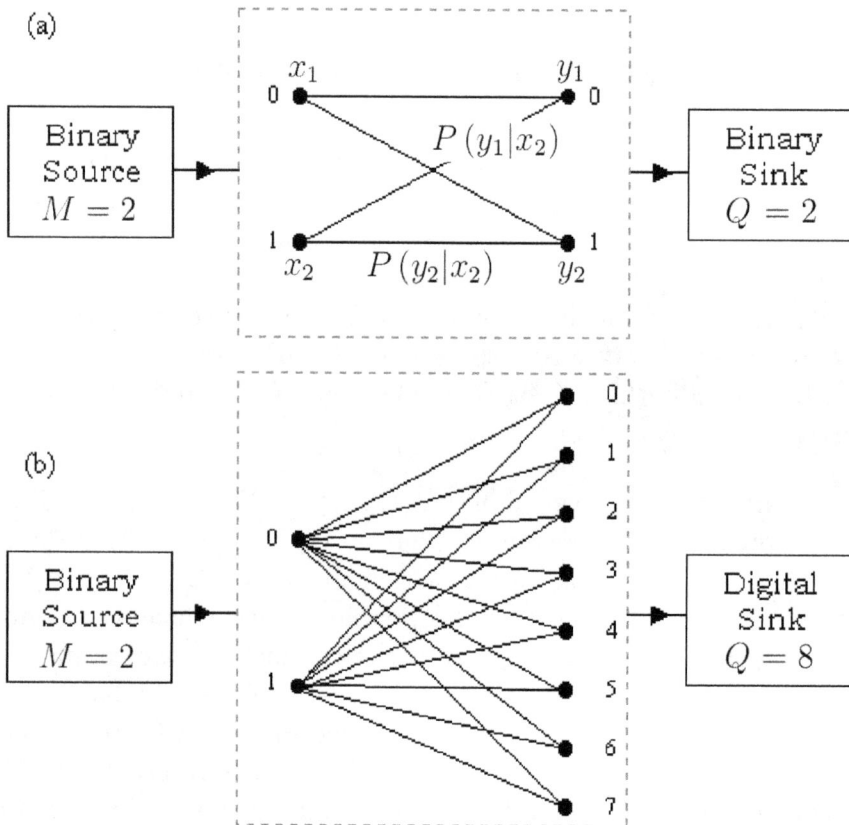

Figure 3.4: Two DMC examples.

The use of the word "memoryless" within DMC indicates the assumption that the successive symbols are statistically independent of each other i.e. if the symbol x_1 is transmitted at time t, then the next symbol to be transmitted at time $t + 1$ does not depend on the transmitted symbol x_1. This is unlike the English alphabet in which the letter u must follow the letter q.

3.2.1 Binary Symmetric Channel (BSC)

In general, let $P(y_j|x_i)$ represent the forward transition probability of receiving symbol y_j given that x_i was transmitted, where $i = 1, 2, \cdots, M$ and $j = 1, 2, \cdots, Q$. Also, let $P(x_i)$ represent the probability of transmitting the symbol x_i. A special case of a DMC is a *binary symmetric channel* (BSC) in which $M = Q = 2$ and $P(y_1|x_2) = P(y_2|x_1)$, as shown in Fig. 3.5. For notational convenience, the forward transition probabilities are labeled P(0|0), P(1|1), P(1|0), P(0|1). For example, $P(1|0) = P(y_2|x_1)$ is the probability of receiving a binary digit 1 at the output of the BSC channel given that a binary digit 0 was transmitted. In practice, most modulation and demodulation processing blocks can be modeled as a BSC.

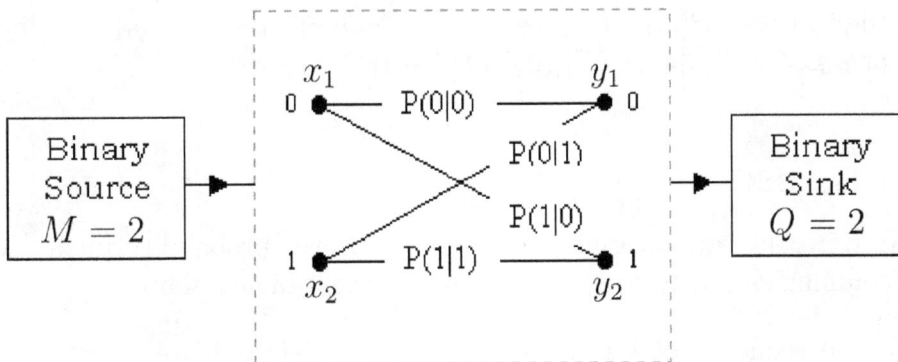

Figure 3.5: Binary symmetric channel (BSC).

Using the forward transition probabilities and simple probability theory, we can establish some very interesting features. For example, for the BSC shown in Fig. 3.5, there are only two possible ways in which a binary digit is output in error from the demodulator. Namely, a binary digit 0 is transmitted and is demodulated as 1 and visa versa. Mathematically, the probability of an error in a binary digit P_e at the output of the BSC is thus given by

$$P_e = P(y_2|x_1) P(x_1) + P(y_1|x_2) P(x_2) \qquad (3.1)$$

The special feature of a BSC is that $P(y_1|x_2) = P(y_2|x_1) = \alpha$, where α is referred to as the *BSC crossover probability* i.e. noise within the channel effects the waveform which represents the binary digit 1 in the same way as it effects the waveform which represents a binary digit 0. Thus we find

$$P_e = \alpha P(x_1) + \alpha P(x_2) = \alpha\left[P(x_1) + P(x_2)\right] = \alpha \qquad (3.2)$$

where we make use of the fact that $P(x_1) + P(x_2) = 1$ i.e. the symbol which is transmitted must be either x_1 or x_2. Of course the probability α depends on the particular modulation scheme, the physical channel and the corresponding demodulator. Notice that P_e does not depend on $P(x_1)$ and $P(x_2)$ only if the channel is symmetric i.e. $P(y_1|x_2) = P(y_2|x_1)$. Finally, given the fact that $P(y_1|x_2) + P(y_2|x_2) = 1$ and $P(y_2|x_1) + P(y_1|x_1) = 1$, we have for a BSC, $P(y_1|x_1) = P(y_2|x_2) = 1 - \alpha$.

Example : BSC

(a) How would you determine the BSC crossover probability for a given communication system ? You may assume a binary source.

(b) If the sequence of binary digits 001 is transferred over a BSC with $\alpha = 0.1$, what is the probability that the received sequence is 100.

Solution

(a) One possible solution is to send an all zero sequence i.e. 0000.... and count the number of ones received. In this case, α is given by the number of ones received divided by the number of zeros transmitted.

(b) Since the channel is a DMC, each output symbol depends only on the corresponding input.

Input sequence	0	0	1
Output sequence	1	0	0
Probability	α	$1 - \alpha$	α

The probability $(100 \mid 001) = P(1|0)P(0|0)P(0|1) = (1 - \alpha)\alpha^2$.

3.2.2 Simulation : DMC

SIMULATION **Bsc:** The DCS shown in Fig. 3.5 is modeled. Binary digits output from a binary source enter a DMC for which $M = Q = 2$. All the forward transition probabilities within this DMC can be controlled. For example, if $P(y_2|x_1)$ is set equal to $P(y_1|x_2)$, then this DMC becomes a BSC. Simulation and theoretical results are compared for $P(y_1)$ and $P(y_2)$, which are respectively given by

$$P(y_2) = P(y_2|x_1)P(x_1) + P(y_2|x_2)P(x_2) \qquad (3.3)$$

and

$$P(y_1) = P(y_1|x_1)P(x_1) + P(y_1|x_2)P(x_2) \qquad (3.4)$$

For $\alpha = 0.1$ and a BSC, simulation results are shown in Fig. 3.6. The simulation results may be improved by increasing the number of binary digits sent through the DCS.

- **Experiment:** What is the condition for maximum noise within the channel ? Modify the code to implement a DMC in which $M = 2$ and $Q = 3$.

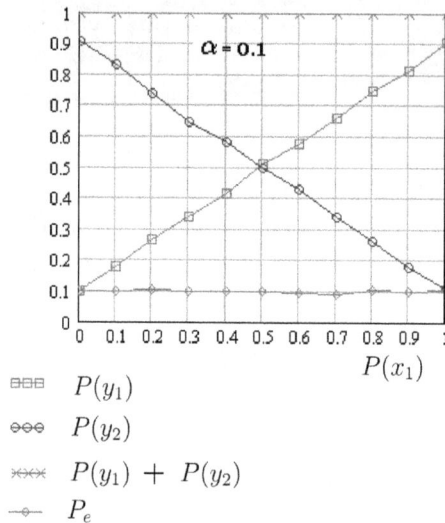

$$\Box\Box\Box \quad P(y_1)$$
$$\ominus\ominus\ominus \quad P(y_2)$$
$$\times\!\!\times\!\!\times \quad P(y_1) \; + \; P(y_2)$$
$$-\!\!\ominus\!\!- \quad P_e$$

Figure 3.6: BSC simulation results.

3.3 Introduction to Error-Control Coding

Error Control Coding

- Introduction to the concept of error-control coding

- Repetition code

- Code rate

In the previous section, it was shown that the probability of an error in a binary digit P_e at the receiver is dictated by the DMC forward transition probabilities. For example, for a BSC, $P_e = \alpha$. To reduce P_e, we could simply increase the transmitter power. But this may not always be desirable. For example, for a mobile (or cell) phone, the necessary battery size and weight may be unacceptable. The most desirable solution is to introduce *error-control coding* (also referred to as *channel coding*) within the DCS, as shown in Fig. 3.7. Error-control coding could be used to only *detect* errors within a given received block. If an error is detected, then an *automatic repeat request* (ARQ) signal is sent back to the transmitter (assuming a duplex link between the transmitter and receiver) to retransmit that block. However, we shall restrict ourselves to *forward error-correction* (FEC) techniques, in which extra binary digits are added by a *channel encoder* prior to transmission, to create a code structure which is resilient to any errors that may occur within the channel. A *channel decoder* is used at the receiver to exploit this code structure to *correct* as many channel errors as possible and subsequently recover the information binary sequence or stream.

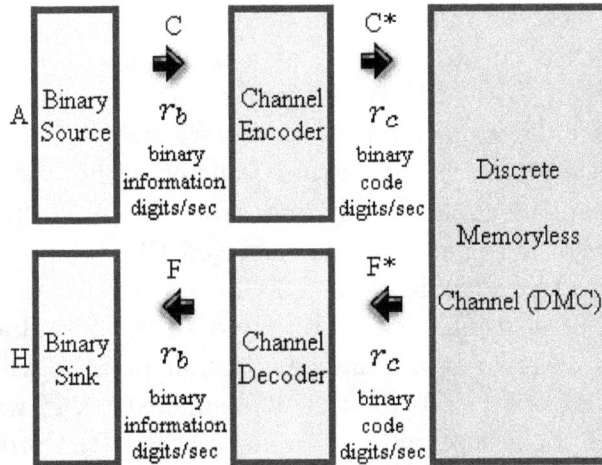

Figure 3.7: Block diagram of a DCS with error-control coding.

In general, k binary digits enter a channel encoder and n binary digits are output. Hence, the *code rate* R_{code} of the encoder is defined by $R_{code} = \frac{k}{n}$. If r_b binary digits/sec enter and r_c binary digits/sec exit from the channel encoder as shown in Fig. 3.7, then the code rate is also given by $R_{code} = \frac{r_b}{r_c}$.

3.3.1 Repetition Code

▶ **Repetition Codes**

- Theoretical and simulated performance
  of a rate $\frac{1}{n}$ repetition codes over a BSC

A *repetition code* is the simplest FEC error control code. The encoding algorithm is to output a *code word* which consists of the input binary digit replicated n times, where n is an odd number greater than or equal to 3 i.e. the code rate $R_{code} = \frac{1}{n}$. For example, if the input to the encoder is a binary digit 1 and $n = 3$, the output code word is 111. A binary digit within this code word is referred to as a *code digit*.

The decoding algorithm is to count the number of binary digit ones and zeros within the received word. If the number of ones is greater than the number of zeros, then output a binary digit one, otherwise output a binary digit zero. Table 3.2 illustrates the encoding and decoding process by example for $n = 3$ for the input information binary stream 1011. Referring to Table 3.2, notice that although five errors occurred within the binary stream that passed through the channel, only one error occurred in the received information binary stream i.e. we have achieved FEC error-control coding. However, if two or more code digits are corrupted, then the decoder will output an incorrect decoded information binary digit. Over a BSC with $n = 1$, the probability of an error in a binary digit P_e at point F will of course be equal to the BSC crossover probability. As n is increased, P_e will be reduced. The rule-of-thumb for acceptable performance is that the probability of an error in a binary digit at point F should be less than 10^{-5}.

Point					Description
C Binary Source Output	1	0	1	1	Information binary digits
C* Code word	111	000	111	111	Code word sequence
F* Channel Output	110	110	011	101	Assumed noisy received words
F Decoder Output	1	1	1	1	Decoded information binary digits

Table 3.2: An example of the binary digits at various points within the DCS.

3.3.2 Information Throughput

▷ Information Throughput

- Throughput concept

- Increase in the BSC crossover probability

Suppose it takes one second for a binary digit to be transferred through the DMC. In this case, without the use of error-control coding, it would take 10 seconds to transfer 10 information binary digits across a DMC. If a rate 1/3 repetition error-control code is used without altering the rate at which code digits are transferred through the DMC, then it would take 30 seconds to transfer the 30 code digits across the DMC. It now takes longer to transfer the information through the DMC, i.e. the *information throughput* has been reduced! A reduction of information throughput is **not** acceptable. As we shall see in section 7.8, the n code digits are forced to occupy the same time-space as the k input information digits to ensure the information throughput remains unchanged i.e. in terms of our repetition code rate 1/3 example, it will still take 10 seconds to transfer 30 code digits through the channel. The penalty incurred is that the level of noise within the channel effectively increases. However, with the use of error-control coding, the overall probability of an error P_e in a binary digit at point F in Fig. 3.7 can easily be reduced below 10^{-5}. It will become apparent in later chapters that to reduce P_e without a reduction in information throughput requires a complex decoder. One consequence of a using a complex decoder is that it does take time to decode. Therefore, in addition to the signal power and channel bandwidth, decoding time is also a precious resource.

3.3.3 Simulation : Repetition Code

⭐ SIMULATION **RepetitionCode:** The DCS shown in Fig. 3.8 is modeled. Binary digits are processed through the DCS until a specified number of errors N_{errors} occur at point F. The information throughput is allowed to be reduced by a factor of R_{code}.

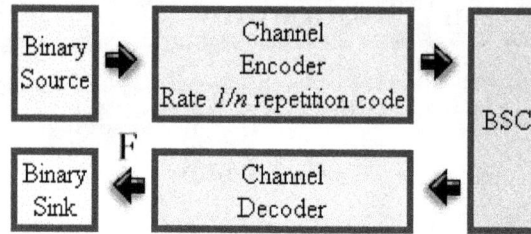

Figure 3.8: Simulated DCS : Rate $1/n$ repetition code over a BSC.

It can be shown (refer to problems section) that P_e at point F is given by

$$P_{rep}(n, \alpha) = \sum_{i=\frac{(n-1)}{2}+1}^{n} \frac{n!}{(n-i)!i!}\alpha^i (1 - \alpha)^{n-i} \qquad (3.5)$$

Using this theoretical expression, Fig. 3.9 shows the correspondence between simulation and theory for two repetition codes over a range of α values. Although the use of a repetition code may at first glance seem to be an appropriate FEC solution, it is rarely used in practice. This is because it can only achieve a low probability of error if $R_{code} \ll 1$ and provided the information throughput of the communication system is dramatically reduced!

- **Experiment**: Modify the simulation code to consider what happens if the information throughput is not allowed to be reduced.

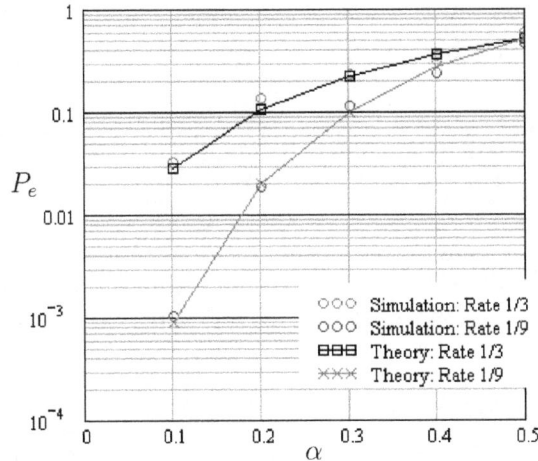

Figure 3.9: Repetition code performance.

3.4 Problems

1. A binary source outputs the following binary sequence:
1010001010001001011101011
(a) What is the probability with which a zero is output from this source based on this sample ?
(b) Check your answer by calculating the probability with which a binary digit one is output and adding this result to your answer of part (a).

2. Consider a digital communication system in which the binary digits 1001 are transferred from a binary source to a binary sink via a BSC channel with crossover probability $\alpha = 0.1$.

(a) What is the probability with which the sequence 1001 is received without error ?
(b) What is the probability with which the received sequence is 0000 ?
(c) What is the probability of one error in any position ?

3. A sequence of N binary digits are transferred over a BSC channel with crossover probability α.
(a) What is the probability of n errors in any position for $n < N$.
(b) What is the probability of n or fewer errors in any position for $n < N$.

4. Consider the DCS shown below.

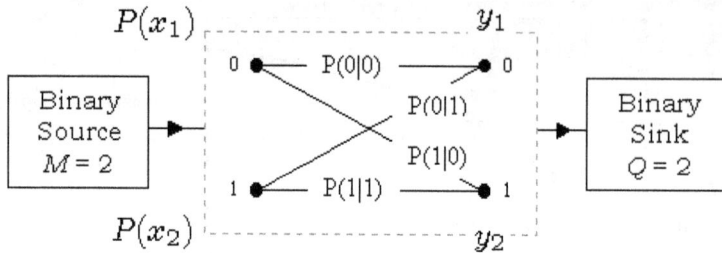

If $P(x_1) = 0.3$, $P(y_1|x_2) = 0.2$, $P(y_2|x_1)= 0.1$, determine the following:

(a) Probability $P(y_1)$.
(b) Probability $P(y_2)$.
(c) Probability that x_1 was transmitted, given that y_1 is received.
(d) Probability that x_2 was transmitted, given that y_1 is received.
(e) Probability that x_1 was transmitted, given that y_2 is received.
(f) Probability that x_2 was transmitted, given that y_2 is received.

5.   In the previous problem if a binary digit 0 is received, what is the probability that this occurred because a binary digit zero was transmitted ?

6. For the DMC shown below, determine an expression for $P(y_1)$, $P(y_2)$ and $P(y_3)$ in terms of $P(x_1)$, $P(x_2)$ and the forward transition probabilities.

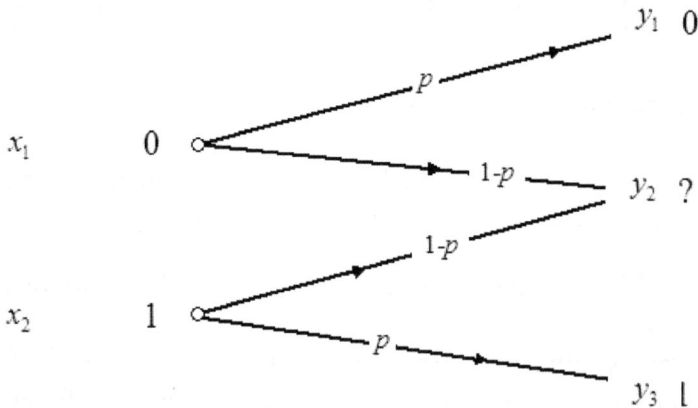

7. Consider the DMC shown below.

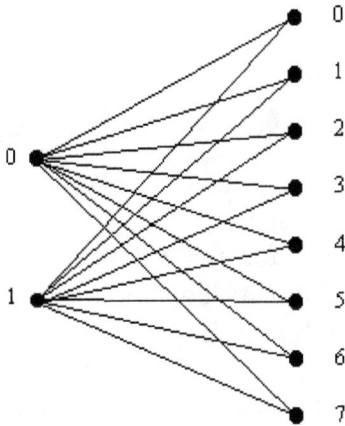

Expressing your answer in terms of the forward transition probabilities (e.g. P(2|0)), determine the following:

(a) Probability that symbol 4 is output from the DMC.

(b) Probability that a binary digit zero was transmitted given that the symbol 4 is received.

8. Consider the use of a rate $\dfrac{1}{n}$ repetition code over a BSC with crossover probability α.

(a) How many channel errors would it take for the decoder to make an error ?

(b) Prove the following formula for the probability of an error in a binary digit over a BSC with crossover probability α.

$$P_{rep}(n, \alpha) = \sum_{i=\frac{(n-1)}{2}+1}^{n} \frac{n!}{(n-i)!i!} \alpha^i (1-\alpha)^{n-i}$$

9. In *Gray* coding, if the input binary word is $(x_1, x_2 \cdots x_n)$, then the output Gray code word $(y_1, y_2 \cdots y_n)$ is given by $y_1 = x_1$ and $y_i = (x_i \oplus x_{i-1})$, for $i = 2 \cdots n$, where $\oplus$ the denotes *modulo-2* addition given by:

Modulo-2 addition		
A	B	A $\oplus$ B
0	0	0
0	1	1
1	0	1
1	1	0

Determine the Gray code for the binary 2^n *tuple* for $n = 2, 3$ and 4. Verify for yourself that each Gray code word differs from its neighbors (previous code word and the next code word) by only one binary digit.

10. A long distance communication system consists of $(N - 1)$ repeaters plus the final receiver, each of which may be modeled as a BSC with a crossover probability α. Determine an expression for the overall probability of an error in a binary digit.

Chapter 4

Information Theory

4.1 Introduction

▷ **Information Theory**

- Source entropy

- Example of a four symbol digital source

- Source information rate

- Predict the performance of an optimum source encoder

In this chapter, we consider the simplified model of a DCS shown in Fig. 4.1. Our aim in this chapter is to establish the minimum average number of binary digits per second required to fully represent the source and the maximum rate at which error-free communication can take place over a given channel.

4.1.1 What is Information ?

A communication system is used to transmit information from a given source to the destination. To utilize the communication channel as efficiently as possible, it is necessary to quantify the amount of "information" conveyed

219

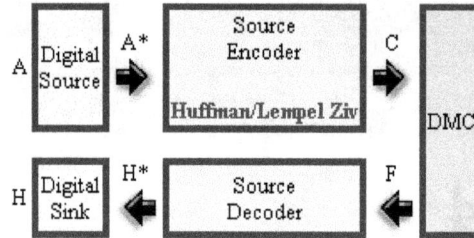

Figure 4.1: Chapter 4 DCS.

by a given source. For example, suppose you attend a lecture in which the speaker gave the same lecture every day, using the exact same words. Although you are listening, information is not being conveyed because you already know what the speaker is going to say! Conversely, more information is conveyed by a less predictable message. By analyzing a digital source in a similar manner, it is possible to establish the average minimum number of binary digits per second required to fully represent an information source and the maximum rate at which reliable communication can take place over a given channel.

4.1.2   Information from a Digital Source

Consider a digital source as shown in Fig. 4.2 which outputs r symbols per second. Let $P(x_i)$ represent the probability with which a symbol x_i is output by the digital source. If the source probabilities $P(x_i)$ remain constant over time and successive symbols are statistically independent of each other, then the source is referred to as a *discrete memoryless source* (DMS). For example, for a binary DMS, the next binary digit to be output does not depend on whether the previous digit output was 1 or 0. Also, the probability with which a binary digit 0 and 1 is output will not change with time. In contrast, the letters of the English alphabet written on this page can be considered to be the output of a digital source with 26 letters if spaces and punctuation are ignored. In this case, the source is not a DMS because for example, in the word "quick", the letter u must follow the letter q.

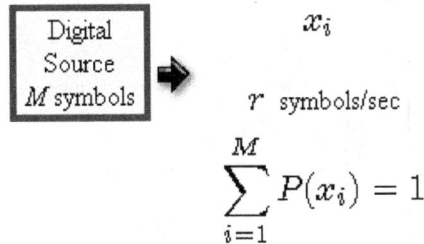

Figure 4.2: Digital source.

Claude Shannon defined the *self-information* I_i associated with a symbol x_i to be given by

$$I_i = \log_2 \frac{1}{P(x_i)} \text{bits} \qquad (4.1)$$

The units of information are "bits", which is not the same as binary digits. The units are however linked to binary digits as follows. Consider a stream of symbols output by a DMS. Any given symbol x_i within this stream conveys $\log_2 \frac{1}{P(x_i)}$ bits of information. If the rate at which symbols are output from the DMS per second is known, then we can determine the average number of bits/sec conveyed by this stream of symbols, referred to as the *information rate R* (bits/sec) of the source. It will become apparent (in the next section) that a source with an information rate of R bits/sec requires a minimum of R binary digits/sec on average to fully represent the DMS.

For a given symbol, say x_a, Fig. 4.3 illustrates how the self-information associated with this symbol depends on $P(x_a)$. Notice how the information conveyed by this symbol goes to zero if it occurs with a probability of 1.

- ⭐ SIMULATION **SelfInformation:** Simulation corresponding to Fig. 4.3. Use this code to understand how to determine the $\log_2(.)$ of a number. Modify the simulation code to discover the information conveyed by a fair dice.

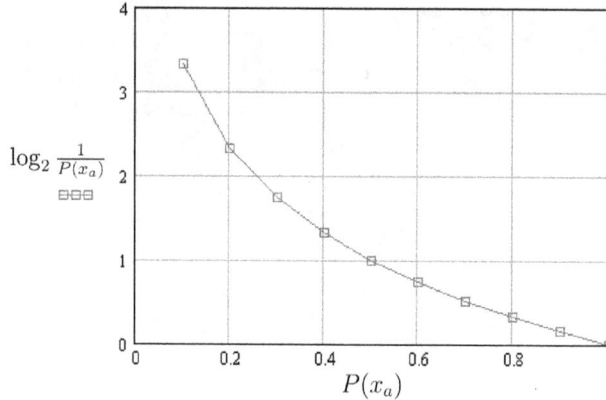

Figure 4.3: Self information.

Based on the definition of self-information, we may state the following properties:

- Since $0 \leq P(x_i) \leq 1$, $I_i \geq 0$ i.e. self information is a positive quantity.

- As $P(x_i) \to 1$, $I_i \to 0$.

- For $I_a > I_b$, $P(x_a) < P(x_b)$.

- For two independent sources, the probability of x_a and x_b being output at the same time is given by $P(x_a)P(x_b)$. Thus $I_{ab} = \log_2 \frac{1}{P(x_a)P(x_b)} = \log_2 \frac{1}{P(x_a)} + \log_2 \frac{1}{P(x_b)} = I_a + I_b$ i.e. as expected, the information conveyed by the joint event of the two independent symbols is equal to the sum of the information conveyed by each symbol.

- If $P(x_1) = P(x_2) = \cdots = P(x_M) = \frac{1}{M}$, then we require a minimum of $\log_2 M$ binary digits to represent each symbol and since $\log_2 M = \log_2 \frac{1}{P(x_i)}$, it is not surprising that the self information of a symbol is defined by $\log_2 \frac{1}{P(x_i)}$. Refer to the example in section 4.1.3 for further details.

⭐ SIMULATION **TwoEvents:** An example of I_{ab}. Experiment with different values for $P(x_a)$ and $P(x_b)$ to get a good feel for the probability of a joint event and I_{ab}.

Our objective is to minimize the number of binary digits required to represent a given source for the efficient utilization of a given communication channel. For a DMS, the average self-information per source symbol, referred to as the *source entropy* $H(X)$, is given by

$$H(X) = \sum_{i=1}^{M} P(x_i)I_i = \sum_{i=1}^{M} P(x_i) \log_2 \frac{1}{P(x_i)} \text{bits per symbol} \qquad (4.2)$$

where X represents the source symbols i.e. $x_1, x_2, x_3, \cdots, x_M$. It will become apparent in the next section that the source entropy indicates the minimum number of binary digits on average required to represent each symbol. If a digital source outputs r symbols per second, then the information rate R of this source is given by

$$R = rH(X) \text{ bits per second} \qquad (4.3)$$

4.1.3 Example: Digital Source

Consider a DMS which can output either one of the digital symbols A, B, C or D. The probability with which each symbol is output and the associated self-information are summarized in Table 4.1. Let $r = 10$ symbols/sec. The source entropy, which is average self-information conveyed per digital symbol, is given by $H(X) = (1/2*1) + (1/4)*2 + (1/8)*3 + (1/8)*3 = 1.75$ bits/symbol. To demonstrate this by simulation, Fig. 4.4 shows a sample of 10 symbols output from the digital source. The self-information conveyed by each symbol is listed next to each symbol. In this simulation, the source entropy is simply the summation of the self-information conveyed by each symbol divided by the number of symbols output. In Fig. 4.4, the source entropy is calculated to be 1.9 bits/symbol. Of course this source entropy estimate will be more accurate if the sample size is increased, as illustrated in Fig. 4.5, in which the source entropy is calculated after n symbols are output by the digital source. As n is increased, the calculated or simulated value converges to the expected value of 1.75 bits/symbol. Given that the source entropy is 1.75 bits/symbol, the information rate $R = rH(X) = 17.5$ bits/sec. If $P(A) = P(B) = P(C) = P(D) = 1/4$, then $H(X) = 2$ bits/symbol. In this case, the information rate $R = 20$ bits/sec.

Symbol	Self-Information
A	1
B	2
C	3
D	3
B	2
A	1
B	2
A	1
D	3
A	1
Total	19
Average	1.9

Source Probabilities

$P(A) = 1/2$
$P(B) = 1/4$
$P(C) = 1/8$
$P(D) = 1/8$

Figure 4.4: Digital source sample output.

Symbol x_i	Probability $P(x_i)$	Self-Information $I_i = \log_2 \frac{1}{P(x_i)}$ bits
A	1/2	1
B	1/4	2
C	1/8	3
D	1/8	3

Table 4.1: Digital Memoryless Source (DMS).

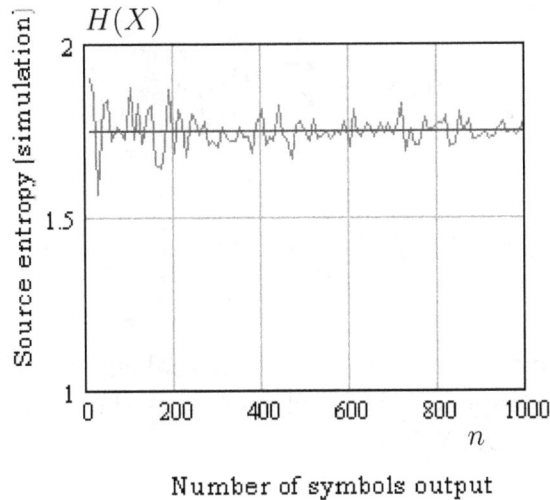

Figure 4.5: Source entropy for a given sample size.

• ⭐ SIMULATION **ABCD source:** The foregoing source entropy calculations are implemented. Try different values for the probability with each symbol is output e.g. what happens if $P(A) = P(B) = P(C) = P(D)$ or if the probability of say symbol A is very much less than any other symbol. Besure to ensure that $P(A) + P(B) + P(C) + P(D) = 1$.

Example : DMS

Show that for a DMS in which all the M symbols are equally likely to be transmitted, the source entropy is given by $\log_2(M)$.

Solution

If all the symbols are equally likely to be output from the source, then $P(x_1) = P(x_2) = \cdots = P(x_M)$. Since $\sum_{i=1}^{M} P(x_i) = 1$, we must have $P(x_i) = \frac{1}{M}$. Hence the source entropy is given by

$$H(X) = \sum_{i=1}^{M} P(x_i) \log_2 \frac{1}{P(x_i)} \tag{4.4}$$

$$= \sum_{i=1}^{M} \frac{1}{M} \log_2 M = M \left(\frac{1}{M} \log_2 M \right) = \log_2 M \text{ bits/symbol} \tag{4.5}$$

Note that the difference between the actual entropy $H(X)$ of a source and the maximum entropy $\log_2 M$ is referred to as the *redundancy* of the source given by $[\log_2 M - H(X)]$.

4.1.4 Source Entropy Bounds

In general, we find that $0 \le H(X) \le \log_2 M$. To prove the lower bound of $H(X) \ge 0$, given that $0 \le P(x_i) \le 1$, we have $\frac{1}{P(x_i)} \ge 1$ and thus $\log_2 \left(\frac{1}{P(x_i)} \right) \ge 0$ as illustrated in Fig. 4.3. Thus $P(x_i) \log_2 \left(\frac{1}{P(x_i)} \right) \ge 0$ and therefore

$$H(X) = \sum_{i=1}^{M} P(x_i) \log_2 \frac{1}{P(x_i)} \ge 0 \tag{4.6}$$

Notice that $P(x_i) \log_2 \frac{1}{P(x_i)} = 0$ if and only if $P(x_i) = 0$ or 1 and that for $P(x_i) = 1$, only the symbol x_i is output by the source for which $H(X) = 0$. To prove the upper bound $H(X) \le \log_2 M$, consider two probability distributions $P(x_i) = P_i$ and $Q(x_i) = Q_i$, where $i = 0, 1, ..., M$ such that $\sum_{i=1}^{M} P_i = 1$ and $\sum_{i=1}^{M} Q_i = 1$. Then

$$\sum_{i=1}^{M} P_i \log_2 \frac{Q_i}{P_i} = \frac{1}{\log_e 2} \sum_{i=1}^{M} P_i \log_e \frac{Q_i}{P_i} \tag{4.7}$$

where we have used the simple relation that $\log_2(z) = \frac{\log_e(z)}{\log_e 2}$. Making use of the inequality $\log_e z \le z - 1$ for $z \ge 0$, where $\log_e z = z - 1$ for $z = 1$ only, and replacing z by $\frac{Q_i}{P_i}$,

$$\sum_{i=1}^{M} P_i \log_e \frac{Q_i}{P_i} \le \sum_{i=1}^{M} P_i \left(\frac{Q_i}{P_i} - 1 \right) = \sum_{i=1}^{M} Q_i - \sum_{i=1}^{M} P_i = 1 - 1 = 0 \quad (4.8)$$

Thus $\sum_{i=1}^{M} P_i \log_2 \frac{Q_i}{P_i} \le 0$, where the equality is true only if $Q_i = P_i$ (i.e. $z = 1$). Now for $Q_i = \frac{1}{M}$, we have

$$\sum_{i=1}^{M} P_i \log_2 \frac{1}{P_i M} = \sum_{i=1}^{M} P_i \log_2 \frac{1}{P_i} + \sum_{i=1}^{M} P_i \log_2 \frac{1}{M} = H(X) + \log_2 \frac{1}{M} \sum_{i=1}^{M} P_i$$
$$(4.9)$$

$$= H(X) + \log_2 \frac{1}{M} = H(X) - \log_2 M \le 0 \quad (4.10)$$

Hence $H(X) \le \log_2 M$, where the equality holds only if the probability of a symbol is $\frac{1}{M}$.

4.1.5 Binary Source

▶ **Information Theory II**

- How to determine the log to the base two of a number

- Closer look at self-information

- Binary source example

- Source entropy of a DMS in which each symbol is equally likely to be output

Consider a binary source which outputs a binary digit zero with probability p. The self-information associated with each symbol is shown in Table 4.2. The source entropy $H(X) = p \log_2\left(\frac{1}{p}\right) + (1-p) \log_2\left(\frac{1}{1-p}\right)$. To concisely represent this expression, it is useful to make use of the *binary entropy function* $\Omega(.)$ defined by

$$\Omega(z) = z \log_2\left(\frac{1}{z}\right) + (1-z) \log_2\left(\frac{1}{1-z}\right) \tag{4.11}$$

Hence, the source entropy of a binary source is simply given by $\Omega(p)$. Figure 4.6 shows how $\Omega(p)$ depends on the probability p.

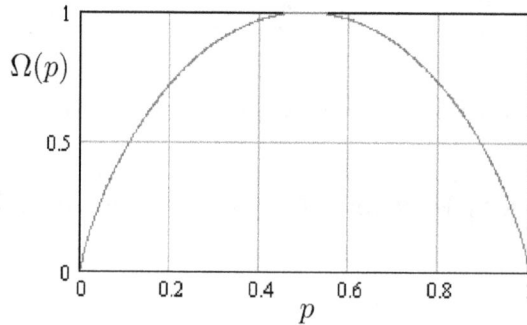

Figure 4.6: Binary source entropy.

Symbol	Probability	Self-Information
0	p	$\log_2\left(\frac{1}{p}\right)$
1	$1-p$	$\log_2\left(\frac{1}{1-p}\right)$

Table 4.2: Binary source.

- ⭐ SIMULATION **BinarySourceEntropy:** Simulation corresponding to Fig. 4.6.

As expected the binary source entropy, which is the average self-information per binary digit, is zero if either an all-zero sequence 00000.... ($p = 1$) or an all-one sequence 11111.... ($p = 0$) comes out of the binary source. An important point to bear in mind is that the binary source entropy is at a maximum for $p = 0.5$. We shall make use of this feature in later sections.

4.1.6 DMS Extension

The second order extension of the DMS in Table 4.1 is shown in Table 4.3. Namely, the symbols are considered in blocks of two. Since the source symbols are statistically independent of each other, the probability of an extended symbol, say AB, is equal to $P(x_A)P(x_B) = 1/2 * 1/4 = 1/8$. In general, for n symbols in a block, the entropy of the extended source is equal to $nH(X)$. In this example, the second order extended source entropy is equal to $2 * 1.75 = 3.5$ bits/extended symbol. Refer to the problems section for another example.

4.1.7 Simulation: Digital Source

⭐ SIMULATION **DigitalSource:** The four symbol digital source with source probabilities as stated in Table 4.1 is modeled. Each symbol output by the digital source is counted and used to confirm the source probabilities. Also the information rate from each source is determined.

- **Experiment:** Modify the simulation code to implement its second order extension.

Extended Symbol	Probability
AA	1/2*1/2
AB	1/2*1/4
AC	1/2*1/8
AD	1/2*1/8
BA	1/4*1/2
BB	1/4*1/4
BC	1/4*1/8
BD	1/4*1/8
CA	1/8*1/2
CB	1/8*1/4
CC	1/8*1/8
CD	1/8*1/8
DA	1/8*1/2
DB	1/8*1/4
DC	1/8*1/8
DD	1/8*1/8

Table 4.3: Extended source.

4.2 Source Coding

Source coding (or *data compression*) is necessary to efficiently utilize a given communication channel. For example, source coding will reduce the time it takes to fax a given page over a telephone line by representing the scanned image with the least number of binary digits as possible. Another example is within your cell phone, where the speech quality is reduced with a focus on clarity and not fidelity. Notice that source coding need not be loss-less. For example, an MP3 file representation of an audio CD song, where the quality has been slightly reduced to efficiently store the song. We shall however, restrict ourselves to loss-less source coding in this chapter.

Recall that in our model of a DCS in Fig. 3.3, a digital source is the combination of the information source and the format block. Although source coding could be considered to be a part of the format block, we shall view this operation to be separate as illustrated in Fig. 4.7. Namely, a source encoder is used to reduce the average number of binary digits required to transmit the digital source output. Notice that the combination of the digital source and source encoder is equivalent to a binary source. In this and the next section, we shall assume that the source probabilities are known. Thereafter, in section 4.4, we shall consider a more practical approach to source coding.

Consider a simple source encoder which replaces each input digital symbol with a binary code word (i.e. string of binary digits). For example, consider a digital source which outputs one of four possible symbols A, B, C or D as listed in Table 4.4. The probability with which each symbol is output from the source, the corresponding self-information and the number of binary digits in per code word L_i are also presented in Table 4.4. For example, if the input to the source encoder is the symbol C, then the output of the source encoder is the binary code word 011. At the receiver, the binary digits must be converted back into the digital symbols by the source decoder. This implies that the code used by the source encoder must be *uniquely decipherable*.

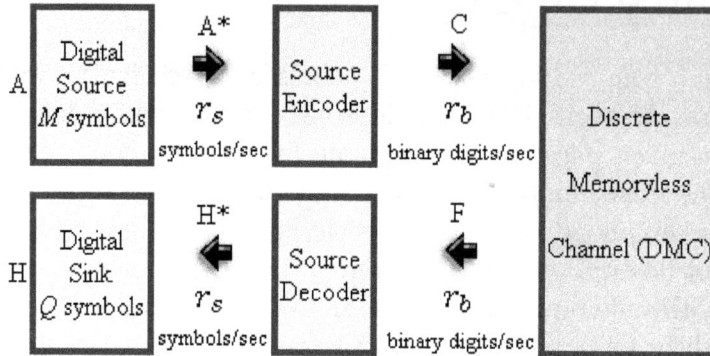

Figure 4.7: DCS with source coding.

The average number of binary digits per symbol output from the source encoder $\overline{L}$ is given by

$$\overline{L} = \sum_{i=1}^{M} P(x_i)L_i \qquad (4.12)$$

i.e. this is the average code word length. In Table 4.4, the average code length $\overline{L} = 1.875$ binary digits. If $r_s = 10$ symbols/sec, then the **average number of binary digits per second output from the source encoder is $r_b = r_s\overline{L}$** = 18.75 binary digits/sec.

Symbol	Probability $P(x_i)$	Self-Information $I_i = \log_2 \frac{1}{P(x_i)}$ bits	Code I	Length L_i
A	1/2	1	0	1
B	1/4	2	01	2
C	1/8	3	011	3
D	1/8	3	0111	4

Table 4.4: Four symbol digital source.

4.2.1   Shannon's Source Coding Theorem

Source Coding

- Example of a simple source encoder

- Average source code word length

- Shannon's source coding theorem

To make efficient use of a communication channel, we wish to minimize the number of binary digits per second necessary to fully represent the source i.e. for a given digital source, we wish to find the code for which $\overline{L}$ is a minimum. This problem is addressed by Shannon's *source coding theorem*, which simply states that $\overline{L}$ must be such that

$$\overline{L} \geq H(X) \tag{4.13}$$

To illustrate this theorem by example, lets refer back to Table 4.4. In this case, the source entropy $H(X) = (1/2) + (2/4) + (3/8) + (3/8) = 1.75$ bits/symbol, however recall that $\overline{L} = 1.875$ binary digits. Clearly, code I is not optimum because ideally $\overline{L} = H(X)$. Hence, we may define the *efficiency* of a source code to be given by

$$\eta = \left(\frac{H(X)}{\overline{L}} 100 \right) \% \tag{4.14}$$

For our ongoing example, the efficiency of code I is $\eta = \left(\frac{H(X)}{\overline{L}} 100 \right) = \frac{1.75}{1.875} 100 = 93\%$. If we multiply both sides of equation 4.13 by r_s and given that $r_b = r_s \overline{L}$ and that the information rate from the source $R = r_s H(X)$ bits/sec, then Shannon's source coding theorem may be expressed as

$$r_b \geq R \tag{4.15}$$

i.e. for an ideal source code, we expect $r_b = R$ binary digits per second on average. The significance of the units "bits" is now clear. For example, recall that the information rate of the DMS summarized in Table 4.1 was 17.5 bits/sec. This means that an optimum source encoder would output on average, 17.5 binary digits/sec.

Example : Source Coding Theorem.

Show that $r_b \geq R$ if the information rate output from a digital source is equal to the information rate output from its corresponding source encoder.

Solution

The combination of a digital source and a source encoder may be viewed as a binary source as illustrated in Fig. 4.8.

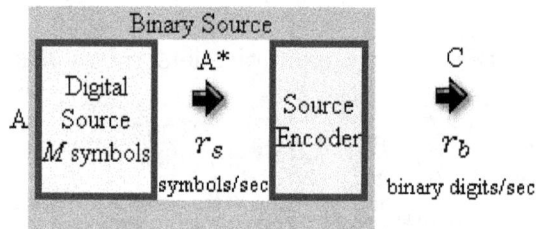

Figure 4.8: Equivalent binary source.

The information rate at point A* = information rate at point C. Equivalently,

$$r_s H(X) = r_b \Omega(p) \qquad (4.16)$$

and therefore

$$\Omega(p) = \frac{r_s H\left(X\right)}{r_b} \tag{4.17}$$

From Fig. 4.6, given that $\Omega(p) \leq 1$, $\frac{r_s H(X)}{r_b} \leq 1$, or equivalently $r_b \geq r_s H(X)$. Given that $R = r_s H\left(X\right)$, we get the expected result of $r_b \geq R$.

4.2.2 Ideal Source Encoder

▶ **Source Coding II**

- Summary of Shannon's source coding theorem

- Source encoder efficiency

The efficiency of an ideal source encoder is 100%. For example, consider a digital source in which each symbol is equally likely to be output as shown in Table 4.5. Using the source code II, the average code word length $\overline{L} = 2$ binary digits is equal to the source entropy $H\left(X\right) = 2$ bits/symbol i.e. efficiency is 100 %.

Symbol	Probability $P\left(x_i\right)$	Self-Information $I_i = \log_2 \frac{1}{P(x_i)}$ bits	Code II	Length L_i
A	1/4	2	00	2
B	1/4	2	01	2
C	1/4	2	10	2
D	1/4	2	11	2

Table 4.5: Ideal source.

4.2.3   Equivalent Binary Source

▶ **Source Coding Example**

- To determine an expression for the probability
  of a binary digit zero at the output of the
  source encoder

Referring back to Fig. 4.8, we shall now determine an expression for the probability of a binary digit zero at point C. Recall that the average code length is given by $\overline{L} = \sum_{i=1}^{M} P(x_i)L_i$. Let N represent the number of symbols output from the digital source. Then, the average number of binary digits output from the source encoder is equal to $N\overline{L} = \sum_{i=1}^{M} NP(x_i)L_i$. However $P(x_i) = \frac{N_i}{N}$, where N_i is the number of times the symbol x_i is output from the digital source. Therefore $N\overline{L} = \sum_{i=1}^{M} NP(x_i)L_i = \sum_{i=1}^{M} N_iL_i$. Hence, the number of binary digit zeros N_{zeros} output from the digital source is given by

$$N_{zeros} = \sum_{i=1}^{M} N_i L_i^{(zeros)} \tag{4.18}$$

where $L_i^{(zeros)}$ is the number of binary digit zeros within the code word for the symbol x_i. Replacing $N_i = NP(x_i)$,

$$N_{zeros} = \sum_{i=1}^{M} NP(x_i) L_i^{(zeros)} \tag{4.19}$$

and therefore, the probability of a binary digit zero p_{zero} output from the equivalent binary source is given by

$$p_{zero} = \frac{\sum_{i=1}^{M} NP(x_i)L_i^{(zeros)}}{N\overline{L}} = \frac{\sum_{i=1}^{M} P(x_i)L_i^{(zeros)}}{\overline{L}} \tag{4.20}$$

- ⭐ SIMULATION **ProbZero:** The four symbol (A, B, C, D) digital source is considered to determine for three different source codes, the source entropy, average code length, information rate, efficiency and the probability p_{zero} for $P(x_i) = \frac{1}{4}$ and for the probabilities in Table 4.4 side-by-side for comparison.

4.3 Huffman Code

The Huffman code is an example of a *compact* code, which minimizes the average number of binary digits required to represent the source symbols. This is achieved by assigning shorter code words to the most frequently occurring symbols.

4.3.1 Huffman Algorithm

▶ **Huffman**

- Huffman source coding algorithm
  explained via an example

The Huffman algorithm is best explained by example and is easy to follow if you watch the corresponding video clip. For a given DMS, the Huffman algorithm generates the optimum source code. For example, the Huffman code for the digital source in Table 4.4 is presented in Fig. 4.9. Notice from Fig. 4.9 that the most probable symbol is assigned the least number of binary digits, which is what we would expect intuitively. Even in the early days of Morse code, the most probable letter "e" was represented by a "dot" and less probable symbols by a combination of "dot" and "dash".

The steps of the Huffman algorithm are illustrated in Figs. 4.10 to 4.14. A brief explanation of the algorithm is as follows.

1. List the symbols with their probabilities in a descending order.

2. Add the probabilities of the last two entries on this list to create the probability p. This probability p will be used in the next step. Copy the list to the adjacent column and delete the last two entries from the list.

3. Starting from the top of the list, insert p at the position for which the probability **below** is either less than or equal to p.

4. Repeat steps 2 and 3 until p is equal to 1. At this stage, we could take the approach of viewing the entries as corresponding to the branch ends of a tree and the deletion of two entries as the merging of two branches. An equivalent view, which is more suited to implementing the algorithm in software, is to create a corresponding number table as in the next step. Steps [1]-[4] to follow are illustrated in Fig. 4.10.

5. List the symbols in the order of descending probabilities. Starting from the top of the list, number each symbol with an integer starting with the number 0. For example, if the symbols A, B, C, D are in the order of descending probabilities, then they are labeled as 0, 1, 2, 3, respectively. These numbers 0 to 3 serve only as labels to keep a track of how the probabilities are manipulated. The value of each number has no significance.

6. Using the table created after step [4], create a similar table using only the integer numbers. For example, if the probabilities p_1 and p_2 (labeled say n_1 and n_2) are added to create $p = p_1 + p_2$, the corresponding integer table entry would be $n_1 + n_2$. Steps [5]-[6] are illustrated in Fig. 4.10.

7. Once the new table is complete, label the last two entries of each column with the binary digits 1 and 0 as illustrated in Fig. 4.11.

8. For each of the last two entries, if the table entry $n_1 + n_2$ corresponds to a binary digit 1, then the code digit of the symbols corresponding to the numbers n_1 and n_2 is 1. Similarly, if the table entry $n_1 + n_2$ corresponds to a binary digit 0, then the code digit of the symbols corresponding to the numbers n_1 and n_2 is 0. By starting from the right most column and working backwards to the first column, the code digits of the code word corresponding to a given symbol are written down. This process is illustrated in Figs. 4.12, 4.13 and 4.14.

- ★ SIMULATION **Huffman:** For an implementation of the Huffman source coding algorithm.

Symbol	Probability	Huffman Code
A	0.5	0
B	0.25	10
C	0.125	111
D	0.125	110

Figure 4.9: Digital source symbols, their probabilities and the corresponding Huffman code.

4.3.2 Simulation : Huffman Source Encoder

★ SIMULATION **InformationRate:** The digital source and source encoder of Tables 4.4 and 4.5 are modeled as shown in Fig. 4.15. Simulation

A	0.5		0.5		0.5	──	1.0
B	0.25		0.25		0.5		
				0.5			
C	0.125		0.25				
		0.25					
D	0.125						

A	0		0		2+3+1	2+3+1+0
B	1		2+3		0	
				2+3+1		
C	2		1			
		2+3				
D	3					

Figure 4.10: Steps [1] to [4].

A	0.5		0.5		0.5
B	0.25		0.25		0.5
C	0.125		0.25		
D	0.125				

A	0		0		2+3+1	1
B	1		2+3	1	0	0
C	2	1	1	0		
D	3	0				

Figure 4.11: Step [5] to [6].

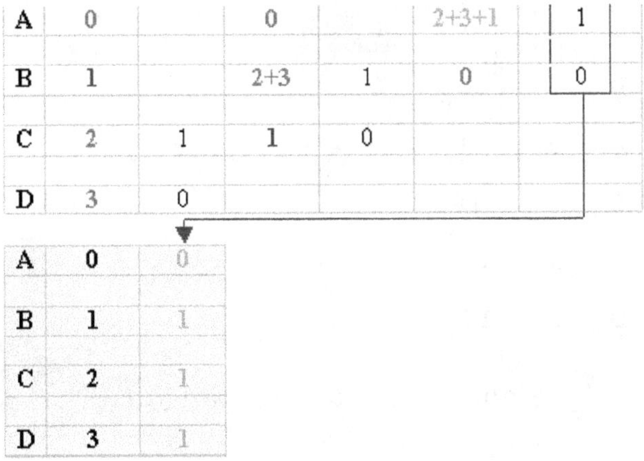

A	0		0		2+3+1	1
B	1		2+3	1	0	0
C	2	1	1	0		
D	3	0				

A	0	0
B	1	1
C	2	1
D	3	1

Figure 4.12: Step [8], part (a).

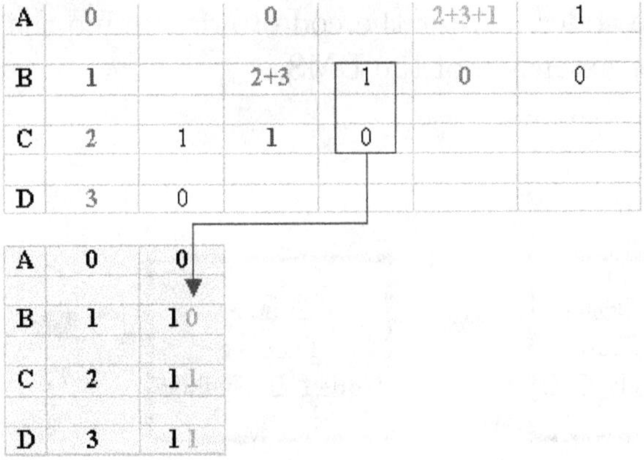

A	0		0		2+3+1	1
B	1		2+3	1	0	0
C	2	1	1	0		
D	3	0				

A	0	0
B	1	1 0
C	2	1 1
D	3	1 1

Figure 4.13: Step [8], part (b).

A	0		0		2+3+1	1
B	1		2+3	1	0	0
C	2	1	1	0		
D	3	0				
A	0	0				
B	1	10				
C	2	111				
D	3	110				

Figure 4.14: Step [8], part (c).

results are compared with the expected theoretical results for the average code length and efficiency of the source code. In both cases, the source probabilities and the corresponding source code may be changed.

- **Experiment:** Verify that the simulation results can be improved by increasing the number of symbols output from the digital source. Try other source codes. Alter the code to determine the efficiency of the second order extension of the DMS.

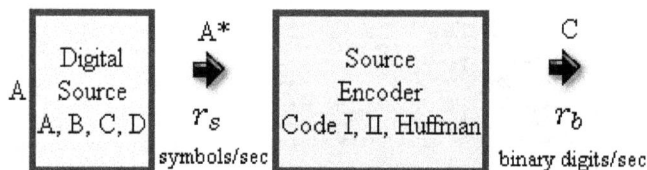

Figure 4.15: Equivalent binary source simulation.

4.4 Universal Source Coding

In the previous section we considered the Huffman source coding algorithm which can only be used if the probability with which each source symbol occurs is known. However, for a real world application, a digital source is typically not a DMS and the source probabilities are difficult to estimate accurately. The solution is to use a *universal source coding algorithm* i.e. one which does not depend on the source statistics. One example of a universal source coding algorithm is the popular *Lempel-Ziv* algorithm. Variations of this algorithm are used to develop software to compress files on a computer. Other applications of the Lempel-Ziv algorithm include the GIF image and TIFF image compression methods.

4.4.1 Lempel-Ziv Algorithm

Lempel-Ziv Algorithm

- The algorithm explained via an example
  in which a sample stream of binary
  digits are encoded and decoded

The first phase of the Lempel-Ziv algorithm is to segment the input binary sequence into binary words which when concatenated together, form the original binary sequence. For example, suppose the input binary sequence to the Lempel-Ziv source encoder is the sequence 01101101110100110 as illustrated in Fig. 4.16. To separate this sequence into words, we begin by storing the binary digits 0 and 1 as shown in step 1. Starting from the beginning of this input sequence, the shortest possible words are 0, 01, 011, 0110, etc. Of these, 0 is already stored. Since the word 01 is not stored, it is extracted from the input sequence and stored, as shown in step 2.

As another example, consider step 4 in which the sequence to be scanned is 01110100110. The shortest possible words are 0, 01, 011, 0111, etc. Of these, only 011 is not stored and so it is extracted and stored as shown in step 5. This process is repeated until all the digits within the input binary sequence have been extracted and stored. For the example in Fig. 4.16, the input sequence 01101101110100110 has been segmented into the binary words 01, 10, 11, 011, 101, 00, 110.

STEP	STORED WORDS	SEQUENCE TO BE SCANNED
1	0, 1	01101101110100110
2	0, 1, 01	101101110100110
3	0, 1, 01, 10	1101110100110
4	0, 1, 01, 10, 11	01110100110
5	0, 1, 01, 10, 11, 011	10100110
6	0, 1, 01, 10, 11, 011, 101	00110
7	0, 1, 01, 10, 11, 011, 101, 00	110
8	0, 1, **01, 10, 11, 011, 101, 00, 110**	

Figure 4.16: Phase I of the Lempel-Ziv algorithm.

The second phase of the algorithm, illustrated in Fig. 4.17, is as follows:

1. Each binary word in step 8 of Fig. 4.16 is labeled with an index number starting from the number one. e.g. the binary word 110 corresponds to the index number 9. The last digit within each binary word is referred to as the *innovation* symbol e.g. within the binary word 110, the innovation symbol is 0.

2. The innovation symbol in a given word is separated e.g. the binary word 110 is split into 11, 0.

3. The index number corresponding to the group of digits other than the innovation symbol is identified e.g. when 110 is split into 11, 0 the newly grouped word is 11, which corresponds to the index number 5.

4. This index number is then replaced with its binary equivalent word and concatenated with the innovation symbol to create the output code word e.g. The binary equivalent of 5 is 101. This is then concatenated with the innovation symbol 0 to create the output code word 1010.

STEP			ENCODER						
1	1	2	3	4	5	6	7	8	9
2	0	1	0 1	1 0	1 1	01 1	10 1	0 0	11 0
			↓	↓	↓	↓	↓	↓	↓
3			1	2	2	3	4	1	5
4			001 1	010 0	010 1	011 1	100 1	001 0	101 0
	code		0011	0100	0101	0111	1001	0010	1010

Figure 4.17: Phase II of the Lempel-Ziv algorithm.

The operation of the source decoder is exactly the inverse operation of the encoder, as illustrated in Fig. 4.18.

1. The innovation symbol is separated from the code word e.g. 0010 is separated into 001, 0.

2. The group of binary digits other than the innovation symbol within the code word are converted into the equivalent decimal number e.g. if 1010 is separated into 101, 0 then the decimal number corresponding to 101 is 5.

3. This decimal number is then used as the index number to identify the corresponding binary word, which is then concatenated with the innovation symbol to recover the original binary word e.g. index 5 corresponds to the binary word 11, which concatenated with 0 recovers the original binary word 110. Notice how the correspondence between the index 5 and the code word 11 is determined from previously decoded binary code words, thereby enabling the source decoder to decode on-the-fly.

Unlike the Huffman code, the Lempel-Ziv creates fixed-length codes e.g. the length of each code word in Fig. 4.17 is 4 binary digits. In Fig. 4.17, 17 binary digits are input and 28 digits are output by the encoder. The

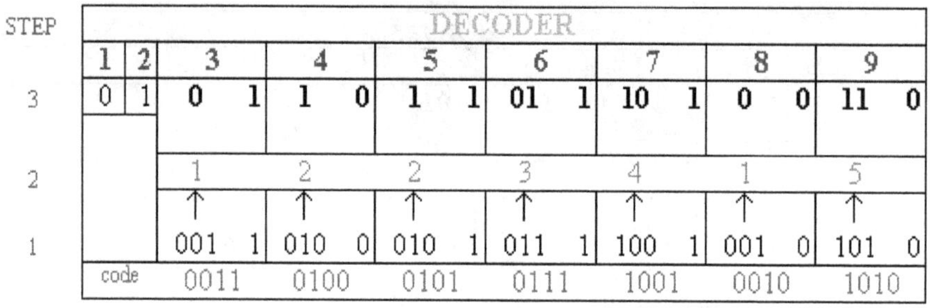

STEP	DECODER															
	1	2	3		4		5		6		7		8		9	
3	0	1	0	1	1	0	1	1	01	1	10	1	0	0	11	0
2			1		2		2		3		4		1		5	
			↑		↑		↑		↑		↑		↑		↑	
1			001	1	010	0	010	1	011	1	100	1	001	0	101	0
code			0011		0100		0101		0111		1001		0010		1010	

Figure 4.18: Lempel-Ziv source decoder.

compression factor in this case of 28/17 is greater than 1. Of course, the compression factor will only be reduced below 1 for a large input sequence. Ideally, most of the redundant binary digits would be removed thereby making the probability of a binary zero and one within the compressed binary stream to be close to 0.5.

4.4.2   Simulation : Lempel-Ziv Algorithm

⭐ SIMULATION **LempelZiv:** The Lempel-Ziv source encoder and decoder are implemented as shown in Fig. 4.19 to illustrate the encoding and decoding process for an input string of binary digits. The number of binary digits input to the encoder, the length of each code word and the probability of a binary digit 0 within the input sequence to the encoder may be altered to observe the trade-offs involved. For example, Fig. 4.20 shows the variation of the compression factor with the probability of a binary digit zero output from the binary source. A total of 4000 binary digits are input to the encoder with 8 binary digits per code word. As expected, maximum compression occurs when the input sequence is an all-ones sequence.

- **Experiment:** Use this code to compress a file. Compare your results with the use of the well known software winzip.

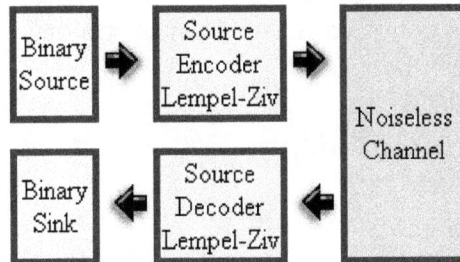

Figure 4.19: Simulated DCS : Lempel-Ziv source coding over a noiseless channel.

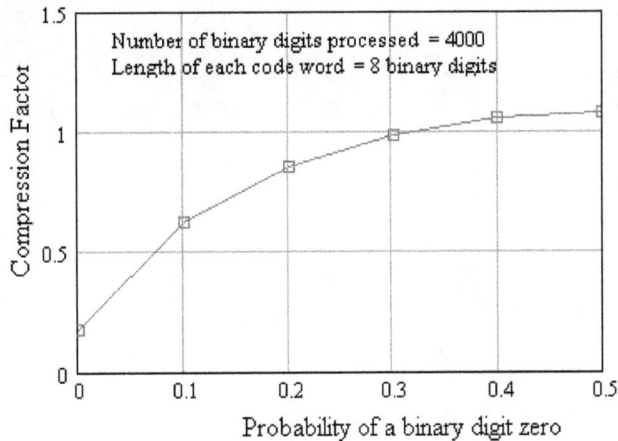

Figure 4.20: Lempel-Ziv source encoder performance.

4.5   Channel Capacity

In section 4.1, the concepts of self-information, source entropy and information rate were used to investigate source coding techniques. In this section,
we continue with more concepts in information theory.

4.5.1   Mutual Information

▷ **Mutual Information**

- Definition of mutual information

- What if the channel is an error free channel

  or a very noisy channel?

Mutual Information $I(x_i, y_j)$ is the information transferred across the
channel when symbol x_i is transmitted and y_j is received and is defined by

$$I(x_i, y_j) = \log_2 \frac{P(x_i|y_j)}{P(x_i)} \text{bits} \qquad (4.21)$$

Using Bayes rule for conditional probabilities, the transition probability
$P(x_i|y_j)$ may be expressed in terms of the corresponding forward transition
probability as

$$P(x_i|y_j) = \frac{P(y_j|x_i)P(x_i)}{P(y_j)} \qquad (4.22)$$

Thus, we may write the mutual information $I(x_i, y_j)$ as

$$I(x_i, y_j) = \log_2 \frac{P(y_j|x_i)}{P(y_j)} \text{ bits} \qquad (4.23)$$

- ⭐ SIMULATION **MutualInformation:** The mutual information of a BSC channel analyzed. Experiment to establish what happens if there is no noise (BSC cross over probability $\alpha = 0$) or maximum noise ($\alpha = 0.5$) within the channel.

For a noiseless channel, $P(x_i|y_j) = 1$ for $i = j$, otherwise $P(x_i|y_j) = 0$ e.g. a noiseless channel for $M = Q = 2$ is illustrated in Fig. 4.21. Hence, $I(x_i, y_j) = \log_2 \frac{1}{P(x_i)}$. As expected, the information transferred across the channel is equal to the self-information of symbol x_i.

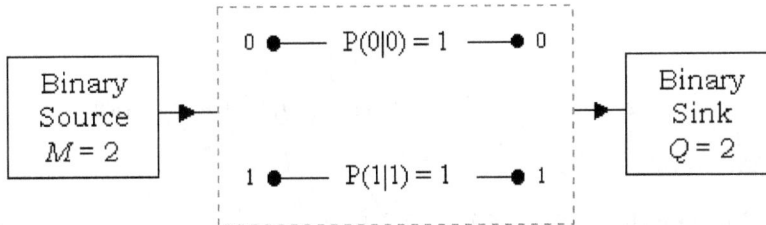

Figure 4.21: Noiseless binary channel.

For a very noisy channel, x_i and y_j are independent. In this case, $P(x_i|y_j) = \frac{P(x_i)P(y_j)}{P(y_j)} = P(x_i)$. Thus $I(x_i, y_j) = \log_2 \frac{P(x_i)}{P(x_i)} = \log_2 1 = 0$. Equivalently in terms of forward transition probabilities, $P(y_j|x_i) = \frac{P(y_j, x_i)}{P(x_i)} = \frac{P(y_j)P(x_i)}{P(x_i)} = P(y_j)$ and therefore $I(x_i, y_j) = \log_2 \frac{P(y_j|x_i)}{P(y_j)} = \log_2 \frac{P(y_j)}{P(y_j)} = 0$. As expected, no information is transferred across the channel.

- ⭐ SIMULATION **Noisy:** The BSC channel is considered once again to show that $P(x_i|y_j) = P(x_i)$ for $\alpha = 0.5$.

4.5.2   Average Mutual Information

▶ **Average Mutual Information**

- Average mutual information in terms of
  the destination and noise entropy

- BSC example

- Channel capacity

The average mutual information $I(X,Y)$ is the average information transferred per symbol given by

$$I(X,Y) = \sum_{i=1,j=1}^{M,Q} P(x_i,y_j)I(x_i,y_j) \text{ bits/symbol} \qquad (4.24)$$

Notice that the definition of $I(X,Y)$ is similar to that of the average self-information (or source entropy) given by $H(X) = \sum_{i=1}^{M} P(x_i)I_i = \sum_{i=1}^{M} P(x_i)\log_2 \frac{1}{P(x_i)}$. It can be shown [Haykin, 2001] as illustrated in Fig. 4.22, that the average mutual information is given by

$$I(X,Y) = H(Y) - H(Y|X) \text{ bits/symbol} \qquad (4.25)$$

where $H(Y)$ is the average information per received symbol (or *destination* entropy) given by

$$H(Y) = \sum_{j=1}^{Q} P(y_j)\log_2 \frac{1}{P(y_j)} \qquad (4.26)$$

and $H(Y|X)$ is the average information per symbol added by the channel (or *noise* entropy) given by

$$H(Y|X) = \sum_{i=1,j=1}^{M,Q} P(y_j|x_i)P(x_i)\log_2 \frac{1}{P(y_j|x_i)} \qquad (4.27)$$

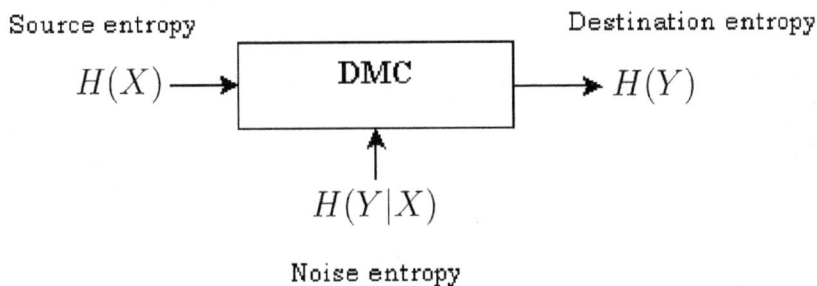

Source entropy   Destination entropy

$$H(X) \longrightarrow \boxed{\text{DMC}} \longrightarrow H(Y)$$

$$H(Y|X)$$

Noise entropy

Figure 4.22: Source, noise and destination entropy.

Some properties of $I(X,Y)$ are [Haykin, 2001]

$$I(X,Y) = I(Y,X), I(X,Y) \geq 0 \qquad (4.28)$$

and

$$I(X,Y) = H(X) + H(Y) - H(X,Y) \qquad (4.29)$$

where

$$H(X,Y) = \sum_{i=1,j=1}^{M,Q} P(x_i,y_j)\log_2 \frac{1}{P(x_i,y_j)} \qquad (4.30)$$

4.5.3   Channel Capacity

▷ **Channel Capacity**

- How to determine the channel capacity of a
  BSC channel

The *channel capacity* C_s is defined as the maximum value of $I(X, Y)$ bits/symbol. Alternatively, by taking into account the rate at which symbols are transmitted into the channel, the channel capacity may also be expressed as the maximum rate C bits/sec at which information can be transferred across the channel. The units of channel capacity are used to distinguish between C and C_s. The following detailed example will illustrate all the concepts of information theory.

4.5.4   Channel Capacity of a Binary Symmetric Channel

The DCS in Fig. 4.23 consists of a binary source, a BSC channel and a binary sink. To determine the channel capacity of the BSC channel, we first have to determine the average mutual information $I(X, Y)$.

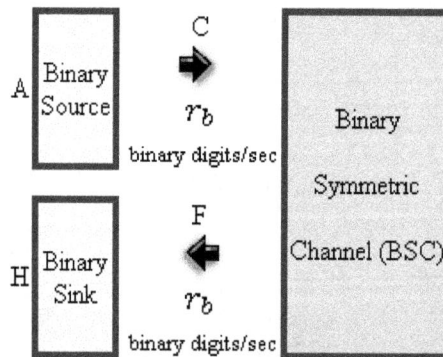

Figure 4.23: Simulated DCS : BSC

The symbols and equations which define the BSC are presented in Table 4.6 from which the source entropy is given by

$$H(X) = P(x_1)\log_2 \frac{1}{P(x_1)} + P(x_2)\log_2 \frac{1}{P(x_2)} = \Omega(p) \text{ bits/symbol} \quad (4.31)$$

Hence, the information rate into the BSC is $r_b\Omega(p)$ bits/sec. The destination entropy is given by

$$H(Y) = \sum_{j=1}^{Q=2} P(y_j)\log_2 \frac{1}{P(y_j)} = P(y_1)\log_2 \frac{1}{P(y_1)} + P(y_2)\log_2 \frac{1}{P(y_2)} \quad (4.32)$$

where $P(y_2) = 1 - P(y_1)$. Hence

$$H(Y) = P(y_1)\log_2 \frac{1}{P(y_1)} + (1-P(y_1))\log_2 \frac{1}{(1-P(y_1))} \quad (4.33)$$

But

$$P(y_1) = P(y_1|x_1)P(x_1) + P(y_1|x_2)P(x_2) \quad (4.34)$$

Let x_1 represent a binary digit zero
Let x_2 represent a binary digit one
α = the BSC crossover probability
p = probability with which x_1 is output
r_b = number of binary digits/sec entering BSC
$P(x_1) = p,\ P(x_2) = 1 - p$
$P(y_2|x_1) = P(y_1|x_2) = \alpha$
$P(y_1|x_1) = P(y_2|x_2) = 1 - \alpha$

Table 4.6: BSC probabilities.

$$= (1 - \alpha)\, p + \alpha\,(1 - p) = p + \alpha - 2\alpha p \qquad (4.35)$$

Therefore

$$H(Y) = \Omega\left(P(y_1)\right) = \Omega\left(p + \alpha - 2\alpha p\right) \ \text{bits/symbol} \qquad (4.36)$$

Similarly, we find that

$$H(Y|X) = \sum_{i=1,j=1}^{2,2} P(y_j|x_i)P(x_i)\log_2 \frac{1}{P(y_j|x_i)} \qquad (4.37)$$

$$= P(y_1|x_1)P(x_1)\log_2 \frac{1}{P(y_1|x_1)} + P(y_1|x_2)P(x_2)\log_2 \frac{1}{P(y_1|x_2)} \qquad (4.38)$$

$$+ P(y_2|x_1)P(x_1)\log_2 \frac{1}{P(y_2|x_1)} + P(y_2|x_2)P(x_2)\log_2 \frac{1}{P(y_2|x_2)} \qquad (4.39)$$

$$\therefore H(Y|X) = (1 - \alpha)\, P(x_1)\log_2 \frac{1}{(1-\alpha)} + \alpha P(x_2)\log_2 \frac{1}{\alpha} + \qquad (4.40)$$

$$\alpha P(x_1)\log_2 \frac{1}{\alpha} + (1 - \alpha)\, P(x_2)\log_2 \frac{1}{(1-\alpha)} \qquad (4.41)$$

$$= P(x_1)\Omega(\alpha) + P(x_2)\Omega(\alpha) = \Omega(\alpha)\left[P(x_1) + P(x_2)\right] = \Omega(\alpha) \ \text{bits/symbol} \qquad (4.42)$$

Therefore,

$$I(X,Y) = \Omega\left(p + \alpha - 2\alpha p\right) - \Omega(\alpha) \ \text{bits/symbol} \qquad (4.43)$$

The results derived for this important example are summarized in Fig. 4.24(a). For $\alpha = 0$, we find $I(X,Y) = \Omega(p)$. This result is as expected because for a noiseless channel, the average information transferred across the channel per symbol must be the same as average information per symbol $\Omega(p)$ transmitted into the channel as illustrated in Fig. 4.24(b).

(a)

$$I(X,Y) = \Omega\,(p + \alpha - 2\alpha p) - \Omega(\alpha) \text{ bits/symbol}$$

$$H(X) = \Omega(p)$$

bits/symbol

$$H(Y|X) = \Omega(\alpha)$$

bits/symbol

$$H(Y) = \Omega\,(p + \alpha - 2\alpha p)$$

bits/symbol

(b)

$$I(X,Y) = \Omega(p) \text{ bits/symbol}$$

$$H(X) = \Omega(p)$$

bits/symbol

$$H(Y|X) = 0$$

bits/symbol

$$H(Y) = \Omega(p)$$

bits/symbol

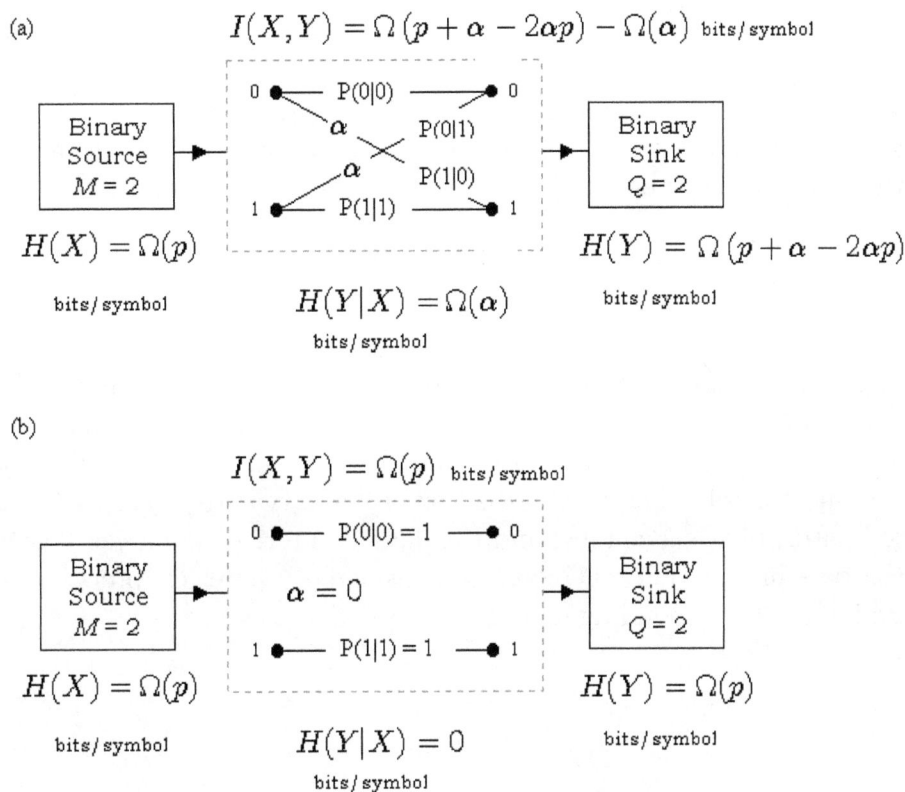

Figure 4.24: Average mutual information over a BSC.

To determine the channel capacity, we have to find the maximum value of $I(X, Y) = \Omega(p + \alpha - 2\alpha p) - \Omega(\alpha)$. By examining this expression carefully, notice that we can control p. Since the binary entropy function $\Omega(z)$ is a maximum when $z = 0.5$, we require $(p + \alpha - 2\alpha p) = 0.5$, which is satisfied when $p = 0.5$. Hence, the BSC channel capacity

$$C_s = 1 - \Omega(\alpha) \text{ bits/symbol} \tag{4.44}$$

A symbol in our example is a binary digit. Alternatively, for r_b binary digits/sec transferred over the BSC, the maximum rate C bits/sec at which information can be transferred across a BSC channel is given by

$$C = [1 - \Omega(\alpha)] \, r_b \text{ bits/sec} \tag{4.45}$$

Figure 4.25 shows how the channel capacity C_s depends on α. Notice that information is not transferred across the channel for the maximum noise situation of $\alpha = 0.5$. For a noiseless channel, $\alpha = 0$ and the channel capacity $C_s = 1$ bit/symbol. The BSC crossover probability α depends on the type of modulator, physical channel and the demodulator. For a given baseband or bandpass modulation scheme to be considered in later chapters, we shall determine the corresponding expression for α.

- ⭐ SIMULATION **AverageMutualInformation:** Simulation corresponding to Figure 4.25.

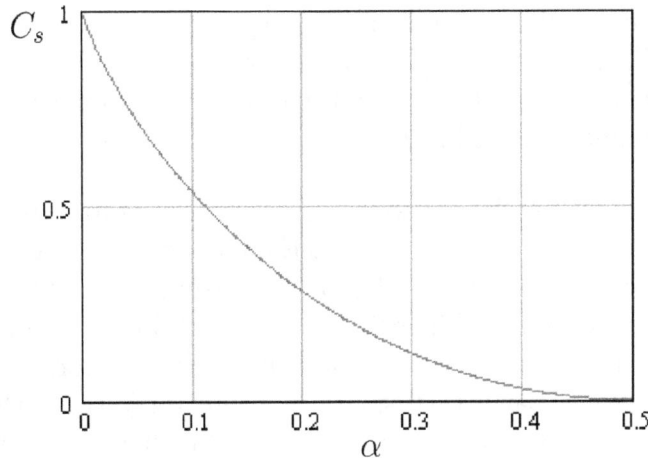

Figure 4.25: Channel capacity of a BSC.

4.6 Channel Coding Theorem

4.6.1 Discrete Memoryless Channel

▷ **Channel Coding Theorem**

- Shannon's channel coding theorem

- Maximum rate at which binary digits can be transferred over a DCS

- Implication which led to 50 years of research in error-control coding

Consider a digital communication system in which a discrete memoryless source (DMS) outputs r_s symbols/sec and a discrete memoryless channel (DMC) through which r_c symbols/sec are transferred. Unlike in Fig. 4.23, where r_b binary digits/sec were transferred over a BSC, we are now considering a more general case. For example, the binary digits output by the binary source in Fig. 4.23 could enter a channel encoder, in which case code digits are transferred over a DMC and r_c symbols/sec would refer to these code digits, where each symbol is a code digit. Let the channel capacity of the DMC be C_s bits/symbol. The rate at which information comes out of the DMS is $R = r_s H(X)$ bits/sec and the rate at which information is passed through the channel is $r_c C_s$ bits/sec.

Shannon's channel coding theorem for a DMC is as follows. If $r_s H(X) \leq r_c C_s$, or equivalently $R \leq r_c C_s$ bits/sec, there exists an error-control coding scheme for which the source output can be transmitted over the channel and be reconstructed with an arbitrarily small probability of error. Conversely if $R > r_c C_s$, it is not possible to transmit information over the channel and reconstruct it with an arbitrary small probability of error. Unfortunately, Shannon's channel coding theorem does not tell us how to create the appropriate channel encoder and decoder. In light of Shannon's source coding theorem, recall that average number of binary digits per second r_b from a source encoder must be such that $r_b \geq R$. However, we require $R \leq r_c C_s$ bits/sec from the channel coding theorem. Thus, the maximum data rate that can be transferred over the channel with an arbitrarily small probability of error is

$$r_b \big|_{\max} = r_c C_s \text{ binary digits/sec} \qquad (4.46)$$

To further explore the implications of the channel coding theorem, consider the DCS shown in Fig. 4.26, in which the channel symbol is a code digit (binary digit), so that r_c is the number of code digits per second transferred over the channel. If p is the probability with which a binary digit zero is output from binary source, then the corresponding source entropy is $\Omega(p)$. For reliable communication over the DMC, Shannon's channel coding theorem requires that $r_b \Omega(p) \leq r_c C_s$. As a quick check for a perfect source encoder, $p = 0.5$, $\Omega(p) = 1$ and $r_b \leq r_c C_s$, from which we recover once again the result that $r_b \big|_{\max} = r_c C_s$. Now, given that $R_{code} = \frac{r_b}{r_c}$, we find the requirement for reliable communication to be

$$R_{code} \Omega(p) \leq C_s \qquad (4.47)$$

The implication of this result is that there is a error-control code with a code rate such that $R_{code}\Omega(p) \leq C_s$, which is capable of achieving an arbitrarily low probability of error. It is not difficult to achieve $\Omega(p) \approx 1$ using an appropriate source encoder. The difficulty is to create a channel encoder and decoder which achieves a very small probability of an error in a binary digit at point F with $R_{code} = C_s$.

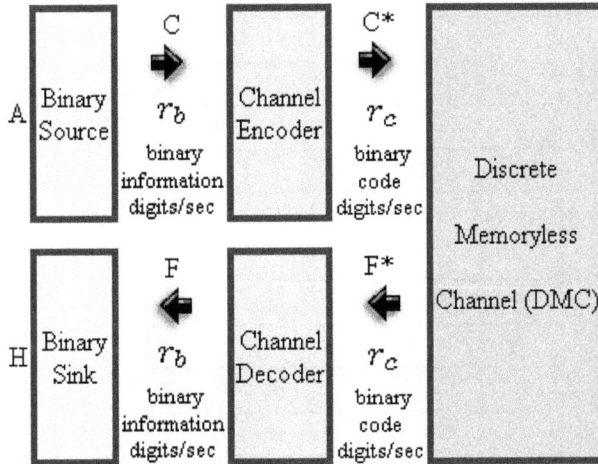

Figure 4.26: Block diagram of a general DCS.

4.6.2   Example: Repetition Code

Repetition Code in Light of Shannon's Channel Coding Theorem

- The error control capability of a rate 1/n repetition code is analyzed with reference to the channel coding theorem

If the DMC is a binary symmetric channel (BSC), then recall that $C_s = 1 - \Omega(\alpha)$ bits/symbol (where is a symbol is a code digit in this case) and thus in this case, we have $R_{code}\Omega(p) \leq 1 - \Omega(\alpha)$. Assuming $p = 0.5$, we have $R_{code} \leq 1 - \Omega(\alpha)$. For a repetition code, Fig. 4.27 shows the variation of the probability of an error in a binary digit P_e versus R_{code}, where $R_{code} = 1, 1/3, 1/5, 1/7, 1/9, 1/11$, for $\alpha = 10^{-2}$ and $\alpha = 10^{-3}$, which correspond to a channel capacity C_s of 0.9192 and 0.98859 bits/symbol, respectively.

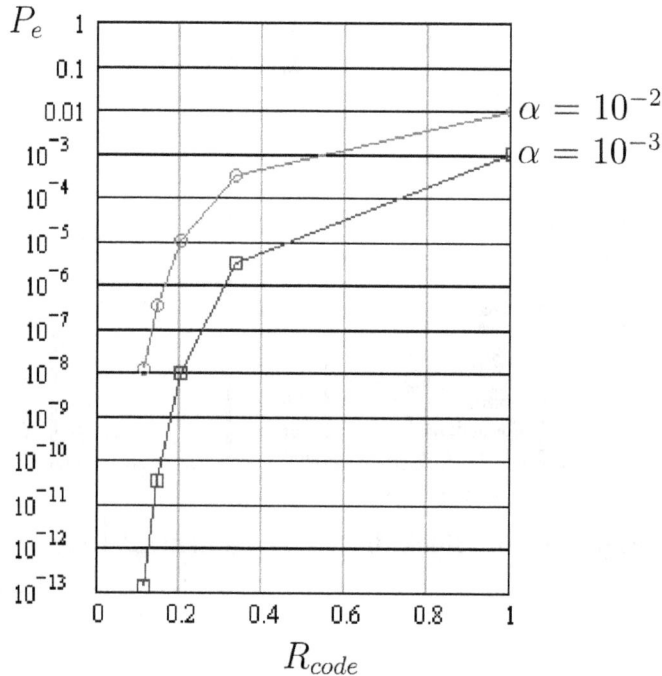

Figure 4.27: Performance of various repetition codes over a BSC for $\alpha = 10^{-2}$ and $\alpha = 10^{-3}$.

- ⭐ SIMULATION **Capacity:** To understand why the channel capacity C_s is 0.9192 and 0.98859 bits/symbol for $\alpha = 10^{-2}$ and $\alpha = 10^{-3}$, respectively. The analysis presented in this simulation code shows the use of the popular *binary phase shift keying* (BPSK) modulation scheme to be considered in chapter 7. Once you understand BPSK, please return to this simulation code.

From Fig. 4.27, we note the following points:

- As expected for $R_{code} = 1$, we find $P_e = \alpha$.

- These curves are only valid if the information throughput is allowed to be reduced by a factor of R_{code}.

- It is necessary for $R_{code} \ll C_s$ to achieve a very small probability of error.

Unlike repetition codes, where the code rate has to approach zero in order to greatly reduce P_e, the channel coding theorem only requires that the code rate be less than or equal to the channel capacity C_s i.e. for a near perfect error-control coding scheme, R_{code} would be just less than C_s and P_e would be very small without requiring a reduction in information throughput. Refer to section 8.1.3 for further details. Motivated by this theorem, ongoing research in this field of error-control coding for approximately the past 50 years has finally led to the development of *turbo codes* and *low density parity check* (LDPC) codes which can achieve a performance close to the Shannon theoretical limit.

4.6.3 Summary of Information Theory

▶ **Summary of Information Theory**

- A summary of the key features that we

  learnt about on information theory

An overview of the information theory covered in the previous sections. Well worth a look to clarify any ambiguities.

4.7   Problems

1. A discrete memoryless source (DMS) consists of three symbols (A, B, C) with probabilities P_A, P_B, P_C.

(a)   If $P_B = P_C$ and $P_A = p$, determine the source entropy in terms of p and the condition under which it is a maximum.

(b)   If $P_A = \dfrac{1}{2}$, what is the amount of information contained in the output sequence CAABBAC.

(c)   How does this compare with the expected average information content ?

(d)   Suppose the symbols are output in blocks of 10, separated by 3 ms spaces. What is the source information rate if each symbol is 1 ms long and $P_A = \dfrac{1}{3}$.

2. A DMS consists of two symbols (A, B) with probabilities P_A, P_B. If the time duration of symbols A and B are T_A and T_B, respectively,

(a) Determine an expression for the information rate.

(b) A telegraph source outputs a "dot" of time duration 0.1 seconds with probability 0.7 and a "dash" of time duration 0.2 seconds. What is the information rate of the source ?

3. The output of an analog to digital converter are eight possible voltage levels $V_1, V_2, \cdots, V_8$, which occur with probability $p_1, p_2, \cdots, p_8$. If $p_1, p_2, \cdots, p_8$ are respectively 0.17, 0.18, 0.15, 0.07, 0.05, 0.17, 0.08, 0.13, calculate the average information conveyed per voltage level.

4. If the 26 letters of the English alphabet are used to communicate information via an online chat

(a)   What is the average information conveyed per character ?

(b)   If each character is to be replaced by a string of binary digits of a fixed length, how many binary digits would be required per character ? Compare your answers to part (a) and (b).

(c)   How often should a person press a key on the keyboard to send information to the receiver at a rate of 5 bits/sec ?

You may assume that each letter is equally likely within the typed message.

5. A DMS consists of four symbols (A, B, C, D) with probabilities P_A, P_B, P_C, P_D as listed in the table below. The output of this source is input to a source encoder which replaces each input symbol by a predefined code word. Three different possible source codes are shown.

Symbol	Probability	Source Code I	Source Code II	Source Code III
A	1/2	0	0	00
B	1/4	01	10	01
C	1/8	011	110	10
D	1/8	0111	111	11

(a)   What is the information content of each letter output from the DMS

(b)   Verify that source code II is uniquely decipherable by encoding and decoding the DMS output sequence BBACDBD.

(c)   What is the efficiency of each source code ?

(d)   Suppose the source encoder utilizes code II and a rate $\frac{1}{2}$ code channel encoder is inserted after the source encoder. If the output of the channel encoder is input to a BSC with a crossover probability α, determine condition on α which satisfies Shannon's channel coding theorem.

6. Consider a binary symmetric channel (BSC). By examining how the average information transferred per symbol depends on the BSC crossover probability α, determine the values for α which corresponds to maximum and minimum noise.

7. Consider the DCS shown below.

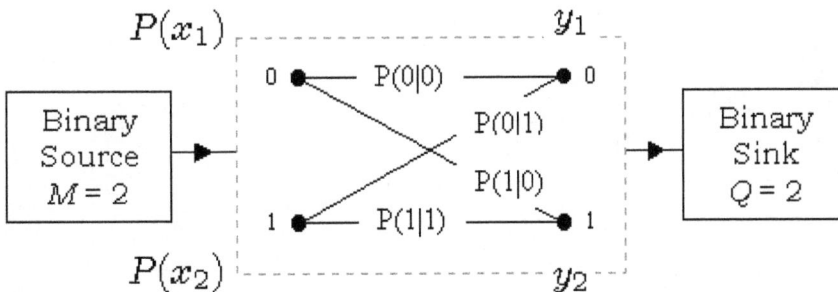

(a)   Under what conditions is the DMC a BSC ?

(b)   For $P(0|1) \neq P(1|0)$, determine an expression for the average mutual information and the probability of an error in a binary digit P_e at the output of the DMC. You may take $P(x_1) = p$, $P(0|1) = \mu$ and $P(1|0) = \lambda$.

(c)   If $P(0|1) = 0.2$, $P(1|0) = 0.1$ and $p = 0.5$, what the average information transferred per symbol and P_e ?

(d)   Under what conditions will the transfer of information go to zero ? What is the value of P_e under these conditions if $p = 0.5$?

(e)   Use your answer to part (b) to determine the channel capacity if $P(0|1) = P(1|0)$.

(f) For part (c), show how the average information transferred per symbol depends on p using a graph. Estimate the value of p from the graph for which this information transfer is maximized.

8. Consider a DMS which is the outcome of a single roll of a fair die. The die is rolled every second. Using a source encoder with a code as shown below and assuming the DMC with transition probabilities as shown below, determine the following:

(a)   Probability of a zero output from the source encoder.

(b)   Source entropy.

(c)   Source information rate.

(d)   Number of binary digits per second output from the source encoder and the minimum possible average number of binary digits per second output from the source encoder. Hence determine the source encoder efficiency.

(e)   The average information transferred per binary digit through this channel.

(f)   The maximum information rate through the channel.

(g)   If the die was biased toward certain outcomes, how can the information transfer through the channel be maximized.

Source code

DIE OUTCOME	Code
1	001
2	010
3	011
4	100
5	101
6	110

Channel transition probabilities

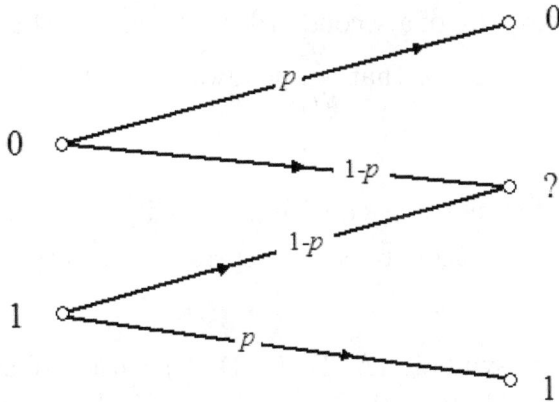

9. Consider a DMC which consists of the serial concatenation of two BSCs, each with crossover probability α. You may assume a binary digit zero is input with probability p.

(a) Determine an expression for the average mutual information and plot its value against α for various values of p.

(b) For $p = 0.5$, what is the channel capacity of the DMC.

(c) If a rate $\frac{1}{2}$ error control code is used to protect the information binary digits, estimate the values of α for which the Shannon's channel coding theorem is satisfied.

10. A digital source outputs one of two possible symbols (*, =). The table below shows the sequence of symbols transmitted into a DMC and the corresponding output.

Input	*	*	=	*	=	=	*	*	*	=	*	=	*	*	*
Output	*	=	*	*	=	*	=	=	*	=	*	=	*	*	=

The duration of the symbol * is 10 ms and the duration of the symbol = is 30 ms. The gap between each symbol is 10 ms. Determine

(a) the source probabilities P_* and $P_=$ and all the forward transitions probabilities.

(b) the average information rate in bits/sec.

11. A DMS consists of three symbols (A, B, C) with probabilities P_A, P_B, P_C. (a) If $P_A = 0.7$, $P_B = 0.2$, $P_C = 0.1$, calculate the source entropy H_1. (b) Find the source entropy H_2 of a second order extension of this source and calculate the ratio $\dfrac{H_2}{H_1}$. (c) Show that $\dfrac{H_2}{H_1}$ is always equal to 2 for any value of P_A, P_B, P_C.

12. A DMS consists of M symbols, each with probability $\dfrac{1}{M}$. Determine the optimum fixed length source encoder code to represent the output of the DMS as a binary stream.

13. DMS consists of five symbols (A, B, C, D, E) with probabilities $P_A = 0.5$, $P_B = 0.15$, $P_C = 0.20$, $P_D = 0.10$ and $P_E = 0.05$. Determine the corresponding Huffman code, average code length and the efficiency of the Huffman code.

14. Consider a DMS which consists of three symbols (A, B, C) with $P_A = 0.7$, $P_B = 0.2$, $P_C = 0.1$.

(a)   Determine the corresponding Huffman code, average code word length and efficiency.

(b)   Determine the Huffman code for the second order extension of the DMS, average code word length and efficiency.

(c)   Compare your answers to part (a) and (b). Is the second order extension more efficient ?

15. Use the Lempel-Ziv source coding algorithm to encode and decode the binary stream 111101110111111101. Determine the compression factor and explain why it is not less than 1.0.

16. If the BSC crossover probability $\alpha = 0.01$ and this channel can transfer a maximum of 10000 symbols per second, what is the maximum data rate that can be supported over this BSC with an arbitrarily low probability of an error.

Chapter 5

Analog to Digital

5.1 Introduction

In this chapter we focus on the information source and the format block within a digital communication system (DCS) as highlighted in Fig. 5.1. Our aim is to convert the analog signal output by the information source (e.g. speech or video signals) into a digital format (via the format block) for transmission over a DCS. This process is known as *analog-to-digital* (A/D) conversion and its inverse operation, required at the receiver, is known as *digital-to-analog* (D/A) conversion. We shall consider the following popular techniques: *pulse code modulation* (PCM), *delta modulation* (DM) and *adaptive delta modulation* (ADM).

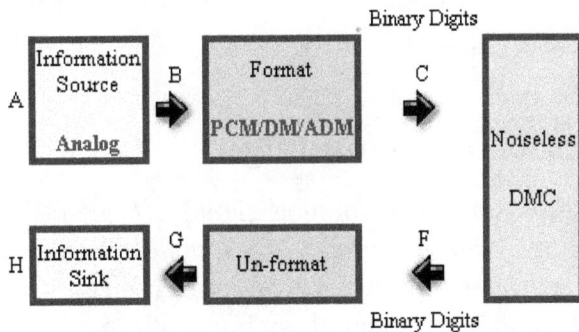

Figure 5.1: Chapter 5 DCS.

To convert an analog signal into a digital signal is a three step process. The first step is to sample the analog signal at uniformly spaced time intervals to obtain a discrete-time signal or a sequence of numbers. The idea of varying the amplitude, width or position of pulses in accordance with the samples of an analog signal is referred to as *pulse modulation*. In *pulse-amplitude modulation* (PAM), the pulse amplitude, obtained via *natural* or *flat-top* sampling (section 5.1.3), represents the analog information. We shall focus on flat-top PAM in which the sample amplitude is held constant for a duration $\tau \leq T_s$, as in Fig 5.9(b), where the amplitudes of pulses of width τ separated in time by T_s seconds are varied in accordance with the sample values. The advantage of using a pulse width $\tau < T_s$ is that we may utilize the time space between pulses to send other similar pulse modulated signals via *time-division multiplexing* (TDM) i.e. we can simultaneously transmit many different signals over a single channel. Briefly, in a TDM system, an anti-aliasing filter is used prior to each input signal. A sample of each input signal is interleaved within T_s. The penalty incurred is an increase in the required bandwidth by a factor equal to the number of signals multiplexed.

The second step is to approximate each sample value to the nearest value in a set of discrete predefined values referred to as *quantization* levels to obtain a discrete-time, discrete-amplitude signal. The third step is to represent each of these predefined values by digital symbols (typically binary digits). We shall consider the first step in this section. The second step is covered in sections 5.2 (uniform quantization) and 5.3 (non-uniform quantization) and finally the third step in section 5.4. In delta modulation (DM), after the first step of PCM, if the difference between successive samples is positive or negative, then a binary digit 1 or 0, respectively is transmitted. DM is covered in section 5.5 together with its adaptive version (ADM).

5.1.1   Instantaneous Sampling

Given that the amplitude of an analog signal $s(t)$ can take on any value in a continuous range, the first step is to *sample* the signal every $T_s = \frac{1}{f_s}$ seconds, where T_s is the sampling interval and f_s is the *sampling* frequency (samples/sec). A sample of a signal is simply its amplitude $s(nT_s)$ at the time of sampling. At all other times in between each sample, the amplitude is set to zero. Let $s_s(t)$ represent the result of multiplying $s(t)$ by an impulse train, in which each impulse is separated by T_s seconds, namely,

$$s_s(t) = s(t) \sum_n \delta\left(t - nT_s\right) = \sum_{n=-\infty}^{\infty} s(nT_s)\delta\left(t - nT_s\right) \qquad (5.1)$$

where the *unit impulse function* is defined by

$$\delta(n) = \begin{cases} 1 & if \quad n = 0 \\ 0 & otherwise \end{cases} \qquad (5.2)$$

For example, consider the analog signal $s(t)$ given by

$$s(t) = \begin{cases} e^{\frac{-t}{T}} \sin\left(\frac{2\pi t}{T}\right) + e^{-0.2t} \cos\left(\pi t\right) & \text{if} \quad t > 0 \\ 0 & \text{otherwise} \end{cases} \qquad (5.3)$$

where T is a constant. A plot of this signal is presented in Fig. 5.2 for $T = 0.5$ secs and its sampled version $s_s(t)$ is presented in Fig. 5.3 using a sampling frequency $f_s = 16$ Hz.

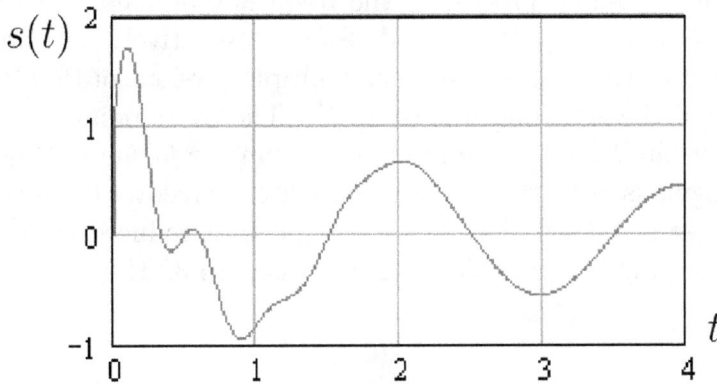

Figure 5.2: Analog signal $s(t)$.

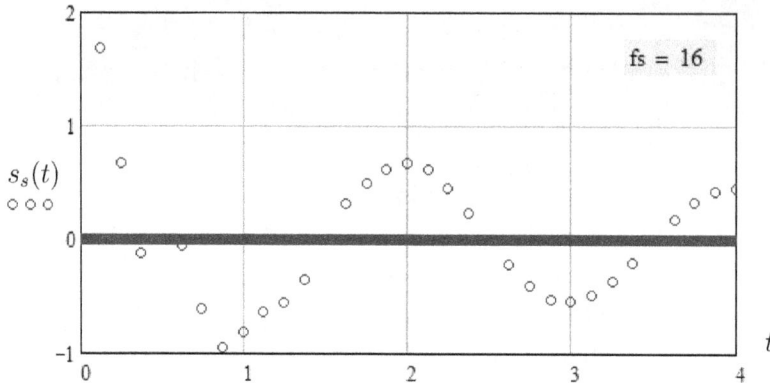

Figure 5.3: Sampled signal $s_s(t) = s(t) \sum_n \delta\left(t - nT_s\right)$.

⭐SIMULATION **SamplingTheorem:** Illustration of the equation
$s_s(t) = \sum_{n=0}^{N-1} s(nT_s)\delta\left(t - nT_s\right)$, where $s(t)$ given by equation 5.3.   Unlike in
section 1.4.3, where the digital signal $x[n] = s\left(\frac{n}{f_s}\right)$ is used to determine the
FFT of a signal $s(t)$ by sampling it every T_s seconds, the function $ss(t)$ is a
function of time and has the value of zero at all times between each sample.
To recover the original signal $s(t)$ from the sampled version $s_s(t)$, we need
to observe the sampling process in the frequency domain. The amplitude
spectra of $s(t)$ and $s_s(t)$, $|S(f)|$ and $|S_s(f)|$ respectively, are presented in
Fig. 5.4.   Notice that $|S_s(f)|$ contains multiple versions of $|S(f)|$, centered
on multiples of the sampling frequency f_s. The overlapping of $|S(f)|$ with
$|S(f \pm f_s)|$ within $|S_s(f)|$ is referred to as *aliasing* or *foldover*. When aliasing
occurs, the signal is distorted and it is impossible to recover the original signal
$s(t)$ from the sampled signal $s_s(t)$. Aliasing can be reduced by increasing f_s
as evident from Fig. 5.5 in which f_s is increased to 32 Hz.

To recover $s(t)$, a low-pass filter is used to remove all the replicas of $S(f)$
from $S_s(f)$ as illustrated in Fig. 5.6, leaving only the one centered on $f = 0$.

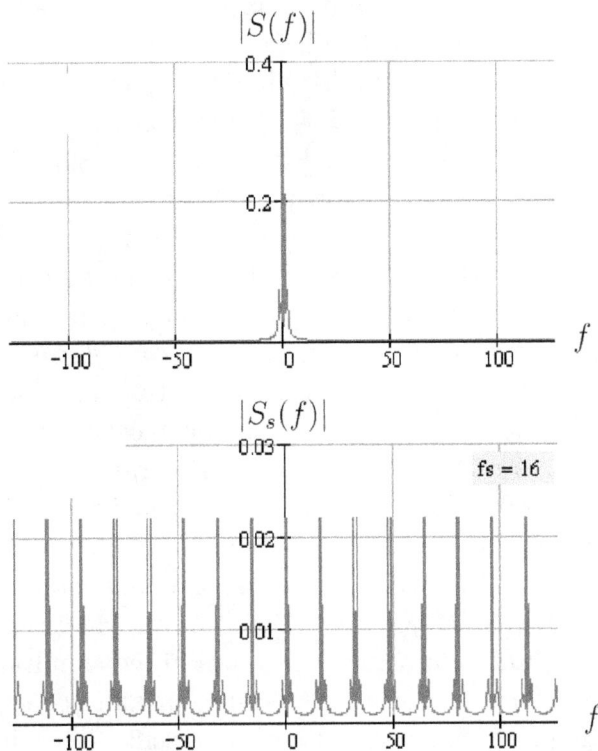

Figure 5.4: FFT amplitude spectrum of the $s(t)$ and $s_s(t)$ signals for $f_s = 16$ Hz.

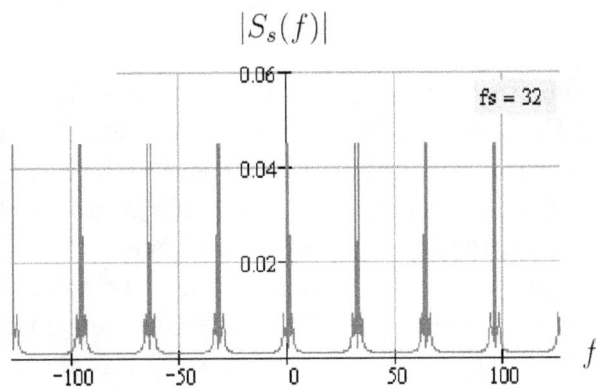

Figure 5.5: FFT amplitude spectrum of $s(t)$ and $s_s(t)$ for $f_s = 32$ Hz.

For clarity, part (b) of Fig. 5.6 is a magnified version of part (a). By taking the inverse Fourier transform of the filtered spectrum, a good approximation of $s(t)$ is recovered from the sampled signal as shown in Fig. 5.7. Since the $|S(f)|$ replicas are separated by f_s, the cut-off frequency f_c of the low-pass filter is set to $f_c = \frac{f_s}{2}$ i.e. in Fig. 5.6, $f_c = 8$ Hz. Of course we have assumed the use of an ideal low-pass filter. In practice, the sampling frequency f_s is increased to further separate the replicas of $|S(f)|$ within $|S_s(f)|$ to accommodate the use of a practical low-pass filter with a gradual cut-off characteristic. Ideally, if $f_{\max}$ is the largest frequency component within $s(t)$, referred to as a *band-limited* signal or a *low-pass* signal, then the *sampling theorem* (later in section 5.1.2) states that provided the sampling frequency $f_s \geq 2f_{\max}$, the signal $s(t)$ can be recovered from its samples $s(nT_s)$, where $2f_{\max}$ is referred to as the *Nyquist* rate or frequency and n is an integer. If the sampling frequency f_s is below the Nyquist frequency $2f_{\max}$, a condition referred to as *under-sampling*, the penalty incurred is aliasing. Equivalently, if we take $f_{\max}$ to be the bandwidth B of the analog signal, then we must ensure $f_s \geq 2B$. Based on the analysis in chapter 1, a signal cannot be of a finite duration and band-limited simultaneously. Given that all practical signals are of a finite duration and consequently are not strictly band-limited, aliasing will always be present. Thus in practice, $f_{\max}$ is usually interpreted as the highest frequency component with a significant spectral amplitude.

If the analog signal $s(t)$ is pre-filtered (i.e. band-limited) *before* sampling using say $f_c = 2.5$ Hz to produce a new signal $s_f(t)$, Fig. 5.8 shows the amplitude spectrum of the sampled signal without and with the use of pre-filtering for $f_s = 16$ Hz. This pre-filter, commonly referred to as the *anti-aliasing* filter, is used to eliminate non-essential high frequency spectral components of the signal. For example, the high frequency content of a speech signal that is not essential for intelligible telephone communication. In practice, the anti-aliasing filter is designed to eliminate any signal components with a frequency greater than $\frac{f_s}{2}$ i.e. for $f_s = 16$ Hz, we would set $f_c = 8$ Hz. Once again we are assuming the use of an ideal low-pass filter.

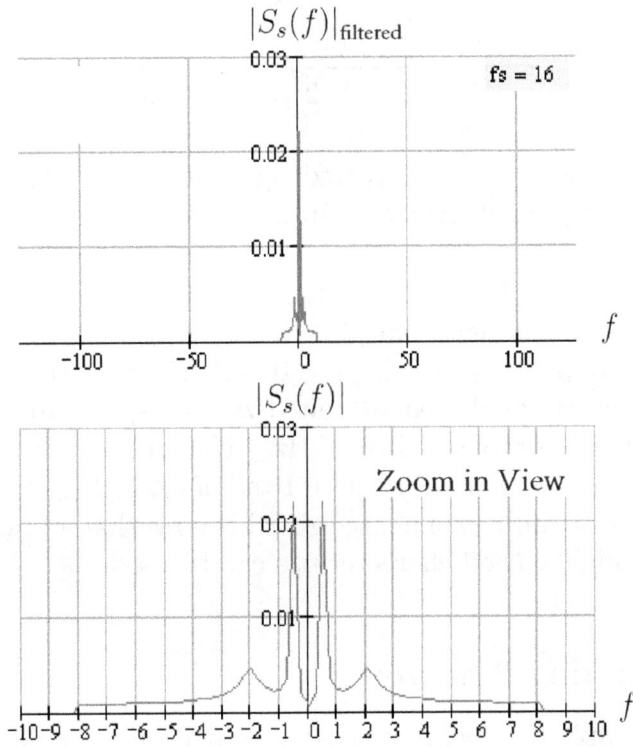

Figure 5.6: Filtered FFT amplitude spectrum of the sampled analog signal with $f_s = 16$ Hz.

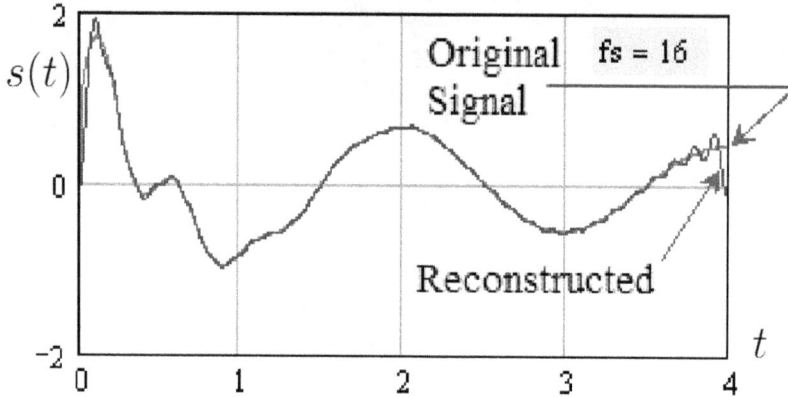

Figure 5.7: Original and reconstructed signal from the filtered spectrum of the sampled analog signal with $f_s = 16$ Hz.

In Fig. 5.8(b), the range from f_{max} to $(f_s - f_{max})$ is called a *guard band*, which in this example is $(f_s - 2f_{max}) = 16 - 2(2.5) = 11$ Hz. In a telephone system, if the speech signal is band-limited via a low-pass filter so that $f_{max} = 3.4$ kHz, then the Nyquist rate is 6.8 kHz. The international standard is to use $f_s = 8$ kHz, which provides a guard band of $(f_s - 2f_{max}) = 8 - 6.8 = 1.2$ kHz. The added advantage of a large guard band is that a practical low-pass filter with a gradual cut-off characteristic can be used.

5.1.2   Sampling Theorem

To prove the sampling theorem, we shall consider for convenience a baseband signal $s(t)$ that has been bandlimited to B Hz (e.g. Fig. 5.6). Now the Fourier series of a periodic impulse train is (refer to section 1.2 as required)

$$\sum_n \delta\left(t - nT_s\right) = A_0 + \sum_{n=1}^{\infty} A_n \cos\left(\frac{2\pi n t}{T_s}\right) + \sum_{n=1}^{\infty} B_n \sin\left(\frac{2\pi n t}{T_s}\right) \qquad (5.4)$$

where

$$A_0 = \frac{1}{T_s} \int_{-T_s/2}^{T_s/2} \delta(t)dt = \frac{1}{T_s} \qquad (5.5)$$

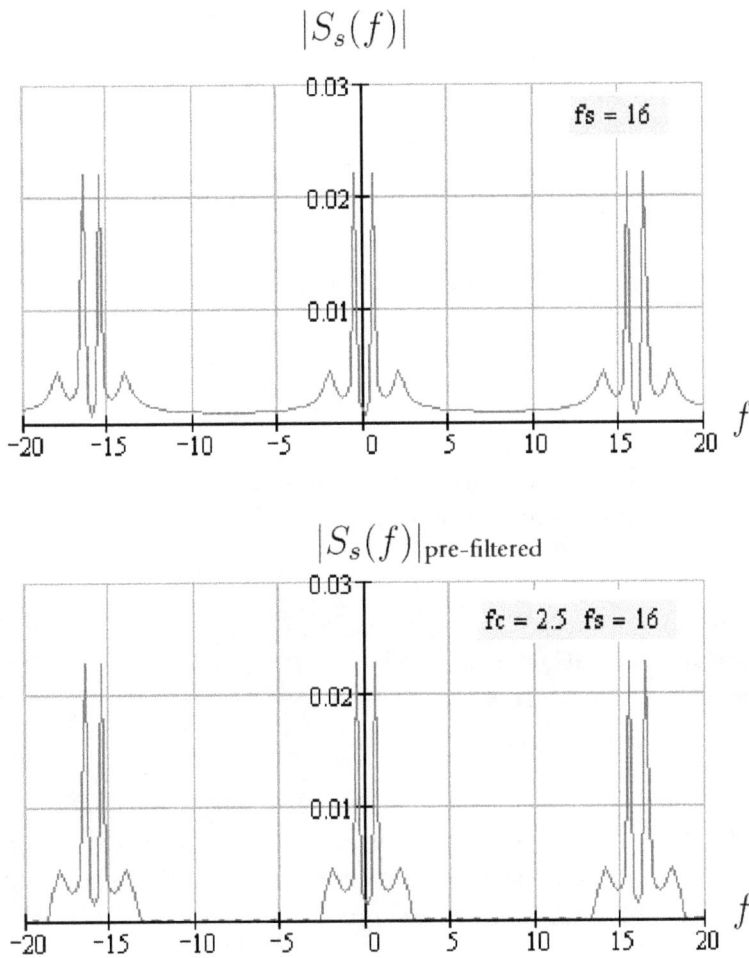

Figure 5.8: FFT amplitude spectrum of the sampled analog signal without (a) and (b) with the use of pre-filtering.

$$A_n = \frac{2}{T_s} \int_{-T_s/2}^{T_s/2} \delta(t) \cos\left(\frac{2\pi nt}{T_s}\right) dt = \frac{2}{T_s} \tag{5.6}$$

$$B_n = \frac{2}{T_s} \int_{-T_s/2}^{T_s/2} \delta(t) \sin\left(\frac{2\pi nt}{T_s}\right) dt = 0 \tag{5.7}$$

so that

$$
\begin{aligned}
s_s(t) &= s(t) \sum_n \delta(t - nT_s) = s(t) \left\{ \frac{1}{T_s} + \frac{2}{T_s} \sum_{n=1}^{\infty} \cos\left(\frac{2\pi nt}{T_s}\right) \right\} \\
&= \frac{1}{T_s} \left[\begin{array}{c} s(t) + 2s(t)\cos(\omega_s t) + 2s(t)\cos(2\omega_s t) \\ +2s(t)\cos(3\omega_s) + \cdots \end{array} \right]
\end{aligned} \tag{5.8}
$$

where for readability, we used the angular frequency $\omega_s = 2\pi f_s = \frac{2\pi}{T_s}$. The Fourier transform of each term on the right-hand side in equation ?? are as follows:

First term $\qquad s(t) \longleftrightarrow S(f)$
Second term $\qquad 2s(t)\cos(\omega_s t) \longleftrightarrow [S(f - f_s) + S(f + f_s)]$
Third term $\qquad 2s(t)\cos(2\omega_s t) \longleftrightarrow [S(f - 2f_s) + S(f + 2f_s)]$
etc.

Thus, given that $[S(f - kf_s) + S(f + kf_s)]$ where $k = 1, 2, 3, \cdots$ is simply the spectrum $S(f)$ shifted to $+kf_s$ and $-kf_s$, the spectrum $S_s(f)$ of $s_s(t)$ consists of $S(f)$ repeating periodically with period T_s as shown already in Figs. 5.5 and Fig. 5.8 and is therefore given by

$$S_s(f) = \frac{1}{T_s} \sum_{n=-\infty}^{\infty} S(f - nf_s) \tag{5.9}$$

Since $s(t)$ is bandlimited to B Hz, we require $f_s \geq 2B$ to avoid an overlap between the repeating $S(f)$ centered on $\pm nf_s$ to recover $S(f)$ from $S_s(f)$, which is equivalent to recovering $s(t)$ from $s_s(t)$.

5.1.3 Natural and Flat-Top Sampling

In practice, *natural* and *flat-top* sampling methods are used as illustrated in Fig. 5.9, where the top of the natural pulse follows the signal. This is because instantaneous sampling, which requires the multiplication of the analog signal by an impulse function every $1/f_s$ seconds, cannot be realized in practice. As mentioned previously, the advantage of using a pulse width $\tau < T_s$ is that we may utilize the time space between pulses to send other similarly pulse modulated signals via time-division multiplexing (TDM). The original signal $s(t)$ can be reconstructed from the natural sampled signal $s_n(t)$ with no distortion by passing $s_n(t)$ through an ideal low-pass filter if the sampling rate $f_s \geq f_{\max}$. However, flat-top sampling introduces amplitude distortion. The main effect is an attenuation of high-frequency components and is referred as the *aperture effect* which can be minimized by ensuring the pulse width τ is very much smaller than the sampling interval T_s.

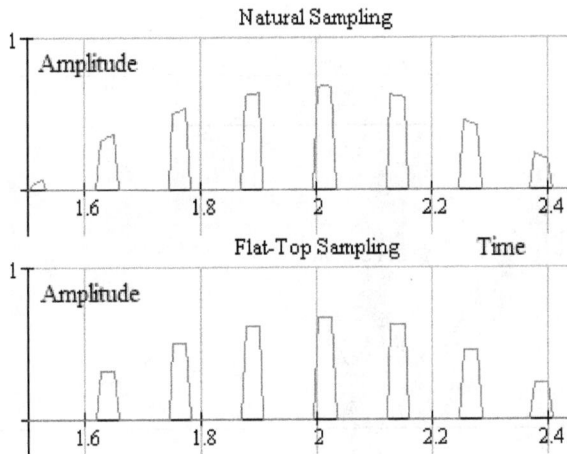

Figure 5.9: (a) Natural and (b) Flat-top sampling.

Sample and Hold

Sample-and-hold is a form of flat-top sampling in which the amplitude is held constant at the previous sample value up to the next sample, instead of setting the amplitude between samples to zero as illustrated in Fig. 5.10. In this case, the sampled signal spectrum does not contain significant replicas of $S(f)$ as shown in Fig. 5.11 because we now have a staircase approximation of the signal. However, it is still necessary to filter the sampled signal using $f_c = \frac{f_s}{2}$ to produce the spectrum shown in Fig. 5.12. Without the filter, the inverse Fourier transform will recover the sample and hold signal presented in Fig 5.10(a). The performance of the sample and hold technique for f_s equal to 16 Hz and 32 Hz is presented in Fig. 5.13.

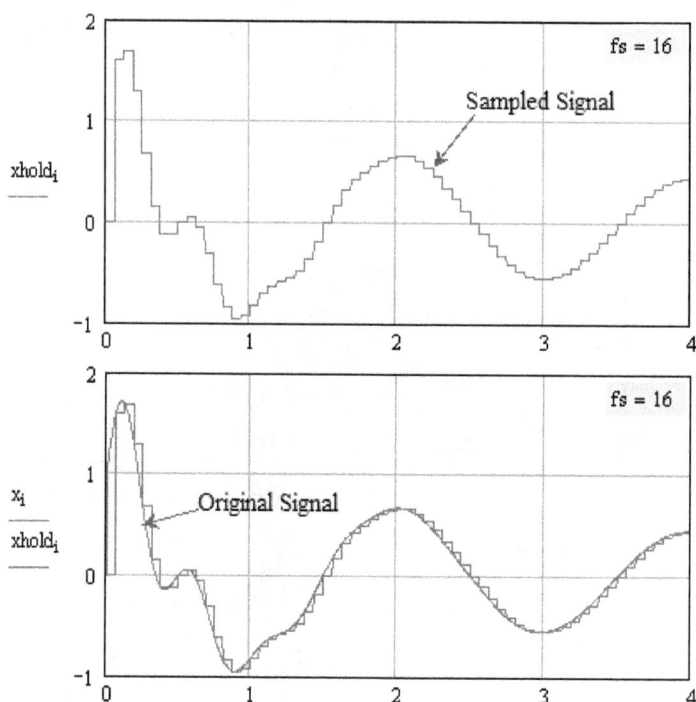

Figure 5.10: Sampled analog signal and its comparison with the original signal.

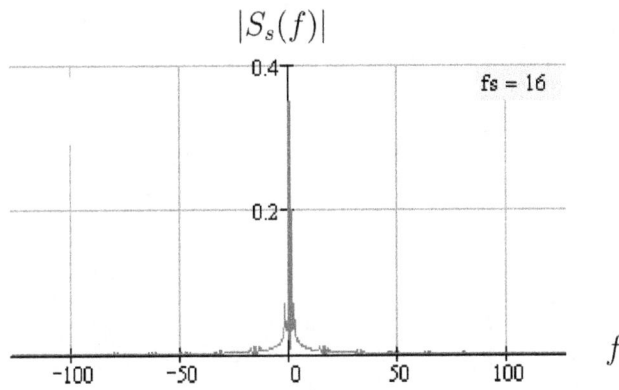

Figure 5.11: FFT amplitude spectrum of the sampled (sample and hold method) signal.

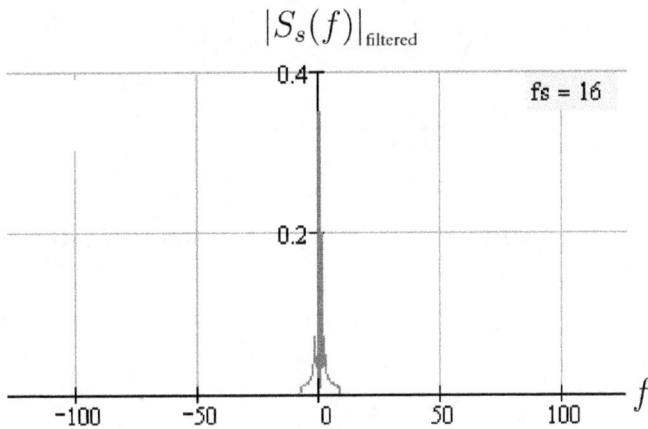

Figure 5.12: FFT filtered amplitude spectrum with $f_c = 8$ Hz.

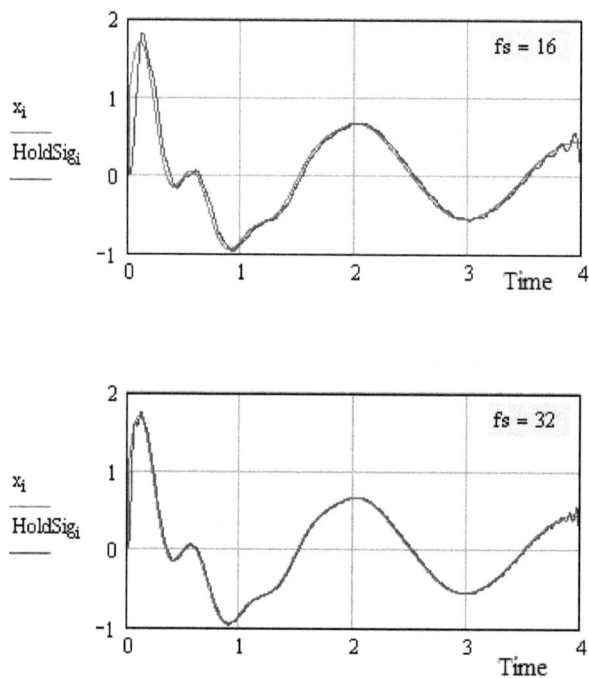

Figure 5.13: Comparison of the reconstructed signal with $s(t)$ for $f_s = 16$ Hz and 32 Hz.

5.1.4   Simulation : Sampling and Filtering

⭐SIMULATION **Sampling:** Instantaneous and sample-hold methods are implemented.

⭐SIMULATION **PrefilterSampling:** The effect of pre-filtering the input signal before sampling is demonstrated.

⭐SIMULATION **NaturalFlatSampling**: The natural and flat-top sampling methods are implemented and compared side-by-side.

- **Experiment:**   Change the input signal and the sampling frequency. For example, verify that we must ensure $f_s > 2f_{max}$ for the special case of sampling a sinusoidal signal $s(t)$ of frequency f_{max}. Also, try the interesting case of sampling a bandpass signal of bandwidth $(f_2 - f_1)$ whose spectrum is such that $S(f) = 0$ except for $\begin{cases} f_1 < f < f_2 \\ -f_2 < f < -f_1 \end{cases}$. First use $f_s \geq 2f_2$ to recover the signal. Now verify that it is not necessary to sample that fast. We can use a sampling frequency less than $2f_2$! It can be shown [Brown, 1980] that we require a minimum sampling frequency of $\frac{2f_2}{n}$, where n is the largest integer not exceeding the ratio $\frac{f_2}{f_2-f_1}$. Verify that not all the sampling frequencies greater than $\frac{2f_2}{n}$ will work and will in addition lead to spectral inversion. There are methods to compensate for spectral inversion, but that is a currently a topic beyond the scope of this book. In practice for linear systems, the bandpass signal spectrum is shifted down to the baseband level for linear baseband signal processing to not only avoid sampling at a high rate, but to also take advantage of readily available hardware pre-designed to process a baseband signal.

5.2   Quantization

After sampling, the amplitude of the samples are still continuous. Given that the transmission of real numbers would require an infinite number of binary digits, the second step in our goal to convert an analog signal $s(t)$ into a digital signal for transmission over a DCS is to *quantize* the signal.

Namely, each sample value is approximated to the nearest value in a set of discrete (uniform or non-uniformly spaced) values. The error introduced by this quantization process makes it impossible to recover an exact replica of the original analog signal. However, for a music CD for example, the errors due to quantization can be reduced to the extent that it is not discernible by the human ear. Uniform and non-uniform quantization will be considered in this and the next section, respectively. For convenience, the input signal to the quantizer will be denoted by $x(t)$. The third and final step of encoding will be presented in section 5.4 within the context of a pulse code modulation system.

5.2.1   Uniform Quantization

A *quantizer* changes the amplitude of the incoming signal $x(t)$ to output an signal $y(t)$ at time t which is constrained to discrete predefined amplitudes. The output signal $y(t)$ range is split into Q_{level} quantization levels. If the spacing Δ between each quantization level is the same, the quantization is said to be *uniform*, as illustrated in Fig. 5.14, for $Q_{level} = 4$. For an infinite number of quantization levels, $y(t) = x(t)$ i.e. the incoming signal is unaffected by the quantizer. A quantizer for which the output is non-zero for values of x close to zero, as in Fig. 5.14, is referred to as a *mid-riser* quantizer. Conversely, a *mid-tread* quantizer has a quantization level set at $y = 0$. To gauge the performance of a given quantizer, we need to make use of the *probability density function* (PDF) of the incoming signal $x(t)$. Please refer to Appendix A.7 for the fundamentals of a PDF. Let the input to the quantizer at time t be the random variable x. Please refer to Appendix A.5 for an introduction to a random variable and A.11 for its expected value.

Suppose the PDF of x, denoted by $p(x)$, is uniform as illustrated in Fig. 5.15. Note that in Fig. 5.15, A is not necessarily the peak amplitude of the signal. The input signal is simply limited to vary only between $-A$ and $+A$ and amplitudes greater than $|A|$ are "chopped-off".

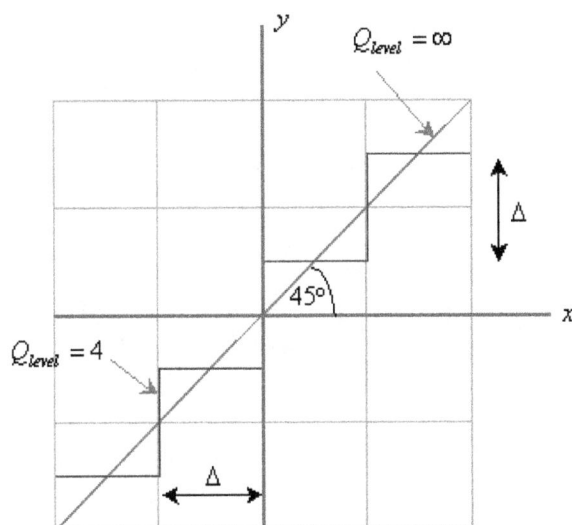

Figure 5.14: Uniform quantization with $Q_{level} = 4$.

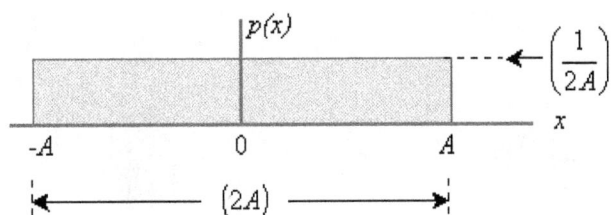

Figure 5.15: Uniform PDF.

For a given PDF, the total area under $p(x)$ is equal to 1 so that

$$\int_{-\infty}^{\infty} p(x)dx = 1 \qquad (5.10)$$

For example, referring to Fig. 5.15 in which $p(x) = \frac{1}{2A}$ for all x,

$$\int_{-\infty}^{\infty} p(x)dx = \int_{-A}^{A} p(x)dx = \frac{1}{2A} \int_{-A}^{A} dx = \frac{1}{2A} [x]_{-A}^{A} = 1 \qquad (5.11)$$

as expected. Alternatively of course, since the PDF is simply a rectangle of height $\left(\frac{1}{2A}\right)$ and width $(2A)$, the area under the PDF is equal to $\left(\frac{1}{2A} * 2A\right) = 1$. Given that x is limited to a minimum and maximum values of $-A$ and A, respectively and $p(x)$ is uniform, the mean value x_m of x is clearly zero, or equivalently,

$$x_m = \int_{-\infty}^{\infty} xp(x)dx = \frac{1}{2A} \int_{-A}^{A} xdx = \frac{1}{2A} \left[\frac{x^2}{2}\right]_{-A}^{A} = 0 \qquad (5.12)$$

Furthermore, the variance σ_m^2 is given by

$$\sigma_m^2 = \int_{-\infty}^{\infty} (x - x_m)^2 p(x)dx = \frac{1}{2A} \int_{-A}^{A} x^2 dx = \frac{1}{2A} \left[\frac{x^3}{3}\right]_{-A}^{A} = \frac{A^2}{3} \qquad (5.13)$$

Having analyzed the PDF presented in Fig. 5.15, consider now the uniform quantizer input x and output y relationship presented in Fig. 5.16. The number of quantization levels Q_{level} in this specific example is 8 although the analysis to follow is valid for any Q_{level}.

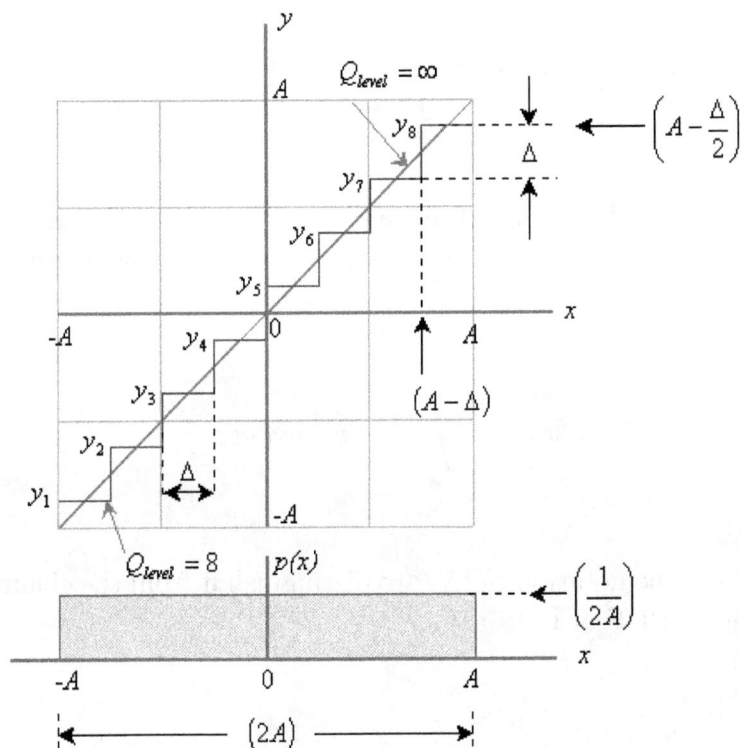

Figure 5.16: Eight level uniform quantization of a random variable.

The difference between the input and the output signals of the quantizer introduces *quantization noise*. Accordingly, the performance of a given quantizer is gauged by the *signal-to-quantization noise ratio (SQNR)* defined by

$$SQNR = \frac{P_s}{\int\limits_{-\infty}^{\infty} (x - y)^2 \, p(x)dx} \tag{5.14}$$

where P_s is the input signal average power and the denominator is the *average quantization noise power*, or equivalently the mean squared error x_{mse} given by

$$x_{mse} = \int\limits_{-\infty}^{\infty} (x - y)^2 \, p(x)dx \tag{5.15}$$

We shall use the notation $SQNR$ to distinguish it from the channel SNR. For the uniform PDF in Fig. 5.16,

$$x_{mse} = \int\limits_{-A}^{A} (x - y)^2 \, p(x)dx$$

$$= \frac{1}{2A} \left[\int\limits_{-A}^{-A+\Delta} (x - y_1)^2 \, dx + \int\limits_{-A+\Delta}^{-A+2\Delta} (x - y_2)^2 \, dx + \cdots + \int\limits_{A-\Delta}^{A} (x - y_{Qlevel})^2 \, dx \right] \tag{5.16}$$

where $y_1, y_2, y_3, \cdots, y_{Qlevel}$ are the quantizer outputs as indicated in Fig. 5.16 for which $y_{Qlevel} = y_8 = A - \frac{\Delta}{2}$. Notice for example in the integral

$\int\limits_{-A}^{-A+\Delta} (x-y_1)^2\, dx$, the term $(x-y_1)$ is the difference between the input and the output of the quantizer. Thus by symmetry

$$\int\limits_{-A}^{-A+\Delta} (x-y_1)^2\, dx = \int\limits_{-A+\Delta}^{-A+2\Delta} (x-y_2)^2\, dx = \cdots = \int\limits_{A-\Delta}^{A} (x-y_{Q_{level}})^2\, dx \quad (5.17)$$

Given that the number of quantization levels $Q_{level} = \frac{2A}{\Delta}$, the x_{mse} can be further simplified to

$$x_{mse} = \frac{1}{2A}\left[\frac{2A}{\Delta}\int\limits_{A-\Delta}^{A} (x-y_{Q_{level}})^2\, dx\right] = \frac{1}{\Delta}\int\limits_{A-\Delta}^{A} (x-y_{Q_{level}})^2\, dx \quad (5.18)$$

Using the integration by substitution method, let $u = (x-y_{Q_{level}})$ for which $\frac{du}{dx} = 1$. The new integration limits are listed in the table below.

	Integration limits	
x	$u = (x-y_{Q_{level}}) = x - \left(A-\frac{\Delta}{2}\right)$	
$A-\Delta$	$A-\Delta-A+\frac{\Delta}{2} = -\frac{\Delta}{2}$	
A	$A-A+\frac{\Delta}{2} = \frac{\Delta}{2}$	

Thus,

$$x_{mse} = \frac{1}{\Delta}\int\limits_{-\frac{\Delta}{2}}^{\frac{\Delta}{2}} u^2\, du = \frac{1}{\Delta}\left[\frac{u^3}{3}\right]_{-\frac{\Delta}{2}}^{\frac{\Delta}{2}} = \frac{\Delta^2}{12} \quad (5.19)$$

As expected, the average quantization noise power $= \frac{\Delta^2}{12} = \frac{A^2}{3Q_{level}^2}$ goes to zero as $Q_{level} \longrightarrow \infty$. Hence, the $SQNR$ is given by

$$SQNR = \frac{P_s}{\int\limits_{-\infty}^{\infty} (x-y)^2 \, p(x) dx} = \frac{P_s}{\frac{\Delta^2}{12}} = \frac{12 P_s}{\Delta^2} = \frac{3 P_s Q_{level}^2}{A^2} \qquad (5.20)$$

which expressed in decibels is given by

$$SQNR(dB) = 10 \log_{10} \left(\frac{3 P_s Q_{level}^2}{A^2} \right) \qquad (5.21)$$

For example, if the input signal is a sinusoidal waveform $s(t) = A \cos \left(\frac{2\pi t}{T} \right)$, then as shown in section 1.1, the average signal power $P_s = \frac{A^2}{2}$. Thus, $SQNR(dB) = 10 \log_{10} \left(\frac{3 Q_{level}^2}{2} \right)$, which for $Q_{level} = 8$ is equal to $SQNR(dB) = 19.8$ dB. Finally, let k represent the number of binary digits per quantization level (or equivalently per sample) so that $Q_{level} = 2^k$, then

$$SQNR(dB) = 10 \log_{10} \left(\frac{3 P_s 4^k}{A^2} \right) = 10 \log_{10} \left(\frac{3 P_s}{A^2} \right) + k 10 \log_{10} (4) \quad (5.22)$$

$$\simeq 10 \log_{10} \left(\frac{3 P_s}{A^2} \right) + 6k \qquad (5.23)$$

If k is increased by 1, the $SQNR$ will increase by 6 dB. This important feature is often referred to as the "6-dB rule".

Peak and Average SQNR

The *peak* signal power A^2 to the average quantization noise power ratio is given by

$$SQNR_{peak} = \frac{A^2}{\frac{\Delta^2}{12}} = 12 \left(\frac{A}{\Delta} \right)^2 = 12 \left(\frac{Q_{level}^2}{4} \right) = 3 Q_{level}^2 \qquad (5.24)$$

For example if $Q_{level} = 8$, then $SQNR_{peak}(dB) = 10 \log_{10} 3Q_{level}^2 = 22.8$ dB. If we take into account the probability of a binary digit error P_e within a PCM word, then it can be shown [Couch, 2001] that

$$SQNR_{peak} = \frac{3Q_{level}^2}{1 + 4\left(Q_{level}^2 - 1\right)P_e} \tag{5.25}$$

A plot of this equation is shown in Fig. 5.17 for various quantization levels versus P_e. As expected, the $SQNR_{peak}$ improves as Q_{level} is increased. The key feature is that provided P_e is less than $\left(\frac{1}{4Q_{level}^2}\right)$, $SQNR_{peak}(dB)$ is essentially independent of P_e, which indicates that the analog signal is corrupted primarily by quantizing noise and not channel noise. Given that we ensure P_e to be at least less than 10^{-6} for any given digital communication system using error-control coding, we may take $SQNR_{peak} = 3Q_{level}^2$. As we shall see in section 5.4, the 56K modem date rate is limited mainly by quantization noise and not by channel errors. Specifically for an input signal with $p(x) = \frac{1}{2A}$, the average signal power $P_s = \int\limits_{-\infty}^{\infty} x^2 p(x) dx = \frac{1}{2A} \int\limits_{-A}^{A} x^2 dx = \frac{A^2}{3}$ to the average quantization noise power ratio is given by

$$SQNR_{average} = \frac{\frac{A^2}{3}}{\frac{\Delta^2}{12}} = \frac{1}{3} SQNR_{peak} = \frac{Q_{level}^2}{1 + 4\left(Q_{level}^2 - 1\right)P_e} \tag{5.26}$$

which for a very small P_e, simply reduces to $SQNR_{average} = Q_{level}^2$.

- ⭐ SIMULATION **SQNRpeak**: Simulation corresponding to Fig. 5.17. Verify for yourself that for $P_e < \left(\frac{1}{4Q_{level}^2}\right)$, the $SQNR_{peak}(dB)$ is essentially independent of P_e.

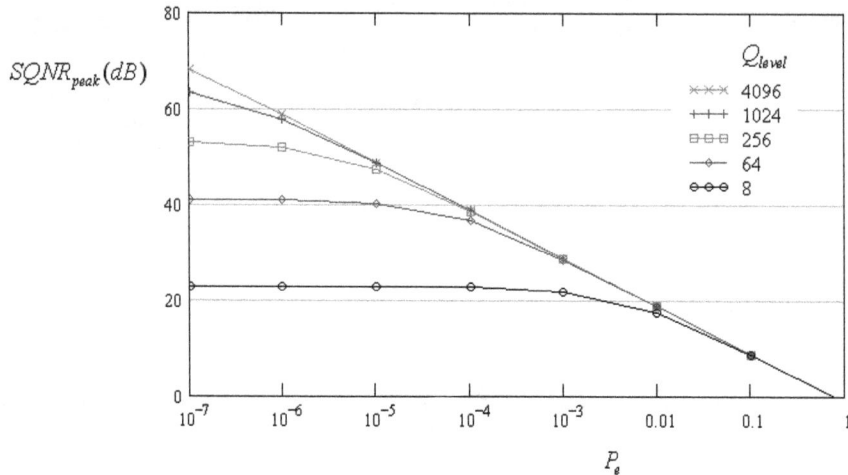

Figure 5.17: $SQNR_{peak}$ versus P_e for various quantization levels.

5.2.2   Simulation : Uniform Quantization of Various Signals

⭐ SIMULATIONS: In the simulation files listed below, the simulation and the theoretical $SQNR$ (where possible) for a variety of input signals are compared over a range of quantization levels. For example, if the input to the quantizer is a uniformly distributed random value x with a mean value $x_m = 0.5$ as in Fig. 5.18, with the corresponding probability histogram presented in Fig. 5.19, we find an excellent correspondence between the theoretical and simulated $SQNR$ values over a range of quantization levels as illustrated in Fig. 5.20.

- **Experiment:** Try other types of input signals.

Input Signal	FileName
Sinusoidal	**QSinusoidal**
Random variable: Uniform PDF	**QUniform**
Random variable: Gaussian PDF	**QGaussian**
Random variable: Large & small amplitudes	**QGaussianMixed**

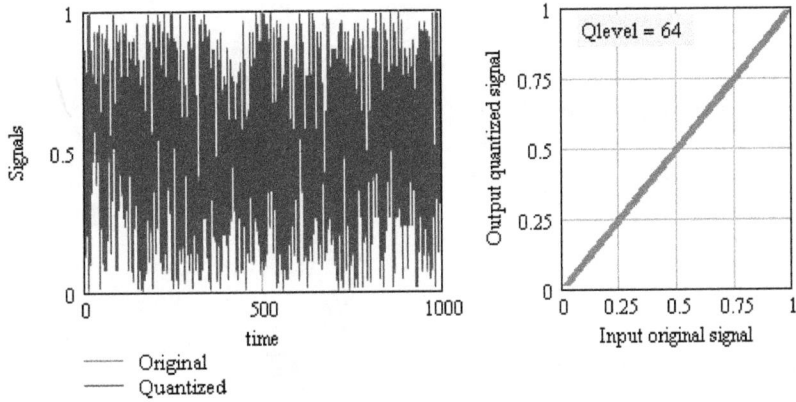

Figure 5.18: Uniform quantization of a random signal with $Q_{level} = 64$.

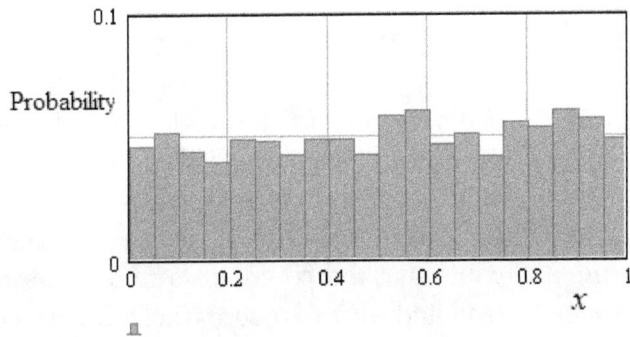

Figure 5.19: Probability histogram of the random signal.

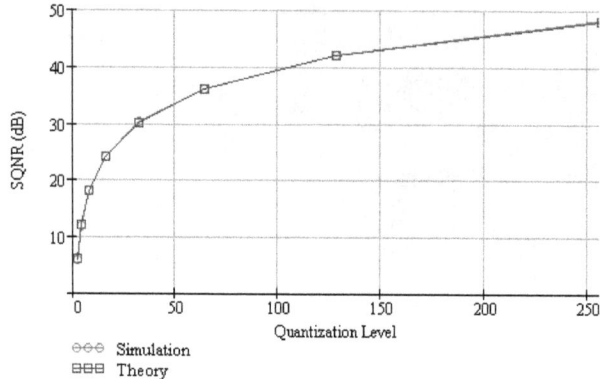

Figure 5.20: Theoretical and simulated $SQNR$ for a input signal with a uniform PDF.

5.3   Nonuniform Quantization

If large amplitudes are relatively rare in the analog signal (e.g. voice), the use of uniform quantization is not efficient. It is better to reduce or increase the spacing Δ between each quantization level when the amplitude of the signal is small or large, respectively. That is, we require the quantization to be *non-uniform*. A flexible method to create a non-uniform quantizer is to make use of a *compressor* and an *expandor*, whose functions respectively we shall denote by $y_{com}(x)$ and $y_{pan}(x)$, on either side of a uniform quantizer as illustrated in Fig. 5.21. The uniform quantizer output is denoted by $y_Q(x)$. The combination of a compressor and an expandor is referred to as a *compandor* and their operation referred to as *companding*.

The expandor performs the inverse operation of a compressor. For example, the North American μ-law standard is as follows:

$$y_{com}(x) = \left\{ \frac{\log_e\left[1 + \mu\frac{|x|}{x_{max}}\right]}{\log_e(1 + \mu)} \right\} \operatorname{sgn}(x) \qquad (5.27)$$

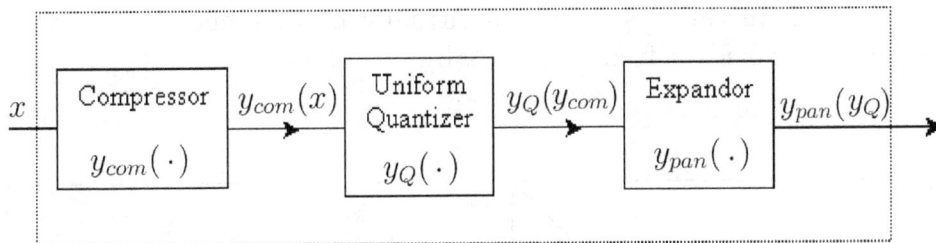

Figure 5.21: Nonuniform quantizer.

$$y_{pan}(x) = x_{\max} \left\{ \frac{(1+\mu)^{|x|} - 1}{\mu} \right\} \mathrm{sgn}(x) \tag{5.28}$$

where $x_{\max}$ is the maximum value of x, μ is a constant that is normally set to 255 and $\mathrm{sgn}(x) = \begin{cases} +1 & \text{if } x \geq 0 \\ -1 & \text{if } x < 0 \end{cases}$. The ratio $\frac{|x|}{x_{\max}}$ simply ensures the input is normalized. Hence the output is within the range $-1 \leq y_{com}(x) \leq 1$. Since $y_{pan}(x)$ is the inverse function of $y_{com}(x)$, if the output of the compandor is input directly into the expandor, then $y_{pan}(y_{com}(x)) = x$. The alternative European A-law standard compressor is given by

$$y_{com}(x) = \begin{cases} \left\{ \dfrac{A\frac{|x|}{x_{\max}}}{1+\log_e(A)} \right\} \mathrm{sgn}(x) & \text{if } 0 \leq \frac{|x|}{x_{\max}} \leq \frac{1}{A} \\[4mm] \left\{ \dfrac{\log_e\left(A\frac{|x|}{x_{\max}}\right)}{1+\log_e(A)} \right\} \mathrm{sgn}(x) & \text{if } \frac{1}{A} \leq \frac{|x|}{x_{\max}} \leq 1 \end{cases} \tag{5.29}$$

where the constant A is normally set to 87.6 to conform with the American standard as illustrated in Fig. 5.22. Referring to Fig. 5.22(a), as μ tends towards the value zero, the μ-law function $y_{com}(x)$ tends towards a straight line at 45 degrees. Similarly, as A tends towards the value 1, the A-law function $y_{com}(x)$ tends towards a straight line at 45 degrees. Indeed for $A = 1$, we find $y_{com}(x) = x$. To ensure compatibility between an American and European telephone system, it was agreed to set $Q_{level} = 256$, $\mu = 255$

and $A = 87.6$. In this case, there is virtually no difference the μ-law and A-law curves as evident from Fig. 5.22(b).

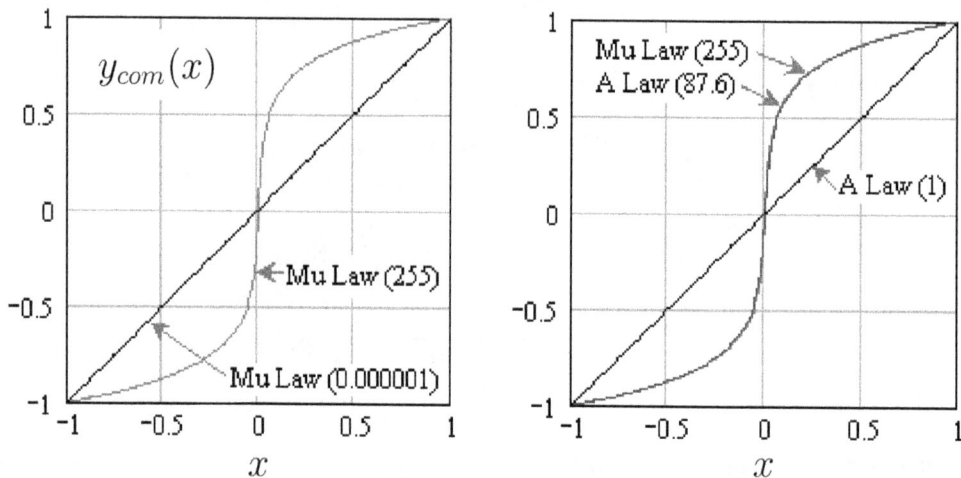

Figure 5.22: (a) North American and (b) European compressor functions.

• ★ SIMULATION **Mulaw:** For the Mathcad/MATLAB code corresponding to Fig. 5.22. Try different values for μ and A to verify for yourself that the international standard values are reasonable.

Using a uniform quantizer of the type presented in the previous section, Fig. 5.23 presents an overlay of the input signal $x = 2\sin(\frac{t}{30})$, the compressor output $y_{com}(x)$ and the quantizer output $y_Q(y_{com})$ with $Q_{level} = 16$. The expandor output signal $y_{pan}(y_Q)$ is presented in Fig. 5.24(a) together with the original input signal x. To highlight the overall non-uniform quantization effect, Fig. 5.24(b) presents a plot of $y_{pan}(y_Q)$ versus x. As a means of comparison, Fig. 5.25 presents the equivalent graphs if the compressor and the expandor were removed, in which case, the input signal x is uniformly quantized to produce the output $y_Q(x)$. Note the flexible nature of the compandor which allows us to increase or decrease the level of non-uniform quantization by simply increasing or decreasing, respectively, the value of μ. Notice also the key feature that the spacing between quantization levels is reduced as the signal amplitude becomes smaller. In summary, the non-uniform and uniform quantized signals are given by $y_{pan}(y_Q(y_{com}(x)))$ and $y_Q(x)$, respectively.

For an input sinusoidal signal, Fig. 5.26 shows the variation of the $SQNR$ in decibels over a range of quantization levels, together with the theoretical $SQNR$ that can be shown to be given by

$$SQNR(dB) \simeq 10 \log_{10} \left(\frac{3Q_{level}^2}{[\ln{(1+\mu)}]^2} \right)$$

$$\simeq 10 \log_{10} \left(\frac{3}{[\ln{(1+\mu)}]^2} \right) + 6k \tag{5.30}$$

Once again if k is increased by 1, the $SQNR$ will increase by 6 dB. Notice that the uniform quantizer outperforms the non-uniform quantizer. However, when the input signal is random, with occasional random large amplitude variations as illustrated in Fig. 5.27, the non-uniform quantizer $SQNR$ is better than the uniform $SQNR$ for a given quantization level as shown in Fig. 5.28. The compression of large input signal variations helps to improve the overall $SQNR$. A key feature of the design philosophy behind a compandor is that the $SQNR$ for a given quantization level is essentially independent of the input signal. That is, the non-uniform curves in Figs. 5.26 and 5.28 are essentially the same even though the input signals are very different. Finally, the difference between the input signal and the quantized signal becomes less, or equivalently the $SNQR$ increases, as we increase the number of quantization levels.

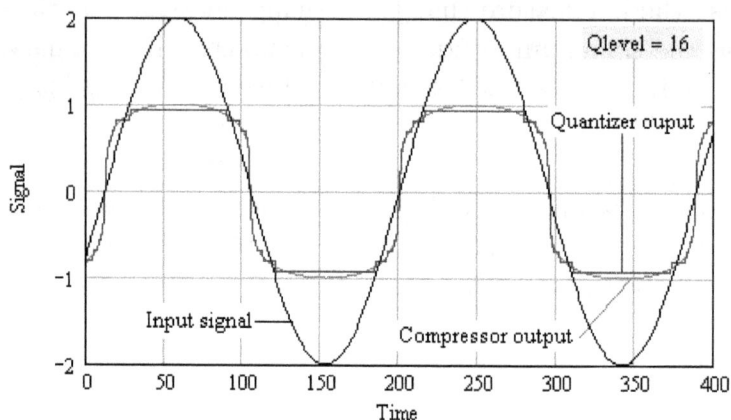

Figure 5.23: Sinusoidal input and the corresponding compressed output, before and after quantization.

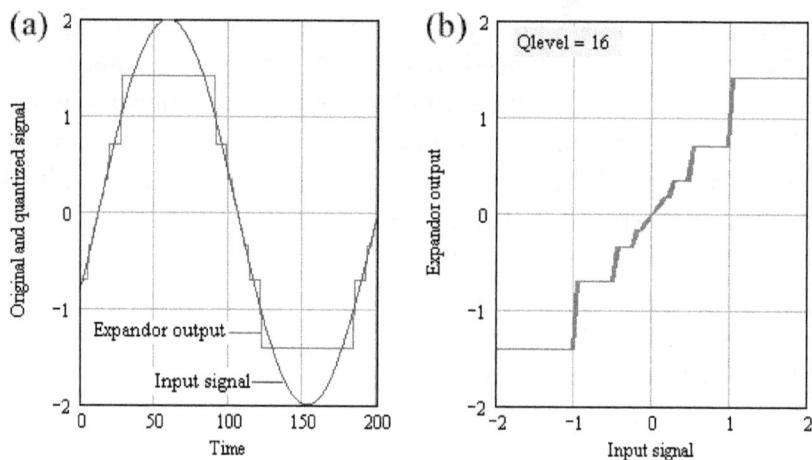

Figure 5.24: Input signal and expandor output signal comparison.

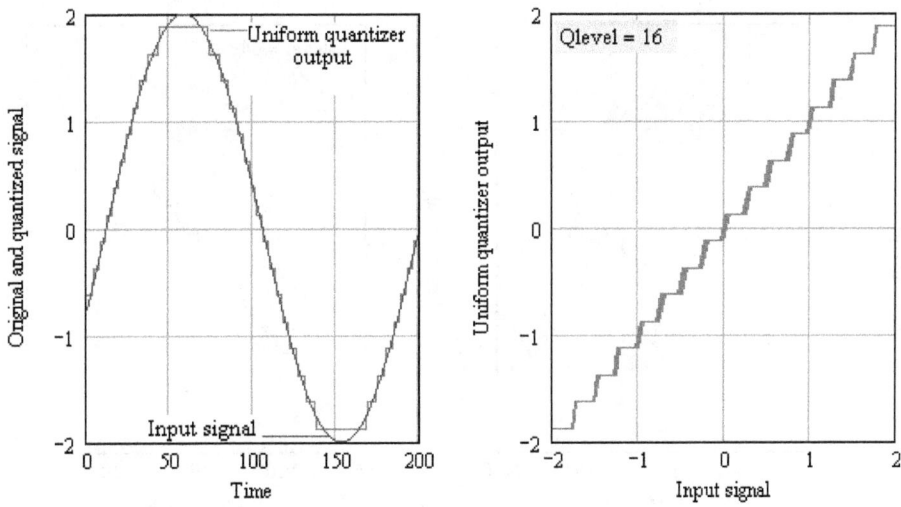

Figure 5.25: Uniform quantization of the input signal.

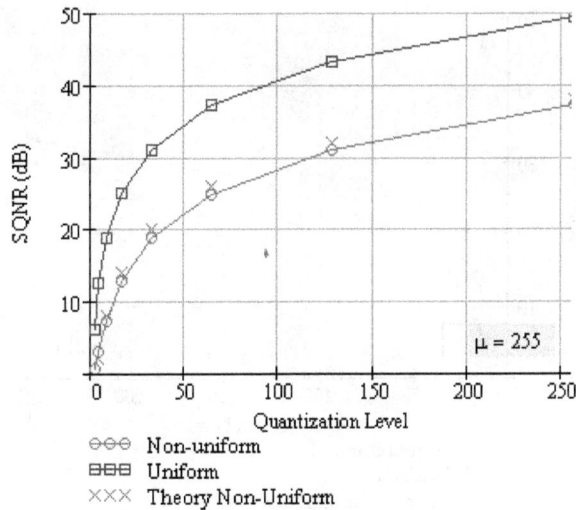

Figure 5.26: Performance of uniform and non-uniform quantizers for an input sinusoidal signal.

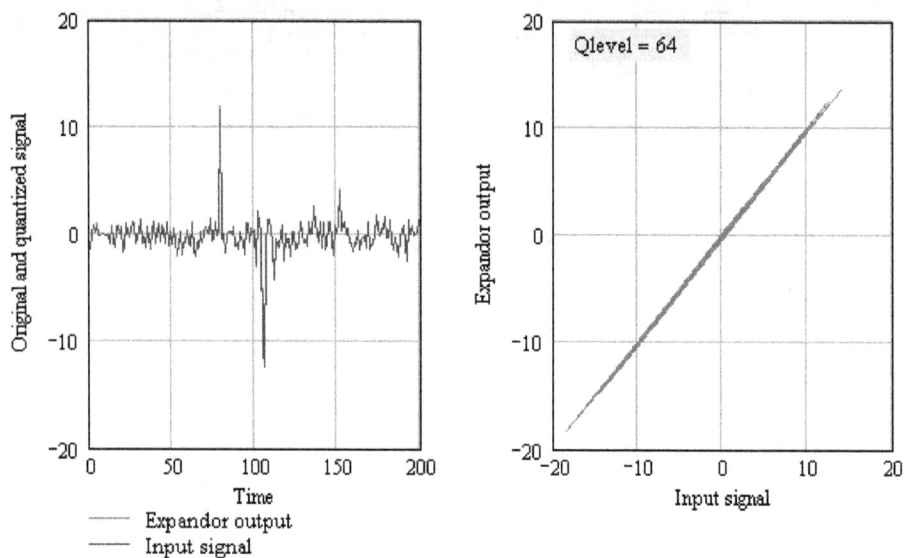

Figure 5.27: Input signal with large occasional amplitude fluctuations.

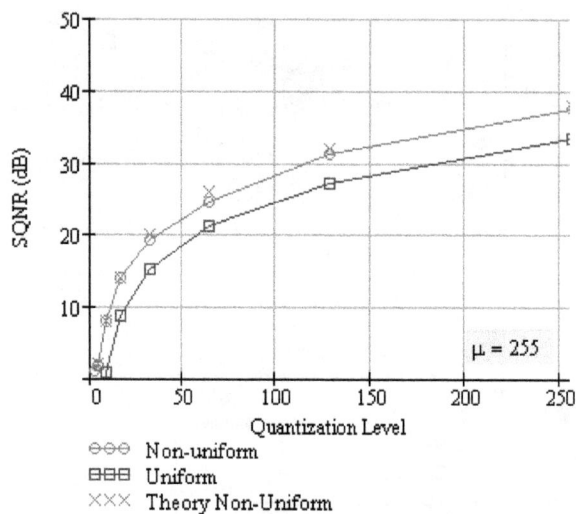

Figure 5.28: Performance of uniform and non-uniform quantizers for an input gaussian random signal with occasional large amplitude fluctuations.

5.3.1 Simulation : Nonuniform Quantization

⭐ SIMULATION **NonUniform:** The processing blocks presented in Fig. 5.21 are implemented.

- **Experiment:** Try a variety of input signals and observe the influence on the non-uniform and uniform $SQNR$ over a range of quantization levels. Verify that the non-uniform $SQNR$ is essentially independent of the input signal for a given quantization level.

5.4 PCM Communication System

Pulse code modulation (PCM) is analog-to-digital conversion in which the samples of an analog signal are represented by binary words to produce a serial bit stream, referred to as a *PCM signal*. PCM signals from all types of analog sources (audio, video, etc.) may be time division multiplexed with data signals (e.g. computers) and transmitted over a high-speed DCS (e.g. long distance telephone communication system). For example, a pulse-amplitude modulated (PAM) signal is quantized and transferred over a DCS by representing a given pulse height by a binary word. For example, using 8 quantization levels as presented in Table 5.1, the pulse amplitudes would be represented by binary words from zero (000) to seven (111). The binary digits are transferred over the channel via a modem as illustrated in Fig. 5.29. At the receiver, the quantized sample values are reconstructed and the analog signal recovered from these samples.

To illustrate a PCM communication system, we shall continue with our ongoing example from the previous section and assume a noiseless channel. The result of quantizing the sampled analog signal using a non-uniform μ-law quantizer is presented Fig. 5.30(a) for $Q_{level} = 8$ with $f_s = 16$ Hz and $\mu = 255$. The sampled signal before quantization is overlaid in this figure for comparison. Notice how the non-uniform quantizer compresses the quantized signal to lie within the range ± 1. The result of passing this sampled quantized signal through a low-pass filter with $f_c = \frac{f_s}{2}$ is presented in Fig. 5.30(b).

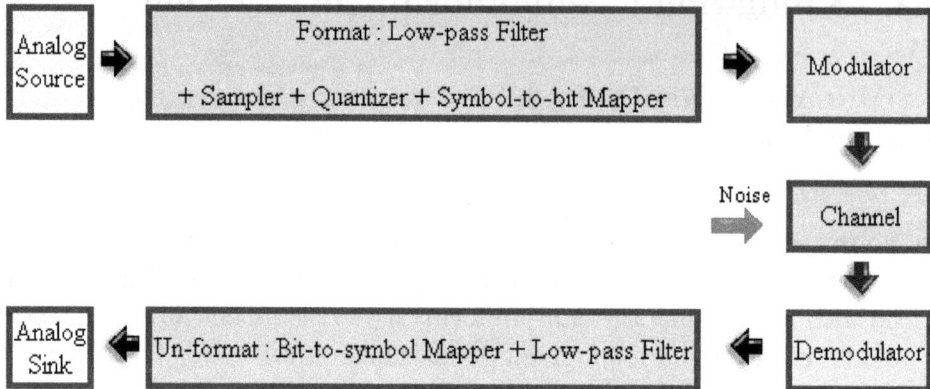

Figure 5.29: PCM communication system.

Performance is clearly improved by increasing the number of quantization levels as evident from Fig. 5.31. Thus by using an appropriate sampling frequency f_s and quantization levels Q_{level}, we can essentially recover the original signal. The symbol-to-bit mapper in Fig. 5.29 may for example, utilize the $Q_{level} = 8$ quantization level to a binary word mapping scheme presented in Table 5.1. To indicate whether the value is positive or negative, a fourth binary digit can be appended. For example, $+3 = 0111$ and $-3 = 0110$, where the appended digit is 1 or 0 to represent a positive or a negative value, respectively.

To prevent aliasing, minimize the required channel bandwidth and improve the $SQNR$, an anti-aliasing (low-pass) filter is used prior to the sampling block as indicated in Fig. 5.29. Although, the anti-aliasing may remove high frequency components from the input signal, the overall advantage (less distortion) is worth this penalty. Since ideal rectangular anti-aliasing filters are impossible to create in practice, the sampling frequency is set to be

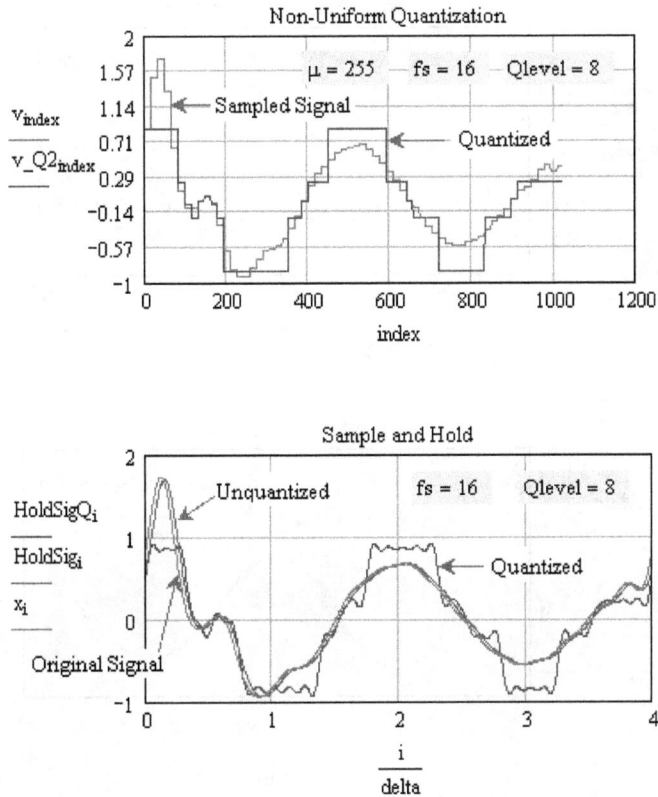

Figure 5.30: (a) Nonuniform quantization of a sampled analog signal with $f_s = 16$ Hz and $Q_{level} = 8$. (b) Signal recovered via a LPF and compared with unquantized case and the original signal.

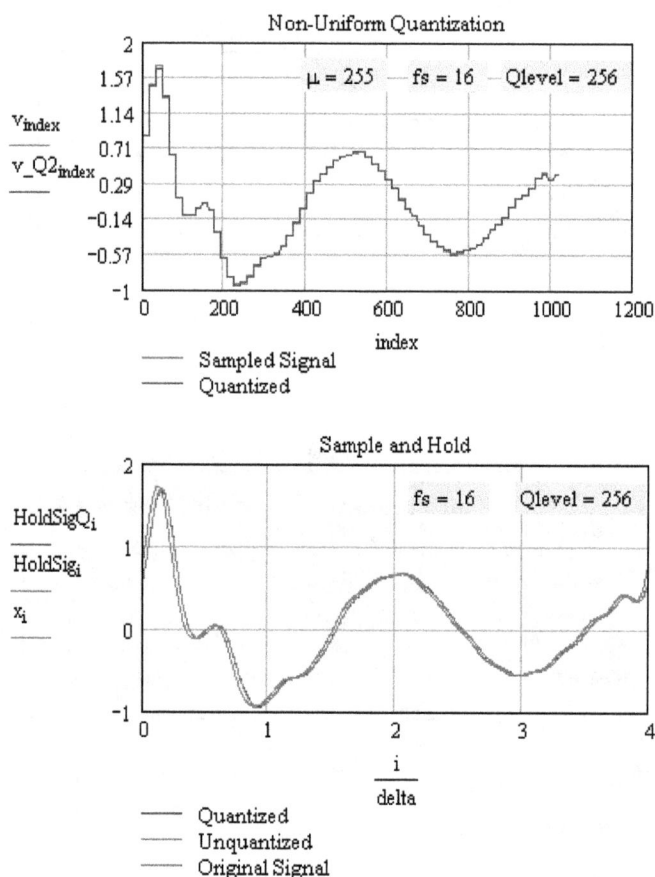

Figure 5.31: (a) Nonuniform quantization of a sampled analog signal with $f_s = 16$ Hz and $Q_{level} = 256$. (b) Signal recovered via a LPF and compared with unquantized case and the original signal.

slightly greater than twice the highest frequency component. For example, the highest frequency component in a voice signal need only be 3400 Hz to ensure intelligible speech. Although this corresponds to a Nyquist sampling rate of 6.8 kHz, the international telephone system standard requires $f_s = 8$ kHz with $Q_{level} = 256$ i.e. 8000 quantized samples per second which corresponds to a 64000 binary digits per second stream for a single one-way telephone conversation. Compare this with a music compact disc (CD) in which the sampling rate is 44.1 kHz with $Q_{level} = 65536$ i.e. 16 binary digits per sample. This high sampling frequency with a large number of quantization levels is necessary to ensure a high quality of sound. The spectrum of the PCM signal is not directly related to the spectrum of the input analog signal. In general, the bandwidth of a PCM signal is *bounded* by

$$B_{PCM} \geq \frac{r_b}{2} \tag{5.31}$$

where r_b is the number of binary digits per second given by

$$r_b = f_s \log_2 Q_{level} \tag{5.32}$$

The actual bandwidth B_{PCM} will depend on the pulse shape used and on the type of line encoding (refer to section 6.1.1). For example, using a rectangular pulse shape with polar non-return-to-zero (NRZ) line coding, the

Quantization level	Word
0	000
1	001
2	010
3	011
4	100
5	101
6	110
7	111

Table 5.1: Quantization level binary words.

first null bandwidth is simply r_b. It will become evident in section 6.3.1 that the lower bound $\frac{r_b}{2}$ (Hz) in equation 5.31 requires the use of $\frac{\sin(x)}{x}$ shaped pulses.

If the analog signal is band-limited to B Hz, then from the sampling theorem we require that $f_s \geq 2B$ and with $Q_{level} = 2^k$ (or $k = \log_2 Q_{level}$), we find that $B_{PCM} \geq kB$ i.e. in PCM, the required channel bandwidth increased by a factor of k. This is the main disadvantage of a PCM system. Given that quantization error is the dominant error in a PCM system, which can be reduced as desired by increasing k, the penalty incurred for improving the quality of a PCM system is an increase in the required channel bandwidth. Recall from the previous sections that for both uniform and non-uniform quantization, the $SQNR$ will increase by 6 dB if k is increased by 1. This is equivalent to an increase by a factor of $10^{0.6} \simeq 4$. Compare this with an increase in the required bandwidth by a factor of $\frac{k+1}{k}$ that is relatively smaller i.e. a small increase in the available bandwidth provides a great advantage.

Note that if there is no input analog signal, or if this signal is nearly constant, the sample values at the quantizer output can oscillate between two adjacent quantization levels. This will result in an unwanted tone of frequency $\frac{1}{2f_s}$ at the output of the PCM system, referred to as *hunting noise*. This can be reduced by using a quantizer which, unlike in Fig. 5.16, has no vertical step at $x = 0$.

5.4.1 Multilevel Signalling

We need not be constrained by the lower bound $\frac{r_b}{2}$ (Hz) if we resort to *baseband multilevel* signaling, in which n binary digits are mapped onto each pulse, which can now range over 2^n levels instead of only two ($M = 2$). For example, for $n = 2$, the mapping scheme could be to map the binary pairs 11, 10, 00 and 01 to the voltage levels +3, +1, -1 and -3, respectively. In section 6.7, the probability of an error in a binary digit for this scheme is determined. Similarly, in *bandpass multilevel* signalling, n binary digits are mapped onto each signal. For example, we could use the QPSK modulation scheme (section 7.5) for $n = 2$. The performance of various bandpass modulation schemes are considered in chapter 7.

5.4.2 Applications

Public Switched Telephone Network (PSTN)

Telephone lines (twisted pair of copper wires) from subscribers (e.g. homes and businesses) in a common geographical area enter the telephone central office (CO). This is referred to as the *local loop* (or the last mile), which in most cases is the only analog part of a digital *Public Switched Telephone Network* (PSTN) network. At the CO, the speech signal is first passed through an anti-aliasing low-pass filter with a bandwidth $B \leq 3400$ Hz, before sampling and quantizing. Specifically, this band-limited signal is sampled at a rate of 8000 samples/sec with each sample converted to an 8-binary digits PCM code word using either the μ-law (North America) or A-law (Europe) quantizer to a produce 64000 binary digits per second signal (64 kb/s or equivalently written 64 kbps), referred to as a *digital signal at level 0* (DS0) signal. The 64 kb/s DS0 PCM signal is time-division-multiplexed (TDM) with the PCM signals from other end-users attached to the CO to create a DS1 signal, which is also referred to as a *T1 signal*. For example in the North American T1 system, 24 voice channels, with each channel producing 64000 binary digits per second ($f_s = 8$ kHz, 8 binary digits/sample), are multiplexed to generate a T1 signal and packaged into frames. A frame consists of 193 binary digits (24 x 8 = 192 binary digits plus 1 framing binary digit) and 8000 frames are sent per second to generate a T1 signal at 1.544 Mb/s. At the same time, the downstream DS1 signal is demultiplexed to extract the 24 PCM signals for the end-users. Given that the time duration of a binary digit $T_b = \frac{1}{1544000}$ secs, the minimum bandwidth required theoretically to transmit a T1 signal is $\frac{1}{2T_b} = 772$ kHz. In practice, $B = \frac{(1+r)}{2T}$, where r is the filter roll-off factor. Taking $r = 1$, we require a bandwidth of $\frac{1}{T_b} = 1544$ kHz.

In the North American transmission hierarchy, four T1 signals are multiplexed to create a DS2 signal at 6.312 Mb/s, seven DS2 signals are multiplexed to create a DS3 signal at 44.736 Mb/s, six DS3 signals are multiplexed to create a DS4 signal at 274.176 Mb/s and finally, two DS4 signals are multiplexed to create a DS5 signal at 560.160 Mb/s. A DS5 signal, which corresponds to 8064 voice channels, is typically sent between switching centres via a fibre optic cable. In contrast, the European hierarchy multiplexes 30 voice channels, with each channel producing 64000 binary digits per second. A further 8 binary digits are used for addressing and 8 binary digits for synchronization. Thus, a frame consists of thirty two 8 binary digits words i.e. 256 binary digits per frame and 8000 frames are sent per second to generate a E1 signal at 256 x 8000 = 2.048 Mb/s. Four E1 signals are then multiplexed to generate a E2 signal at 8.448 Mb/s, four E2 signals are multiplexed to generate a E3 signal at 34.368 Mb/s, four E3 signals are multiplexed to generate a E4 signal at 139.264 Mb/s and finally four E4 signals are multiplexed to generate a E5 signal at 565.148 Mb/s, which corresponds to 7680 voice channels.

56K Modem

In the digital PSTN network, the μ-law or A-law quantization noise limits the signal-to-noise power ratio for the telephone channel to be approximately 30 dB. The quantization noise can be approximated as additive white gaussian noise (AWGN) provided the transmitted signal contains independent, identically-distributed random symbols, the symbol timing is independent of the sampling clock at the PCM encoder and the number of quantization levels is large. If we take the channel bandwidth to be 3.5 kHz, then from Shannon's channel capacity theorem (section 8.11.2), the channel capacity C is given by

$$C = B \log_2 \left(1 + \frac{E_b r_b}{N_o B} \right) = B \log_2 \left(1 + \frac{P}{N} \right) \tag{5.33}$$

where $P = E_b r_b$ is the signal power and $N = N_o B$ is noise power. For a signal-to-noise power ratio $\left(\frac{P}{N} \right) = 30$ dB, $C = 3500 \log_2 \left(1 + 10^{30/10} \right) \simeq 35$ kb/s. In practice, using the *quadrature amplitude modulation* (QAM) scheme (section 7.9) with *trellis-coded modulation* (TCM) (section 8.9), this limit is 33.6 kb/s.

If the Internet service providers (ISP) is digitally connected both to the Internet and to a telephone company's central office (CO), then in the downstream direction, the ISP sends an 8-bit digital signal, which is sent without A/D or D/A conversion (and therefore avoid quantization noise) over the PSTN to the CO. This 8-bit signal, which arrives at a rate of 8000 times per second, undergoes a single D/A conversion and is sent over the local loop as one of the 256 (0 to 255) analog signal levels using pulse amplitude modulation (PAM, section 6.7). Note that the output of a D/A converter is a Q_{level} digital signal and strictly an analog signal. Given that digital-to-analog conversion is unaffected by quantization noise and consequently there is no loss of information transmitted from the ISP's digital modem to the user's analog modem, we could argue that it should then be possible to transmit data from the ISP to the user at 64 kb/s because each pulse represents 8-bits and 8 bits/pulse x 8000 pulses/sec = 64 kb/s. However, the non-uniformly spaced 256 levels which make it difficult to distinguish between the more closely spaced levels, as well a bandwidth constraint of about 3500 Hz, limits the maximum achievable rate to 56,000 b/s. This is because the maximum Nyquist rate with no inter-symbol interference is $2B$ pulses/sec over a bandwidth of B and for a channel bandwidth of 3500 Hz with each pulse representing 8 bits, the maximum rate is 2(3500)(8) = 56,000 b/s. The solution is to make use of the most robust 128 amplitude levels. For 56 kb/s transmission, each symbol is maximally coded from 7 bits of each 8-bit PCM word. For approximately 52 kb/s transmission, 92 levels are used and so forth. Using fewer levels decreases the probability of an error at a cost of lower data rate.

In the V.90 standard modem, the upstream path includes an A/D converter and therefore, quantization noise limits the maximum data rate in the upstream direction to 35 kb/s. However, in the V.92 standard modem, the upstream direction also uses the PCM method and supports up to 48 kb/s using sophisticated cancellation methods to convey data simultaneously in both directions (i.e. full-duplex). Notice, the asymmetrical maximum speeds for end users to download and upload data from an ISP digitally connected to the PSTN. This is acceptable because users typically download information from the internet most of the time. Bear-in-mind that a web browser normally indicates the download speed in kB/s, which is kBytes/s and not kb/s, where 1 byte is 8 binary digits. A good approximation to the average download speed you can expect in kB/s for a given modem connection speed is to divide this number by 10 (instead of 8), to take into account the start, stop and parity bits e.g. 4.9 kB/s for a modem connection speed of 49000 b/s. Finally, by compressing the data immediately before sending it and having a matching decompression on the other side of the link, we can maximize the amount of data transferred via the 56K modem. The current data compression standard used in a 56K modem is the V.44, which provides a particularly noticeable improvement when browsing the web, because HTML files are highly compressible.

DSL

Digital subscriber line (DSL) is a very clever method to overcome the 56 kb/s bottleneck by utilizing the bandwidth between 26 kHz to 1100 kHz of the twisted-pair copper wire to transfer (full-duplex) digital data directly to the computer over the local loop. There are many DSL variations [Gitlin *et al.*, 1992] - *Asymmetrical*-DSL (ADSL), *high-bit-rate*-DSL (HDSL), *very-high-bit-rate*-DSL (VDSL), *integrated service digital network* (ISDN). Of these, ADSL has superseded 56K dial-up internet access throughout the world. ADSL provides a continuously available ("always on") connection and simultaneously accommodates analog (voice) information on the same line. A splitter is used, both at the CO and the user end, to separate the analog voice (below 4 kHz) from the download and upload passband transmissions (above 25 kHz). The bandwidth above the 4 kHz voice band is used to provide download data rates from typically 256 kb/s to a maximum of 6.1 Mb/s and upload data rates from 128 kb/s up to 640 kb/s, over the local loop to the customer via multiple subcarriers (up to 256) over a length of approximately 12,000 feet. ADSL is called "asymmetric" because of the asymmetrical manner in which the download data rate is very much greater than the upload data rate. The subcarriers used for upload and download occupy the 26-138 kHz band and the 138-1100 kHz band, respectively. Each subcarrier, separated by 4.3125 kHz, behaves as a virtual modem using *orthogonal frequency division multiplexing* (OFDM) (refer to section 7.10) with variable size QAM signal constellations for each of the subcarriers. This is also referred to as *discrete multitone* (DMT) modulation. The aggregate of the virtual modems dictates the total available speed.

Digital Audio

The left and right audio signals of a stereo recording are sampled and uniformly quantized using a sampling rate of 44.1 kHz with $Q_{level} = 2^{16}$ to ensure excellent sound quality. Although the human upper limit for audible sound is approximately 20 kHz, which corresponds to a Nyquist sampling rate of 40 kHz, the sampling rate is 44.1 kHz to prevent aliasing. A given digital audio can be compressed by removing sounds which either the human ear cannot hear, or which are disguised by other sounds e.g. via the *MPEG-1 Layer 3* (MP3) format which has revolutionized the way people listen and store music.

5.4.3 Simulation : PCM

SIMULATION **PCMSampleandHold:** The PCM system presented in Fig. 5.32 is modeled. In addition to verifying the figures presented in this section, you may experiment with the pre-filter cut-off frequency f_c, sampling frequency f_s, quantization level Q_{level} and quantizer type (uniform and non-uniform), the input signal $s(t)$, etc. and view the influence on the spectral content of the signal in between any processing block within the PCM system.

SIMULATION **PCMAudio**: An audio signal is sent through the PCM system. A *.wav file is imported as the input signal and the output signal with uniform, non-uniform and un-quantized processing is saved as a *.wav file. The audio file is written to the hard drive at various stages within the communication system.

- **Experiment:** Alter the numerous variables in this simulation and listen to the effect of your changes. e.g. the sampling rate. Record your own voice and send it through the PCM system. For example, the voice recording of my wife Renuka is stored in the file renuka.wav. Figures 5.33 to 5.37 show this signal and its spectrum at various points (A to E) within the noiseless communication system using $Q_{level} = 256$, $\mu = 255$, $f_s = 8000$ Hz, $f_c = 4000$ Hz at the transmitter and $f_c = 2000$ Hz at the receiver. Convince yourself why a telephone line works well with a bandwidth of approximately 3.5 kHz. Finally, if you are feeling adventurous, represent the quantized signal with binary digits and send these over a BSC. This final step is easy. I have left it out for your benefit!

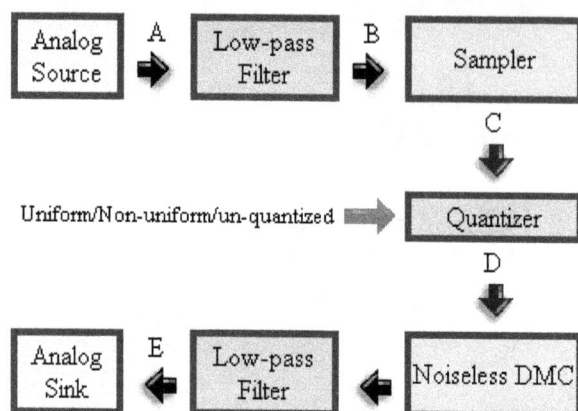

Figure 5.32: Simulated PCM communication system.

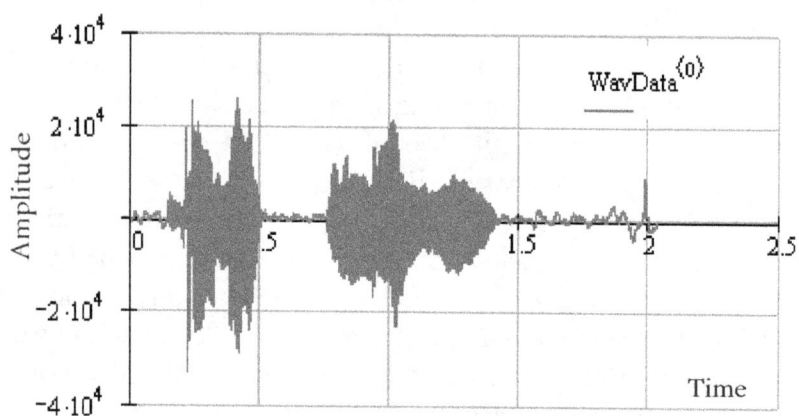

Figure 5.33: Input voice signal at point A.

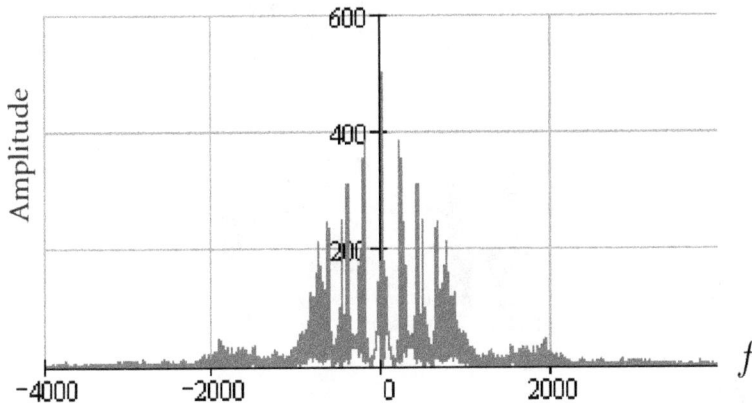

Figure 5.34: FFT amplitude spectrum of the input signal with $f_c = 4000$ Hz at point B.

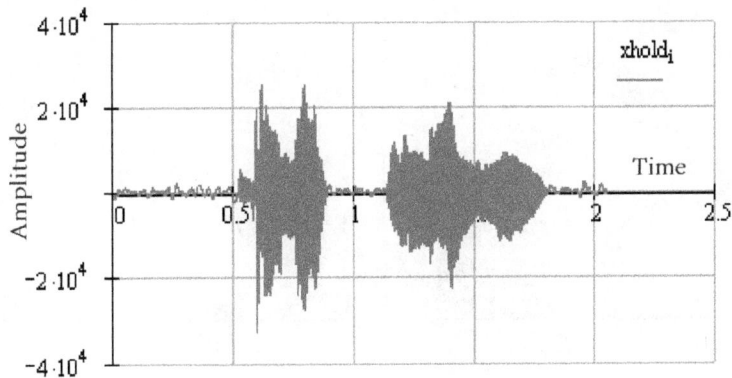

Figure 5.35: Sample and hold output using $f_s = 8000$ Hz at point C.

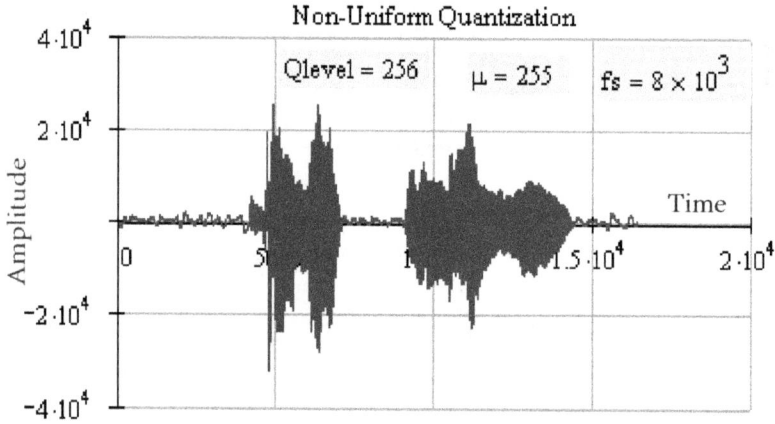

Figure 5.36: Quantized signal at point D.

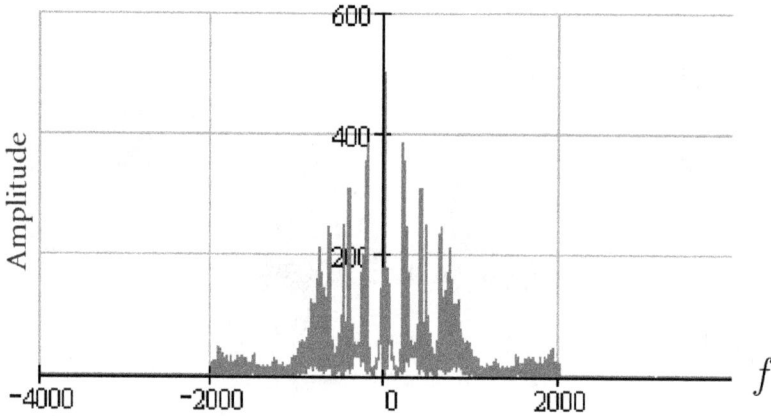

Figure 5.37: FFT amplitude spectrum of the output signal with $f_c = 2000$ Hz at point E.

5.5 Delta Modulation (DM)

Delta modulation is a simple analog-to-digital technique in which the input signal is sampled at a much higher rate than the Nyquist rate to exploit the correlation between adjacent samples. The basic concept is to create a staircase approximation to the input signal as follows. At each sample time, if the current sample value x_i is larger then the *previous* predicted value $x_{i-1}^{(p)}$, then the *current* predicted value $x_i^{(p)}$ is increased by a small fixed amount Δ (delta). Conversely, if x_i is less then $x_{i-1}^{(p)}$, then $x_i^{(p)}$ is decreased by a small fixed amount Δ (delta). Equivalently,

$$x_i^{(p)} = x_{i-1}^{(p)} + \Delta \left[signum \left(x_i - x_{i-1}^{(p)} \right) \right] \tag{5.34}$$

where the function $signum\,(z)$ returns $+1$ if $z \geq 0$, otherwise -1. Thus $x^{(p)}$ is a staircase signal which either increases or decreases by Δ every $\frac{1}{f_s}$ seconds as illustrated in Fig. 5.38. Notice that $x^{(p)}$ is held constant at $x_{i-1}^{(p)}$ between samples.

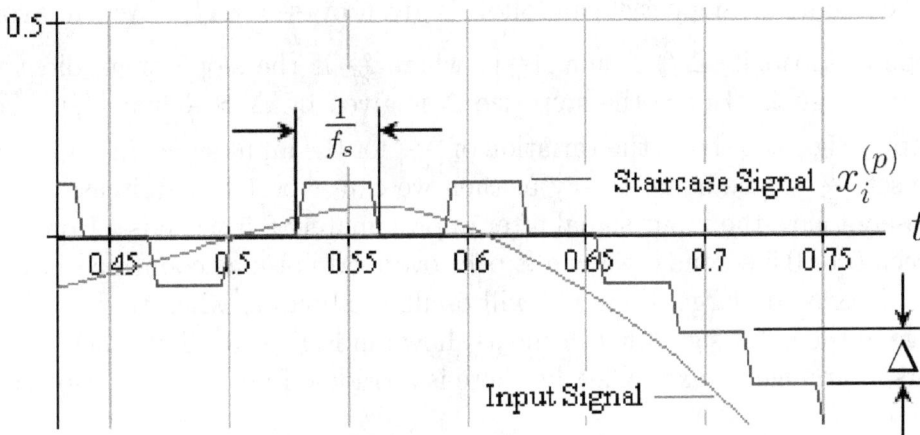

Figure 5.38: Input signal and its staircase approximation.

Given that $x^{(p)}$ can only rise or fall by an amount Δ in $\frac{1}{f_s}$ seconds, it may not be able to closely follow the input signal as illustrated in Fig. 5.39. This phenomena is referred to as *slope overload* which leads to *overload noise*. In Fig. 5.39 and in all the figures to be shown in this section, f_s was set to $4f_{max}$ to make the staircase clearly visible.

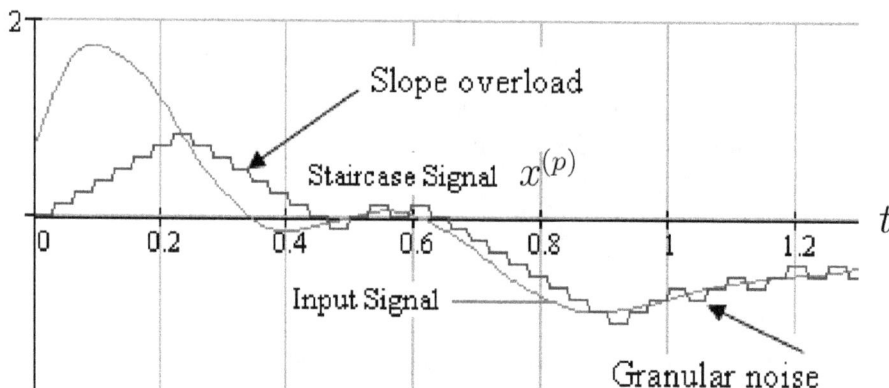

Figure 5.39: Slope overload and Granular noise.

The maximum slope $x^{(p)}$ can follow is given by $\frac{\Delta}{\frac{1}{f_s}} = \Delta f_s$. We can avoid a slope overload if $\Delta f_s > \max\left|\frac{dx}{dt}\right|$, where $\frac{dx}{dt}$ is the slope or gradient of the input signal. Hence the step size Δ is given by $\Delta > \frac{1}{f_s}\max\left|\frac{dx}{dt}\right|$. For example, Fig. 5.40 shows the variation of $\left|\frac{dx}{dt}\right|$ for the input signal in Fig. 5.39. If we set $\left|\frac{dx}{dt}\right|$ to be equal to say 5, then we can expect the staircase signal to closely follow the input signal after approximately 0.5 seconds. However, between 0 to 0.5 seconds, we can expect overload noise as confirmed in Fig. 5.41. If Δ is too large, then $x^{(p)}$ will oscillate about x_i when there is little change in the input signal with time as shown in Fig. 5.39. This phenomena results in *granular noise*. Clearly, there is a trade-off between the two types of noise.

Having created a staircase signal which closely follows the input signal, the output of the delta modulator is a binary sequence created as follows. For each step rise, the output is a binary digit 1 and for each step fall, the

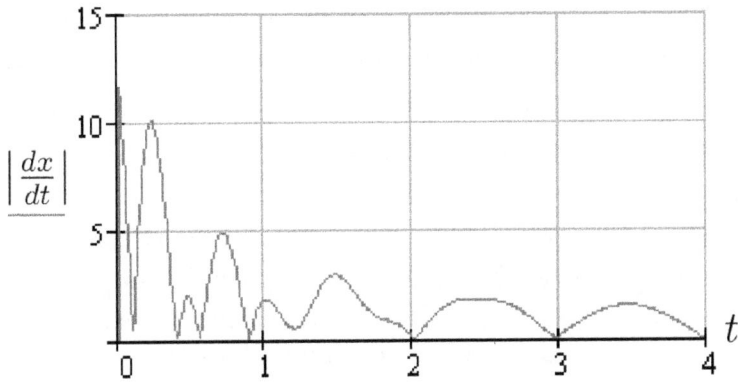

Figure 5.40: Magnitude of the slope variation with time.

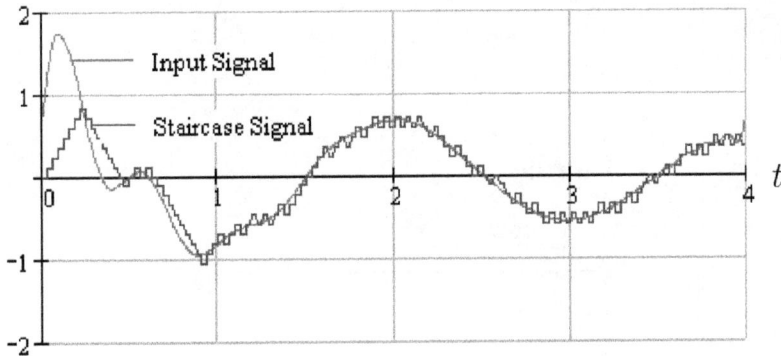

Figure 5.41: Input and staircase signals for $\Delta = \frac{5}{4f_{\max}}$.

output is a binary digit 0. Thus each sample corresponds to only either a single binary digit 1 or 0. In this case, the number of binary digits per second is given by $r_b = f_s \log_2 Q_{level} = f_s \log_2 2 = f_s$. Compared with PCM, r_b is not necessarily reduced with the use of DM because f_s is very much larger than in PCM. Note that we can use a quantizer with $Q_{level} > 2$ to map the error $\left(x_i - x_{i-1}^{(p)} \right)$ on a quantization level which is then transmitted via binary digits as in PCM. But given that the sampling rate is very high to ensure the error $\left(x_i - x_{i-1}^{(p)} \right)$ is very close to zero, it is feasible to set $Q_{level} = 2$. At the receiver, the delta demodulator simply reconstructs the staircase signal by increasing or decreasing the output by Δ depending on whether a binary 1 or 0 is received, respectively. The reconstructed staircase signal is then passed through a low-pass filter to smooth out the sharp edges as illustrated in Fig. 5.42 for f_s equal to $4f_{\max}$ and $8f_{\max}$.

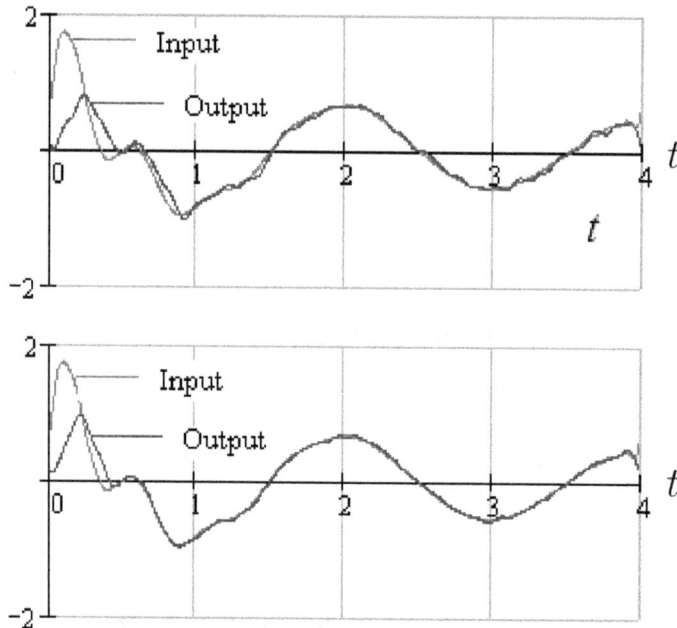

Figure 5.42: Input and demodulated signal with (a) $\Delta = \frac{5}{4f_{\max}}$ and (b) $\Delta = \frac{5}{8f_{\max}}$.

A key advantage of DM is that unlike PCM in which the severity of the error depends on the location of binary digit errors within the $\log_2 Q_{level}$ binary digits per quantization level, each binary digit in DM has equal importance. However, PCM will outperform DM provided the probability of an error in a binary digit P_e is sufficiently low enough (typically below 10^{-6}). This requirement is easily achieved with the use of error-control coding.

5.5.1 Adaptive Delta Modulation

In *adaptive delta modulation* (ADM), the magnitude of the step size Δ is allowed increase if successive values of $\left[signum \left(x_i - x_{i-1}^{(p)} \right) \right]$ are of the same polarity. Conversely, Δ is reduced if successive values of $\left[signum \left(x_i - x_{i-1}^{(p)} \right) \right]$ are of opposite polarity. Consider the following simple algorithm:

Step 1:
Initialize a variable *previous* $= 1$
Set the minimum step size Δ min

Step 2:
$current = \left[signum \left(x_i - x_{i-1}^{(p)} \right) \right]$

$$\Delta = \begin{cases} \frac{\Delta prev}{current} \left[current + 0.5 previous \right] & \text{if} \quad \Delta prev \geq \Delta \min \\ \\ \Delta \min & \text{otherwise} \end{cases}$$

$x_i^{(p)} = x_{i-1}^{(p)} + \Delta \left[signum \left(x_i - x_{i-1}^{(p)} \right) \right]$
previous $=$ *current*
$\Delta prev = \Delta$

Step 3:
Go back to step 2.

For example, if *previous* $=$ *current* $= 1$, then $\Delta = 1.5 \Delta prev$ i.e. Δ increases by a factor of 1.5 above its previous value. Not surprisingly, ADM typically outperforms DM. This feature can be traded for a reduction in r_b and therefore the bandwidth requirement.

5.5.2   Simulation : DM and ADM

⭐SIMULATION **DeltaModulation:** The DCS shown in Fig. 5.43 is implemented.

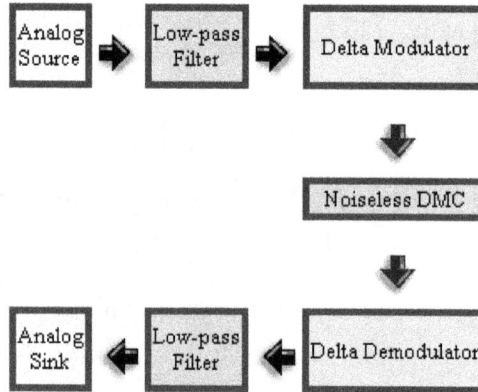

Figure 5.43: Delta Modulation and demodulation over a noiseless channel.

⭐SIMULATION **AdaptiveDeltaModulation:** The DM in Fig. 5.43 is replaced by the ADM. Given that the difference between the input signal and the staircase signal is no longer either $\pm\Delta$, the demodulator has to be adaptive to ensure the step size changes to match the modulator. For simplicity, this feature is inherently assumed within the simulation code. Simulation results are presented in Fig. 5.44 to compare DM and ADM over the same time period. Unlike DM, the ADM staircase step size increases from time $t = 0$ in order to catch the input signal.

- **Experiment:** Observe how the levels of overload and granular noise are effected by the sampling frequency and the step size Δ. Try various other input signals (e.g. sinusoidal waveform). Modify the simulation code to use a quantizer with $Q_{level} > 2$ to map the error $\left(x_i - x_{i-1}^{(p)}\right)$ on a quantization level which is then transmitted via binary digits. Verify that the ADM scheme improves the $SQNR$.

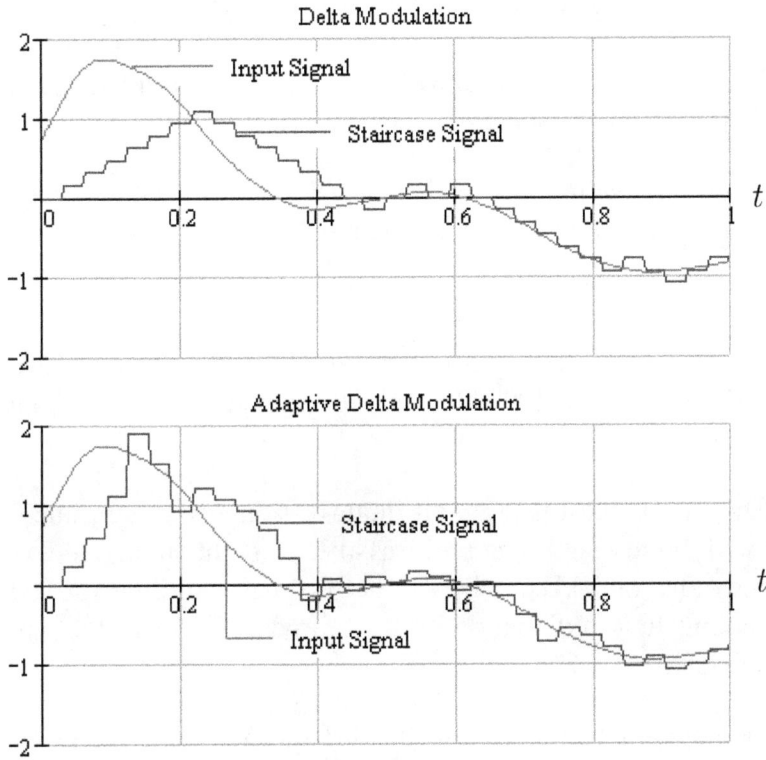

Figure 5.44: DM and ADM comparison.

5.6 Problems

1. Determine the Nyquist frequency f_s of sinc pulse given by $s(t) = \dfrac{\sin(At)}{\pi t}$ for $A = 2\pi$. Verify your answer by sampling and reconstructing this signal.

2. Suppose noise is added to the pulse in problem 1 such that

$$s(t) = \frac{\sin(At)}{\pi t} + 0.2\cos(2\pi 50 t) \text{ with } A = 2\pi.$$

Show that the noise can be removed by passing the sampled signal through a low-pass filter with a cut-off frequency $f_c = 1$ Hz.

3. If a sinusoidal signal $s(t) = A\cos(2\pi f_o t)$ is sampled at a frequency of $f_s = 1.5 f_o$, show that the reconstructed signal is given by $s(t) = A\cos(\pi f_o t)$ with the use of a low-pass filter with $f_c = \dfrac{f_s}{2}$.

4. An analog signal is transmitted via PCM using 256 quantization levels. Given that the bit rate is 100000 binary digits/sec, what is the bandwidth of the analog signal.

5. (a) Show that PCM using 8-bit quantization and a sampling frequency of 8000 Hz with binary signaling requires at least eight times the voice analog channel bandwidth of 4 kHz. (b) At the Nyquist sampling rate, show that binary signalling first null bandwidth required is $2\log_2(Q_{level})$ times the analog bandwidth.

6. A voice signal band-limited to 3800 Hz is to be sent via a binary PCM communication system. For virtually error-free transmission, we require r_b to be less than or equal to 56000 binary digits per second.

(a) Determine the appropriate quantization level Q_{level} and the sampling frequency f_s.
(b) The time duration T_b of a binary digit.
(c) The bandwidth (first null and minimum) of the binary PCM signal.
(d) The average signal-to-quantization noise power.

7. Verify the 6 dB rule which says that an additional 6 dB improvement in the $SQNR$ is obtained for each binary digit added to the PCM word.

8. (a) Approximately how many minutes worth of CD quality music (44.100 kHz, 16 bit stereo) music can be stored on a 700 MB CD-R ? (b) Analyze the audio file CdQuality.wav.

9. In a PCM communication system, suppose the channel can be modeled as a BSC channel with a crossover probability of 10^{-6}. What is quantization level required to ensure that the peak signal power to the average quantization noise power is greater than 40 dB ?

10. It can be shown the if P_e is the probability of an error in a binary digit in the binary PCM signal, then the average $SQNR = \dfrac{Q_{level}^2}{1 + 4\left(Q_{level}^2 - 1\right)P_e}$.

(a) For $P_e = 10^{-4}$ what is quantization level required to ensure the $SQNR$ is greater than 25 dB ?

(b) For $P_e = 10^{-3}$, what is the maximum possible $SQNR$ that can be achieved ?

(c) What happens as $P_e \to 0$?

11. Determine an expression for the time duration T_b of a binary digit necessary to ensure that the information throughput is not reduced by transmitting an analog signal via PCM system in which the sampling frequency is f_s and the number of quantization levels is Q_{level}.

12. A requirement for the magnitude of the quantization distortion error $|e|$ in terms of the peak-to-peak analog voltage $(2A)$ is $|e| \le \left(\dfrac{p}{100}\right)(2A)$ where p is a percentage.

(a) Given that $|e|_{\max} = \dfrac{x}{2}$, determine an expression for the minimum quantization level.

(b) If the quantization distortion must not exceed ± 2 % of the peak-to-peak analog signal, determine the required minimum number of bits per PCM word.

13. The frequency response of a low-pass Butterworth filter is such that $|H(f)| = \dfrac{1}{\sqrt{1 + \left(\dfrac{f}{f_c}\right)^{2n}}}$, where n is the number of capacitors or inductors used to create this filter and f_c is the cut-off frequency.

(a) Verify that as n is increased, the frequency response becomes closer to that of an ideal low-pass filter with a cut-off frequency f_c.

(b) If a voice signal is pre-filtered using a Butterworth filter with $n = 8$ and $f_c = 3000$ Hz, what is the sampling rate necessary to ensure that aliasing is reduced to the -40 dB point in the power spectrum ?

14. Simulate a random variable with a PDF as shown below. Determine the signal to quantization ratio in decibels $SQNR$ (dB) for a uniform quantizer with $Q_{level} = 8$. Compare your simulation result with the expected value given by $10 \log_{10} (Q_{level}^2)$.

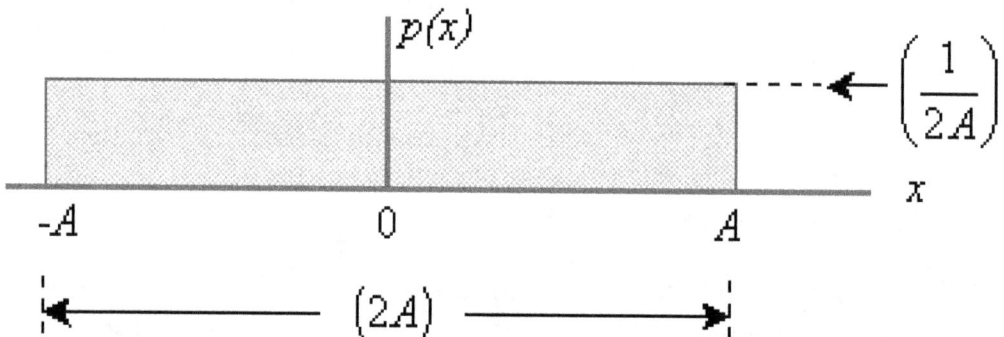

15. Consider a random variable with a PDF as shown below. Derive an expression for the signal to quantization ratio in decibels $SQNR$ (dB) for a uniform quantizer with $Q_{level} = 8$. Verify your theoretical result by simulation.

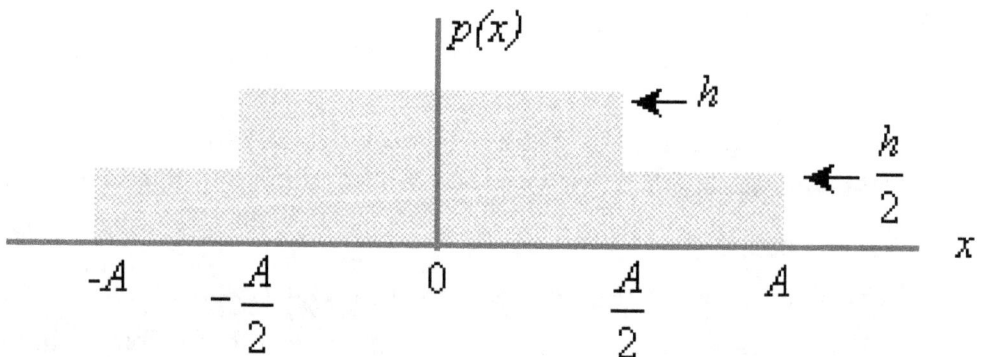

16. An ideal anti-aliasing filter with a cut-off frequency 3.4 kHz is used for a PCM voice communication system.

(a)   If a uniform quantization is employed, then the $SQNR(dB)$ may be expressed as $SQNR(dB) = 4.8 + 6k - \alpha_{dB}$ where $\alpha_{dB} = 10 \log_{10} \left(\dfrac{A^2}{P_s} \right)$. If $\alpha_{dB} = 12$ dB for the voice signal, how many binary digits per second would be required to represent this voice signal if we wish to ensure the $SQNR$ greater than 38 dB ?

(b) If μ-law compandor is used instead, with $\mu = 255$, determine the $SQNR$ in decibels with $Q_{level} = 256$.

17. Consider a signal $s(t)$ whose Fourier transform is given by

$$S(f) = \begin{cases} 1 & if & |f| \le 1 \\ 0 & otherwise \end{cases}.$$

If a "mains-hum" noise signal $n(t) = 0.2 \cos(2\pi 50t)$ is added to $s(t)$, show that by sampling the resultant signal $(s(t) + n(t))$ at 8 Hz and using an ideal low-pass filter with a cut-off frequency of 1 Hz, the mains-hum noise signal can be eliminated.

I am enough of an artist to draw freely upon my imagination. Imagination is more important than knowledge. Knowledge is limited. Imagination encircles the world.

— *Albert Einstein*

Chapter 6

Baseband Signaling

6.1   Introduction

In this chapter we focus on the DCS shown in Fig. 6.1 in which binary digits
are transferred over a *baseband channel*, namely, a channel with a frequency
passband that typically includes zero frequency ($f = 0$).   Notice that the
combination of the information source and the format block is modeled as
a binary source.   This is very typical for a DCS.   For example, the binary
source in Fig. 6.1 could represent the front-end of the PCM system which
outputs $r_b = f_s \log_2 Q_{level}$ binary digits per second.   Referring to Fig. 6.1,
the modulator maps the binary digits (at point C) onto electrical pulses
(at point D) that are transferred over a physical baseband channel.   This
mapping process is referred to as *line coding* or *transmission coding*.

A demodulator is used to retrieve the binary digits from the received
baseband signal with an aim to minimize the probability of an error in a
binary digit P_e at point F. Errors that occur at point F are the result of two
main effects.   Firstly, intersymbol interference (ISI) to be covered in section
6.3.1, which is due to the inevitable filtering that takes place in a practical channel, that causes the pulse shape to be distorted and interfere with
neighboring pulses.   Fortunately, with the use of appropriate filters at the
transmitter and receiver, it is possible to overcome this problem.   Secondly,
thermal noise which cannot be eliminated is due to the motion of electrons
within a conductor.   However, with the use of a matched filter (section 6.2)
within the demodulator to maximize the received signal-to-noise power ratio
and by employing error-control coding (chapter 8), the P_e at point F can be

reduced to an acceptable low value.

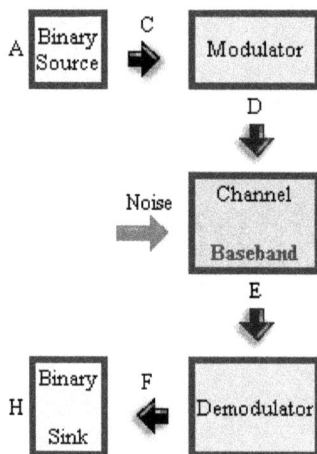

Figure 6.1: Chapter 6 DCS.

6.1.1   Line Codes

There are many possible ways in which binary digits can be mapped onto (or equivalently assigned to) electrical pulses as illustrated in Figs. 6.2 to 6.7 for the binary sequence 1100101.   These waveforms (pulses), which are called *line codes,* are only a sample of the many possible line codes [Sklar, 2001] and are often referred to by more than one name.   The selection of a given line code depends typically on which combination of the following features are most important for a given application:

- the power spectral density, in particular at 0 Hz

- bandwidth

- immunity to noise

- inherent synchronization features

- inherent error detection properties

Line codes can be classified as *return-to-zero* (RZ) or *non-return-to-zero* (NRZ) depending on whether the voltage returns to zero volts or not within its symbol interval, respectively. In Figs. 6.2 to 6.5, the mapping between a given binary digit and a pulse is obvious. For the bipolar RZ or *alternative mark inversion* (AMI) line code in Fig. 6.6 (typically used for PCM), a binary digit 1 corresponds to an alternating pulse and a binary digit 0 is sent as 0 volts. For the Miller line code in Fig. 6.7, a binary digit 1 is represented by either $+A$ or $-A$ for a duration of T_b seconds and a binary digit 0 is represented by either a transition from $+A$ to $-A$ or a transition from $-A$ to $+A$ over a duration of T_b seconds. The selection of pulses is made to ensure there is always a transition from $+A$ to $-A$ or $-A$ to $+A$ at the start of each binary digit. A transition every T_b seconds ensures that even if an all-ones sequence is transmitted, the receiver is able to easily synchronize to the incoming baseband signal. Such a baseband signal is referred to as self-clocking. An additional sought advantage is an inherent error detection capability. For example, if the pulse corresponding to a binary digit does not alternate within the bipolar baseband signal in Fig. 6.6, then an error must have occurred.

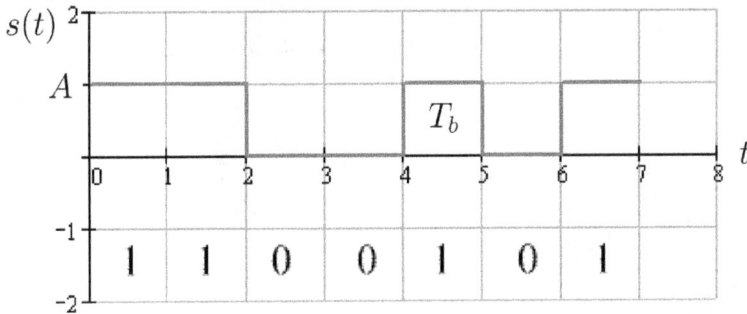

Figure 6.2: On-Off NRZ or Unipolar NRZ line code baseband signal.

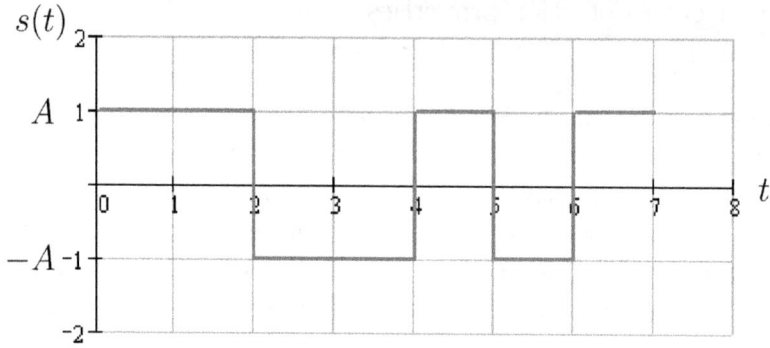

Figure 6.3: Polar NRZ or NRZ-L line code baseband signal.

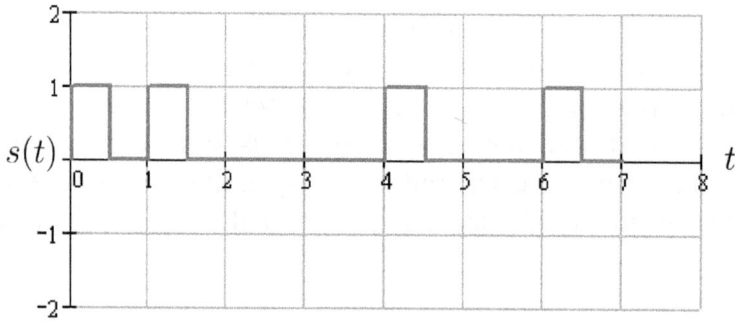

Figure 6.4: Unipolar RZ line code baseband signal.

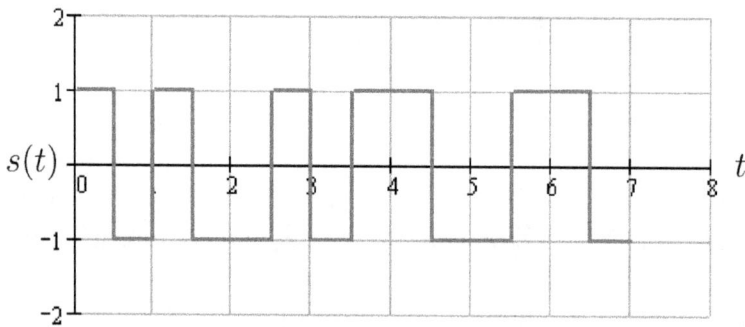

Figure 6.5: Manchester or Split phase line code baseband signal.

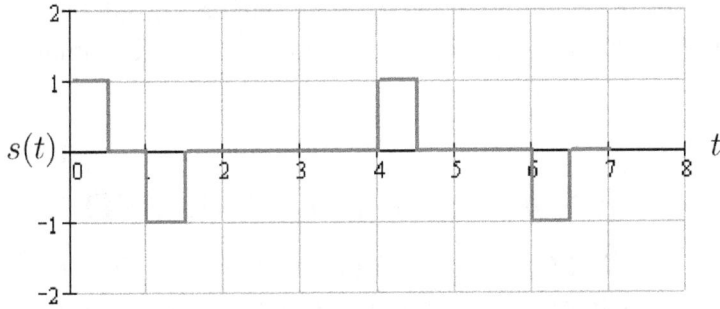

Figure 6.6: Bipolar RZ or RZ-AMI line code baseband signal.

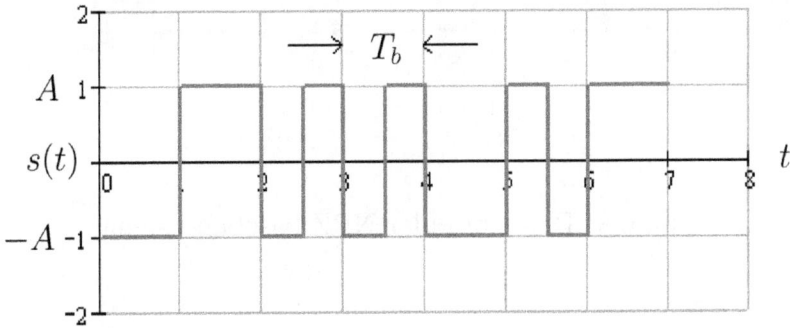

Figure 6.7: Miller line code baseband signal.

• ⭐ SIMULATION **BaseBandSignals:** Simulation corresponding to Figs. 6.2 to 6.7. You may experiment with an input binary stream of any length and observe the corresponding line code signals.

6.1.2   Random Polar NRZ Line Code PSD

Consider the polar NRZ baseband signal is shown in Fig. 6.8 in which a binary digit 1 and 0 is represented by a pulse of amplitude A and $-A$, respectively. Each pulse is of duration T_b seconds.

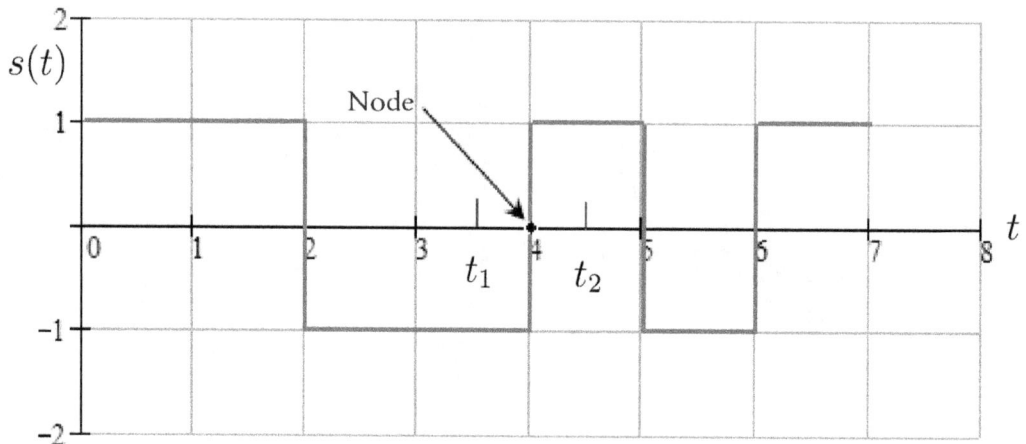

Figure 6.8: Random polar NRZ baseband signal.

From Fig. 6.8, the autocorrelation function $R_X(t_1, t_2)$ is given by

$$R_X(t_1, t_2) = \overline{x_1 x_2} = E[X(t_1)X(t_2)] \qquad (6.1)$$

If $(t_2 - t_1) > T_b$, then t_1 and t_2 will be in different pulse intervals and therefore, the random variables $X(t_1)$ and $X(t_1)$ will be independent. In this case,

$$R_X(t_1, t_2) = (\overline{x_1})(\overline{x_2}) \tag{6.2}$$

where the amplitude x_1 of $X(t_1)$ will either be A or $-A$ with probability $\frac{1}{2}$. Thus

$$\overline{x_1} = \sum_{x_1} x_1 P(x_1) = A\frac{1}{2} + (-A)\frac{1}{2} = 0 \tag{6.3}$$

Similarly, $(\overline{x_2}) = 0$. Hence, $R_X(t_1, t_2) = (\overline{x_1})(\overline{x_2}) = 0$ for $(t_2 - t_1) > T_b$. If $(t_2 - t_1) < T_b$, then t_1 and t_2 may or may not be in the same pulse interval. In this case,

$$R_X(t_1, t_2) = \sum_{x_1} \sum_{x_2} x_1 x_2 P(x_1, x_2) \tag{6.4}$$

$$
\begin{aligned}
&= A^2 P(x_1 = A, x_2 = A) + A^2 P(x_1 = -A, x_2 = -A) \\
+ (-A^2)\, P(x_1 &= -A, x_2 = A) + (-A^2) P(x_1 = A, x_2 = -A) \tag{6.5}
\end{aligned}
$$

By symmetry, $P(x_1 = A, x_2 = A) = P(x_1 = -A, x_2 = -A)$ and $P(x_1 = -A, x_2 = A) = P(x_1 = A, x_2 = -A)$. Thus

$$R_X(t_1, t_2) = 2A^2 \left[P(x_1 = A, x_2 = A) - P(x_1 = -A, x_2 = A) \right] \tag{6.6}$$

From Bayes rule,

$$P(x_1 = A, x_2 = A) = P(x_1 = A | x_2 = A)P(x_2 = A) = \frac{P(x_1 = A | x_2 = A)}{2}$$

$$(6.7)$$

$$P(x_1 = -A, x_2 = A) = P(x_1 = -A | x_2 = A)P(x_2 = A) = \frac{P(x_1 = -A | x_2 = A)}{2}$$

$$(6.8)$$

where

$$P(x_1 = A | x_2 = A) + P(x_1 = -A | x_2 = A) = 1 \qquad (6.9)$$

Hence

$$R_X(t_1, t_2) = A^2 \left[1 - 2P(x_1 = -A | x_2 = A) \right] \qquad (6.10)$$

The probability $P(x_1 = -A | x_2 = A)$ is given by the joint probability of the following two events:

- $Event1 = P(t_1 \text{ and } t_2 \text{ are in adjacent pulse intervals})$

  $= P(\text{of a node between } t_1 \text{ and } t_2)$

- $Event2 = P(\text{amplitude change from } - A \text{ to } A) = \frac{1}{2}$

so that

$$P(x_1 = -A | x_2 = A) = P(Event1) P(Event2) = \frac{1}{2}P(Event1) \qquad (6.11)$$

Now $P(Event1)$ is equal to the area under the PDF of the node between t_1 and t_2, where the node is the point of transition from $-A$ to A.   Given that this node is equally likely within a T_b interval, the height of the PDF is simply $\frac{1}{T_b}$ and thus,

$$P\left(Event1\right) = (t_2 - t_1)\frac{1}{T_b} \tag{6.12}$$

So that

$$P(x_1 = -A | x_2 = A) = \frac{1}{2}\left(\frac{t_2 - t_1}{T_b}\right) \tag{6.13}$$

and therefore

$$R_X(t_1, t_2) = A^2\left[1 - \left(\frac{t_2 - t_1}{T_b}\right)\right] \tag{6.14}$$

Let $\tau = t_2 - t_1$ and given that $R_X(\tau)$ is an even function of τ, we have

$$R_X(\tau) = \begin{cases} A^2\left(1 - \frac{|\tau|}{T_b}\right) & \text{if} \quad |\tau| < T_b \\ 0 & \text{otherwise} \end{cases} \tag{6.15}$$

Since $\overline{x_1} = \overline{x_2} = \cdots = \overline{x_N} = 0$ and the autocorrelation function depends only on τ, this binary random process is wide-sense stationary. Thus from the Wiener-Khintchine theorem, the power spectral density $S_X(f)$ is given by the FT of the $R_X(\tau)$, which from problem 12 in chapter 1, is given by

$$S_X(f) = A^2 T_b \operatorname{sinc}^2(f T_b) \tag{6.16}$$

6.1.3 Comparison of Line Code PSDs

The power spectral density (Watts/Hz) of the polar (NRZ), Bipolar (RZ) and Manchester (RZ) lines codes are presented in Fig. 6.9. To ensure a fair comparison, the time duration of each binary digit T_b has been fixed to 1 second and the amplitude of a given baseband signal selected to ensure the average signal power is 1 watt i.e. the area under any of three curves shown in Fig. 6.9 is 1.

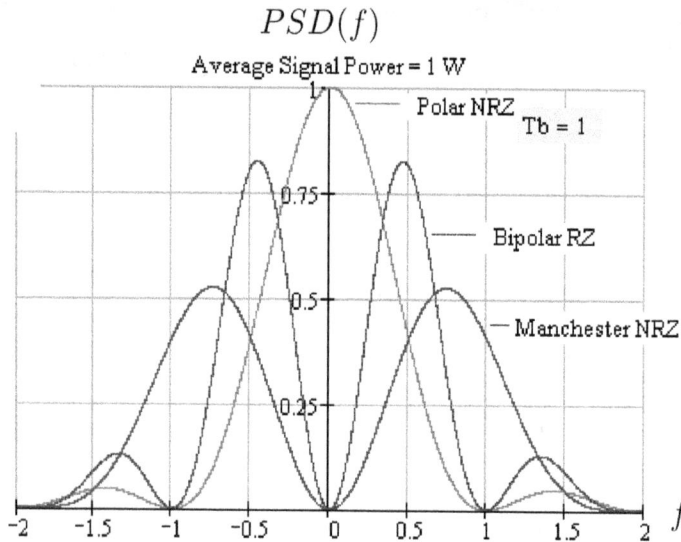

Figure 6.9: Examples of line codes PSD.

- ⭐ SIMULATION **PSD:** For the theoretical expressions of the PSDs corresponding to Fig. 6.9.

Referring to Fig. 6.9, it is clear that such baseband signals have a significant power at low frequencies and are therefore suited for transmission over low-pass (i.e. baseband) channels. For example, the first null bandwidth for a binary polar NRZ and the bipolar RZ signal is equal to $\frac{1}{T_b}$, whereas its $\frac{2}{T_b}$ for the Manchester NRZ signal. An important feature evident from 6.9 is that both the bipolar and Manchester baseband signals have no d.c. component. This is an advantage because the d.c. component and even very low frequency components (i.e. near 0 Hz) are typically heavily attenuated across a given channel.

After line coding, the pulses are filtered (i.e. shaped) to improve their spectral efficiency and/or improve their immunity to intersymbol interference (section 6.3.1). The *spectral efficiency* η of a digital signal, defined by $\eta = \frac{r_b}{B}$, is the number of binary digits per second that can be supported by each hertz of bandwidth i.e. the units of η are bits/sec/Hz, where if each pulse that represents a single binary digit occupies T_b seconds, then the number of binary digits per second $r_b = \frac{1}{T_b}$. If rectangular pulses are used, then the spectral efficiency is 0 bits/sec/Hz in principle because the bandwidth of each pulse is infinite! However in practice with the use of filters, we may use the first-null bandwidth from Fig. 6.9, for which the spectral efficiencies of Polar NRZ, Bipolar RZ and Manchester NRZ are given by

$$\eta_{PolarNRZ} = \frac{r_b}{\frac{1}{T_b}} = 1 \text{ bits/s/Hz} \tag{6.17}$$

$$\eta_{BipolarRZ} = \frac{r_b}{\frac{1}{T_b}} = 1 \text{ bits/s/Hz} \tag{6.18}$$

$$\eta_{ManchesterNRZ} = \frac{r_b}{\frac{2}{T_b}} = \frac{1}{2} \text{ bits/s/Hz} \tag{6.19}$$

6.1.4 Multilevel Polar NRZ

The spectral efficiency can be improved by using a multilevel signal (section 5.4.1). The null bandwidth of a multilevel polar NRZ signal is given by

$$B_{null} = \frac{1}{T_s} \tag{6.20}$$

where T_s is the time duration of a pulse $T_s = nT_b$ and there are 2^n possible levels of each pulse. Thus, the multilevel polar NRZ spectral efficiency is given by

$$\eta_{MultilevelPolarNRZ} = \frac{r_b}{B_{null}} = \frac{\frac{1}{T_b}}{\frac{1}{nT_b}} = n \text{ bits/s/Hz} \tag{6.21}$$

The spectral efficiency cannot be improved indefinitely by simply increasing n, because the receiver must now distinguish between the 2^n possible levels of each pulse. Refer to section 6.7 for further details. It will become evident in section 8.11.2 that the spectral efficiency for essentially error-free communication is limited by Shannon's channel capacity theorem.

6.1.5   Simulation I : Random Binary Line Code PSD

⭐SIMULATION **RandomPolarPSD:** A polar NRZ signal is created with 100 binary digits that are generated with the probability of a zero and one equal to 0.5. For example, the first 14 digits of a typical random binary polar NRZ signal are shown in Fig. 6.10. This specific signal's autocorrelation function is shown in Fig. 6.11 together with the theoretical function given by equation 6.15. A plot $S_X(f)$ is shown in Fig. 6.12 in which we note the excellent agreement between simulation and theory.

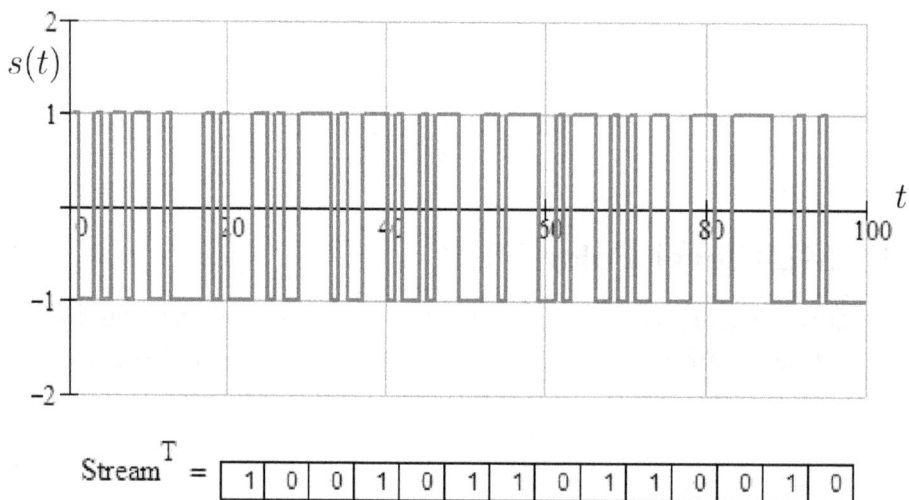

Figure 6.10: Random binary digit polar NRZ signal.

⭐SIMULATIONS **RandomBipolarPSD** and **RandomManPSD:** The above steps are repeated for the bipolar and Manchester line codes. The random binary Manchester line code PSD is presented in Fig. 6.13.

⭐SIMULATION **PSD2AutoCorrelation:** The inverse FFT of the theoretical PSD for a given line code is taken to determine the autocorrelation function. The autocorrelation function for the Manchester line code is shown in Fig. 6.14.

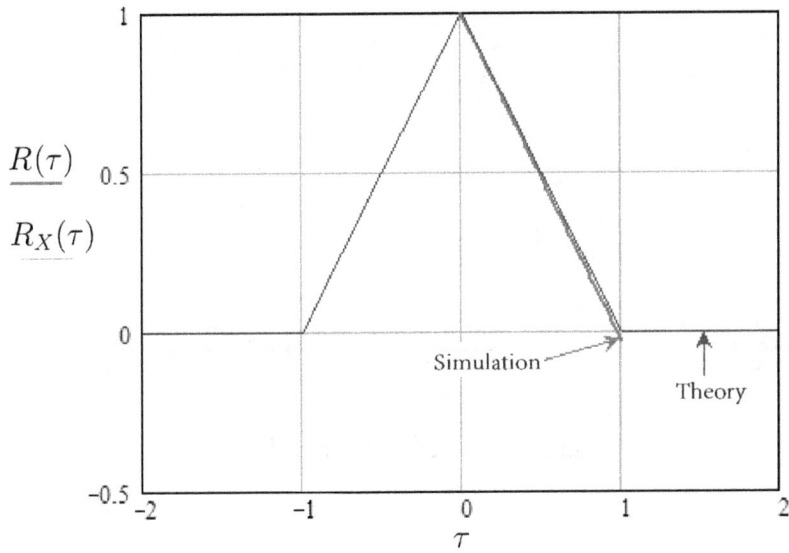

Figure 6.11: Autocorrelation function of a random binary polar (NRZ) signal.

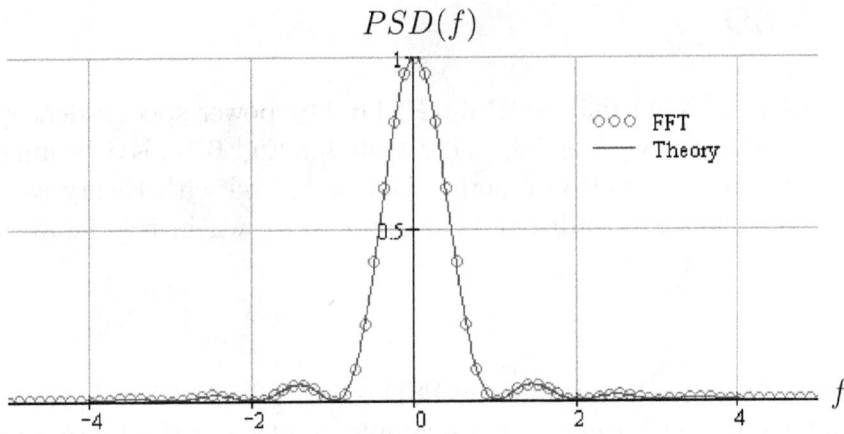

Figure 6.12: Polar (NRZ) PSD.

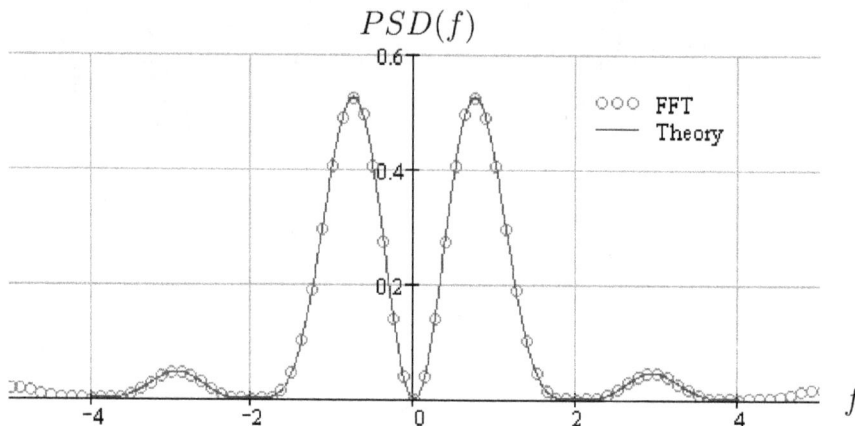

Figure 6.13: Manchester PSD.

- **Experiment:** Change the pulse width to determine its influence on the PSD.

6.1.6   Simulation II : Random Multilevel Polar NRZ PSD

⭐ SIMULATION **MultilevelPolarPSD:** The power spectal density of a random 8-level polar NRZ signal, as illustrated in Fig. 6.15, is determined via the FT of its autocorrelation function and compared with theory as shown in Fig. 6.16. The first null bandwidth is also shown in this figure and is given by $B_{null} = \frac{1}{3T_b}$.

- **Experiment:** Determine the PSD of other line codes via simulation and confirm with theory. For example, implement the *2B1Q line code* used in the North American digital subscriber line (DSL) standard in which pairs of binary digits are mapped to amplitude levels as shown in the Table 6.1 to generate a 4-ary PAM signal. Notice the use of Gray coding (refer to problems section in chapter 3).

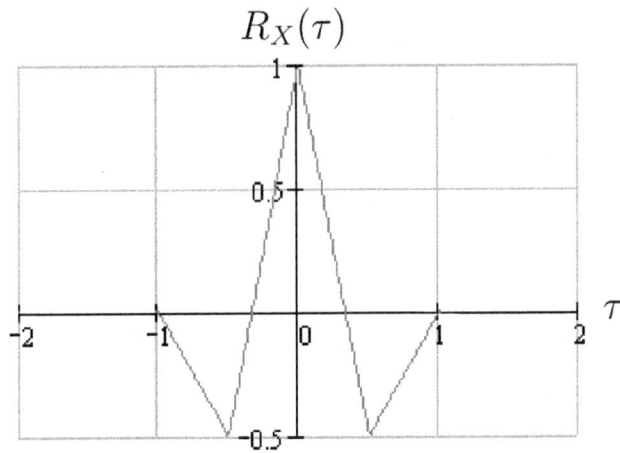

Figure 6.14: Autocorrelation function via the inverse FFT of the theoretical Manchester PSD.

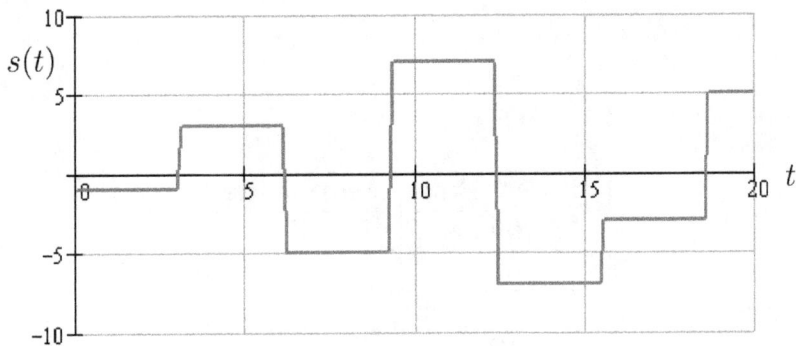

Figure 6.15: Random 8-level polar NRZ baseband signal.

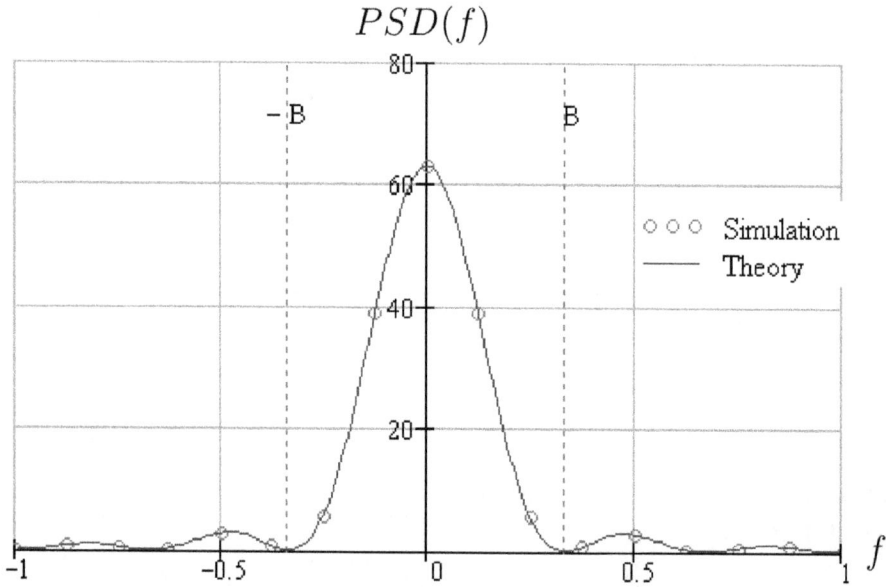

Figure 6.16: Power spectral density of an 8-level polar NRZ baseband signal.

Binary Pair	Amplitude
00	-3
01	-1
11	+1
10	+3

Table 6.1: 2B1Q line code.

6.2 Signal Detection

We now focus our attention on how to optimally detect a **single** pulse in the presence of additive white gaussian noise and consider a **sequence** of pulses later in section 6.3, where we shall encounter the additional complexity of adjacent pulses smearing (overlap in time) leading to what is referred to as *intersymbol interference* (ISI).

6.2.1 Matched Filter

For a given input signal, the *matched filter* is a linear filter that maximizes the ratio of the instantaneous output signal power to the average output noise power. Let $r(t)$ represent the received noisy pulse from an ideal *distortionless* channel given by

$$r(t) = s(t) + n(t) \qquad \text{for } 0 \leq t \leq T \tag{6.22}$$

where $s(t)$ is the transmitted pulse (or signal) of time duration T and $n(t)$ is a zero-mean AWGN signal with a power spectral density of $\frac{N_o}{2}$ W/Hz. This received signal $r(t)$, which is free of intersymbol interference, is input to a filter (linear time invariant) with a transfer function (or equivalently frequency response) $H(f)$, and the corresponding output $y(t)$ is sampled at time T to produce $y(T)$ as illustrated in Fig. 6.17.

The output $y(t)$ is given by

$$y(t) = s_o(t) + n_o(t) \tag{6.23}$$

where $s_o(t)$ is the signal component $s_o(t)$ and $n_o(t)$ is the noise component. Thus **without** noise, the output $y(t) = s_o(t)$ is given by the inverse FT of $H(f)S(f)$, where $S(f)$ is the FT of $s(t)$ i.e.

$$s_o(t) = \int_{-\infty}^{\infty} H(f)S(f)e^{j2\pi ft}df \tag{6.24}$$

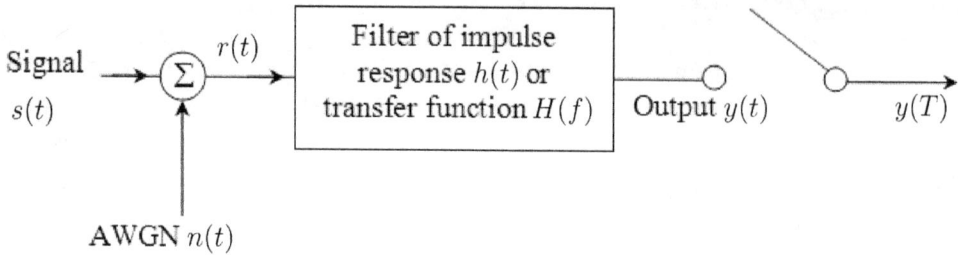

Figure 6.17: Matched filter with impulse response $h(t)$.

The **instantaneous** output signal power P_{out} at time t is given by $s_o^2(t)$. Note that this is **not** the average output signal power. The *average* noise power N_{out} for **white noise** is given by

$$N_{out} = E\left[n_o^2(T)\right] = \int_{-\infty}^{\infty} \frac{N_o}{2} |H(f)|^2\, df = \frac{N_o}{2} \int_{-\infty}^{\infty} |H(f)|^2\, df \qquad (6.25)$$

where the relation $PSD_{out}(f) = PSD_{in}(f)\,|H(f)|^2$ from section 1.8.11 was used. To minimize the probability of incorrectly identifying the symbol associated with the received signal in the presence of noise, we require the filter to **maximize** the instantaneous output signal power to the average noise power $\left(\frac{P_{out}}{N_{out}}\right)_T$ ratio at time T, given by

$$\left(\frac{P_{out}}{N_{out}}\right)_T = \frac{s_o^2(T)}{N_{out}} \qquad (6.26)$$

$$= \frac{\left|\int_{-\infty}^{\infty} H(f)S(f)e^{j2\pi fT}\, df\right|^2}{\frac{N_o}{2}\int_{-\infty}^{\infty}|H(f)|^2\, df} \le \frac{\int_{-\infty}^{\infty}|H(f)|^2\, df \int_{-\infty}^{\infty}|S(f)|^2\, df}{\frac{N_o}{2}\int_{-\infty}^{\infty}|H(f)|^2\, df} \qquad (6.27)$$

where we have made use of the Cauchy-Schwarz's inequality (Appendix A.12.1) that

$$\left| \int_{-\infty}^{\infty} A(f)B(f)df \right|^2 \leq \int_{-\infty}^{\infty} |A(f)|^2 \, df \int_{-\infty}^{\infty} |B(f)|^2 \, df \qquad (6.28)$$

with $A(f)$ and $B(f)$ being replaced by $H(f)$ and $S(f)e^{j2\pi fT}$, respectively and $\left| e^{j2\pi fT} \right| = 1$. Thus

$$\left(\frac{P_{out}}{N_{out}} \right)_T \leq \frac{2}{N_o} \int_{-\infty}^{\infty} |S(f)|^2 \, df \qquad (6.29)$$

or equivalently

$$\left(\frac{P_{out}}{N_{out}} \right)_T \leq \frac{2E}{N_o} \qquad (6.30)$$

where E is energy of the input signal given by $E = \int_{-\infty}^{\infty} |S(f)|^2 \, df$ using Parseval's theorem. It can be shown [Couch, 2001] that the Cauchy-Schwarz's equality holds if $A(f) = cB^*(f)$ where c is an arbitrary constant and $*$ denotes the complex conjugate. Thus, the maximum value of $\frac{2E}{N_o}$ will be true if $H(f)$ is equal to the optimum transfer function $H_{opt}(f)$ given by

$$H_{opt}(f) = cS^*(f)e^{-j2\pi fT} \qquad (6.31)$$

Its important to note the maximum value of $\left(\frac{P_{out}}{N_{out}} \right)_T$ does **not** depend on the shape of the input signal but rather on its energy E and the single-sided noise power spectral density N_o. With $H_{opt}(f) = S^*(f)e^{-j2\pi fT}$ (and taking $c = 1$), the output of the matched filter at time T is given by

$$s_o(T) = \int_{-\infty}^{\infty} H_{opt}(f)S(f)e^{j2\pi fT} df$$

$$= \int_{-\infty}^{\infty} S^*(f)e^{-j2\pi fT} S(f)e^{j2\pi fT} df$$

$$= \int_{-\infty}^{\infty} |S(f)|^2 \, df = E \qquad (6.32)$$

Furthermore, the average noise power output is given by

$$N_{out} = \frac{N_o}{2} \int_{-\infty}^{\infty} |H_{opt}(f)|^2 \, df = \frac{N_o}{2} E \qquad (6.33)$$

Thus as expected, we find that

$$\left(\frac{P_{out}}{N_{out}}\right)_T = \frac{\left| \int_{-\infty}^{\infty} H_{opt}(f) S(f) e^{j2\pi fT} df \right|^2}{\frac{N_o}{2} \int_{-\infty}^{\infty} |H_{opt}(f)|^2 \, df} = \frac{E^2}{\frac{N_o}{2} E} = \frac{2E}{N_o} \qquad (6.34)$$

- ⭐ SIMULATION **Schwartz**: The matched filter in Fig. 6.17 is modeled under a noiseless channel ($n(t) = 0$) with various types of input pulses, including the pulse for which $S(f) = H_r(f)$. The corresponding matched filter frequency response $H_{opt}(f)$ is set equal to $S^*(f)e^{-j2\pi fT}$ to verify that in each case $\left(\frac{P_{out}}{N_{out}}\right)_T = \frac{2E}{N_o}$. The variation of the input and output signal with time are determined from the inverse FT under a noiseless environment to confirm also that $y(T) = s_o(T) = E$.

Having found the optimum frequency response $H_{opt}(f)$, the corresponding impulse response $h_{opt}(t)$ is given by the inverse FT of $H_{opt}(f)$, namely

$$h_{opt}(t) = \int_{-\infty}^{\infty} cS^*(f)e^{-j2\pi fT} e^{j2\pi ft} df = c \int_{-\infty}^{\infty} S^*(f)e^{-j2\pi f(T-t)} df \qquad (6.35)$$

For a real signal, $S(-f) = S^*(f)$ and therefore

$$h_{opt}(t) = c \int_{-\infty}^{\infty} S(-f)e^{-j2\pi f(T-t)} df = c \int_{-\infty}^{\infty} S(f)e^{j2\pi f(T-t)} df \qquad (6.36)$$

$$= \begin{cases} cs(T-t) & 0 \le t \le T \\ \\ 0 & otherwise \end{cases} \tag{6.37}$$

where we have made use of the time shifting FT property (section 1.5). Taking $c = 1$, the matched filter impulse response is given by $s\,(T-t)$ i.e. the impulse response of the filter is simply a delayed mirror image of the signal $s(t)$ we wish to detect i.e. $h_{opt}(t)$ is "matched" to $s(t)$. The output $y(t)$ is then given by

$$y(t) = r(t) * h_{opt}(t) = \int_0^t r(\tau) h_{opt}(t-\tau) d\tau \tag{6.38}$$

For $h_{opt}(t) = s\,(T-t)$,

$$h_{opt}(t-\tau) = s\,(T-(t-\tau)) = s\,(T-t+\tau) \tag{6.39}$$

and therefore

$$y(t) = \int_0^t r(\tau) s\,(T-t+\tau)\,d\tau \tag{6.40}$$

At time $t = T$

$$y(T) = \int_0^T r(\tau) s\,(T-T+\tau)\,d\tau = \int_0^T r(\tau) s\,(\tau)\,d\tau \tag{6.41}$$

This result has a very interesting implication! Namely, a matched filter is equivalent to a *correlator* in which the signals $r\,(t)$ and $s(t)$ are multiplied and integrated over the time duration T as summarized in Fig. 6.18. Specifically, the received signal $r(t) = s(t) + n(t)$ is correlated with $s(t)$ such that the output at time T is given by

$$y(T) = \int_0^T r(t)s(t)dt \qquad (6.42)$$

where $y(T)$ will have a gaussian distribution with a mean value equal to $\int_0^T s(t)s(t)\,dt = E$, the energy of the signal $s(t)$, and the variance σ^2 of $y(T)$ is given by (refer to problems section)

$$\sigma^2 = E\left[n_o^2\right] = N_{out} = \frac{N_o}{2}E \qquad (6.43)$$

In practice, the received signal $r(t)$ is correlated with a *basis function* (to be introduced later in section 6.5) instead of $s(t)$. In this case, given that a basis function has unit energy, the variance σ^2 is given by $\sigma^2 = \frac{N_o}{2}$ (refer to problems section).

(a)

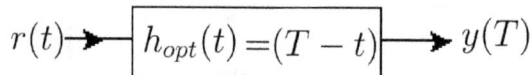

(b)

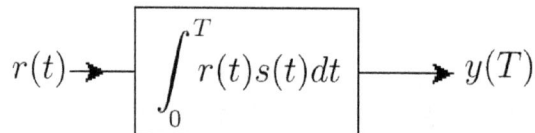

Figure 6.18: (a) Matched Filter (b) Correlator.

To illustrate the foregoing theory by example, consider the signal $s(t)$ presented in Fig. 6.19 for which $T = 2$ secs. The corresponding impulse response $h(t)$ is shown in Fig. 6.20.

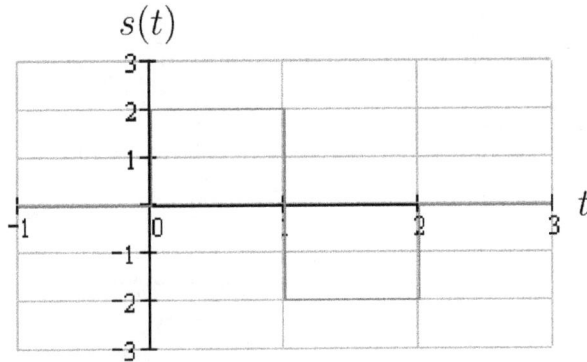

Figure 6.19: Input signal.

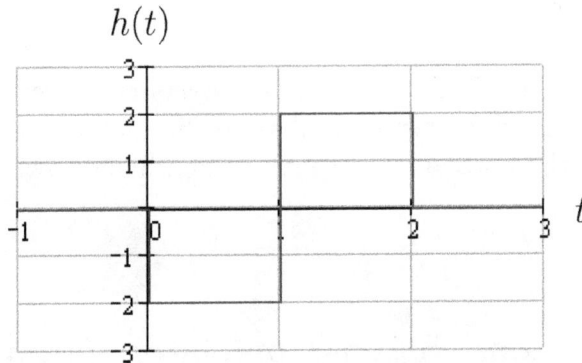

Figure 6.20: Impulse response.

Under a noiseless channel, the results of *convolving* the input signal $s(t)$ with the impulse response $h(t)$, overlaid with the result of *correlating* the input signal $s(t)$, is shown in Fig. 6.21. As expected, the output of the matched filter and the correlator are identical at time $T = 2$ secs (only). Furthermore, under this noiseless scenario, $y(T)$ is equal to the energy of the signal (8 Joules) as expected.

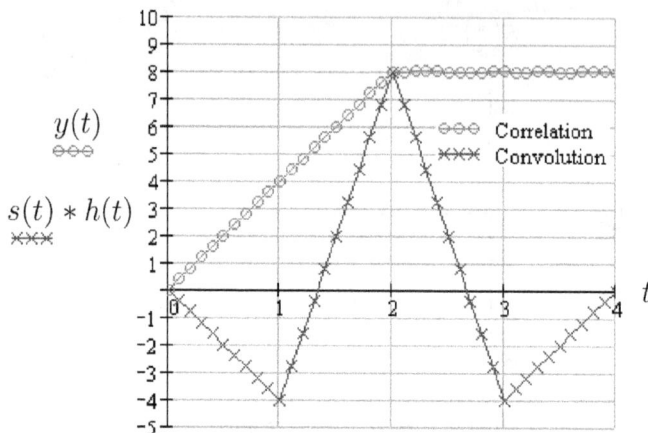

Figure 6.21: Comparison of the correlator and matched filter output.

- ⭐ SIMULATION **MatchedFilterCorrelator:** Select an input signal from six different types to observe the matched filter impulse response and the correlator output. Experiment with other types of input signals.

6.2.2   Simulation : Signal Correlator Variance

⭐ SIMULATION **SignalCorrelator:** The signal $s(t) = A$ for $0 \le t \le T$ is corrupted an additive white gaussian noise signal $n(t)$ to create $r(t) = s(t) + n(t)$ and correlated with $s(t)$ as illustrated in Fig. 6.18(b). A typical example of the input signal $r(t) = s(t) + n(t)$ to the correlator is presented in Fig. 6.22.

The output $y(T)$ is expected to have a gaussian distribution with a mean value equal to $\int_0^T s(t)s(t)\,dt = A^2T = E$ and variance $\sigma^2 = \frac{N_o}{2}E = \frac{N_oA^2T}{2}$. This is verified by simulation in Figs. 6.23 and 6.24. The simulation results are slightly inaccurate because the noise signal is only an approximation to a true AWGN signal.

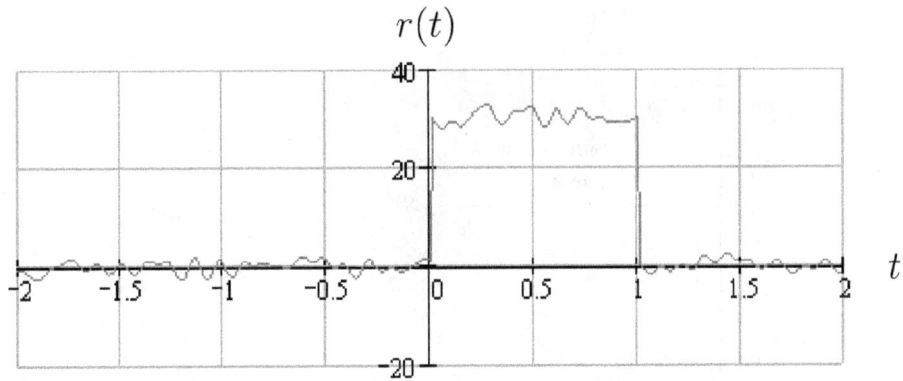

Figure 6.22: Received signal from an AWGN channel.

- **Experiment:** Try other types of pulses (e.g. shape the pulse to remove the sharp corners, triangular pulse, etc.)

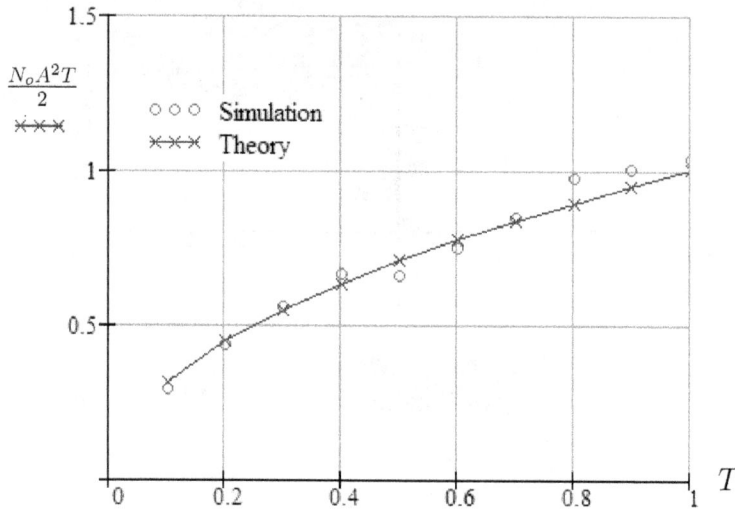

Figure 6.23: Simulated and theoretical standard deviation σ of the correlator output.

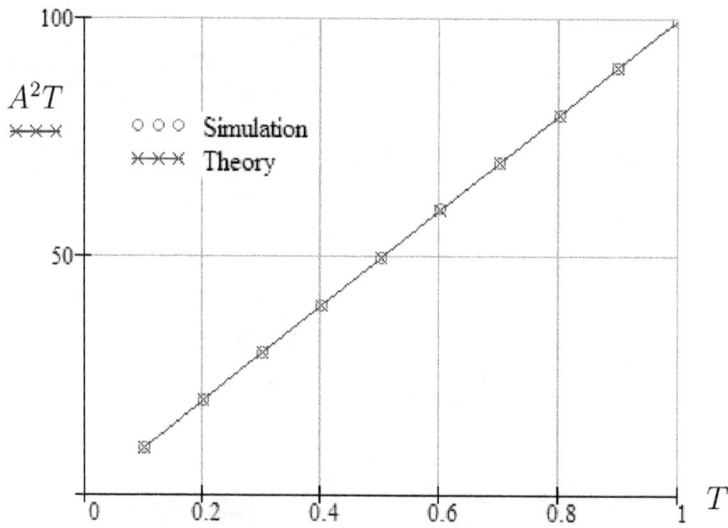

Figure 6.24: Simulated and theoretical mean of the correlator output.

6.3 BandLimited AWGN Channel

In this section, we consider digital transmission through a bandlimited AWGN channel (e.g. telephone, microwave, satellite, underwater acoustic,etc.) modeled as a linear filter with a bandwidth limitation. We cannot use rectangular shaped pulses (of infinite bandwidth) through a *dispersive* channel with a finite bandwidth, because the high frequency components which contribute to the sharp edges, would be eliminated. Furthermore, the pulse would be distorted because those components which passed through would suffer amplitude and phase distortion. This leads to *intersymbol interference* (ISI) i.e. interference between adjacent symbols, which in turn, leads to an increase in the probability of an error at the sink.

6.3.1 Intersymbol Interference (ISI)

A simple and effective method to gauge the level of ISI (and channel noise) in practice is to use an oscilloscope to superimpose each pulse on top of each other as shown in Fig. 6.25. This figure is referred to as an *eye diagram* because it resembles an eye. The partial closing of the eye is to due to both distortion and noise. If the intersymbol interference is high, the vertical opening between some, or all, signal levels may disappear altogether. In that case, the eye is said to be closed. If there are wide vertical spacings between signal levels, which indicates immunity to additive noise, then the eye is said to be open. A wide or small horizontal opening indicates that a large or small timing offset can be tolerated, respectively. The slope of the inner eye indicates sensitivity to timing jitter or variance in the timing offset. For example, a steep slope means that the eye closes rapidly as the timing offset increases.

- ★ SIMULATION **EyeDiagram**: For further details in which amplitude and phase distortion is introduced within a periodic rectangular pulse signal. Experiment with different levels of distortion. Try a different pulse shape to change the eye shape.

A transmitter and receiver that contains reactive elements (inductors, capacitors) may also introduce ISI. For example, in section 1.8.1 we found

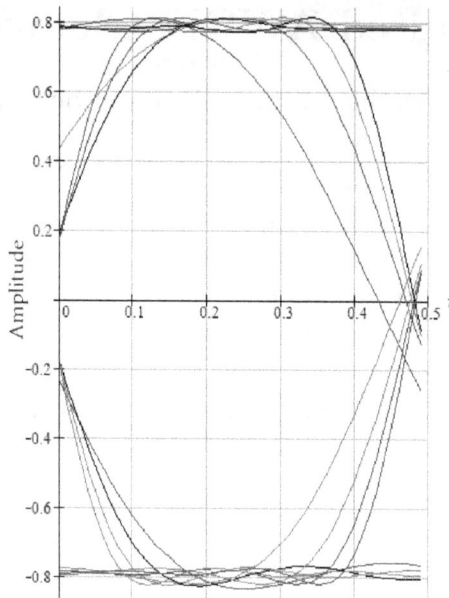

Figure 6.25: Eye diagram.

that the rectangular pulse input to a RC filter is spread in time. Although our goal is not to reproduce the transmitted pulse but to establish which pulse has been received from a finite set of possible pulses, ISI may lead to symbol errors even in a noiseless channel because energy from neighboring symbols may lead to an incorrect decision. This is because, as it will become evident from the sections to follow, it is the energy of the received pulse which directly influences the probability of an error in a symbol (e.g. binary digit). Thus, unlike the rectangular shaped pulses used in Figs. 6.2 to 6.5 for a given line code, the solution is to *shape* each pulse appropriately to minimize ISI as we shall discover in section 6.3.3. Before we consider this in detail, we must first address the important question of what is the maximum rate at which information can be transmitted over a channel of bandwidth B Hz ?

6.3.2   Maximum Information Rate

Recall from the sampling theorem, that a signal of bandwidth B can recon- structed using $2B$ samples/sec. Assuming an error-free channel, the required channel bandwidth to transmit this signal is simply B Hz. Therefore, the channel should be able to transmit $2B$ samples/sec without errors. Indeed, Nyquist showed that one pulse generated every T seconds (i.e. $\frac{1}{T}$ pulses per second) can be detected given a baseband bandwidth of $\frac{1}{2T}$ without ISI provided each pulse of a received sequence is of the form $\text{sinc}\left(\frac{t}{T}\right)$. Thus, a maximum of $2B$ pulses per second (*Nyquist rate*) can be transmitted over a channel of bandwidth B (*Nyquist bandwidth*). For example, the binary stream 1100101 represented with sinc(.) pulses is shown in Fig. 6.26 and its corresponding resultant signal in Fig. 6.27.

Although the focus of this chapter is the transfer of binary digits from the source to the sink via a baseband channel, we shall not restrict a pulse to represent only a single binary digit in this section. Thus, the symbol T in this section refers to the time duration of a possible multilevel pulse $M \geq 2$.

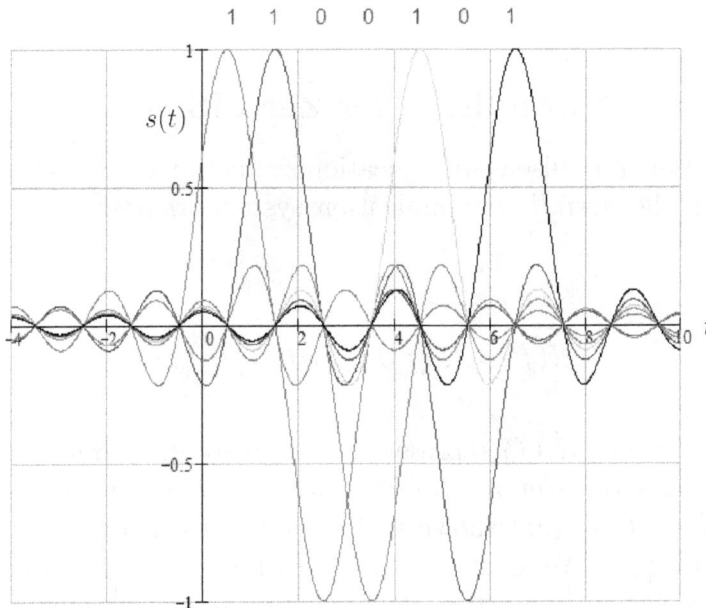

Figure 6.26: The individual sinc pulses of the binary stream 1100101.

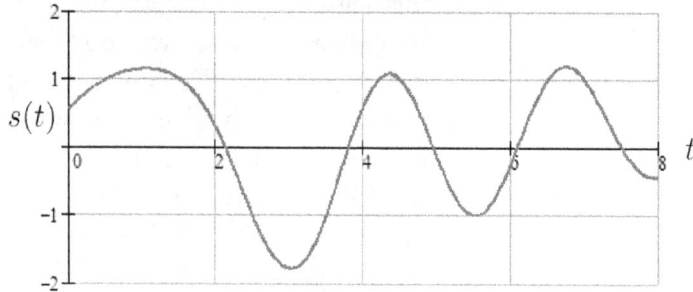

Figure 6.27: Sinc pulse baseband signal.

• ⭐ SIMULATION **SincPulse:** To see how the tails of the pulses cancel each other. You may alter the binary stream as required.

6.3.3 Nyquist Condition for Zero ISI

Consider a baseband pulse communication system modeled as shown in Fig. 6.28 in which the *overall* communication system transfer function $H(f)$ is given by

$$H(f) = H_t(f)H_c(f)H_e(f)H_r(f) \qquad (6.44)$$

where $H_t(f)$ and $H_r(f)$ represent respectively, the *transmitter* and *receiver* filters and the *channel transfer function* $H_c(f)$ is equalized via an *equalizing filter* $H_e(f)$ (not shown in Fig. 6.28 for convenience) that is ideally equal to $\frac{1}{H_c(f)}$. We shall come back to the equalizing filter in section 6.3.5. For now, simply note that the channel acts as a filter plus noise $n(t)$.

Let the input to the transmitting filter be the impulse train $\sum\limits_{n=-\infty}^{\infty} a_n\delta(t - nT)$, where $\{a_n\}$ is the sequence of amplitudes (e.g. 0 and 1 for binary)

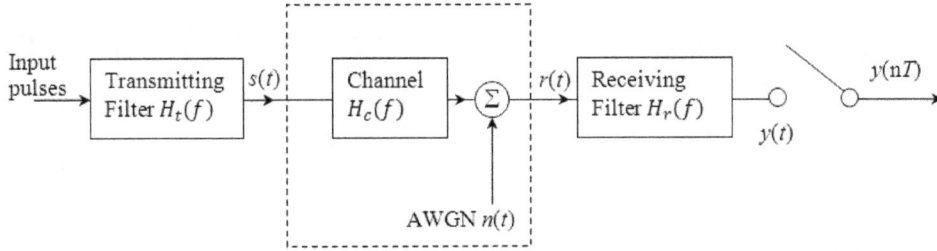

Figure 6.28: Bandlimited baseband communication system.

output from a data source and $\delta(t)$ is the unit impulse function. Then the baseband signal $s(t)$ at the output of the transmitting filter is given by

$$s(t) = \sum_{n=-\infty}^{\infty} a_n \delta(t - nT) * h_t(t) = \sum_{n=-\infty}^{\infty} a_n h_t(t - nT) \tag{6.45}$$

where $*$ denotes convolution, $h_t(t)$ is the impulse response of the transmitting filter and T is the symbol interval, given by $T = \frac{n}{r_b}$, where r_b is the binary data rate. Notice that $h_t(t)$ is the pulse shape which controls the spectral characteristics of $s(t)$. The channel output $r(t)$ is given by

$$r(t) = \sum_{n=-\infty}^{\infty} a_n h_{tc}(t - nT) + n(t) \tag{6.46}$$

where $h_{tc}(t) = h_t(t)*h_c(t)$ is the impulse response of the transmitting filter $h_t(t)$ and channel filter $h_c(t)$ combination and $n(t)$ represents the AWGN. Finally, the output $y(t)$ of the receiving filter is given by

$$y(t) = \sum_{n=-\infty}^{\infty} a_n h_{tcr}(t - nT) + w(t) \tag{6.47}$$

where $h_{tcr}(t) = h_t(t) * h_c(t) * h_r(t)$ is the cascaded impulse response and $w(t) = n(t) * h_r(t)$ is the noise at the output of the receiving filter. Given that the output $y(t)$ is sampled every T seconds,

$$y(mT) = \sum_{n=-\infty}^{\infty} a_n h_{tcr}(mT - nT) + w(mT) \qquad (6.48)$$

$$= h_{tcr}(0)a_m + \sum_{n \neq m} a_n h_{tcr}(mT - nT) + w(mT) \qquad (6.49)$$

The first term on the right-hand side of equation 6.49 is the desired symbol a_m scaled by $h_{tcr}(0)$. The second term represents the intersymbol interference due to the other symbols at the sampling instant $t = mT$. If the receiving filter $h_r(t)$ is matched to $h_{tc}(t)$, then

$$h_{tcr}(0) = h_{tc}(t) * h_r(t) \qquad (6.50)$$

$$= \int_{-\infty}^{\infty} h_{tc}^2(t)dt = \int_{-\infty}^{\infty} |H_t(f)|^2 |H_c(f)|^2 df = E_{tc} \qquad (6.51)$$

where E_{tc} is the energy in the channel output $h_{tc}(t)$.

- ★ SIMULATION **ImpulseTrain:** An illustration of the baseband communication system in Fig. 6.28, in which ten binary digits are generated with probability 0.5 to create an unit impulse train $\sum_{n=-\infty}^{\infty} a_n \delta(t - nT)$, that is subsequently input to the transmitting filter, with an impulse response $h_t(t)$ set to be the root raised cosine filter $h_{root}(t)$ (equation 6.65) to be covered later in this section. The channel is assumed to be an ideal channel, so that $y(t) = \sum_{n=-\infty}^{\infty} a_n h_{tr}(t - nT)$. We shall take $h_{tr}(t)$ to be $h_{raised}(t)$ presented in equation 6.66. The choice for $h_t(t)$ and $h_{tr}(t)$ will become evident once you have covered this entire section. In this ideal communication system model, it is shown that $y(mT) = h_{tr}(0)a_m$ and equation 6.51 is verified. Please do take the time for research with this simulation. For example, what happens if the channel has a non-ideal frequency response ?

The third term in equation 6.49 is the noise component with a power spectral density of $\frac{N_o}{2}|H_{tc}(f)|^2$ where $H_{tc}(f) = H_t(f)H_c(f)$. Specifically, a zero mean gaussian random variable with variance

$$\sigma^2 = \frac{N_o}{2}\int\limits_{-\infty}^{\infty} |H_t(f)|^2\, |H_c(f)|^2\, df = \frac{N_o E_{tc}}{2} \qquad (6.52)$$

If an equalizer is used such that $H_e(f) = \frac{1}{H_c(f)}$ (e.g. a telephone modem will use a 'training sequence' in the handshaking to equalize the non-ideal channel response) or similarly, if we assume that there is no channel distortion, we may write $H_c(f) = ke^{-j2\pi f t_d}$, where k is a constant and t_d is the time-delay. If $|H_c(f)|$ is not constant, then we shall have *amplitude distortion* and if $\theta_c(f)$ is not a linear function of frequency, then we shall have *phase distortion*. Refer back to section 1.8.10 for further details. For convenience, we shall take $k = 1$ and $t_d = 0$. Thus $H_{tc}(f) = H_t(f)$ and the matched filter has a frequency response $H_r(f) = H_t^*(f)$ so that

$$y(mT) = h_{tr}(0)a_m + \sum_{n \neq m} a_n h_{tr}(mT - nT) + w(mT) \qquad (6.53)$$

To remove the effect of ISI, the overall communication system must be designed such that $h_{tr}(mT - nT) = 0$ for $n \neq m$ and $h_{tr}(0) \neq 0$. Without any loss of generality, we may take $h_{tr}(0) = 1$, so that the overall communication system must be designed such that

$$h_{tr}(nT) = \begin{cases} 1 & if \quad n = 0 \\ 0 & \quad n \neq 0 \end{cases} \qquad (6.54)$$

The corresponding impulse response $h_{tr}(t)$ is simply the inverse FT of $H_{tr}(f)$ so that

$$h_{tr}(t) = \int\limits_{-\infty}^{\infty} H_{tr}(f)e^{j2\pi f t}df \qquad (6.55)$$

The range of integration in the above equation can be divided into segments as

$$h_{tr}(t) = \sum_{m=-\infty}^{\infty} \int_{(2m-1)/2T}^{(2m+1)/2T} H_{tr}(f)e^{j2\pi fnT} df \qquad (6.56)$$

and thus at the sampling instants $t = nT$, we find that $h_{tr}(nT)$ given by

$$h_{tr}(nT) = \sum_{m=-\infty}^{\infty} \int_{(2m-1)/2T}^{(2m+1)/2T} H_{tr}(f)e^{j2\pi fnT} df \qquad (6.57)$$

$$= \sum_{m=-\infty}^{\infty} \int_{-1/2T}^{1/2T} H_{tr}\left(f + \frac{m}{T}\right) e^{j2\pi fnT} df \qquad (6.58)$$

Assuming the integration and summation can be interchanged, we have

$$h_{tr}(nT) = \int_{-1/2T}^{1/2T} \left[\sum_{m=-\infty}^{\infty} H_{tr}\left(f + \frac{m}{T}\right) \right] e^{j2\pi fnT} df \qquad (6.59)$$

Finally, if $\sum_{m=-\infty}^{\infty} H_{tr}\left(f + \frac{m}{T}\right) = T$, then

$$h_{tr}(nT) = \int_{-1/2T}^{1/2T} T e^{j2\pi fnT} df = \frac{\sin(n\pi)}{n\pi} = \begin{cases} 1 & if \quad n = 0 \\ 0 & \quad n \neq 0 \end{cases} \qquad (6.60)$$

which shows that $h_{tr}(t)$ with $H_{tr}(f)$ satisfying

$$\sum_{m=-\infty}^{\infty} H_{tr}\left(f + \frac{m}{T}\right) = T \qquad (6.61)$$

produces zero ISI. If the channel has a bandwidth B, so that $H_c(f) = 0$ for $|f| > B$, then the replicas of $H_{tr}\left(f + \frac{m}{T}\right)$ in equation 6.61 will not overlap if $T < \frac{1}{2B}$, and therefore $\sum_{m=-\infty}^{\infty} H_{tr}\left(f + \frac{m}{T}\right) \neq T$ implying we cannot remove ISI. If $T = \frac{1}{2B}$, then it is only possible to satisfy equation 6.61 if

$$H_{tr}\left(f\right) = \begin{cases} T & if \quad |f| < B \\ 0 & otherwise \end{cases} \qquad (6.62)$$

which via the inverse FT corresponds to

$$h_{tr}(t) = T\int_{-B}^{B} e^{j2\pi ft}df = \frac{1}{2B}\left[\frac{e^{j2\pi ft}}{j2\pi t}\right]_{-B}^{B} = \mathrm{sinc}\,(2Bt) \qquad (6.63)$$

or equivalently $\mathrm{sinc}\left(\frac{t}{T}\right)$, as discussed in section 6.3.2. To generate a sinc(.) pulse in practice is impossible. Even if we could, the timing would have to be perfect to ensure no ISI because each pulse would then extend into every pulse in the entire sequence. But what if $T > \frac{1}{2B}$? In this case, we can easily satisfy equation 6.61 using well known practical filters. For example, a popular choice for $H_{tr}(f)$ is the *raised cosine transfer function*, denoted by $H_{raised}(f)$, such that $H_{raised}(f) = H_t(f)H_r(f)$, where $H_t(f)$ and $H_r(f)$ are *root raised cosine filters* given by

$$H_t(f) = H_r(f) = \begin{cases} 1 & if \quad |f| < f_1 \\ \cos\left[\frac{\pi}{4}\left(\frac{|f|+B-2f_0}{B-f_0}\right)\right] & if \quad f_1 \le |f| \le B \\ 0 & if \quad |f| > B \end{cases} \qquad (6.64)$$

The frequency $f_0 = \frac{1}{2T}$ is the *6-dB bandwidth*, $B = \frac{(1+r)}{2T}$ is the *absolute bandwidth*, $f_1 = \frac{(1-r)}{2T}$ is the frequency at which the roll-off begins and r is the filter roll-off factor. The variation of $H_{raised}(f)$ is shown in Fig. 6.29 and $H_t(f)$ for $r = 0$, 0.5 and 1 and $T = 1$ sec is shown in Fig. 6.30, where the Nyquist minimum bandwidth case corresponds to $r = 0$.

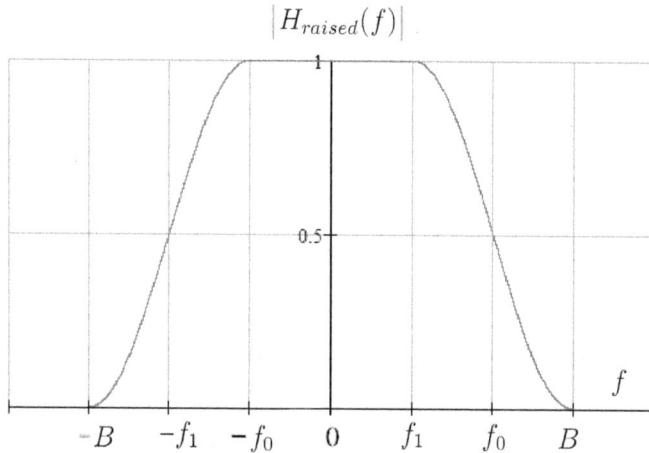

Figure 6.29: Raised cosine frequency response.

• ⭐ SIMULATION **RootRaisedCosine:** The raised and root-raised cosine characteristics are analyzed side-by-side.

Splitting the raised cosine transfer function $H_{raised}(f)$ into $H_t(f)$ and $H_r(f)$ helps to minimize the bandwidth of the transmitted signal and to filter out noise at the receiver. But more importantly, since the complex conjugate $H_t^*(f) = H_r(f)$, the requirement for matched filtering to maximize the received signal-to-noise power ratio is satisfied (missing only a linear phase factor) for an additive white gaussian noise (AWGN) channel. The transmitted pulse shape is the impulse response $h_{root}(t)$ of the *root* raised cosine filter given by the inverse Fourier transform of $H_t(f)$ to be

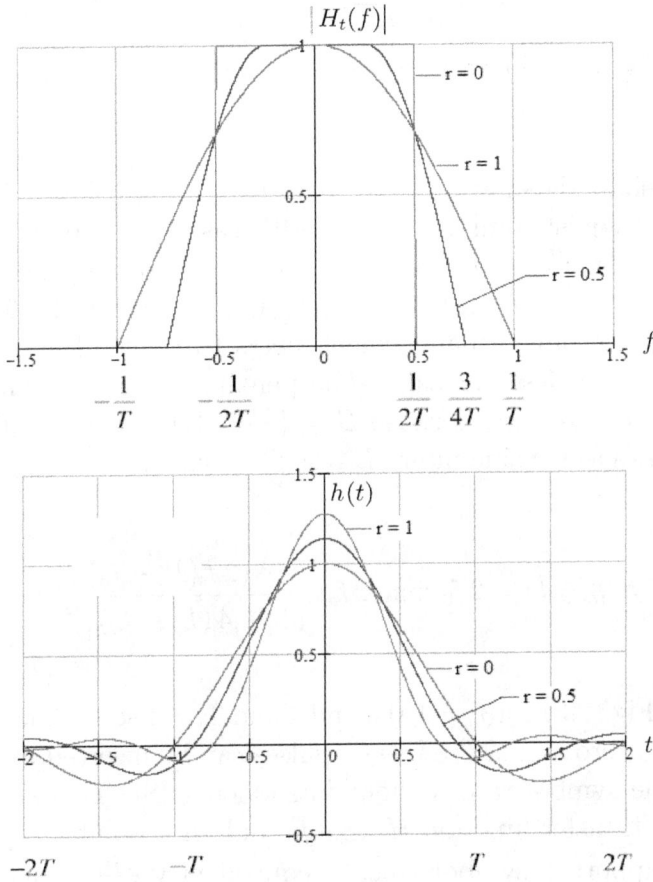

Figure 6.30: (a) Frequency response and (b) Impulse response of the root raised cosine filter.

$$h_{root}(t) = \begin{vmatrix} \left(1 - r + \frac{4r}{\pi}\right)\frac{1}{T} & if & t = 0 \\[2ex] \frac{4r}{\pi T}\dfrac{\cos\left(\frac{\pi t}{T}(1+r)\right)+\frac{T}{4rt}\sin\left(\frac{\pi t}{T}(1-r)\right)}{\left[1-\left(\frac{4rt}{T}\right)^2\right]} & elseif & \left(\frac{4rt}{T}\right)^2 \neq 1 \\[3ex] \frac{r}{T\sqrt{2}}\left[\left(1 - \frac{2}{\pi}\right)\cos\left(\frac{\pi}{4r}\right) + \left(1 + \frac{2}{\pi}\right)\sin\left(\frac{\pi}{4r}\right)\right] & elseif & \left(\frac{4rt}{T}\right)^2 = 1 \end{vmatrix}$$

$$(6.65)$$

This pulse shape is shown in Fig. 6.30 for $r = 0$, 0.5 and 1 and $T = 1$ sec, where the Nyquist minimum bandwidth case corresponds to $r = 0$ i.e. the pulse shape for $r = 0$ in this case is identical to the pulse $\text{sinc}(t)$. A timing error will introduce ISI. However, as r is increased, the amplitude of the pulse tail becomes smaller which helps to reduce ISI errors because timing then becomes less critical. The penalty incurred is an increase in the bandwidth requirement because $B = \left(\frac{1+r}{2T}\right)$ Hz. The impulse response $h_{raised}(t)$ of the raised cosine filter $H_{raised}(f)$ given by

$$h_{raised}(t) = 2f_0 \,\text{sinc}\,(2f_o t)\,\frac{\cos\left(2\pi\left(B - f_o\right)t\right)}{1 - \left[4\left(B - f_0\right)t\right]^2} \qquad (6.66)$$

is shown in Fig. 6.31 for $r = 0$, 0.5 and 1 and $T = 1$ sec. Since the impulse response goes to zero at $T, 2T, 3T, \cdots$, pulses can be inserted at these points to maximize the symbol rate without introducing ISI. Therefore, a system with an overall transfer function $H_{raised}(f)$ and an absolute bandwidth $B = \frac{(1+r)}{2T}$ Hz can support $\frac{1}{T}$ symbols/sec, or equivalently $\frac{2B}{1+r}$ symbols/sec. For $r = 0$, we recover the Nyquist's discovery that a bandwidth of $B = \frac{1}{2T}$ Hz can support $\frac{1}{T}$ symbols/sec or equivalently $2B$ symbols/sec.

- ⭐ SIMULATION **RootRaisedCosine:** For further details, where in addition, the inverse Fourier transform of $H_t(f)$ and $H_{raised}(f)$ are shown to correspond to $h(t)$ and $h_{raised}(t)$, respectively.

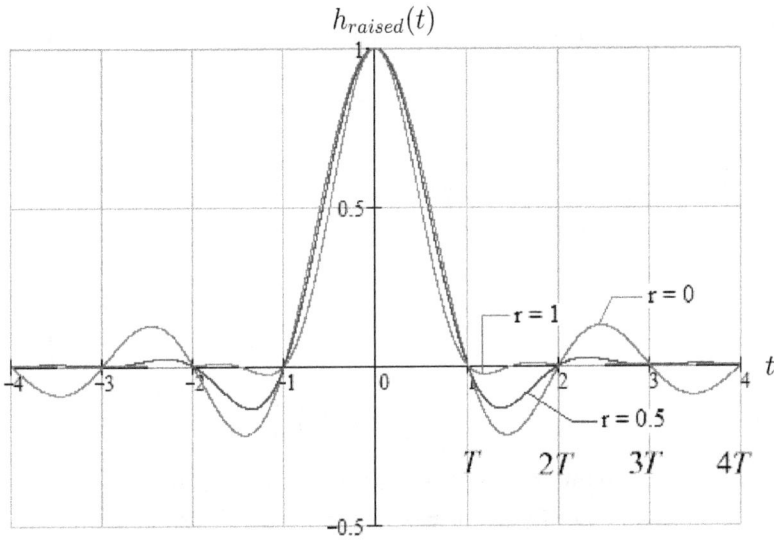

Figure 6.31: Raised cosine filter impulse response.

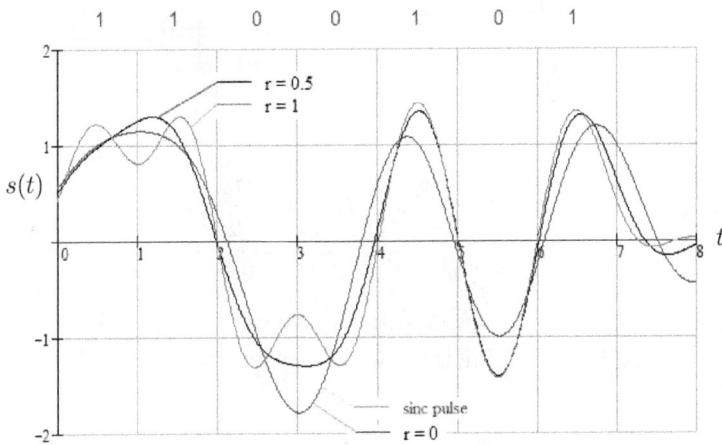

Figure 6.32: Root raised cosine shaped pulses compared with sinc pulses.

Finally in Fig. 6.32 we compare the signal created with a root raised cosine filter with the sinc pulse signal in Fig. 6.27 for $r = 0$, 0.5 and 1 and $T = 1$ sec. As expected, the $r = 0$ curve overlays the "sinc pulse" curve.

- ⭐ SIMULATION **ShapedPulse:** Simulation corresponding to Fig. 6.32. You may experiment with different values for r and change the binary sequence.

6.3.4   Partial-Response Signaling

The analysis so far suggests that we must ensure that $T > \frac{1}{2B}$ to build a practical system with no ISI. However, it is possible to achieve $T = \frac{1}{2B}$, or equivalently $2B$ symbols per second within a bandwidth of B Hz, by using *partial-response signaling* (also known as *correlative-level coding*) in which intersymbol interference is added and removed in a controlled manner. Specifically, we relax the condition for zero ISI (equation 6.54) to be

$$h_{tr}(nT) = \begin{cases} 1 & if \quad n = 0, 1 \\ 0 & otherwise \end{cases} \tag{6.67}$$

Then, let us define $Z(f)$ as

$$Z(f) = \sum_{m=-\infty}^{\infty} H_{tr}\left(f + \frac{m}{T}\right) \tag{6.68}$$

Given that $Z(f)$ is a periodic function with period $\left(\frac{1}{T}\right)$, it can be expressed in terms of an exponential Fourier series (section 1.2.1) as

$$Z(f) = \sum_{n=-\infty}^{\infty} z_n e^{\frac{j2\pi nf}{1/T}} = \sum_{n=-\infty}^{\infty} z_n e^{j2\pi nfT} \tag{6.69}$$

where z_n is the complex coefficient given by

$$z_n = \frac{1}{1/T} \int_{1/T}^{1/2T} Z(f) e^{\frac{-j2\pi nf}{1/T}} df = T \int_{-1/2T}^{1/2T} Z(f) e^{-j2\pi nfT} df \qquad (6.70)$$

Comparing equation 6.70 with equation 6.59, we have

$$z_n = Th_{tr}(-nT) \qquad (6.71)$$

To satisfy $h_{tr}(nT) = \begin{cases} 1 & if \quad n = 0 \\ 0 & n \neq 0 \end{cases}$, we require $z_n = \begin{cases} T & if \quad n = 0 \\ 0 & n \neq 0 \end{cases}$,
which corresponds $Z(f) = T$ i.e. the Nyquist condition for zero ISI presented
in equation 6.61. However, to satisfy $h_{tr}(nT) = \begin{cases} 1 & if \quad n = 0, 1 \\ 0 & otherwise \end{cases}$, we
now require

$$z_n = \begin{cases} T & if \quad n = 0, -1 \\ 0 & otherwise \end{cases} \qquad (6.72)$$

which corresponds to

$$Z(f) = \sum_{n=-\infty}^{\infty} z_n e^{j2\pi nfT} = T + Te^{-j2\pi fT} \qquad (6.73)$$

which is equal to $\left(\frac{1}{2B} + \frac{1}{2B}e^{-\frac{j\pi f}{B}} \right)$ for $T = \frac{1}{2B}$. Thus

$$H_{tr}(f) = \begin{cases} \frac{1}{2B}\left[1 + e^{-\frac{j\pi f}{B}}\right] & if \quad |f| < B \\ 0 & otherwise \end{cases} \qquad (6.74)$$

and taking the inverse FT,

$$h_{tr}(t) = \text{sinc}(2Bt) + \text{sinc}(2Bt - 1) \qquad (6.75)$$

Notice that $h_{tr}(t)$, referred to as a *duobinary* pulse, is simply the combination of two sinc pulses time-displaced by T seconds with respect to each

other as shown in Fig. 6.33, together with its magnitude spectrum $|H_{tr}(f)|$ for $T = 1$ sec. Furthermore, notice in Fig. 6.33, $|H_{tr}(f)|$ is not equal to zero at $f = 0$. This is undesirable because many communication channels cannot transmit a d.c. component. The solution is to use a *modified* duobinary pulse using two sinc pulses separated by $2T$ seconds such that

$$h_{tr}(nT) = \text{sinc}\,(2Bt) + (-1)\,\text{sinc}(2Bt - 2) \qquad (6.76)$$

In general, we have a class of bandlimited pulses, referred to as *partial response signals*, given by

$$h_{tr}(nT) = \sum_{n=0}^{N-1} w_n \,\text{sinc}(2Bt - n) \qquad (6.77)$$

where for example, $w_0 = w_1 = 1$ for a duobinary pulse and $w_0 = 1, w_1 = 0, w_2 = -1$ for the modified duobinary pulse.

- ⭐ SIMULATION **DuoBinaryPulse:** An implementation of the analysis presented in this section and a plot of the pulses. You can experiment with other weights. For example, try $w_0 = -1, w_1 = 0, w_2 = 2$, $w_3 = 0, w_4 = -1$. Verify for yourself that magnitude spectrum of the modified duobinary pulse is given by $H_{tr}(f) = \frac{j}{B}\sin\left(\frac{\pi f}{B}\right)$ for $|f| \leq B$, otherwise its zero.

6.3.5   Equalization

In section 6.3.3, we assumed that

$$H_e(f) = \frac{1}{H_c(f)} = \frac{1}{|H_c(f)|}e^{-j\theta_c(f)} \quad \text{for } |f| \leq B \qquad (6.78)$$

where the equalizing filter $H_e(f)$ is placed after the receiving filter $H_r(f)$ in Fig. 6.28. To achieve $H_c(f)H_e(f) \simeq 1$, we could use a *finite-impulse*

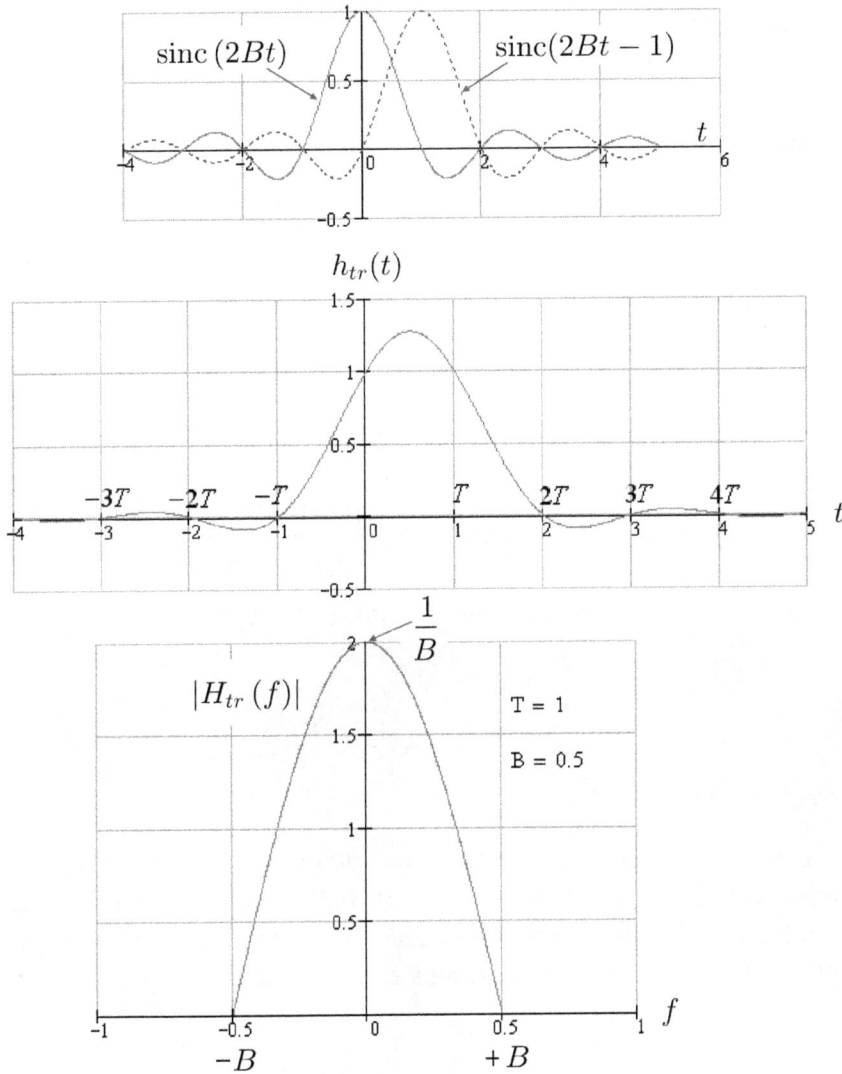

Figure 6.33: Duobinary pulse and its magnitude spectrum.

response (FIR) filter with adjustable tap coefficients c_n as shown in Fig. 6.34. The input to this filter are the sampled values output from the receiving filter in Fig. 6.28 in the previous section. This type of filter is referred to as *tapped delay-line* or *transversal filter*. Lets begin by taking a look at a simple equalizer.

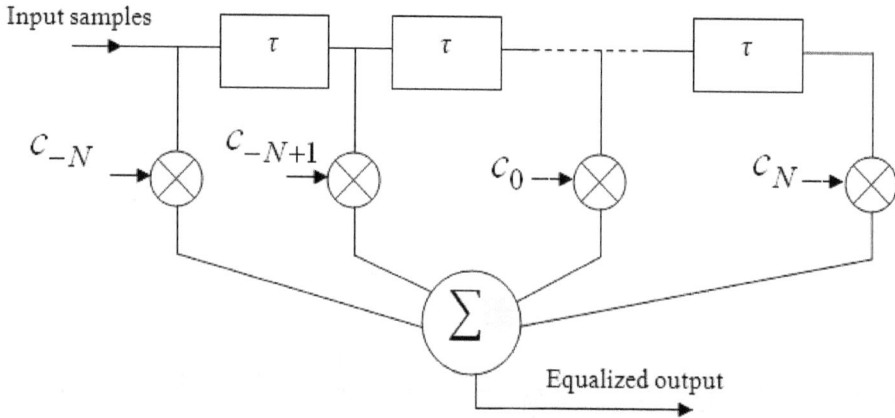

Figure 6.34: Linear transversal filter.

Zero-Forcing Equalizer

Suppose a single impulse is transmitted as a training signal and the received set of pulse samples are 0.2, 0.1, -0.2, 1.0, 0.3, -0.1, 0.3, where 0.2 is first value to enter the FIR filter and the value 1.0 corresponds to the main lobe of the pulse. Let the vector $\mathbf{x}$ represent these input samples, so that

$$\mathbf{x} = \begin{bmatrix} x_{-3} \\ x_{-2} \\ x_{-1} \\ x_0 \\ x_1 \\ x_2 \\ x_3 \end{bmatrix} = \begin{bmatrix} 0.2 \\ 0.1 \\ -0.2 \\ 1 \\ 0.3 \\ -0.1 \\ 0.3 \end{bmatrix} \tag{6.79}$$

Now suppose we create a square matrix $\mathbf{X}$ with $(2N+1)$ columns using $\mathbf{x}$ and a coefficients vector $\mathbf{c}$ with $(2N+1)$ *equalizer coefficients* such that

$$\mathbf{X} = \begin{bmatrix} x_0 & x_{-1} & x_{-2} & \cdots & x_{-(N+1)} \\ x_1 & x_0 & x_{-1} & x_{-2} & \vdots \\ x_2 & x_1 & x_0 & x_{-1} & x_{-2} \\ \vdots & x_2 & x_1 & x_0 & x_{-1} \\ x_{N+1} & \cdots & x_2 & x_1 & x_0 \end{bmatrix} \quad \text{and} \quad \mathbf{c} = \begin{bmatrix} c_{-N} \\ \vdots \\ c_0 \\ \vdots \\ c_N \end{bmatrix} \quad (6.80)$$

so that for the input vector in equation 6.79 with $N = 1$, we have

$$\mathbf{X} = \begin{bmatrix} 1 & -0.2 & 0.1 \\ 0.3 & 1 & -0.2 \\ -0.1 & 0.3 & 1 \end{bmatrix} \quad \text{and} \quad \mathbf{c} = \begin{bmatrix} c_{-1} \\ c_0 \\ c_1 \end{bmatrix} \quad (6.81)$$

where we are only making use of the samples $x_{\pm 1}$ and $x_{\pm 2}$ on either side of the main-lobe x_0. Now to force the samples on either side of equalizer output to be zero, we require

$$\mathbf{X}\mathbf{c} = \mathbf{q} \tag{6.82}$$

where $\mathbf{q} = \begin{bmatrix} 0 \\ 1 \\ 0 \end{bmatrix}$. Therefore, the tap coefficients must be given by

$$\mathbf{c} = \mathbf{X}^{-1}\mathbf{q} \tag{6.83}$$

so that in our ongoing example, we have

$$\mathbf{X}^{-1} = \begin{bmatrix} 0.9339 & 0.2026 & -0.0529 \\ -0.2467 & 0.8899 & 0.2026 \\ 0.1674 & -0.2467 & 0.9339 \end{bmatrix} \quad (6.84)$$

and hence

$$\mathbf{c} = \begin{bmatrix} 0.2026 \\ 0.8899 \\ -0.2467 \end{bmatrix} \tag{6.85}$$

In general, all the row entries in the vector $\mathbf{q}$ are zero except the $(N+1)$ row, which is set equal to 1. If we set the coefficients in Fig. 6.34 to those equation 6.84, insert the samples $\mathbf{x}$ into the transversal filter and flush the filter with zeros, then the output samples $\mathbf{y}$ are given by

$$\mathbf{y} = \begin{bmatrix} 0.0405 \\ 0.1982 \\ -0.0009 \\ 0.0 \\ 1.0 \\ 0.0 \\ -0.1022 \\ 0.2916 \\ -0.0740 \\ 0.0000 \end{bmatrix} \tag{6.86}$$

As expected, the FIR filter output is forced to be zero on either side of the desired pulse. This type of equalizer is referred to as a *zero-forcing equalizer*. Taking a closer look at $\mathbf{x}$, the sum of all the ISI magnitude contributions before equalization is equal to 1.2 and 0.7075 after equalization.

- ⭐ SIMULATION **TransversalFilter:** Implementation of the matrix analysis, transversal filter, ISI analysis and to view the input and output samples on a graph. Experiment with other input sample vectors $\mathbf{x}$. How could we force two samples on either side of $\mathbf{y}$ to be zero ?

Minimum Mean-Square Error (MMSE) Equalization

The zero-forcing equalizer is not used in practice because the coefficient calculations do not take noise into account. Indeed, the effects of noise are
actually enhanced wherever $H_c(f)$ is small, because the equalizer compen-
sates with a large gain to ensure $H_c(f)H_e(f) = 1$. A better solution is
to relax the zero-forcing condition in favour of a *minimum mean-square error* (MMSE) equalizer in which the coefficients are chosen to minimize the
mean-square error between the desired output and the actual noisy output.
Specifically, from equation 6.82, we multiply both sides by $\mathbf{X}^T$, so that

$$\left(\mathbf{X}^T\mathbf{q}\right) = \left(\mathbf{X}^T\mathbf{X}\right)\mathbf{c} \tag{6.87}$$

from which the transversal filter coefficients $\mathbf{c}$ are then given by

$$\mathbf{c} = \left(\mathbf{X}^T\mathbf{X}\right)^{-1}\left(\mathbf{X}^T\mathbf{q}\right) \tag{6.88}$$

where $\left(\mathbf{X}^T\mathbf{q}\right)$ and $\left(\mathbf{X}^T\mathbf{X}\right)$ are referred to as the *cross-correlation* and
auto-correlation matrix, respectively [Qureshi, 1985]. Finally, the output of
the filter $\mathbf{y}$ is given by

$$\mathbf{y} = \mathbf{X}\mathbf{c} \tag{6.89}$$

The cross and auto-correlation matrices are unknown *a priori*, but can be
approximated by transmitting a test signal and using time average estimates.
For our ongoing example, the matrix $\mathbf{X}$ is created by staggering $\mathbf{x}$ such that

$$\mathbf{X} = \begin{bmatrix}
0.2 & 0 & 0 & 0 & 0 & 0 & 0 \\
0.1 & 0.2 & 0 & 0 & 0 & 0 & 0 \\
-0.2 & 0.1 & 0.2 & 0 & 0 & 0 & 0 \\
1 & -0.2 & 0.1 & 0.2 & 0 & 0 & 0 \\
0.3 & 1 & -0.2 & 0.1 & 0.2 & 0 & 0 \\
-0.1 & 0.3 & 1 & -0.2 & 0.1 & 0.2 & 0 \\
0.3 & -0.1 & 0.3 & 1 & -0.2 & 0.1 & 0.2 \\
0 & 0.3 & -0.1 & 0.3 & 1 & -0.2 & 0.1 \\
0 & 0 & 0.3 & -0.1 & 0.3 & 1 & -0.2 \\
0 & 0 & 0 & 0.3 & -0.1 & 0.3 & 1 \\
0 & 0 & 0 & 0 & 0.3 & -0.1 & 0.3 \\
0 & 0 & 0 & 0 & 0 & 0.3 & -0.1 \\
0 & 0 & 0 & 0 & 0 & 0 & 0.3
\end{bmatrix} \qquad (6.90)$$

for which we find

$$\mathbf{c} = \begin{bmatrix}
-0.1674 \\
0.0145 \\
0.1910 \\
0.9607 \\
-0.1948 \\
0.0189 \\
-0.2495
\end{bmatrix} \qquad (6.91)$$

- ⭐ SIMULATION **MMSE**: Implementation of the MMSE equalizer in which the output $\mathbf{y}$ given by equation 6.89 and implementing a transversal filter with the coefficients given in equation 6.91, are shown to produce the same output $\mathbf{y}$. The maximum single ISI magnitude before and after equalization and the sum of all the ISI magnitude contributions before and after equalization are compared. What happens to the filter output $\mathbf{y}$ if the size of $\mathbf{X}$ is reduced by using fewer entries from the input vector $\mathbf{x}$?

In practice, the filter coefficients **c** are found using an iterative algorithm which avoids the need to determine the inverse of a matrix as in equation 6.88.

6.3.6 Simulation I: Random Sinc Pulse Signal PSD

⭐ SIMULATION **SincPSD:** A sinc pulse baseband signal similar to Fig. 6.27 is generated from a 100 random binary digits i.e. each binary digit is represented by a sinc pulse. From the power spectral density (PSD) of this signal, which is given by the Fourier transform of the signal's autocorrelation function, the expected minimum bandwidth of $\frac{1}{2T}$ Hz with the use of sinc(.) pulses is verified.

- **Experiment:** Try a different type of pulse (e.g. the duobinary pulse) and compare its bandwidth with $\frac{1}{2T}$ Hz.

6.3.7 Simulation II: Baseband Filtering

⭐ SIMULATION **Nyquist:** An input sinc pulse and a rectangular pulse are processed side-by-side through the processing blocks shown in Fig. 6.35. Please do allow the software to complete its task. This is a computer intensive simulation. Let $H_{signal}(f)$ be the Fourier transform (FT) of the input pulse. The transmitted pulse at point D is given by the inverse FT of $H_{signal}(f)H_t(f)$. This pulse is sent through a channel with a frequency response $H_c(f) = |H_c(f)|\,e^{j\theta_c(f)}$. The received pulse at point E is given by the inverse FT of $H_{signal}(f)H_t(f)H_c(f)$ and the output pulse is given by the inverse FT of $H_{signal}(f)H_t(f)H_c(f)H_r(f)$.

Recall that for distortionless transmission, we require $|H_c(f)|$ equal to a constant and $\theta_c(f) = -2\pi f t_d$ as shown in Fig. 6.36(a). To illustrate a channel with distortion by example, let $H_c(f) = \frac{1}{1+j2\pi f t_c}$ (low-pass filter) or equivalently, $|H_c(f)| = \frac{1}{\sqrt{1+(2\pi f t_c)^2}}$ and $\theta_c(f) = -\tan^{-1}(2\pi f t_c)$, as shown in

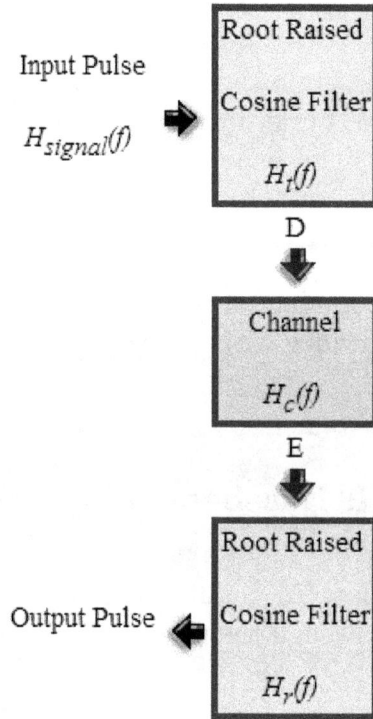

Figure 6.35: Pulse filtering.

Fig. 6.36(b). The result of processing the two pulses summarized in Table 6.2 through the filters in Fig. 6.35 are presented in Fig. 6.37 for $H_c(f) = e^{-j2\pi f t_d}$ (distortionless) and $H_c(f) = \frac{1}{1+j2\pi f t_c}$ (distortion) with $r = 1$, $t_c = t_d = T = 1$ sec.

- **Experiment:** Try $r = 0.1$, or input a unit impulse function $\delta(t)$, to verify for yourself that the pulse shapes are as expected at every stage of the communication system. Change the width of a given input pulse and the characteristics of the filters to gain an appreciation of filtering within a baseband channel. Modify the code to consider other types of pulses and have some fun by introducing other types of distortion within the channel.

PulseType	Input pulse	Fourier transform $\frac{H_{signal}}{H_t(f)}$
Root raised cosine	$h_{root}(t)$	
Rectangular	$s(t) = \begin{cases} 1 & if \quad \lvert t \rvert < \tau \\ 0 & otherwise \end{cases}$	$2\tau \operatorname{sinc}(2\tau f)$

Table 6.2: Pulses for experimentation.

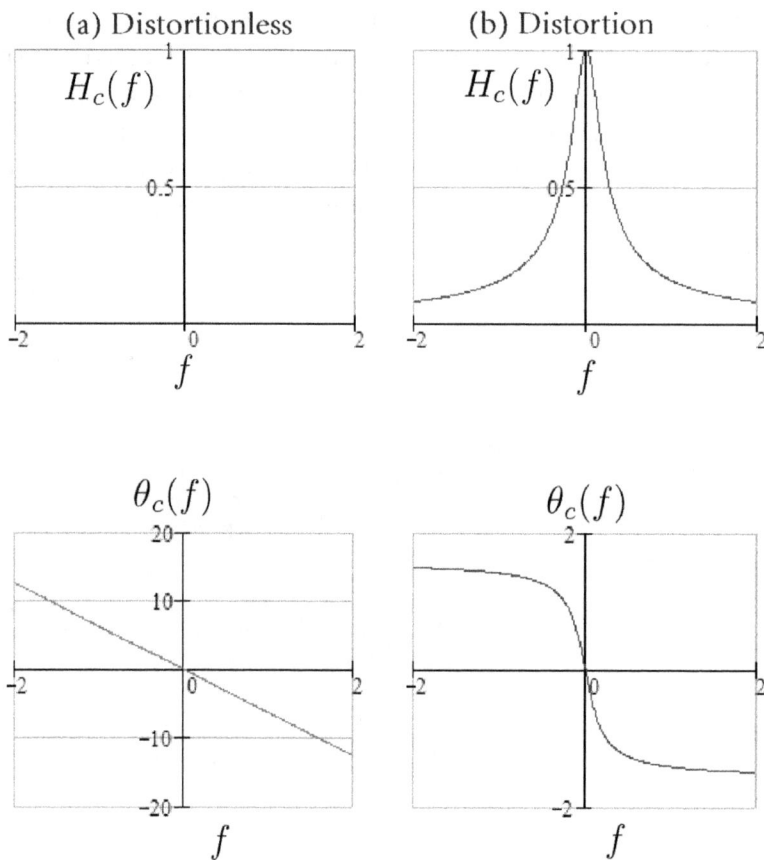

Figure 6.36: Channel transfer function for no distortion and an example of distortion.

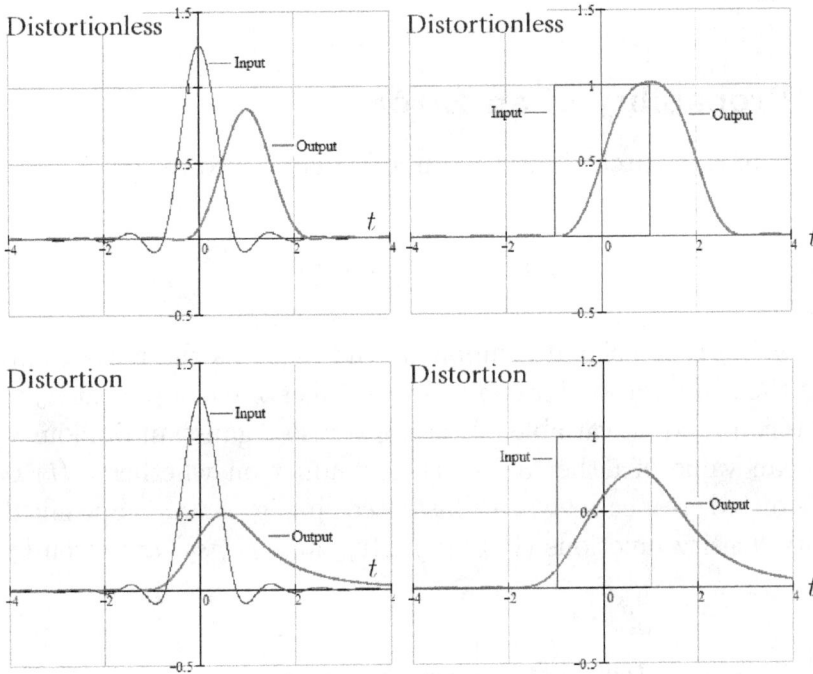

Figure 6.37: Input and output sinc and rectangular pulse shapes with $r = 1$.

6.4   Performance

In this section, we shall determine an expression for P_e using the matched filter in Fig. 6.17 of section 6.2. We are once again assuming an ideal distortionless channel (unconstrained bandwidth with no ISI) to reduce the complexity of the analysis to follow in this and the remaining sections of this chapter. For a binary communication system, given that we have not yet employed error-control coding, this P_e over an AWGN channel is simply the BSC crossover probability (refer to section 3.2).

6.4.1   Probability of an Error

The output of the matched filter at time T may be expressed as

$$y(T) = s_o(T) + n_o(T) \tag{6.92}$$

where $s_o(T)$ is the signal component and $n_o(T)$ is the noise component. Dropping the (T) notation for convenience, this noise component n_o is a zero mean gaussian random variable. Hence y is also a gaussian random variable with a mean value of either a_1 or a_2 depending on whether $s_1(t)$ or $s_2(t)$ was transmitted, respectively, as illustrated in Fig. 6.38. Specifically, the probability density functions (PDFs) $f(y|s_1)$ and $f(y|s_2)$ are given by

$$f(y|s_1) = \frac{1}{\sigma\sqrt{2\pi}} \exp\left[-\frac{(y-a_1)^2}{2\sigma^2}\right] \tag{6.93}$$

$$f(y|s_2) = \frac{1}{\sigma\sqrt{2\pi}} \exp\left[-\frac{(y-a_2)^2}{2\sigma^2}\right] \tag{6.94}$$

where the mean correlator outputs a_1 and a_2 correspond respectively to $s_1(t)$ and $s_2(t)$. Recall from the previous section that for matched filtering, $s_o(T) = E$ and $\sigma^2 = \frac{N_o}{2}E$. Thus for example with polar NRZ signaling, $a_1 = E_b$ and $a_2 = -E_b$.

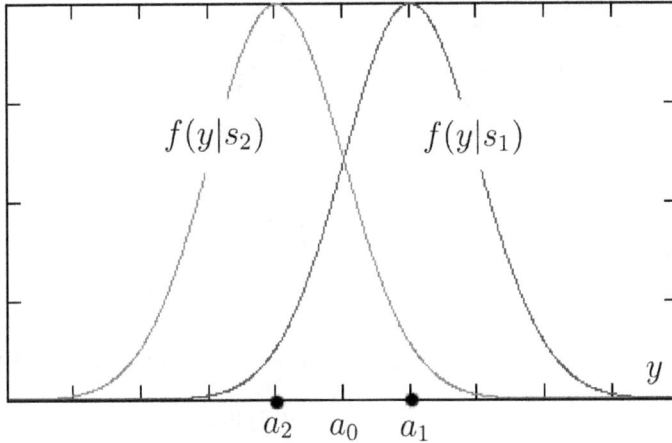

Figure 6.38: Probability density functions of $y(T)$.

• ⭐ SIMULATION **Gaussian:** To learn how a gaussian probability density function (PDF) is completely described by its mean and standard deviation σ.

By symmetry, we may invoke that the optimum threshold a_0 is set at the point of intersection of the two PDFs i.e. $a_0 = \frac{a_1 + a_2}{2}$. In general however, it can be shown (refer to the problems section) that the optimum threshold

$$a_0 = \frac{a_1 + a_2}{2} + \frac{\sigma^2}{a_1 - a_2} \ln \left(\frac{P(s_2)}{P(s_1)} \right) \qquad (6.95)$$

where $P(s_1)$ is the a priori probability that $s_1(t)$ was transmitted and $P(s_2) = 1 - P(s_1)$ is the a priori probability that $s_2(t)$ was transmitted. For $P(s_1) = P(s_2)$, we find that $\ln \left(\frac{P(s_2)}{P(s_1)} \right) = 0$ and hence $a_0 = \frac{a_1 + a_2}{2}$.

Since $s_1(t)$ and $s_2(t)$ correspond to a binary digit 1 and 0, respectively, for convenience let $P(1|0)$ represent the probability that $s_2(t)$ is incorrectly demodulated. Similarly, let $P(0|1)$ be the probability that $s_1(t)$ is incorrectly demodulated. To determine expressions for $P(1|0)$ and $P(0|1)$, refer to the PDF shown in Fig. 6.39. The shaded area encompassed between any two points, say a and b, along the horizontal axis is equal to the probability

$$P(a < y \le b) = \int_a^b f(y|s_2)dy \qquad (6.96)$$

with which y lies between a and b. Since the area under a PDF is equal to 1, we have $\int_{-\infty}^{\infty} f(y|s_1)dy = 1$ and $\int_{-\infty}^{\infty} f(y|s_2)dy = 1$.

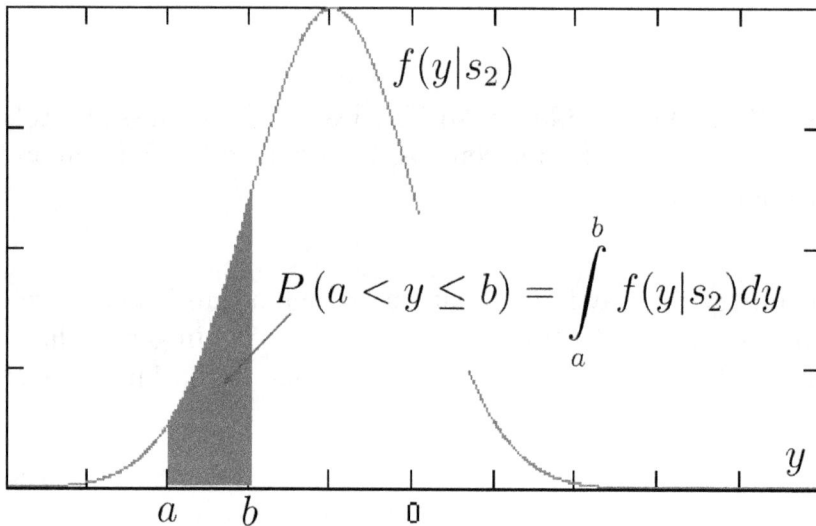

Figure 6.39: Area under a probability density function.

Thus referring to Fig. 6.38, $P(1|0) = P(y > a_0|0)$ and $P(0|1) = P(y \le a_0|1)$. Hence

$$P(1|0) = \int_{a_0}^{\infty} f(y|s_2)dy = \int_{a_0}^{\infty} \frac{1}{\sigma\sqrt{2\pi}} \exp\left[-\frac{(y-a_2)^2}{2\sigma^2}\right] dy \qquad (6.97)$$

Using the integration method of substitution, let $z = \frac{(y-a_2)}{\sigma}$, so that $\frac{dz}{dy} = \frac{1}{\sigma}$. Changing the integration limits, we find that for $y = a_0$, $z = \frac{a_0-a_2}{\sigma}$. But $a_0 = \frac{a_1+a_2}{2}$, and therefore $z = \frac{\frac{a_1+a_2}{2}-a_2}{\sigma} = \frac{a_1-a_2}{2\sigma}$. For the integration limit $y = \infty$, we get $z = \infty$. Hence

$$P(1|0) = \int_{\frac{a_1-a_2}{2\sigma}}^{\infty} \frac{1}{\sigma\sqrt{2\pi}} \exp\left[-\frac{z^2}{2}\right] (\sigma dz) \qquad (6.98)$$

which simplifies to

$$P(1|0) = \int_{\frac{a_1-a_2}{2\sigma}}^{\infty} \frac{1}{\sqrt{2\pi}} \exp\left[-\frac{z^2}{2}\right] dz \qquad (6.99)$$

We may rewrite this integral in terms of a zero mean, unit standard deviation gaussian PDF $f(z)$ as shown in Fig. 6.40, where

$$f(z) = \frac{1}{\sqrt{2\pi}} \exp\left[-\frac{z^2}{2}\right] \qquad (6.100)$$

so that

$$P(1|0) = \int_{\frac{a_1-a_2}{2\sigma}}^{\infty} f(z)dz \qquad (6.101)$$

The integral over $f(z)$ can only be solved numerically. A convenient solution is to express it in terms of the *Q-function* defined as

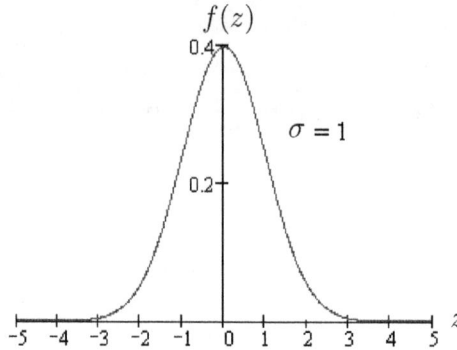

Figure 6.40: Gaussian probability density function with zero mean and unit standard deviation.

$$Q(u) = \int\limits_{u}^{\infty} \frac{1}{\sqrt{2\pi}} \exp\left(-\frac{z^2}{2}\right) dz \qquad (6.102)$$

Alternatively, in terms of the *complementary error function,* which is defined by

$$\text{erfc}(u) = \frac{2}{\sqrt{\pi}} \int\limits_{u}^{\infty} \exp\left(-z^2\right) dz \qquad (6.103)$$

the Q-function is given by

$$Q(u) = \frac{1}{2} \text{erfc}(\frac{u}{\sqrt{2}}) \qquad (6.104)$$

where as shown in Fig. 6.41, $Q(u)$ is simply the area to the right of the coordinate u, on a unit standard deviation, zero mean gaussian PDF. The difference $Q(u_1) - Q(u_2)$ is illustrated in Fig. 6.41. Since the total area under the PDF is 1 and its a symmetrical function about the zero axis, $Q(0) = 0.5$, $Q(-\infty) = 1$ and $Q(\infty) = 0$.

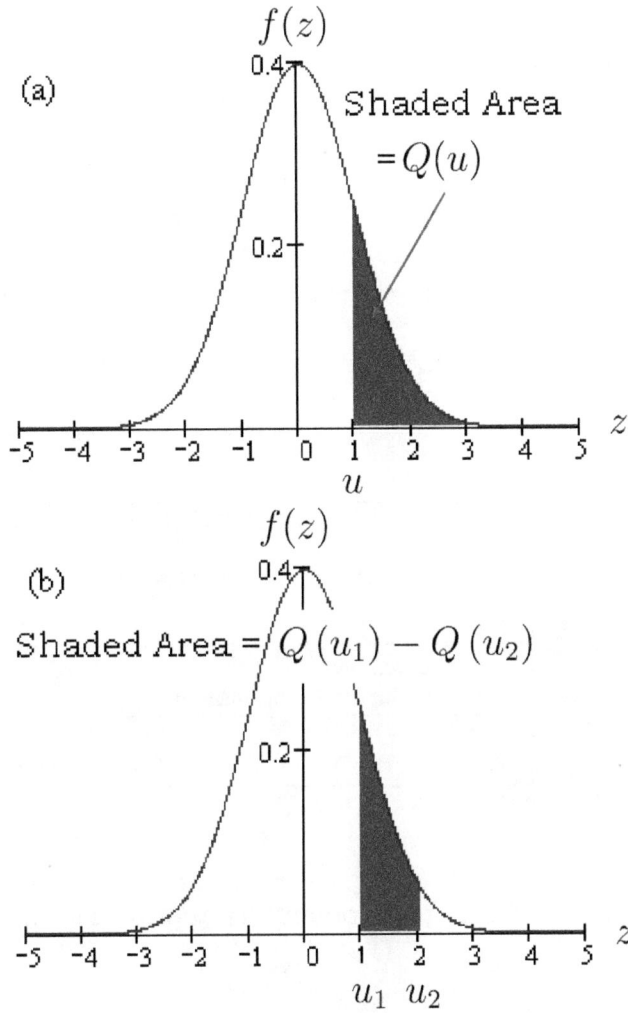

Figure 6.41: Q-function defined in terms of a Gaussian probability density function with zero mean and unit standard deviation.

The values of $Q(u)$ are normally tabulated (Appendix B.1), or made available within a mathematical software package such as Mathcad. For large values of u, we can approximate the Q-function as

$$Q(u) \approx \frac{1}{u\sqrt{2\pi}} \exp\left(-\frac{u^2}{2}\right) \tag{6.105}$$

which as evident from Fig. 6.42 is a good approximation for $u > 3$.

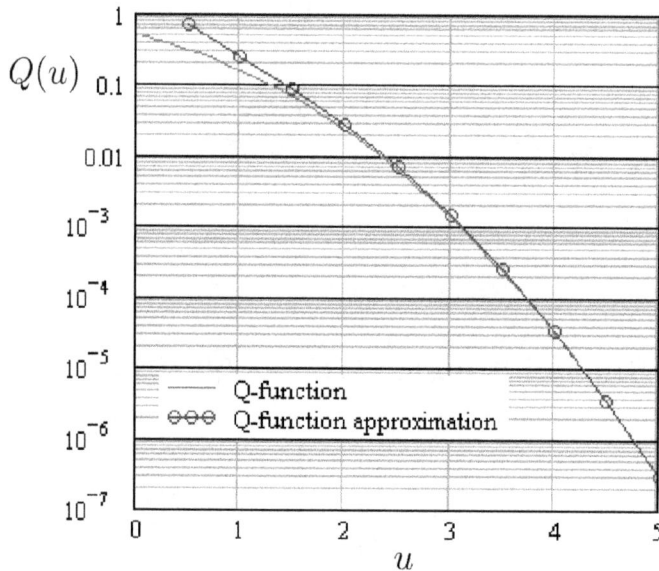

Figure 6.42: Q function and its approximation.

• ⭐ SIMULATION **Qfunction:** Simulation corresponding to Fig. 6.42. Alter the expression in equation 6.105 to try your own approximation.

Expressing the probability $P(1|0)$ in terms of the Q-function, we find

$$P(1|0) = \int_{\frac{a_1-a_2}{2\sigma}}^{\infty} \frac{1}{\sqrt{2\pi}} \exp\left[-\frac{z^2}{2}\right] dz = Q\left(\frac{a_1-a_2}{2\sigma}\right) \tag{6.106}$$

By symmetry from Fig. 6.38, $P(1|0) = P(0|1)$ i.e. the equivalent model of this binary communication system over an AWGN channel in which each transmitted signal is equally likely to be received in error is shown in 6.43.

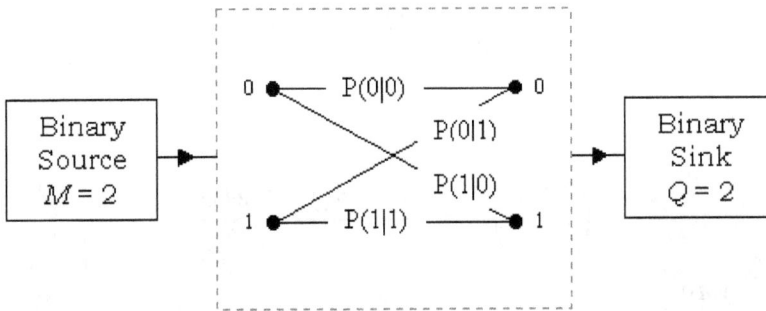

Figure 6.43: Binary communication system model over an AWGN channel.

Therefore as in section 3.2, the probability of an error in a binary digit P_e at the output of the demodulator is given by $P_e = P(1|0)P(0) + P(0|1)P(1)$, where $P(0)$ and $P(1)$ are respectively the probabilities with which a binary digit zero and one are transmitted. Since $P(0) + P(1) = 1$, we find

$$P_e = P(1|0)\left[P(0)+P(1)\right] = P(1|0) = Q\left(\frac{a_1-a_2}{2\sigma}\right) \tag{6.107}$$

Given the nature of the $Q(.)$ function, to minimize P_e, we need to determine the linear filter that maximizes $\frac{a_1-a_2}{2\sigma}$ or equivalently maximize $\frac{(a_1-a_2)^2}{\sigma^2}$, where $(a_1 - a_2)^2$ is the instantaneous power of the difference signal $[s_1(t) - s_2(t)]$

and σ^2 is the average noise power. Recall that for a matched filter, $\left(\frac{P_{out}}{N_{out}}\right)_T =$ $\frac{2E}{N_o}$ where E is the energy of the input signal. Thus, for a filter matched to the input difference signal $[s_1(t) - s_2(t)]$,

$$\left(\frac{P_{out}}{N_{out}}\right)_T = \frac{2E_d}{N_o} \qquad (6.108)$$

where E_d is the energy of the difference signal at the filter input given by

$$E_d = \int_0^{T_b} [s_1(t) - s_2(t)]^2 \, dt \qquad (6.109)$$

so that

$$\left(\frac{P_{out}}{N_{out}}\right)_T = \frac{(a_1 - a_2)^2}{\sigma^2} = \frac{2E_d}{N_o} \qquad (6.110)$$

and therefore

$$P_e = Q\left(\frac{a_1 - a_2}{2\sigma}\right) = Q\left(\frac{1}{2}\sqrt{\frac{2E_d}{N_o}}\right) = Q\left(\sqrt{\frac{E_d}{2N_o}}\right) \qquad (6.111)$$

For example, for unipolar NRZ signaling in which $s_1(t) = +A$ and $s_2(t) = 0$ for $0 < t \leq T_b$, we find $E_d = \int_0^{T_b} [A]^2 \, dt = A^2 T_b$ and thus

$$(P_e)_{UnipolarNRZ} = Q\left(\sqrt{\frac{A^2 T_b}{2N_o}}\right) = Q\left(\sqrt{\frac{E_b}{N_o}}\right) \qquad (6.112)$$

where the average energy per binary digit $E_b = \frac{A^2 T_b}{2}$. Similarly, for polar NRZ signaling in which $s_1(t) = +A$ and $s_2(t) = -A$ for $0 < t \leq T_b$, we find $E_d = \int_0^{T_b} [2A]^2 \, dt = 4A^2 T_b$ and thus

$$(P_e)_{PolarNRZ} = Q\left(\sqrt{\frac{4A^2T_b}{2N_o}}\right) = Q\left(\sqrt{\frac{2E_b}{N_o}}\right) \qquad (6.113)$$

where $E_b = A^2T_b$. Alternatively, since $a_1 = E_b$, $a_2 = -E_b$ and $\sigma^2 = \frac{N_o}{2}E_b$,

$$(P_e)_{PolarNRZ} = Q\left(\frac{a_1 - a_2}{2\sigma}\right) = Q\left(\frac{2E_b}{2}\sqrt{\frac{2}{N_oE_b}}\right) = Q\left(\sqrt{\frac{2E_b}{N_o}}\right) \quad (6.114)$$

6.4.2 Signal-to-Noise Ratio (SNR)

A given theoretical expression for P_e is usually expressed in terms of the $\frac{E_b}{N_o}$ ratio because the specific values of E_b and N_o are not as important as the ratio which dictates the value of P_e. This ratio $\frac{E_b}{N_o}$, referred to as the *signal-to-noise ratio* (SNR), is usually expressed in *decibels*. Expressing a number in decibels is simply the conversion of a number, say x, onto a logarithm scale on which the units are decibels (dB) as shown in Fig. 6.44, so that the SNR

$$\left(\frac{E_b}{N_o}\right) dB = 10\log_{10}\left(\frac{E_b}{N_o}\right) \qquad (6.115)$$

To distinguish between the actual ratio $\frac{E_b}{N_o}$, and the ratio $\left(\frac{E_b}{N_o}\right)$dB, let SNR represent the signal-to-noise ratio in decibels i.e. $SNR = 10\log_{10}\left(\frac{E_b}{N_o}\right)$. To calculate P_e, if the SNR is specified, we need to recover the actual ratio $\frac{E_b}{N_o}$. For example, for $SNR = 9.6$ dB, the ratio $\left(\frac{E_b}{N_o}\right) = 10^{\frac{SNR}{10}} = 10^{\frac{9.6}{10}} = 9.12$. For example, for polar NRZ baseband signaling,

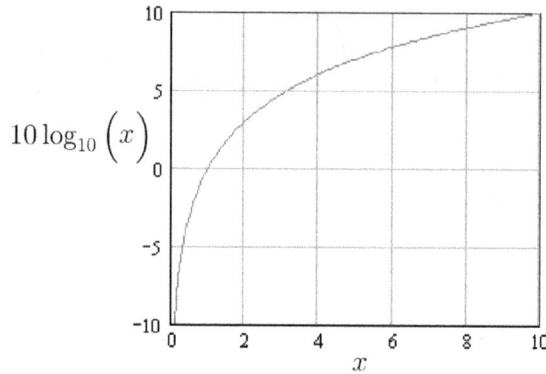

Figure 6.44: Decibel scale.

$$P_e = Q\left(\sqrt{\frac{2E_b}{N_o}}\right) = Q\left(\sqrt{2 * 9.12}\right) = Q\left(4.27\right) \approx 1.03 \text{ x } 10^{-5} \qquad (6.116)$$

i.e. for every one million binary digits processed through the modulator, AWGN channel and demodulator, we can expect approximately 10 errors on average if the channel SNR is 9.6 dB. This specific example is worth remembering because the required SNR to achieve a P_e of approximately 10^{-5} is typically used as a point of reference for the performance of a DCS.

- ⭐ SIMULATION **Decibels:** A plot of equation $\left(\frac{E_b}{N_o}\right) dB = 10 \log_{10}\left(\frac{E_b}{N_o}\right)$ which corresponds to Fig. 6.44. Use the code to determine what is the value of $\frac{E_b}{N_o}$ corresponding to 9.6 dB ?

Example : Q-function

Show that $Q(-u) = 1 - Q(u)$ and show that for $u > 0$, the area $\int\limits_{-\infty}^{u} \frac{1}{\sqrt{2\pi}} \exp\left(-\frac{z^2}{2}\right) dz = 1 - Q(u)$.

Solution

The curves below illustrate the variation of $Q(u)$, $Q(-u)$ and $1-Q(u)$. By inspection we see that $Q(-u) = 1-Q(u)$ and the integral $\int_{-\infty}^{u} \frac{1}{\sqrt{2\pi}} \exp\left(-\frac{z^2}{2}\right) dz$

is simply the area shown in green which is $1-Q(u)$. Note that $\int_{-\infty}^{-u} \frac{1}{\sqrt{2\pi}} \exp\left(-\frac{z^2}{2}\right) dz = 1 - Q(-u) = Q(u)$.

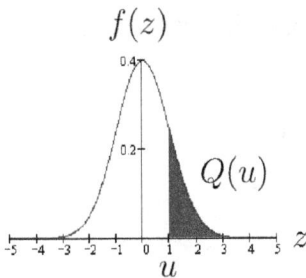

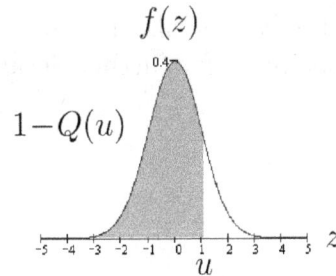

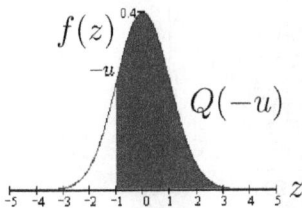

The total area under a PDF is 1
In each of the three examples shown $u = 1$.

6.4.3 Simulation I: Gaussian Random Number Generator

SIMULATION **Histogram:** A gaussian random number generator is implemented. Given that we shall require the use of this generator for all the simulations with an AWGN channel, it is important to check it works as expected. With the mean value set to $+1$, 1000 noise samples are presented in Fig. 6.45, together with the corresponding probability histogram. Refer to Appendix A for further details. Overlaid on this histogram is the expected height of each solid-bar, that is calculated using the theoretical expression for a gaussian PDF. The excellent agreement between simulation and theory implies that the gaussian random number generator model is sufficiently accurate.

- **Experiment:** Alter the number of samples and try different values for the mean and standard deviation. Of course, there are several other methods to check the gaussian nature of the generator. A good opportunity for you to test your programming skills e.g. determine the cumulative distribution function and compare it with theory. Refer to Appendix A for further details.

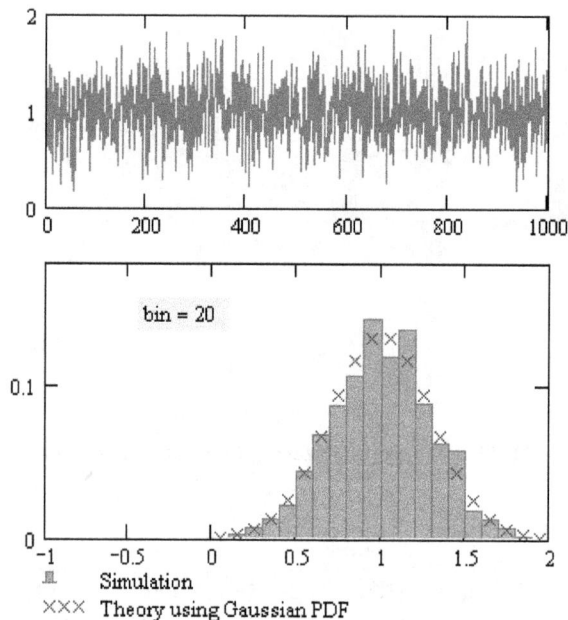

Figure 6.45: Probability histogram.

6.4.4   Simulation II: Polar NRZ

⭐SIMULATION **PolarSignal:** The DCS shown in Fig. 6.46 is modeled using a gaussian random number generator with the mean set to E_b or $-E_b$, depending on whether the binary source output is a binary digit 1 or 0 respectively and the variance is set to $\sigma^2 = \frac{N_o E_b}{2}$. In terms of the SNR which is equal to $10 \log_{10} \left(\frac{E_b}{N_o} \right)$ in decibels, $\sigma = \frac{E_b}{\sqrt{2 \left(10^{SNR/10} \right)}}$. If N_{rand} represents a given random number generator output, then depending on whether the binary source output is a binary digit 0 or 1, an error counter is incremented if $N_{\text{rand}} > a_0$ or $N_{\text{rand}} \leq a_0$, where $a_0 = 0$. The probability P_e is then equal to the number of errors divided by the number binary digits output from the binary source. The simulation results for P_e versus the channel SNR (dB) (equivalently $\left(\frac{E_b}{N_o} \right) dB$) are presented in Fig. 6.47 and compared with the theoretical expression $(P_e)_{PolarNRZ} = Q \left(\sqrt{\frac{2E_b}{N_o}} \right)$. Notice the excellent agreement between simulation and theory. The theoretical performance of unipolar NRZ signaling is also shown in Fig. 6.47 for comparison. A key feature is that polar NRZ can achieve a given desired P_e at a lower SNR. Specifically, there is approximately a 3 dB difference between the unipolar and polar schemes at high $SNRs$.

- **Experiment:** Try different values of E_b to convince yourself that it makes no difference. This is because σ will change accordingly to provide the same P_e for a given SNR (dB). It is the ratio $\frac{E_b}{N_o}$ that matters ! Alter the code to investigate other types of line coding schemes.

6.4.5   Simulation III: Polar NRZ : Digital Filter Technique

⭐SIMULATION **DigitalConvolution:** The simulation **ImpulseTrain** in section 6.3.3 is modified to show how a sampled version (i.e. digital signal) of the transmitted signal baseband signal

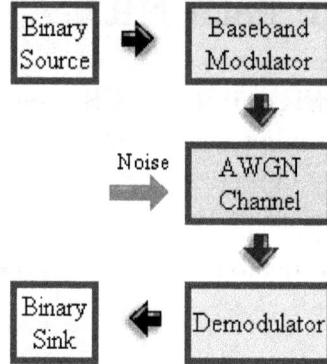

Figure 6.46: Polar NRZ baseband signaling over an AWGN channel.

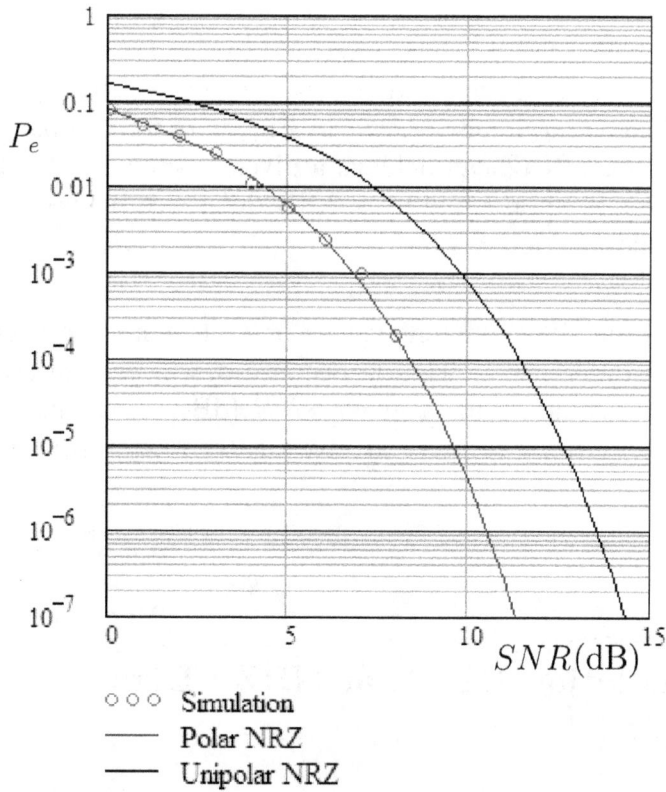

○ ○ ○   Simulation
——   Polar NRZ
——   Unipolar NRZ

Figure 6.47: Polar NRZ baseband signaling performance.

$$s(t) = \sum_{n=-\infty}^{\infty} a_n \delta(t - nT) * h_t(t) = \sum_{n=-\infty}^{\infty} a_n h_t(t - nT) \qquad (6.117)$$

can be alternatively determined via the convolution of the input digital signal and the transmitting filter impulse response $h_t(t)$ set to be the root raised cosine filter $h_{root}(t)$ (equation 6.65). Having demonstrated the equivalence of these two techniques, the convolution method is used in the next simulation.

6.4.6 Simulation IV: Polar NRZ via Unit Impulse Functions

SIMULATION **DigitalPolarNRZ:** The communication system in Fig. 6.46 is modeled once again to determine the performance of polar NRZ signaling over an AWGN channel. In this case however, the binary digits are represented as ± 1 unit impulse functions sent through a transmitting filter with the impulse response $h_{root}(t)$ of equation 6.65. Noise is added to the digital signal $x[n]$ output from the transmitter filter using a gaussian random number generator with zero mean and standard deviation $\sigma = \sqrt{\frac{E_b}{2\left(10^{\frac{SNR}{10}}\right)}}$, where SNR is the channel signal-to-noise ratio in decibels (dB). Given that $\sum |x[n]|^2$ is the total energy within the non-periodic digital signal $x[n]$ representing N binary digits (1.4), the energy per binary digit $E_b = \frac{1}{N} \sum |x[n]|^2$.

6.5   Basis functions

In section 6.2, it was shown that the matched filter output at time T is equal $y(T) = \int_0^T r(t)s(t)dt$. Instead of correlating $r(t)$ with $s(t)$, we shall develop a demodulator which correlates $r(t)$ with a *basis function* for *antipodal* signals for which $s_2(t) = -s_1(t)$. Basis functions allow us to geometrically represent signal **waveforms** as signal **points** on a *signal constellation diagram*. Specifically, recall that in the previous section we considered *binary modulation*, in which the modulator mapped a binary digit 1 and 0 to the signals

$\dot{s}_1(t)$ and $s_2(t)$, respectively. In general, we have *M-ary modulation,* in which n binary digits are mapped onto each signal using a set of $M = 2^n$ signals. We shall now consider a method to create a set of $N \leq M$ orthonormal waveforms, referred to also as *basis functions,* that will allow us to geometrically represent the M signals as signal points within a N-dimenional signal constellation diagram.

6.5.1   Gram-Schmidt Orthogonalization Procedure

Consider first the concept of *orthonormal basis functions,* which is similar to the well known right-handed coordinate system shown in Fig. 6.48. Namely, the unit vectors i, j, k each have a magnitude of 1 and their directions are those of the x, y and z axis, respectively. For example, a vector $\underline{v}$ shown in Fig. 6.48 can be written in terms of the **unit** vectors as $\underline{v} = 4i + 3j + 2k$.

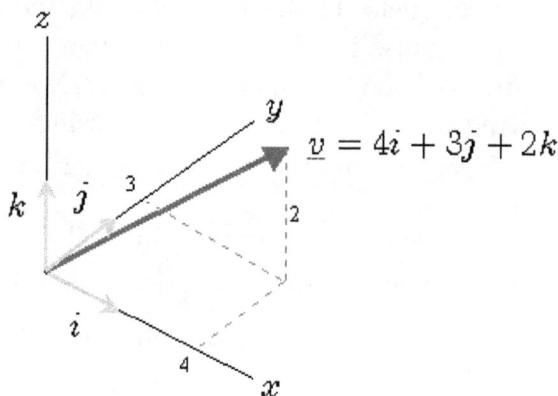

Figure 6.48: Right-handed coordinate system example.

In a similar manner, a set of M energy signals $\{s_1(t), s_2(t), \cdots, s_M(t)\}$ each of duration T seconds can be represented by a linear combinations of a set of N *orthonormal basis functions* $\{\phi_1(t), \phi_2(t), \cdots, \phi_N(t)\}$ where $N \leq M$, so that

$$s_1(t) = s_{11}\phi_1(t) + s_{12}\phi_2(t) + \cdots + s_{1N}\phi_N(t) \qquad (6.118)$$

$$s_2(t) = s_{21}\phi_1(t) + s_{22}\phi_2(t) + \cdots + s_{2N}\phi_N(t) \qquad (6.119)$$

$$\vdots$$

$$s_M(t) = s_{M1}\phi_1(t) + s_{M2}\phi_2(t) + \cdots + s_{MN}\phi_N(t) \qquad (6.120)$$

where for $i = 1, 2, \ldots M$ and $j = 1, 2, \ldots N$. The *coefficients* s_{ij} and the basis functions $\phi_j(t)$ are determined using the *Gram-Schmidt* orthogonalization procedure as follows. The first basis function $\phi_1(t)$ is defined by

$$\phi_1(t) = \frac{s_1(t)}{\sqrt{E_1}} = \frac{s_1(t)}{\sqrt{\int_0^T s_1^2(t)dt}} \qquad (6.121)$$

where

$$E_1 = \int_0^T s_1^2(t)dt \qquad (6.122)$$

is the energy of the signal $s_1(t)$. Thus $s_1(t) = \sqrt{E_1}\phi_1(t) = s_{11}\phi_1(t)$ where $s_{11} = \sqrt{E_1}$ and the energy of $\phi_1(t)$ given by $\int_0^T \phi_1^2(t)dt$ is 1 joule. For $s_2(t)$, we similarly define (via an intermediate function $g_2(t)$) the second basis function $\phi_2(t)$ by

$$\phi_2(t) = \frac{g_2(t)}{\sqrt{\int_0^T g_2^2(t)dt}} = \frac{s_2(t) - s_{21}(t)\phi_1(t)}{\sqrt{E_2 - s_{21}^2}} \qquad (6.123)$$

where E_2 is the energy of the signal $s_2(t)$ and

$$g_2(t) = s_2(t) - s_{21}(t)\phi_1(t) \tag{6.124}$$

is orthogonal to $\phi_1(t)$ over the time durtation $0 \leq t \leq T$, where the coefficient

$$s_{21} = \int_0^T s_2(t)\phi_1(t)dt \tag{6.125}$$

The basis functions $\phi_1(t)$ and $\phi_2(t)$ are said to be *orthonormal* because

$$\int_0^T \phi_1(t)\phi_2(t)dt = 0, \int_0^T \phi_2(t)\phi_2(t)dt = 1, \int_0^T \phi_1(t)\phi_1(t)dt = 1 \tag{6.126}$$

In general, we may define

$$g_i(t) = s_i(t) - \sum_{j=1}^{i-1} s_{ij}\phi_j(t) \tag{6.127}$$

where

$$s_{ij} = \int_0^T s_i(t)\phi_j(t)dt \tag{6.128}$$

for $j = 1, 2, ..., i - 1$.

- ⭐ SIMULATION **Orthonormal:**   A verification of equation 6.126 with a simple example.

To clarify the forgoing concepts by an example, consider binary modulation ($M = 2$) in which the two signals $s_1(t)$ and $s_2(t)$ shown in Fig. 6.49 represent a binary digit '1' and '0', respectively. The time duration of each signal is T_b seconds. Using the Gram-Schmidt orthogonalization procedure, the first basis function

$$\phi_1(t) = \frac{s_1(t)}{\sqrt{E_1}} = \frac{A}{\sqrt{A^2 T_b}} = \frac{1}{\sqrt{T_b}} \text{ for } 0 \leq t \leq T_b \qquad (6.129)$$

The second basis function is not required because

$$s_{21} = \int_0^{T_b} s_2(t)\phi_1(t)dt = \int_0^{T_b} -A\frac{1}{\sqrt{T_b}}dt = \frac{-AT_b}{\sqrt{T_b}} = -A\sqrt{T_b} \qquad (6.130)$$

and therefore

$$\phi_2(t) = \frac{s_2(t) - s_{21}(t)\phi_1(t)}{\sqrt{E_2 - s_{21}^2}} = \frac{(-A) - (-A\sqrt{T_b})\left(\frac{1}{\sqrt{T_b}}\right)}{\sqrt{E_2 - s_{21}^2}} = \frac{-A + A}{\sqrt{E_2 - s_{21}^2}} = 0$$
$$(6.131)$$

Hence, the signals $s_1(t)$ and $s_2(t)$ in terms of a single ($N = 1$) basis function $\phi_1(t)$ are given by

$$s_1(t) = \sqrt{E_b}\phi_1(t) \quad \text{for } 0 \leq t \leq T_b \quad \text{Binary digit '1'}$$
$$(6.132)$$
$$s_2(t) = -\sqrt{E_b}\phi_1(t) \quad \text{for } 0 \leq t \leq T_b \quad \text{Binary digit '0'}$$

where $E_b = A^2 T_b$ is the transmitted signal energy per binary digit. As a quick check, $s_1(t) = \sqrt{E_b}\phi_1(t) = \sqrt{A^2 T_b}\left(\frac{1}{\sqrt{T_b}}\right) = A$ for $0 \leq t \leq T_b$ and the

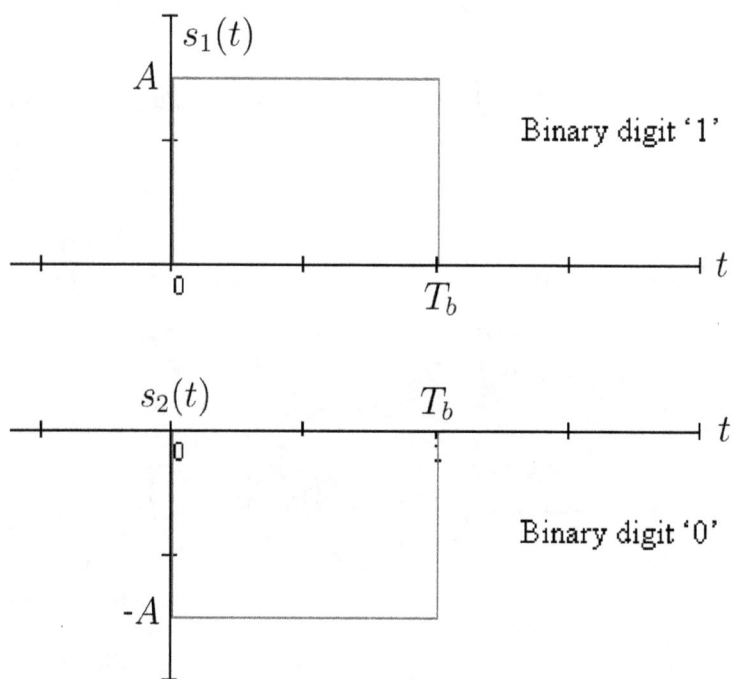

Figure 6.49: Baseband signals representing a binary digit '1 and '0'.

energy in signal $s_1(t)$ is given by $\int\limits_0^{T_b} s_1^2(t)dt = \int\limits_0^{T_b} E_b\phi_1^2(t)dt = E_b\int\limits_0^{T_b} \phi_1^2(t)dt = E_b = A^2T_b$ as expected.

In Fig. 6.48, the vector $\underline{v} = 4i + 3j + 2k$ is visually represented as a single dot in the three-dimensional space. Similarly, a signal $s_i(t)$ can be represented as a dot within a N-dimensional *signal space or signal constellation diagram* using a set of N orthonormal basis functions $\{\phi_1(t), \phi_2(t), \cdots, \phi_N(t)\}$. Continuing with our example in which we require only a single basis function $\phi_1(t)$, a visual representation of $s_1(t)$ and $s_2(t)$ within the one dimensional signal space diagram is presented in Fig. 6.50. The two dots are referred to as *signal points*.

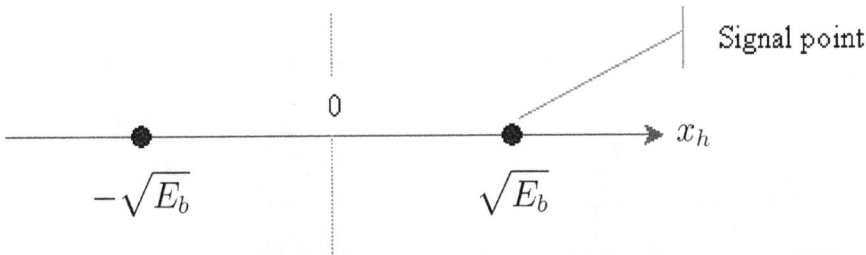

Figure 6.50: Signal constellation diagram.

The coordinate x_h along the horizontal axis of $\phi_1(t)$ is given by

$$x_h = \int\limits_0^{T_b} s_i(t)\phi_1(t)dt \qquad (6.133)$$

where $s_i(t)$ is either $s_1(t)$ or $s_2(t)$. Thus for $s_1(t)$,

$$x_h = \int\limits_0^{T_b} s_1(t)\phi_1(t)dt = \sqrt{E_b}\int\limits_0^{T_b} \phi_1^2(t)dt = \sqrt{E_b} \qquad (6.134)$$

and for $s_2(t)$,

$$x_h = \int_0^{T_b} s_2(t)\phi_1(t)dt = -\sqrt{E_b}\int_0^{T_b}\phi_1^2(t)dt = -\sqrt{E_b} \qquad (6.135)$$

6.5.2 Demodulator

Having introduced the concept of basis functions, consider once again the matched filter, but now in terms of correlating with a basis function ϕ_j. From Fig. 6.17 (section 6.2), recall the output $y(t)$ is given by

$$y(t) = r(t) * h_{opt}(t) = \int_0^t r(\tau)h_{opt}(t-\tau)d\tau \qquad (6.136)$$

For $h_{opt}(t) = \phi_j\left(T_b - t\right),$

$$y(t) = \int_0^t r(\tau)\phi_j\left(T_b - t + \tau\right)d\tau \qquad (6.137)$$

At time $t = T_b$

$$y(T_b) = \int_0^{T_b} r(\tau)\phi_j\left(\tau\right)d\tau \qquad (6.138)$$

For a noiseless channel, let the received signal $r(t) = s_1(t) = \sqrt{E_b}\phi_1$ where E_b is the average energy per binary digit. In this case,

$$y(T_b) = \int_0^{T_b} s_1(\tau)\phi_1\left(\tau\right)d\tau = \sqrt{E_b}\int_0^{T_b}\phi_1^2(\tau)d\tau = \sqrt{E_b} \qquad (6.139)$$

is equal to the coordinate $x_h = \int\limits_0^{T_b} s_1(t)\phi_1(t)dt$ of the signal point. Thus
the demodulator for a binary communication (antipodal signals) is as shown
in Fig. 6.52. The output $y(T_b)$ of the correlator at time T_b will be a gaussian
random variable with a mean value equal to x_h and variance $\sigma^2 = \frac{N_o}{2}$ (refer
to problem 19). For the signals in Fig. 6.49, the output of correlator under a
noiseless channel will be either $+\sqrt{E_b}$ or $-\sqrt{E_b}$ depending on whether $s_1(t)$ or
$s_2(t)$ was transmitted. Notice how its energy content of the signals and
not the shape that matters. With noise, the probability density functions
(PDFs) of the signal points will be as illustrated in Fig. 6.51. If $y(T_b) > 0$, we
decide that the received signal is $s_1(t)$ and output a binary digit 1, otherwise
output a binary digit zero.

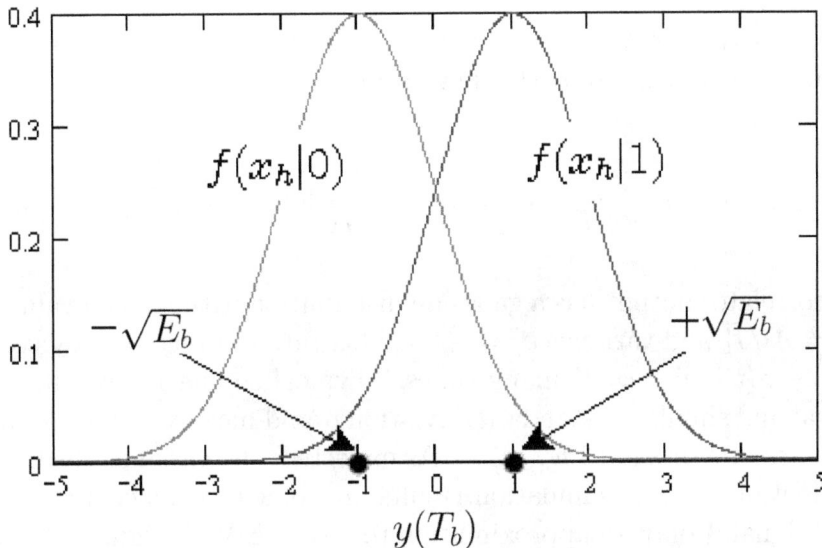

Figure 6.51: Signal points probability density functions.

6.5.3   Simulation : Correlation Variance

⭐ SIMULATION **BasisCorrelator:** The signal $s_1(t) = A$ for $0 \le t \le T_b$

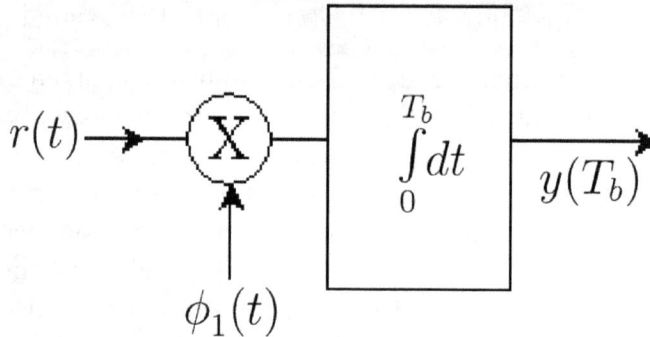

Figure 6.52: Demodulator.

is corrupted with an AWGN signal $n(t)$ and correlated with the corresponding basis function $\phi_1(t)$, where $\phi_1(t)$ is given by

$$\phi_1(t) = \frac{s_1(t)}{\sqrt{E_1}} = \frac{A}{\sqrt{A^2 T}} = \frac{1}{\sqrt{T_b}} \text{ for } 0 \leq t \leq T_b \qquad (6.140)$$

The correlator output has a gaussian distribution with a mean value equal to $\sqrt{E_b} = A\sqrt{T_b}$ and variance $\sigma^2 = \frac{N_o}{2}$ as illustrated in Fig. 6.53 by correlating $[s_1(t) + n(t)]$ with $\phi_1(t)$ many times. Typical simulation results for the theoretical and simulated standard deviation σ and mean value for a range of T_b values are presented in Fig. 6.54. As expected, the variance is essentially independent of T_b. The simulation results are somewhat inaccurate because the noise signal is only an approximation to a true AWGN signal. A simpler and more accurate simulation method will be presented in the next section.

- **Experiment:** Try other types of pulses an compare with theory. Be sure to change the expression for E_1 accordingly.

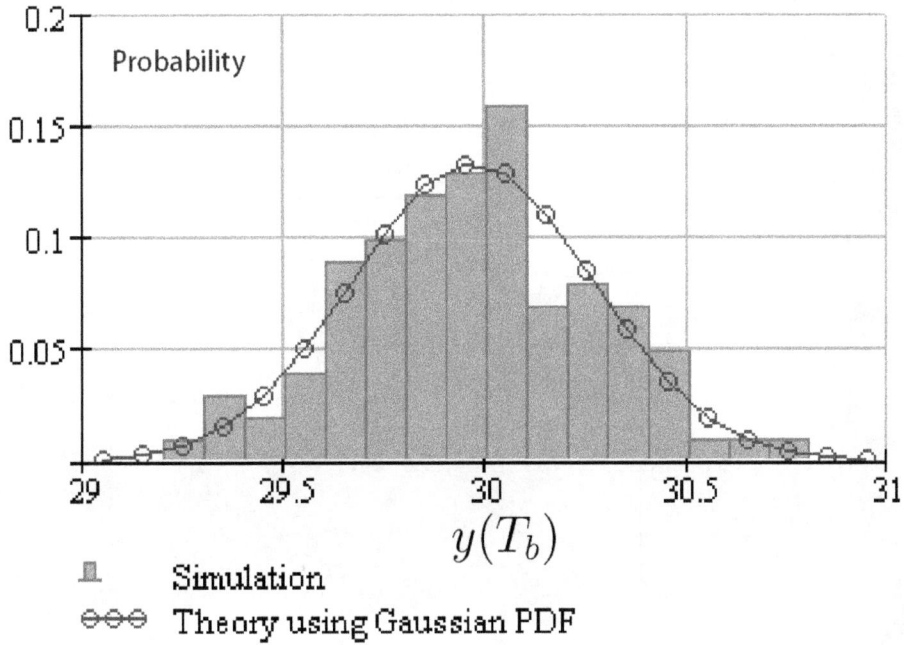

Figure 6.53: Correlator output probability histogram for $A = 30$ volts and $T_b = 1$ second.

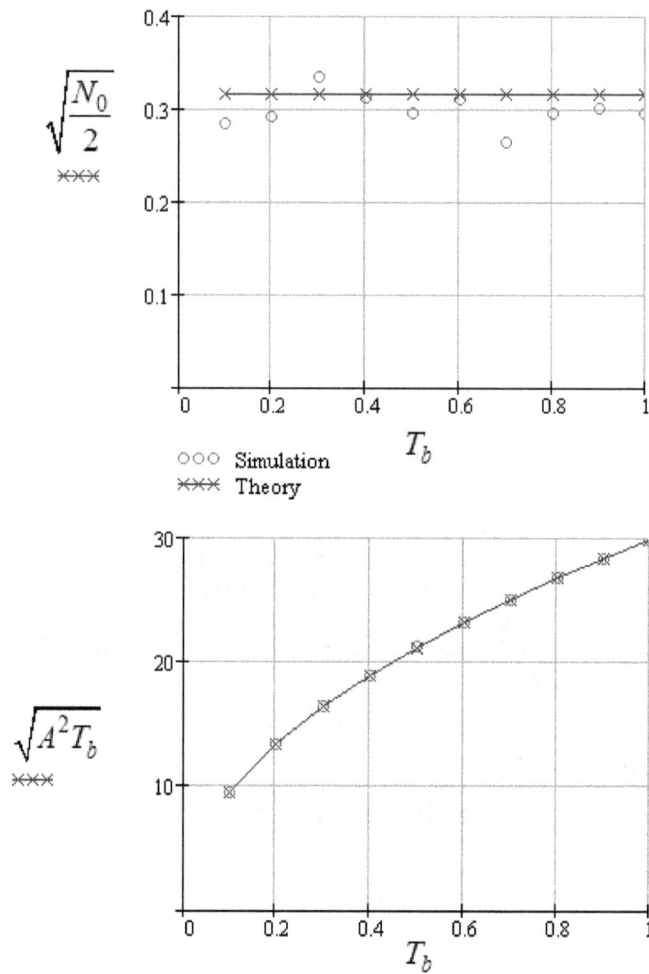

Figure 6.54: Simulated and theoretical standard deviation mean of the correlator output.

6.6   Correlator Performance

▶**Performance in a AWGN Channel** (Refer to section **7.3**)

▶**Q-Function** (Refer to section **7.3**)

For binary communication with antipodal signals, the incoming signal $r(t)$ is correlated with a basis function $\phi_1(t)$ so that $y(T_b) = \int_0^{T_b} r(t)\phi_1(t)\,dt$, where $r(t) = s_i(t) + n(t)$. Based on the analysis presented in the previous section, the probability density functions (PDFs) $f(y|s_1)$ and $f(y|s_2)$ are given by

$$f(y|s_1) = \frac{1}{\sigma\sqrt{2\pi}} \exp\left[-\frac{(y-a_1)^2}{2\sigma^2}\right] \tag{6.141}$$

$$f(y|s_2) = \frac{1}{\sigma\sqrt{2\pi}} \exp\left[-\frac{(y-a_2)^2}{2\sigma^2}\right] \tag{6.142}$$

where the mean correlator outputs a_1 and a_2 correspond respectively to $s_1(t)$ and $s_2(t)$ and $\sigma = \sqrt{\frac{N_o}{2}}$. Unlike in section 6.4, with antipodal signals, $a_1 = \sqrt[4]{E_b}$ and $a_2 = -\sqrt{E_b}$. Following the analysis presented in section 6.4,

$$P_e = Q\left(\frac{a_1 - a_2}{2\sigma}\right) \tag{6.143}$$

Since $\sigma = \sqrt{\frac{N_o}{2}}$, we find

$$P_e = Q\left(\frac{(a_1 - a_2)}{2}\sqrt{\frac{2}{N_o}}\right) = Q\left(\sqrt{\frac{(a_1 - a_2)^2}{2N_o}}\right) \tag{6.144}$$

Hence for polar NRZ signalling in which the binary digits 1 and 0 are represented by $s_1(t) = +A$ and $s_2(t) = -A$ for a duration of T_b seconds, respectively, so that the average energy per binary digit $E_b = A^2 T_b$, we find $a_1 = \sqrt{E_b}$ and $a_2 = -\sqrt{E_b}$ so that

$$(P_e)_{PolarNRZ} = Q\left(\sqrt{\frac{2A^2 T_b}{N_o}}\right) = Q\left(\sqrt{\frac{2E_b}{N_o}}\right) \qquad (6.145)$$

as expected.

6.6.1 Simulation : Polar NRZ Demodulation via a Basis Function

☆SIMULATION **PolarBasis:** The DCS shown in Fig. 6.46 , section 6.4 is modeled once again.   The difference now is that the gaussian random number generator mean is set to $\sqrt{E_b}$ or $-\sqrt{E_b}$, depending on whether the binary source output is a binary digit 1 or 0 respectively and the variance is set to $\sigma^2 = \frac{N_o}{2}$. The corresponding simulation results confirm the curves presented earlier in Fig. 6.47, section 6.4.

- **Experiment:** Modify the simulation code to investigate for example, the performance of bipolar NRZ signalling.

6.7   M-ary Pulse Amplitude Modulation (PAM)

In M-ary *pulse amplitude modulation* (PAM), there are M possible amplitudes of the multilevel signal (i.e. $M = 2$ for polar NRZ) and each PAM signal represents $\log_2 M$ binary digits. In the popular Fast Ethernet 100BASE-T2 communication standard, 5 level PAM $\{-2, -1, 0, +1, +2\}$ with 25,000000 pulses per second with 4 bits per symbol (expanded into two 3-bit symbols) over an Ethernet cable provide a data rate of 100 Mb/s.

The incoming signal $r(t) = s_i(t) + n(t)$ is correlated with a basis function $\phi(t)$ so that

$$y(T) = \int_0^T r(t)\phi(t)\, dt \qquad (6.146)$$

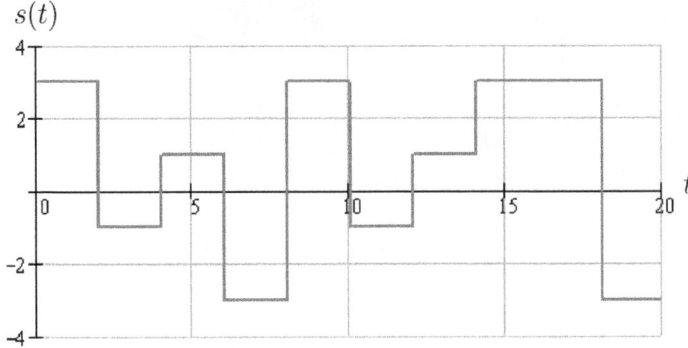

Figure 6.55: 4-ary PAM signal.

where T is the time duration of signal (or pulse). For a rectangular pulse, the basis function is given by

$$\phi(t) = \frac{1}{\sqrt{T}} \quad \text{for } 0 \leq t \leq T \tag{6.147}$$

An example of a 4-level signal is shown in Fig. 6.55 with $A = 1$ volt and $T = 2$ s for clarity.

Assuming each level is equally likely, the average signal power is given by $P_s = \frac{(-3A)^2 + (-A)^2 + (3A)^2 + (A)^2}{4} = 5A^2$. Given that power is the rate of flow of energy, the average transmitted signal energy is given by

$$E_s = P_s T = 5A^2 T \tag{6.148}$$

In general,

$$P_s = \frac{1}{M} \sum_{i=1}^{M} A_i^2 \tag{6.149}$$

where A_i is the amplitude of the signal $s_i(t)$ and $E_s = P_s T$. To minimize P_s and avoid a d.c. component, the M levels are selected to be symmetric about the origin. In general, $A_i = (2i - 1 - M)A$ for $i = 1, 2, 3, ..., M$. For example, for $M = 4$, $A_1 = -3A$, $A_2 = -A$, $A_3 = A$, $A_4 = 3A$. Thus,

$$P_s = \frac{A^2}{M} \sum_{i=1}^{M} (2i - 1 - M)^2 = \frac{A^2 (M^2 - 1)}{3} \qquad (6.150)$$

and

$$E_s = \frac{A^2 T (M^2 - 1)}{3} \qquad (6.151)$$

Let $x_h^{(i)}$ denote the horizontal coordinate of a signal point on the M-ary PAM signal constellation diagram. Then

$$x_h^{(i)} = \int_0^T A_i \frac{1}{\sqrt{T}} dt = A_i \frac{1}{\sqrt{T}} (T) = A_i \sqrt{T} \qquad (6.152)$$

Thus for $M = 4$, the signal point coordinates are $-3A\sqrt{T}$, $-A\sqrt{T}$, $+A\sqrt{T}$, $+3A\sqrt{T}$ as shown in Fig. 6.56. Given that the energy of a pulse of amplitude A_i is given by $E_i = A_i^2 T$, the signal point coordinates are $-\sqrt{E_1}$, $-\sqrt{E_2}$, $+\sqrt{E_3}$, $+\sqrt{E_4}$, where $E_1 = E_4 = (3A)^2 T$ and $E_2 = E_3 = A^2 T$.

Notice that for $M = 2$, the signal point coordinates are $-\sqrt{E_1}$ and $+\sqrt{E_2}$, where $E_1 = E_2 = A^2 T$ as in section 6.5. Indeed, referring to Fig. 6.56, the probability of an error for the $(M - 2)$ inner levels (i.e. excluding the minimum and maximum levels) is now twice the error for the $M = 2$ binary case, because an error can occur on either side of a level and the probability of an error for either the maximum or the minimum level is the same as the binary case. Thus, based on the analysis in section 6.6, the probability of an error P_e for $M = 2$ is given by

$$P_e = Q\left(\frac{a_1 - a_2}{2\sigma}\right) = Q\left(\frac{\sqrt{A^2 T} - (-\sqrt{A^2 T})}{2\sigma}\right) = Q\left(\frac{\sqrt{A^2 T}}{\sigma}\right) \qquad (6.153)$$

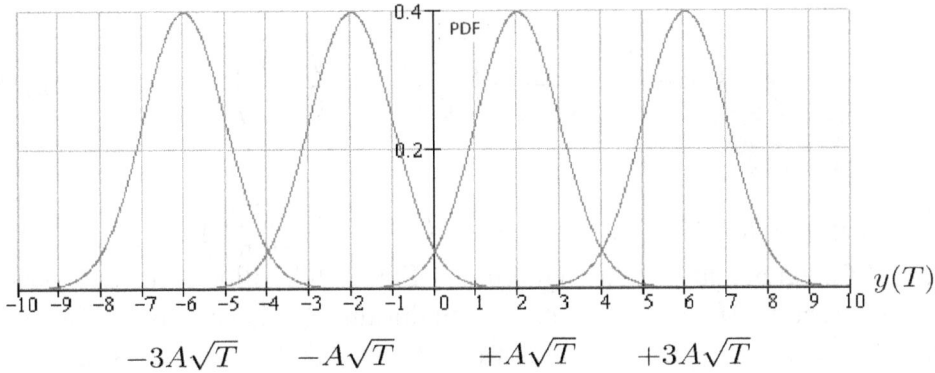

Figure 6.56: Probability density functions of the four signal points of 4-ary PAM over an AWGN channel.

where $\sigma = \sqrt{\frac{N_o}{2}}$, so that

$$P_e = Q\left(\sqrt{\frac{2A^2T}{N_o}}\right) \tag{6.154}$$

Hence, the average probability of a symbol error P_M for M-ary PAM is given by

$$
\begin{aligned}
P_M &= \frac{1}{M}\left[(M-2)(2P_e) + (2)(P_e)\right] && \text{(6.155)} \\
&= \frac{2(M-1)P_e}{M} = \frac{2(M-1)}{M}Q\left(\sqrt{\frac{2A^2T}{N_o}}\right) && \text{(6.156)}
\end{aligned}
$$

Given that each pulse represents $\log_2 M$ binary digits, the average energy per binary digit

$$E_b = \frac{E_s}{\log_2 M} = \frac{A^2T\left(M^2 - 1\right)}{3\log_2 M} \tag{6.157}$$

and consequently,

$$P_M = \frac{2(M-1)}{M} Q \left(\sqrt{\frac{(6 \log_2 M) \, E_b}{(M^2 - 1) \, N_o}} \right) \tag{6.158}$$

For example, for $M = 2$, $P_{M=2} = \frac{2(2-1)}{2} Q \left(\sqrt{\frac{(6 \log_2 2) E_b}{(4-1) N_o}} \right) = Q \left(\sqrt{\frac{2E_b}{N_o}} \right)$ as expected for binary antipodal signalling. Figure 6.57 illustrates the performance of various M-ary schemes. To maintain P_M at say 10^{-5}, Fig. 6.58 shows the $\frac{E_b}{N_o}(dB)$ required to maintain P_M at say 10^{-5} as M is increased from 2 to 256. For each additional binary digit transmitted per level as M increases from 2 (1 binary digit) to 256 (8 binary digits), the required increase in $\frac{E_b}{N_o}(dB)$ is shown in Fig. 6.59. For example, the first point in Fig. 6.59 indicates that an increase of 4.16 dB is required to maintain $P_M = 10^{-5}$ as M goes from 2 to 4.

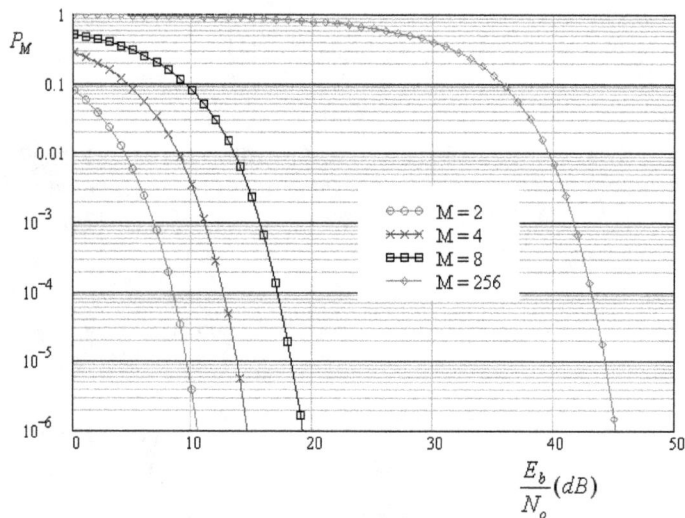

Figure 6.57: M-ary PAM performance over an AWGN channel.

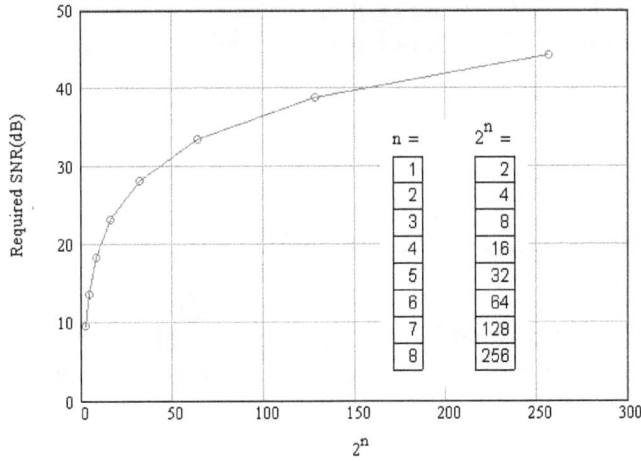

Figure 6.58: $\frac{E_b}{N_o}(dB)$ required to maintain $P_M = 10^{-5}$.

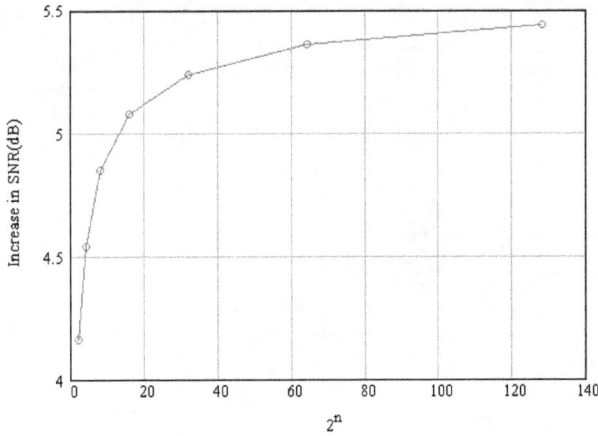

Figure 6.59: Increase in $\frac{E_b}{N_o}(dB)$ required for each additional binary digit per pulse while maintianing $P_M = 10^{-5}$.

- SIMULATION **MaryPam:** Simulation corresponding to Figs. 6.57 and 6.58.   Determine the required increase in $\frac{E_b}{N_o}(dB)$ for large M to achieve a specific P_M. What is the corresponding required transmitter power increase ?

6.7.1   Simulation : 4-ary PAM

SIMULATION **FourPAM:** 4-aryPAM is simulated over an AWGN channel. The simulation results, together with theoretical curve are presented in Fig. 6.60.

- **Experiment:**   Verify that the performance will improve if the probability with which a binary zero is output from the binary source is greater than 0.5. Can you explain why? Try various values for A and T. Why does the performance remain the same ?   Finally, alter the simulation code to simulate 8-ary PAM.

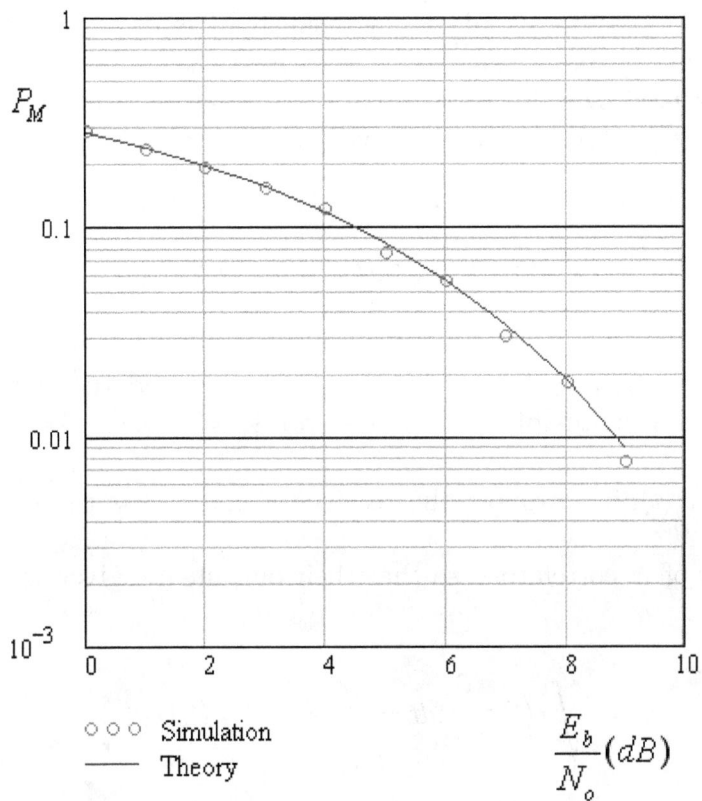

Figure 6.60: Simulated and theoretical 4-ary PAM performance over an AWGN channel.

6.8   M-ary Optimum Coherent Receiver in AWGN

In this section, we shall rise-above the analysis presented in the previous sections to establish the optimum receiver for an M-ary signal in AWGN. The receiver operation can be separated into that of a *demodulator*, which converts the received signal $r(t)$ into an N-dimensional vector $\mathbf{y} = (y_1, y_2, \cdots, y_N)$ and a *detector* (also referred to as a *signal transmission decoder*) which decides which of the M possible signals was transmitted based on $\mathbf{y}$. Recall that a set of M energy signals $\{s_1(t), s_2(t), \cdots, s_M(t)\}$ each of duration T seconds can be represented by a linear combinations of a set of N orthonormal basis functions $\{\phi_1(t), \phi_2(t), \cdots, \phi_N(t)\}$ such that

$$s_i(t) = \sum_{j}^{N} s_{ij}\phi_j(t) \qquad 0 \le t \le T \tag{6.159}$$

where $N \le M$, $i = 1, 2, \ldots M$, $j = 1, 2, \ldots N$ and the coefficients s_{ij} provide the coordinates of the *signal vector* $\mathbf{s}_i = (s_{i1}, s_{i2}, \cdots, s_{iN})$, where $s_{ij} = \int_0^T s_i(t)\phi_j(t)dt$. Suppose the received signal $r(t)$ is passed through a parallel bank of N correlators, so that their outputs are given by

$$y_j = \int_0^T r(t)\phi_j(t)dt = \int_0^T \left[s_i(t) + n(t)\right]\phi_j(t)dt \tag{6.160}$$

$$= \int_0^T s_i(t)\phi_j(t)dt + \int_0^T n(t)\phi_j(t)dt = s_{ij} + n_j \tag{6.161}$$

where the output vector $\mathbf{y} = (y_1, y_2, \cdots, y_N)$, the *noise vector* $\mathbf{n} = (n_1, n_2, \cdots, n_N)$ and $\{n_j\}$ are random variables with

$$E\left[n_j\right] = \int_0^T E\left[n(t)\right]\phi_j(t)dt = 0 \tag{6.162}$$

and convariances (refer to A.12)

$$E\left[n_j n_i\right] = \int_0^T \int_0^T E\left[n(t)n(\tau)\right]\phi_j(t)\phi_i(t)dt d\tau \qquad (6.163)$$

$$= \int_0^T \int_0^T \frac{N_o}{2}\delta(t-\tau)\phi_j(t)\phi_i(t)dt d\tau \qquad (6.164)$$

$$= \frac{N_o}{2}\int_0^T \phi_j(t)\phi_i(t)dt = \frac{N_o}{2}\delta_{ij} \qquad (6.165)$$

where $\delta_{ij} = \begin{cases} 1 & for & i=j \\ 0 & otherwise \end{cases}$. Thus $\{n_j\}$ are zero-mean uncorrelated statistically independent gaussian random variables with the same variance $\frac{N_o}{2}$ so that $E\left[y_j\right] = E\left[s_{ij}+n_j\right] = s_{ij}$ with variance $\sigma = \frac{N_o}{2}$. Hence, the conditional probability density $f(\mathbf{y}|\mathbf{s}_i)$ is given by

$$f(\mathbf{y}|\mathbf{s}_i) = \prod_{j=1}^N \frac{1}{\sigma\sqrt{2\pi}}\exp\left[-\frac{(y_j-s_{ij})^2}{2\sigma^2}\right] \qquad (6.166)$$

$$= \frac{1}{\sigma^N(2\pi)^{N/2}}\exp\left[-\sum_{j=1}^N \frac{(y_j-s_{ij})^2}{2\sigma^2}\right] \qquad (6.167)$$

where $\sum_{j=1}^N (y_j-s_{ij})^2$ is the squared *Euclidean distance* $d_{Euc}(\mathbf{y},\mathbf{s}_i) = \|\mathbf{y}-\mathbf{s}_i\|^2$ between the N-dimensional vector $\mathbf{y} = (y_1, y_2, \cdots, y_N)$ and the signal vector $\mathbf{s}_i = (s_{i1}, s_{i2}, \cdots, s_{iN})$. This is not surprising because, for example for $N=3$, the noise vector $\mathbf{n}$ would have a spherical distribution about the signal point in the 3-dimensional space. Indeed, simple intuition leads us to the optimum detector criterion, called the *maximum a posteriori probability* (MAP) criterion, which is to choose the $\mathbf{s}_i$ which maximizes the probability $P(\mathbf{s}_i|\mathbf{y})$. Using Bayes's rule, we have

$$P\left(\mathbf{s}_i|\mathbf{y}\right) = \frac{f(\mathbf{y}|\mathbf{s}_i)P\left(\mathbf{s}_i\right)}{f\left(\mathbf{y}\right)} \tag{6.168}$$

where $P\left(\mathbf{s}_i\right)$ is the *a priori probability* of ith signal being sent. Given that typically that $P\left(\mathbf{s}_i\right) = \frac{1}{M}$ and $f\left(\mathbf{y}\right)$ is independent of $\mathbf{s}_i$, we conclude that our equivalent decision rule is to select $\mathbf{s}_i$ which maximizes $f(\mathbf{y}|\mathbf{s}_i)$. Taking a closer look at this decision criterion, referred to as the *maximum-likelihood* (ML) criterion, we have from equation 6.167 that

$$\log_e\left(f(\mathbf{y}|\mathbf{s}_i)\right) = \log_e\left(\frac{1}{\sigma^N\left(2\pi\right)^{N/2}}\right) - \sum_{j=1}^{N}\frac{(y_j - s_{ij})^2}{2\sigma^2} \tag{6.169}$$

Thus to maximize $f(\mathbf{y}|\mathbf{s}_i)$, we simply need to minimize the *squared Euclidean distance*

$$d_{Euc}\left(\mathbf{y}, \mathbf{s}_i\right) = \|\mathbf{y} - \mathbf{s}_i\|^2 = \sum_{j=1}^{N}(y_j - s_{ij})^2 \tag{6.170}$$

i.e. select $\mathbf{s}_i$ that is closest to $\mathbf{y}$. For an interesting insight, which leads to a practical implementation, suppose we expand equation 6.170 so that

$$d_{Euc}\left(\mathbf{y}, \mathbf{s}_i\right) = \sum_{j=1}^{N}(y_j)^2 - 2\sum_{j=1}^{N}y_j s_{ij} + \sum_{j=1}^{N}(s_{ij})^2 \tag{6.171}$$

We can ignore the term $\sum_{j=1}^{N}(y_j)^2$ because its common to all $d_{Euc}\left(\mathbf{y}, \mathbf{s}_i\right)$ and we can also ignore the $\sum_{j=1}^{N}(s_{ij})^2$ if all the signals have the same energy, which leaves the $-2\sum_{j=1}^{N}y_j s_{ij}$. Therefore, to minimize $d_{Euc}\left(\mathbf{y}, \mathbf{s}_i\right)$ is equivalent to simply maximizing $\sum_{j=1}^{N} y_j s_{ij}$ as done in practice. We conclude that the optimum receiver, commonly referred to as *correlation receiver*, consists of a demodulator and a detector, where the demodulator part of the optimum receiver can be implemented using a parallel bank of N correlators or matched filters.

6.9 Problems

1. Sketch the On-off RZ, On-Off NRZ, Polar RZ, Polar NRZ, Unipolar RZ, Bipolar RZ, Manchester and Miller line code baseband signal for the binary stream 11001011101.

2. The diagram below shows a polar RZ signal representing the binary stream 1010 in which the pulse height $A = 1$ volt, width is 0.5 s and $T_b = 1$s. For a random binary stream in which the probability of a zero is equal to 0.5, verify that the PSD is given by

$$PSD_{polarRZ}(f) = \frac{A^2 T_b}{4} \text{sinc}\left(\frac{fT_b}{2}\right)$$

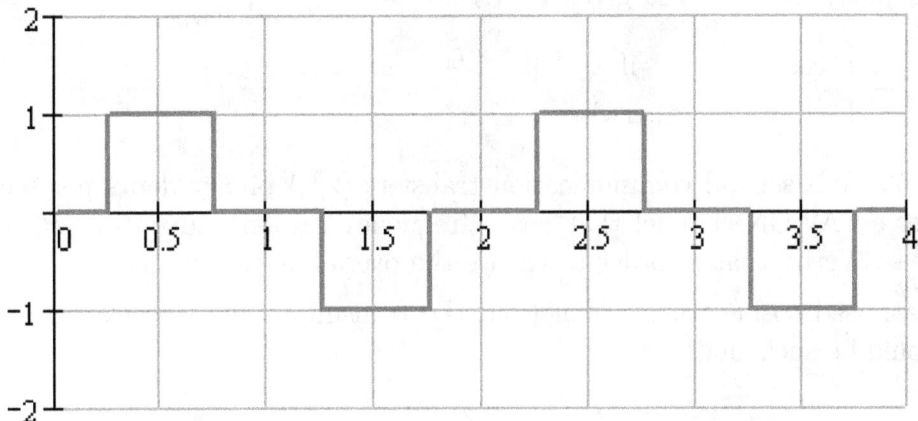

3. A binary PCM signal, which conveys 56000 binary digits/sec, is to be transmitted using a 16 level multilevel signal. What is the minimum channel bandwidth required ?

4. Experiment with the multilevel polar NRZ simulation code in section 6.1.6 to determine how the PSD of a 8 multilevel signal depends on the range of the levels and the time duration of a multilevel pulse. Determine the average signal power via the

(a) levels, assuming each level is equally likely.
(b) amplitudes of the specific baseband signal generated.
(c) area under the theoretical PSD.
(d) area under the PSD of the specific baseband signal generated.

5. Sketch the polar signal representing the binary stream 0111001101 in which a binary digit 1 and 0 is represented by the pulse sinc (t) and $-$ sinc (t), respectively with $T_b = 1$ s.

6. (a) Generate a random binary stream of digits and represent this stream as a polar signal in which a binary digit 1 and 0 is represented by the pulse sinc (t) and $-$ sinc (t), respectively with $T_b = 1$ s. By finding the PSD of this signal, verify that $B_{PCM}^{(min)} = \dfrac{1}{2T_b}$ Hz.

(b) The PSD of a polar NRZ line code is given by $PSD(f) = \dfrac{|S(f)|^2}{T_b}$, where $S(f)$ is the Fourier transform of the pulse of height 1. Determine $PSD(f)$ if a pulse shape $s(t)$ is given by (i) $s(t) = \begin{cases} 1 & \text{if} & |t| \leq \frac{T_b}{2} \\ 0 & \text{otherwise} \end{cases}$ (ii)

$s(t) = \begin{cases} \cos\left(\frac{\pi t}{T_b}\right) & \text{if} & |t| \leq \frac{T_b}{2} \\ 0 & \text{otherwise} \end{cases}$.

7. A baseband communication transfers $(2f_o)$ binary digits per second over an AWGN channel that has a frequency response given by $H_c(f) = \dfrac{1}{1+j\frac{f}{f_o}}$. Verify that in order to ensure the overall system frequency response is a raised cosine transfer function, the transmitter and receiver functions should be such that

$$|H_t(t)| = |H_r(t)| = \begin{cases} \left(1+\left(\frac{f}{f_o}\right)^2\right)^{1/4} \cos\left(\frac{\pi f}{4f_o}\right) & if & 0 \leq |f| \leq 2f_o \\ \\ 0 & otherwise \end{cases}$$

where the roll-off factor $r = 1$.

8. Consider a unipolar NRZ baseband communication system over an additive white gaussian noise (AWGN) channel in which the amplitude of a pulse is 1 mV and of duration 10^{-4} seconds. If the double sided noise power spectral density is 5 x 10^{-12} W/Hz, what is the performance of this signalling scheme ?

9. Consider the use of a baseband signalling scheme over an AWGN in which

$s_1(t) = A$ for $0 \leq t \leq T$ for a binary digit 1
$s_2(t) = -A$ for $0 \leq t \leq T$ for a binary digit 0.

If $A = 1mV$ and the single-sided noise power spectral density $N_o = 10^{-9}$ W/Hz,

(a) what is the maximum rate r_b at which binary digits can be transmitted to ensure that the probability of an error in a binary digit $P_e \leq 10^{-6}$.

(b) If $r_b = 100$ binary digits/sec, what is the corresponding P_e. How does this compare with your answer to part (a).

10. (a) For a binary communication system in which there is no ISI, what is the required bandwidth for a data rate of 1024 bits/sec ?
(b) If the time duration of a baseband signal pulse is 10^{-6} seconds and the system bandwidth is 0.9 MHz, what is the required roll-off filter factor ?

11. (a) Show that the maximum analog signal bandwidth that can be transferred via a binary polar NRZ PCM signal over a channel of bandwidth $B_{channel}$ without ISI is given by $\frac{B_{channel}}{(1+r)\log_2 Q_{level}}$. State any necessary assumptions.
(b) What is the maximum analog signal bandwidth that can be accommodated if a multilevel signal with M levels is used instead.

12. If N analog signals are time-division multiplexed and sent over a channel with a raised cosine overall transfer function, determine an expression for the required absolute channel bandwidth in terms of the sampling frequency f_s, the roll-off factor r and N.

13. Consider the use of a 2^k - level PAM signal to transfer 10^6 binary digits per second. Compare the theoretical minimum bandwidth required for no ISI and the bandwidth required using a filter roll-off factor $r = 0.5$ for $k = 1, 2, \cdots 6$.

14. For binary communication over an AWGN channel,

(a) show that the optimum threshold a_0 to minimize the probability of an error in a binary digit is given by $a_o = \frac{a_1 + a_2}{2} + \frac{\sigma^2}{a_1 - a_2} \ln \left(\frac{P(s_2)}{P(s_1)} \right)$.

(b) determine a_0 for binary antipodal signalling, with $a_1 = \sqrt{E_b}$, $a_2 = -\sqrt{E_b}$ and $\sigma^2 = \frac{N_o}{2}$.

15. In the previous problem for $a_1 = -1$, $a_2 = 1$, $\sigma = 0.793$ and $P(s_2) = 0.3$,

(a) plot a graph of P_e versus a_0 and determine the value of a_0 which minimizes P_e. Verify your answer using the optimum threshold formula $a_o = \frac{a_1 + a_2}{2} + \frac{\sigma^2}{a_1 - a_2} \ln \left(\frac{P(s_2)}{P(s_1)} \right)$.

(b) plot a graph of a_0 versus $P(s_2)$ over the range 0 to 1 and verify that $a_0 = \frac{a_1 + a_2}{2}$ for $P(s_2) = 0.5$.

(c) repeat parts (a) and (b), but now with $a_1 = 3$, $a_2 = 1$.

16. If a rectangular pulse of amplitude A and duration T is input to an RC filter with a frequency response $H(f) = \frac{1}{1 + j\frac{f}{f_o}}$, where $f_o = \frac{1}{2\pi RC}$, determine the following given that the single-sided noise power spectral density $N_o = 10^{-11}$ W/Hz.

(a) The input signal energy if $A = 2$ volts and $T = 0.5$ seconds.

(b) The average output noise power N_{out} and the output signal power P_{out} at time $t = T$ if the RC filter time constant t_c is 0.3 seconds. Hence determine the signal-to-noise power ratio $\frac{P_{out}}{N_{out}}$.

(c) If a matched filter is used instead of the RC filter, determine $\frac{P_{out}}{N_{out}}$.

(d) Repeat part(b) for values of t_c ranging from 0.1 to 4. From a graph of $\frac{P_{out}}{N_{out}}$ versus t_c, determine the value of t_c which maximizes $\frac{P_{out}}{N_{out}}$, and determine the ratio $\frac{\left(\frac{P_{out}}{N_{out}} \right)_{RCFilter}}{\left(\frac{P_{out}}{N_{out}} \right)_{MatchedFilter}}$. Does this ratio change as you use set different values for A and T ?

17. Consider a binary baseband communication system in which a binary digit 1 and 0 are represented by $s_1(t) = +A$ and $s_2(t) = -A$, respectively for $0 \le t \le T$. If the receiver is the well known integrate and dump receiver, in which the received signal $r(t)$ is integrated over a time period T and then sampled at time $t = T$, determine an expression for the probability of an error in a binary digit over an AWGN channel given that the standard deviation σ of the output is given by $\sigma = \sqrt{\frac{N_oT}{2}}$.

18. In the previous problem of a binary polar NRZ baseband communication system over an AWGN channel with a single-sided noise power spectral density N_o, it was stated the standard deviation σ of the integrate and dump output is given by $\sigma = \sqrt{\frac{N_oT_b}{2}}$, where T_b is pulse duration. (a) Prove this expression and (b) verify it by simulation.

19. Consider a *correlator* in which the received signal $r(t) = s(t) + n(t)$ is correlated with $s(t)$ such that the output at time T is given by $y(T) = \int_0^T r(t)s(t)dt$.

(a) Prove that the variance σ^2 of $y(T)$ is given by $\sigma^2 = \frac{N_o}{2}E_b$, where E_b is the energy of the signal $s(t)$.

(b) If $r(t)$ is correlated with its basis function, so that $y(T) = \int_0^T r(t)\phi(t)dt$, show that $\sigma^2 = \frac{N_o}{2}$.

20. Consider a binary baseband communication system in which a binary digit 1 and 0 are represented by the signals $s_1(t)$ and $s_2(t)$, respectively, each of time duration T_b. If the received signal $r(t)$ is correlated with $s_1(t)$ such that output $y(T) = \int_0^T r(t)s_1(t)dt$, plot a graph of $y(T)$ versus T for $T = 0, 0.1 \cdots T_b$ under the following conditions for the two possible situations where $r(t) = s_1(t)$ or $r(t) = s_2(t)$:

(a) $s_1(t) = a(t)$ and $s_2(t) = b(t)$
(b) $s_1(t) = a(t)$ and $s_2(t) = -a(t)$
(c) $s_1(t) = b(t)$ and $s_2(t) = -b(t)$

where $a(t) = \begin{cases} A & if \quad 0 \le t \le T_b \\ 0 & otherwise \end{cases}$ and

$b(t) = \begin{cases} A & if \quad 0 \le t \le \frac{T_b}{2} \\ -A & if \quad \frac{T_b}{2} < t \le T_b \\ 0 & otherwise \end{cases}$

and $T_b = 2$ seconds and $A = 1$ volt.

(d) If the channel is an AWGN channel and the correlator receiver is as shown in the figure below, derive an expression for the probability of an error in a binary digit P_e for the signal assignments of parts (a), (b) and (c).

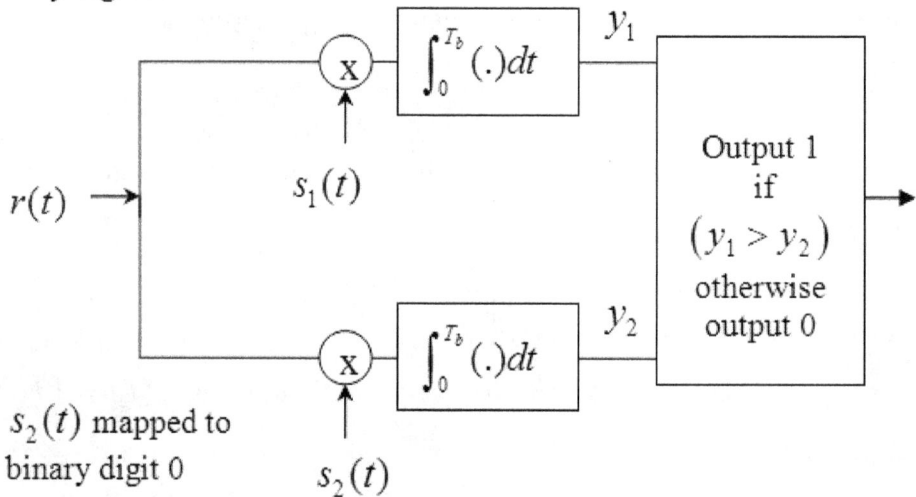

21. Consider a PCM system in which the $SQNR_{peak}(dB)$ is required to be at least 50 dB. If the probability of a binary digit error P_e within a PCM word due to channel noise is 10^{-7}, determine the minimum number of quantization levels required and the minimum bandwidth of the PCM signal if the input analog signal is band-limited to 3.4 kHz. What is the increase the required bandwidth if the polar NRZ line code is used ?

Chapter 7

Bandpass Signaling

7.1 Introduction

In this chapter we shall focus on the modulator, channel and the demodulator for communication over a *bandpass channel*, in which the frequency passband does not include zero frequency ($f = 0$) as illustrated in Fig. 7.1. Unlike in a baseband communication system (previous chapter) where the baseband signal power is at low frequencies, the power in a bandpass signal is at high frequencies. Specifically, the double-sided baseband signal spectrum is shifted to be centered on a frequency other than zero. Refer to the frequency shift property in section 1.5 for further details. The binary digits are transferred via a combination of high frequency sinusoidal waveforms. For linear systems, the bandpass signal spectrum can be shifted down to the baseband level for linear baseband signal processing. Thus, we can expect the performance of certain bandpass and baseband modulation schemes to be equivalent. For example, it will be shown that with antipodal signalling, the performance of polar NRZ is the same as that of binary phase shift keying (BPSK).

The ability to shift the signal spectrum permits communication over those channels (e.g. satellite, microwave link, etc.) over which it is either impossible or impractical to transfer a baseband signal. For example, an impractically large antenna (many miles long) would be required to transmit a voice signal via electromagnetic waves through space. The solution is to "carry" the baseband signal using a high frequency sinusoidal waveform, referred to as a *carrier*. Of the many advantages in using a carrier, some of the key advantages are as follows:

- a carrier can be matched to the characteristics of the channel.

- a given channel can be used for the simultaneous communication of several bandpass signals by appropriately sharing the available channel bandwidth e.g. frequency or time division multiplexing.

- a high frequency carrier can easily penetrate the atmosphere.

- the size of an antenna (e.g. cell phone) need not be large.

- the available channel bandwidth can be regulated to minimize interference between different users e.g. radio spectrum.

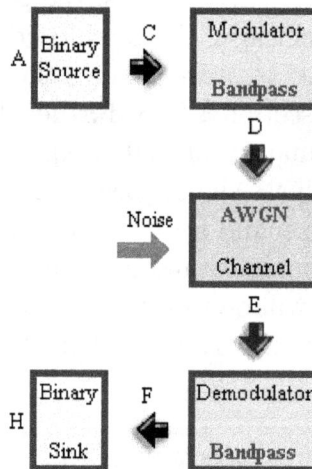

Figure 7.1: Chapter 7 DCS.

7.1.1 Fundamentals

Given that a sinusoid has only three features (amplitude, frequency, phase) which can be altered, there are three basic bandpass modulation techniques. *Amplitude shift keying* (ASK), *frequency shift keying* (FSK) and *phase shift keying* (PSK), in which respectively, the amplitude A, frequency f_c or phase θ of a sinusoidal waveform $s(t) = A \cos(2\pi f_c t + \theta)$ is changed (referred to as *switching* or *keying*) in accordance with the input binary digits. For example, if the input symbols to the modulator are binary digits, then the modulator transmits either one of two different of signals $s_1(t)$ or $s_2(t)$ for a duration of T_b seconds as summarized in Table 7.1 and illustrated in Fig. 7.2 for the input binary sequence 1100101. Of course the mapping scheme used between a given binary digit and signal is ensured to be the same within the modulator and demodulator.

In general, for M input symbols, the amplitude, frequency or phase for ASK, FSK or PSK is selected from the sets $A \in \{A_1, A_2, \cdots A_M\}$, $f \in \{f_1, f_2, \cdots f_M\}$ and $\theta \in \{\theta_1, \theta_2, \cdots \theta_M\}$, respectively. The modulation schemes in this case are referred to as M-ary ASK, M-ary FSK, or M-ary PSK digital modulation. Notice that both PSK and FSK signals have a constant envelope, making them resilient to channel noise that effects the amplitude of a signal. This is because the information is not embedded within the amplitude, but within its phase or frequency. Finally, we shall assume a linear AWGN channel with a bandwidth large enough so as not to distort the transmitted signal.

Modulation Scheme	Signals	Switched
ASK	$s_1(t) = A_1 \cos(2\pi f_c t + \theta)$	Amplitude
	$s_2(t) = A_2 \cos(2\pi f_c t + \theta)$	
FSK	$s_1(t) = A \cos(2\pi f_1 t + \theta)$	Frequency
	$s_2(t) = A \cos(2\pi f_2 t + \theta)$	
PSK	$s_1(t) = A \cos(2\pi f_c t + \theta_1)$	Phase
	$s_2(t) = A \cos(2\pi f_c t + \theta_2)$	

Table 7.1: Fundamental bandpass modulation schemes.

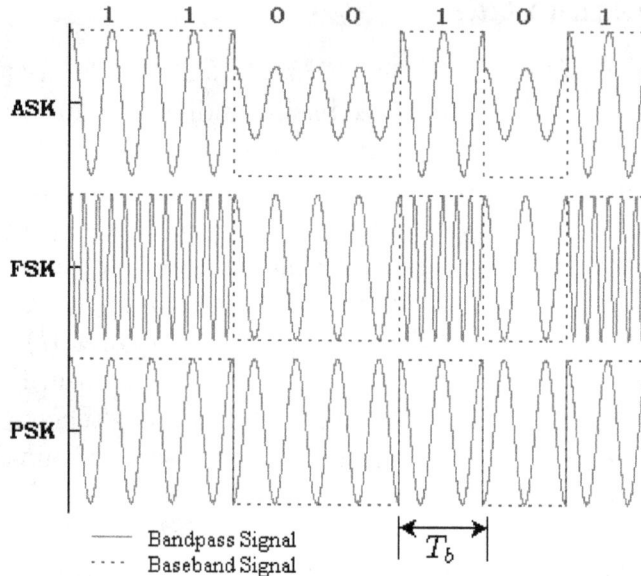

Figure 7.2: ASK, FSK and PSK modulation scheme examples.

Demodulators are classified as *coherent* or *noncoherent* depending on whether a phase-recovery technique is used or not, respectively. Coherent demodulators outperform their corresponding noncoherent variety with the penalty of increased complexity. In this chapter, we begin with coherent binary PSK and M-ary PSK and take into account the influence of error-control coding on the information throughput. Thereafter, *quadrature amplitude modulation* (QAM) in which both the amplitude and phase are switched. Then we consider coherent binary FSK and M-ary FSK. Finally, *differential coherent* PSK and noncoherent FSK, before comparing the coherent modulation schemes from the perspective of power and bandwidth requirements for a given desired binary digit error probability.

7.1.2   Applications

Differential BPSK (section 7.13) and QPSK (section 7.5) are used within the wireless LAN IEEE 802.11b-1999 standard to provide data rates ranging from 1 Mbit/s to 11 Mbit/s. In the wireless LAN standard IEEE 802.11g-2003, OFDM (section 7.10) in combination with BPSK (section 7.2), QPSK or

QAM (section 7.9) are used to provide data rates of 6, 9, 12, 18, 24, 36, 48 and 54 Mbit/s. OFDM is the means by which high speed internet access (ADSL) is provided into the home via copper wires over the "last mile". However, the WiMAX IEEE 802.16 standard could in near future, provide wireless high-speed internet access over this last mile in addition to connecting Wi-Fi hotspots to the Internet. OFDM is now a part of the wireless LAN standards IEEE 802.11a, g, n and HIPERLAN/2 and has been adopted for numerous modern wideband audio and video broadcast applications e.g. digital audio broadcasting (DAB), digital video broadcasting for terrestrial (DVB-T) and handheld (DVB-H) for mobile phones. There are of course numerous other applications of the modulation schemes presented in this chapter e.g. satellite communications, bluetooth, etc.

7.2 Binary Phase-Shift Keying (BPSK)

▷ **BPSK**

- Fundamentals

- BPSK Signal Energy

- Bandpass and baseband signals

In *binary phase shift keying* (BPSK), the phase of a high frequency carrier is switched by 180 degrees in accordance with the input binary digits, as summarized in Table 7.2 i.e. if the input to the BPSK modulator is a binary digit '1', then the signal $s_1(t)$ is transmitted into the channel. For a binary digit '0', the signal $s_2(t)$ is transmitted, which differs in phase by 180 degrees. Notice that $s_1(t)$ and $s_2(t)$ are antipodal signals.

Figure 7.3 illustrates the output of a BPSK modulator for the input binary sequence 1100101. The corresponding baseband signal is also shown for convenience. Each binary digit occupies a time-space of T_b seconds because a given signal is transmitted for a time duration of T_b seconds. Within this time duration, the carrier frequency f_c is given by $f_c = \frac{n_c}{T_b}$ to ensure that an integer number of cycles n_c are completed within the time duration T_b. For example in Fig. 7.3, $n_c = 2$.

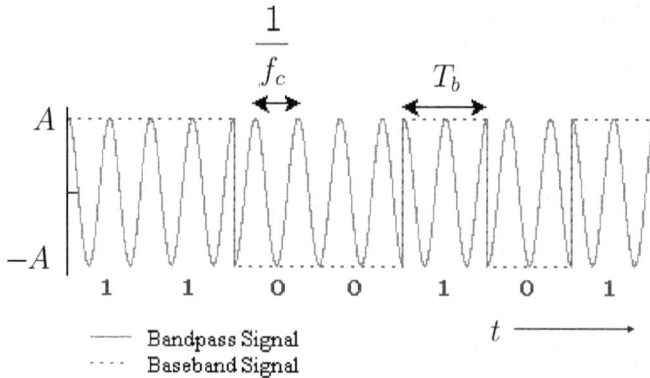

Figure 7.3: Output of a BPSK modulator.

- In section 7.8, it will be shown how the time duration of the transmitted signal $s_i(t)$ has to be reduced to maintain the *information throughput* if error-control coding is used.

7.2.1 BPSK Signal Energy

For the signal $s_1(t) = A\cos(2\pi f_c t)$, the transmitted signal energy per binary digit E_b is given by

$s_1(t)$ represents binary digit 1	$s_1(t) = A\cos\left(2\pi f_c t\right)$
$s_2(t)$ represents binary digit 0	$s_2(t) = A\cos\left(2\pi f_c t + \pi\right) = -A\cos\left(2\pi f_c t\right)$
where A is the amplitude of the carrier and f_c is the carrier frequency.	

Table 7.2: BPSK signals.

$$E_b = \int_0^{T_b} [s_1(t)]^2 \, dt = A^2 \int_0^{T_b} \cos^2(2\pi f_c t) dt \qquad (7.1)$$

Using $\cos^2 \theta = \frac{1+\cos 2\theta}{2}$,

$$E_b = \frac{A^2}{2} \int_0^{T_b} (1 + \cos(4\pi f_c t)) \, dt = \frac{A^2}{2} \left[t + \frac{\sin(4\pi f_c t)}{4\pi f_c} \right]_0^{T_b} \qquad (7.2)$$

Recall that the time duration of a BPSK signal $T_b = n_c T_c = \frac{n_c}{f_c}$, or equivalently $n_c = f_c T_b$. Thus $\sin(4\pi f_c T_b) = \sin(4\pi n_c) = 0$ because n_c is an integer and therefore

$$E_b = \frac{A^2 T_b}{2} \qquad (7.3)$$

Notice that E_b can be increased either by increasing the duration T_b of the signal and/or the amplitude A of the signal. Of these two quantities, the amplitude A will have a greater influence on E_b. Since $A = \sqrt{\frac{2E_b}{T_b}}$, the expressions for $s_1(t)$ and $s_2(t)$ may be written as

$$s_1(t) = \sqrt{\frac{2E_b}{T_b}} \cos(2\pi f_c t) \qquad (7.4)$$

and

$$s_2(t) = -\sqrt{\frac{2E_b}{T_b}} \cos(2\pi f_c t) \qquad (7.5)$$

Given that the time duration of a BPSK signal is T_b, the transmitted signal power P_t is equal to

$$P_t = E_b r_b \tag{7.6}$$

where $r_b = \frac{1}{T_b}$ is the number of binary digits per second r_b transferred via this modulation scheme.

7.2.2   BPSK Signal Constellation Diagram

▶ **BPSK Demodulator**

- Energy in a BPSK signal

- Concept of a basis function

- BPSK signal constellation diagram

- Decision device

The basis function $\phi_1(t)$ for $s_1(t) = \sqrt{\frac{2E_b}{T_b}} \cos\left(2\pi f_c t\right)$ and $s_2(t) = -\sqrt{\frac{2E_b}{T_b}} \cos\left(2\pi f_c t\right)$ can be shown to be given by

$$\phi_1(t) = \sqrt{\frac{2}{T_b}} \cos\left(2\pi f_c t\right) \tag{7.7}$$

and the BPSK signals $s_1(t)$ and $s_2(t)$ in terms of $\phi_1(t)$ are

$$s_1(t) = \sqrt{E_b}\phi_1(t) \text{ and } s_2(t) = -\sqrt{E_b}\phi_1(t) \tag{7.8}$$

Let $r(t)$ represent the signal received from the channel given by

$$r(t) = s_i(t) + n(t) \tag{7.9}$$

where $s_i(t)$ is either $s_1(t)$ or $s_2(t)$ and $n(t)$ is the additive noise signal. As in section 6.5, the coordinate x_h along the horizontal axis of the signal constellation diagram is given by

$$x_h = \int_0^{T_b} r(t)\phi_1(t)dt \tag{7.10}$$

If there is no noise in the channel i.e. $n(t) = 0$, then $r(t)$ is either $s_1(t)$ or $s_2(t)$. For $r(t) = s_1(t)$, $x_h = \int_0^{T_b} s_1(t)\phi_1(t)dt$. But $s_1(t) = \sqrt{E_b}\phi_1(t)$, so that $x_h = \sqrt{E_b}\int_0^{T_b} \phi_1^2(t)dt$ further simplies to $x_h = \sqrt{E_b}$ because $\int_0^{T_b}\phi_1^2(t)dt = E_\phi = 1$. Similarly, for $r(t) = s_2(t)$, we find that $x_h = -\sqrt{E_b}$. The corresponding signal constellation diagram is shown in Fig. 7.4 for $E_b = 1$.

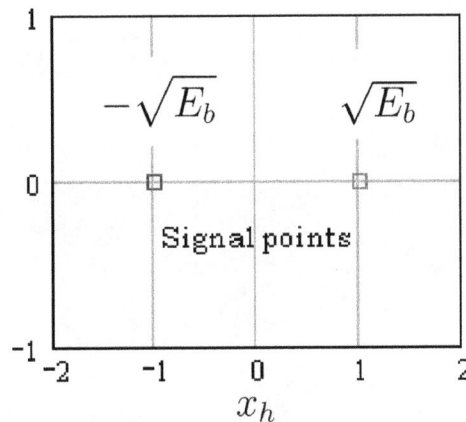

Figure 7.4: BPSK signal constellation diagram.

• ⭐ SIMULATION **BPSKSignalSpace:** Equation 7.10 is implemented to determine the BPSK signal points for a given E_b. Alter the expressions for $s_1(t)$ and $s_2(t)$ to consider what happens if there is a phase difference with respect to $\phi_1(t)$.

Under noisy conditions, the received signal $r(t) = s_i(t) + n(t)$ will be corrupted. In this case, the coordinate x_h for a given $r(t)$, will not lie on $\pm\sqrt{E_b}$. It may lie on either side of $\pm\sqrt{E_b}$ depending on the amount of noise within the channel. For example, Fig. 7.5 illustrates the received signal point under noisy conditions assuming the signal $s_1(t)$ is transmitted with $E_b = 1$.

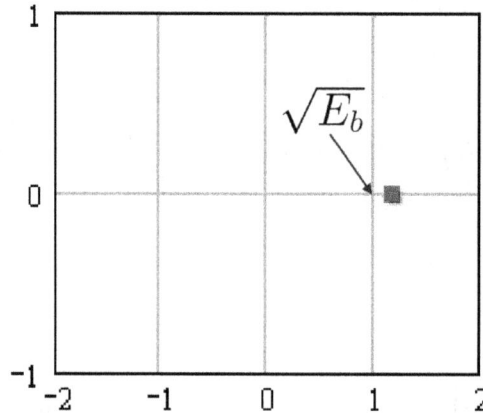

Figure 7.5: Coordinate x_h under noisy conditions when transmit $s_1(t)$.

Thus under noisy conditions,

$$x_h = \int_0^{T_b} r(t)\phi_1(t)dt = \int_0^{T_b} s_i(t)\phi_1(t)dt + \int_0^{T_b} n(t)\phi_1(t)dt \qquad (7.11)$$

$$= \left\{ \begin{array}{ll} \sqrt{E_b} + n_1 & if \quad s_i(t) = s_1(t) \\ -\sqrt{E_b} + n_2 & if \quad s_i(t) = s_2(t) \end{array} \right\} \qquad (7.12)$$

where for an AWGN, n_1 and n_2 are gaussian random variables with mean zero and standard deviation $\sigma = \sqrt{\frac{N_o}{2}}$.

▶ **BPSK Signal Point PDF**

- A closer look at the signal point probability density
  function to see how it can be used to determine
  the BSC crossover probability

Figure 7.6 shows the PDF $f(x_h|s_1)$ curve of the received signal point if only $s_1(t)$ is transmitted, overlaid with the PDF $f(x_h|s_2)$ curve if only $s_2(t)$ is transmitted.

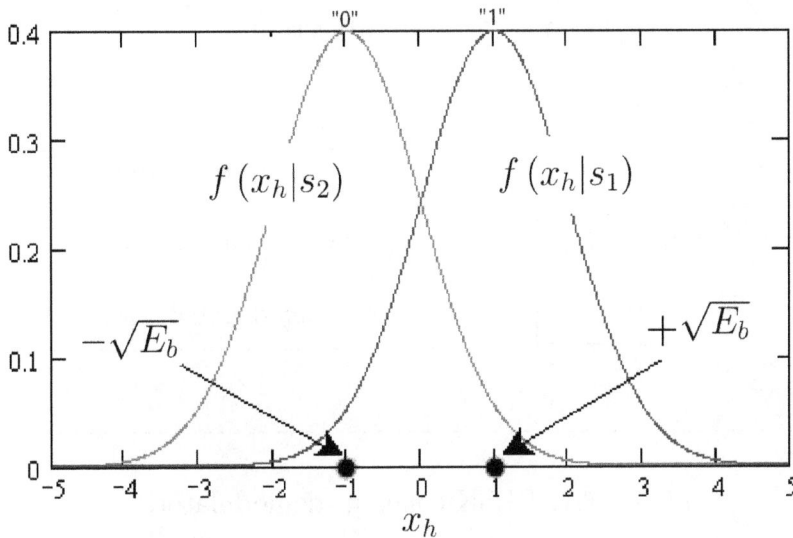

Figure 7.6: Probability density functions of the BPSK signal points.

Figure 7.7 shows a block diagram of the corresponding BPSK demodulator. At the receiver, the incoming signal $r(t)$ is multiplied by $\phi_1(t)$ and the product of these two sinusoidal waveforms is integrated over the time duration T_b of the signal. The output x_h of the integrator is then compared with the threshold value 0. This method of forcing a decision on whether the incoming signal corresponds to a binary 0 or 1 is known as *hard-decision*. Referring to Fig. 7.6, notice that as E_b is increased, the two signal points will move further apart and we can expect the probability of a demodulation

error to decrease i.e. a reduction in the probability of an error in a binary
digit at the demodulator output.

The BPSK demodulator shown in Fig. 7.7 assumes that the phase difference between the incoming signal and the basis function is zero. If the phase
difference between $r(t)$ and $\phi_1(t)$ is θ, it is easily shown (refer to problems section) that the signal point coordinates under a noiseless channel will now be
$\pm\sqrt{E_b}\cos\theta$. As θ increases from zero, the signal points come closer together.
For $\theta = \frac{\pi}{2}$, $x_h = 0$ for both signal points under a noiseless environment!

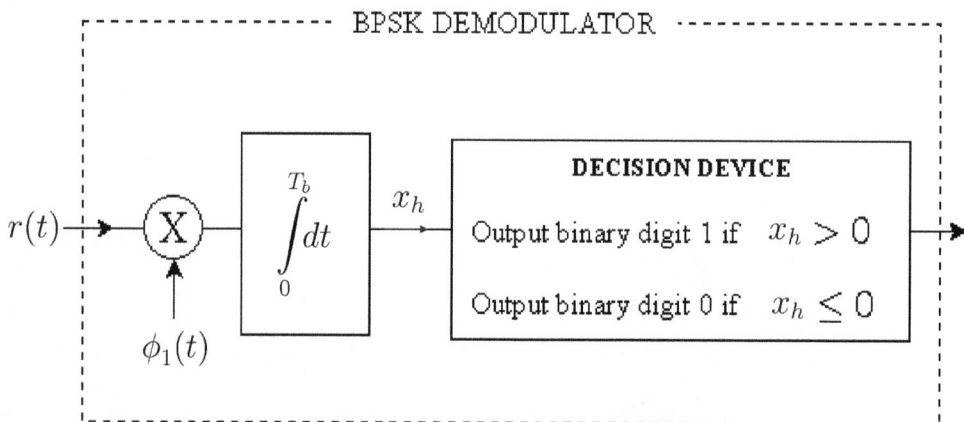

Figure 7.7: BPSK coherent demodulator.

Example : Basis Function Energy

Show that the energy of the signal $\phi_1(t)$ is 1 joule.

Solution

The energy in $\phi_1(t) = \sqrt{\frac{2}{T_b}}\cos\left(2\pi f_c t\right)$ is given by

$$E_\phi = \int_0^{T_b} \left[\sqrt{\frac{2}{T_b}}\cos(2\pi f_c t)\right]^2 dt = \frac{2}{T_b}\int_0^{T_b}\cos^2(2\pi f_c t)dt$$

Using $\cos^2 \theta = \frac{1+\cos 2\theta}{2}$,

$$E_\phi = \frac{1}{T_b} \int_0^{T_b} (1 + \cos(4\pi f_c t)) \, dt = \frac{1}{T_b} \left[t + \frac{\sin(4\pi f_c t)}{4\pi f_c} \right]_0^{T_b}$$

$$= \frac{1}{T_b} \left(T_b + \frac{\sin(4\pi f_c T_b)}{4\pi f_c} \right). \quad \text{But } \sin(4\pi f_c T_b) = \sin(4\pi n_c) = 0$$

Hence $E_\phi = \frac{1}{T_b} (T_b) = 1$ joule.

7.2.3   Simulation : BPSK Signal Point Variance

⭐ SIMULATION **BandpassCorrelator:** The signal $s_1(t) = \sqrt{\frac{2E_b}{T_b}} \cos(2\pi f_c t)$ for $0 < t \leq T_b$ is corrupted with an additive white gaussian noise signal $n(t)$ and correlated with $\phi_1(t) = \sqrt{\frac{2}{T_b}} \cos(2\pi f_c t)$ as in Fig. 7.7. The probability histogram of the correlator output x_h for $\frac{E_b}{N_o} = 10$ dB is shown in Fig. 7.8, highlighting the good correspondence between simulation and theory.

If the time duration $T_b = \frac{n_c}{f_c}$ of a signal is increased by increasing the number of cycles n_c whilst keeping the carrier frequency f_c constant, the corresponding simulated mean and standard deviation of x_h are presented in Fig. 7.9. As expected, $\sigma = \sqrt{\frac{N_o}{2}}$ and is independent of E_b.

- **Experiment:** Correlate the input signal with $\phi_1(t) = \sqrt{\frac{2}{T_b}} \sin(2\pi f_c t)$.

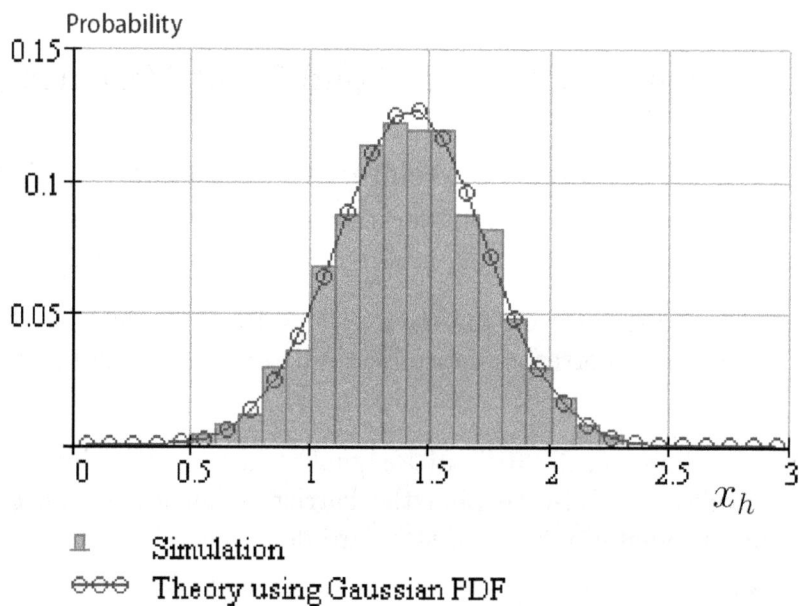

Figure 7.8: Probability histogram of the correlator output. (Mean, Standard deviation) : Simulation (1.416, 0.311), Theory (1.414, 0.316).

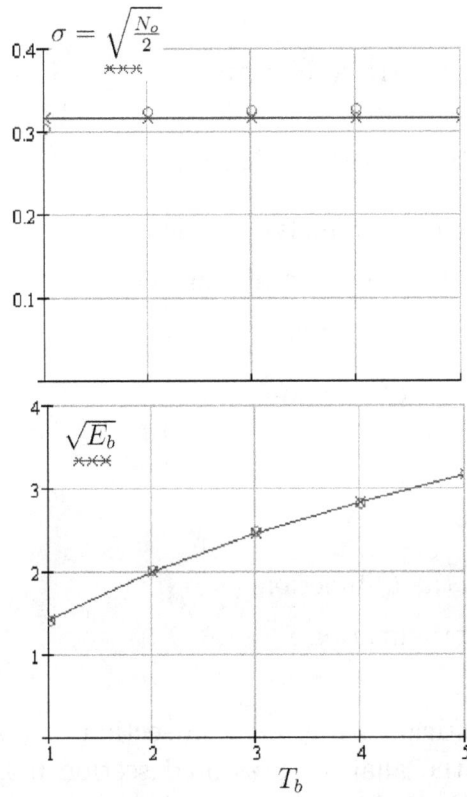

Figure 7.9: Simulated and theoretical standard deviation mean of the correlator output.

7.3   BPSK Performance

The BPSK signal bandwidth, power spectral density and signal power are considered later in section 7.15. For now, we simply consider the probability of an error in a binary digit at the output of a BSC channel (section 3.2) over which the binary digits are transferred via BPSK.

7.3.1   Probability of a Error

Performance in a AWGN Channel

- A general mathematical analysis to determine the probability of an error by finding the area under a signal point probability density function

- Introduction to the Q-function

Q-Function

- A closer look at the Q-function including its approximation

Figure 7.6 is identical to Fig. 6.51 in section 6.5, with $a_1 = \sqrt{E_b}$ and $a_2 = -\sqrt{E_b}$. From the analysis presented section 6.6, the probability of incorrectly demodulating a binary digit is given by

$$P_e = Q\left(\sqrt{\frac{(a_1 - a_2)^2}{2N_o}}\right) = Q\left(\sqrt{\frac{2E_b}{N_o}}\right) \qquad (7.13)$$

Note that with the use of error-control coding, the noiseless signal points a_1 and a_2 will lie at $\pm\sqrt{E_b R_{code}}$ as we shall discover in section 7.8.1.

BPSK Performance

- Determine an expression for the probability P_e of a binary digit error using BPSK over AWGN channel

- Performance curve of P_e versus the channel signal-to-noise ratio in decibels

- Why P_e is the BSC crossover probability

Figure 7.10 shows the performance of the uncoded BPSK modulation scheme over an AWGN over the channel SNR range of 0 to 10 dB. The interesting feature is that the performance of BPSK is the same as that of polar NRZ baseband signalling. This is not surprising because the demodulation process depends on the pulse energy and not its shape.

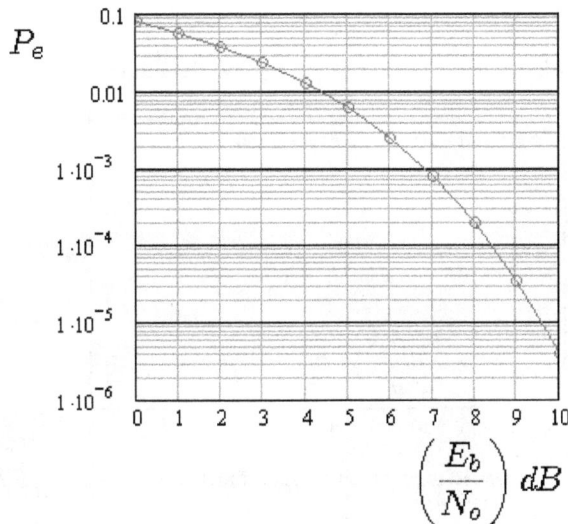

Figure 7.10: BPSK performance curve.

- ⭐ SIMULATION **BPSKTheory:** A plot of equation 7.13 which corresponds to Fig. 7.10. Use this code to determine P_e for a channel SNR of 9.6 dB.

7.3.2   Energy per Binary Digit

For a given channel SNR, the energy per binary digit E_b can be taken to be 1 joule without any loss of generality. To demonstrate this feature, suppose an all-zero sequence is transmitted over an AWGN channel with $SNR = $ -1 dB using BPSK. The standard deviation σ of the probability density function $f(x_h|0)$ is given by $\sigma = \sqrt{\frac{N_o}{2}} = \sqrt{\frac{E_b}{2\left(10^{SNR/10}\right)}}$, where $SNR = \left(\frac{E_b}{N_o}\right)$dB $= 10\log_{10}\left(\frac{E_b}{N_o}\right)$. As E_b is altered, the mean and standard deviation σ of $f(x_h|0)$ is effected as shown in Fig. 7.11. However, the shaded area, which is the probability $P(1|0) = \alpha$, remains constant for a given channel SNR. Clearly α depends only on the specified SNR for any value of E_b because σ will alter accordingly.

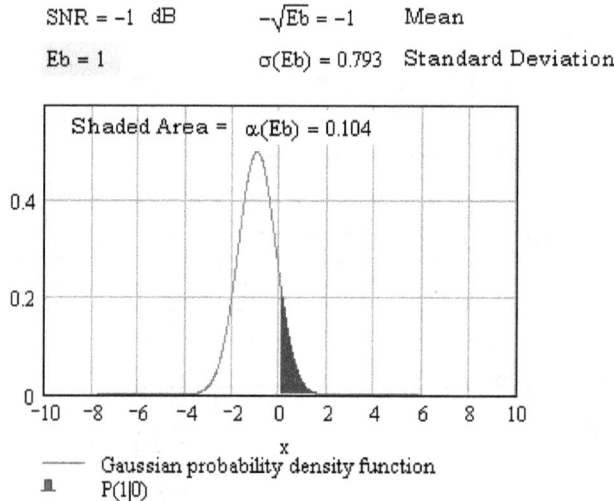

Figure 7.11: Influence of E_b on α having specified the operating SNR.

- ⭐ SIMULATION **Eb:** The shaded area in Fig. 7.11 is determined for a given E_b. What happens as E_b is changed ?

7.3.3 Random BPSK Signal PSD

The PSD of a random BPSK signal is easily determined by making use of the property that the **bandpass** power spectral density $PSD_{BP}(f)$ is the frequency-shifted version of the random polar NRZ **baseband** (section 6.1.2) power spectral density $PSD_B(f)$ such that

$$PSD_{BP}(f) = \frac{1}{4}[PSD_B(f - f_c) + PSD_B(f + f_c)] \qquad (7.14)$$

where $PSD_B(f) = A^2 T_b \, \text{sinc}^2(fT_b)$ (equation 6.16) and therefore

$$PSD_{BP}(f) = \frac{A^2 T_b}{4} \text{sinc}^2\left[T_b(f - f_c)\right] + \frac{A^2 T_b}{4} \text{sinc}^2\left[T_b(f + f_c)\right] \qquad (7.15)$$

The corresponding autocorrelation function (problem 19 in chapter 7) is given by

$$R_X(\tau) = \begin{cases} \frac{A^2}{2}\left(1 - \frac{|\tau|}{T_b}\right)\cos\left(2\pi f_c \tau\right) & \text{for} \quad |\tau| \leq T_b \\ 0 & \text{for} \quad |\tau| > T_b \end{cases} \qquad (7.16)$$

7.3.4 Simulation : BPSK

⭐ SIMULATION **BPSK:** The DCS shown in Fig. 7.12 is simulated using a gaussian random number generator to add noise with variance $\sigma^2 = \frac{N_0}{2}$ to the signal points in Fig. 7.4.

If the binary source outputs a binary digit zero, then a gaussian noise sample is added to $-\sqrt{E_b}$. Conversely, for a binary digit one output, a gaussian noise sample is added to $\sqrt{E_b}$. If the result is greater than zero, then a binary digit one is output, otherwise a binary digit zero is output. By counting the

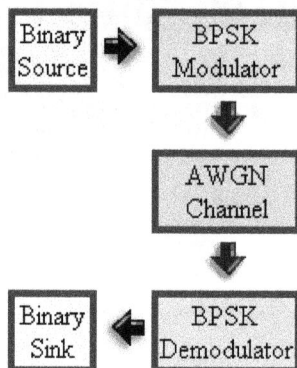

Figure 7.12: Simulated DCS : BPSK over an AWGN channel.

number of binary digit errors, the probability of an error in a binary digit
at the output of the demodulator is calculated for a given SNR.   The simulation points and the theoretical performance curve are presented in Fig.
7.13. The simulation has only been carried out over the SNR range of 0 to 6
dB, because at higher $SNRs$, it is very time consuming to generate sufficient
binary digit errors to estimate P_e accurately. As a general rule of thumb,
any given simulation point should correspond to at least 10 or more binary
digit errors.

- **Experiment**:   Rewrite simulation **BPSK** to generate binary digits
  until the binary digit error count increases to a predefined number,
  say N_{errors}.   Experiment with $N_{errors} = 5$, 10 and 100 to verify for
  yourself how the accuracy of the simulation results improves as N_{errors}
  is increased.

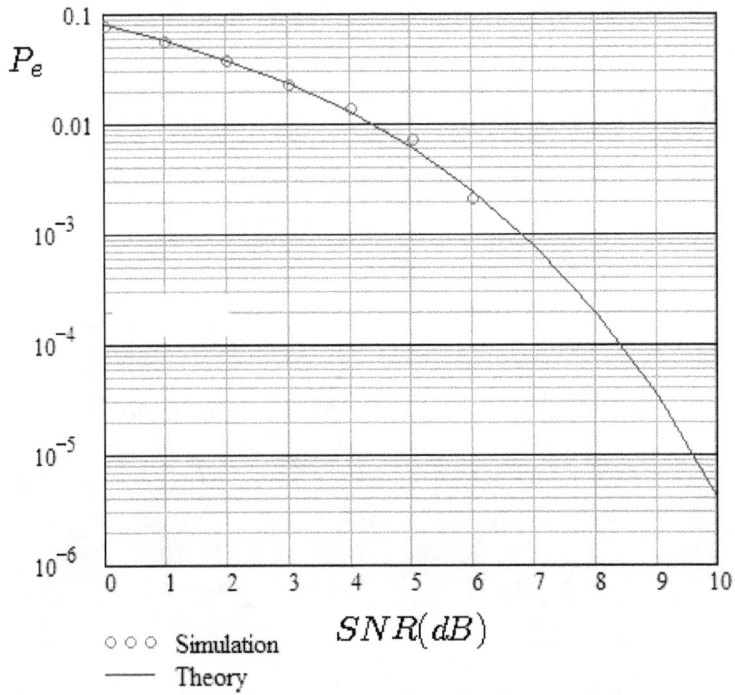

Figure 7.13: Comparison between simulation and theory for uncoded BPSK over an AWGN channel.

7.4   Fading Multipath Channel

A typical example of a fading multipath channel is a mobile (cell-phone) communication system in which the base station signal arrives at the cell phone from multiple paths via reflections from surronding buildings, etc. Specifically, consider the transmission of the carrier $s(t) = \cos(2\pi f_c t)$ through a noiseless multipath channel. The received signal $r(t)$, that is the combination of n propagation paths each with a corresponding attenuation factor $\alpha_n(t)$, a process referred to as *fading*, may be expressed as

$$r(t) = \sum_n \alpha_n(t) \cos\left[2\pi f_c\left(t - \tau_n(t)\right)\right] \qquad (7.17)$$

where $\tau_n(t)$ is the propagation delay on the nth path. Expressing $r(t)$ in complex form,

$$r(t) = \mathrm{Re}\left[\sum_n \alpha_n(t)e^{j2\pi f_c t}e^{-j2\pi f_c \tau_n(t)}\right] = \mathrm{Re}\left[c(t)e^{j2\pi f_c t}\right] \qquad (7.18)$$

where

$$c(t) = \sum_n \alpha_n(t)e^{-j2\pi f_c \tau_n(t)} = \sum_n \alpha_n(t)e^{-j\theta_n(t)} \qquad (7.19)$$

is the sum of phasors varying in amplitude $\alpha_n(t)$ and phase $\theta_n(t)$ with time.   Separating $c(t)$ into its real $c_r(t)$ and imaginary $c_i(t)$ parts and assuming a large number of propagation paths, $c(t)$ can be modelled as

$$c(t) = c_r(t) + jc_i(t) \qquad (7.20)$$

where $c_r(t)$ and $c_i(t)$ are zero-mean gaussian random variables (from the central-limit theorem). Thus we may write $c(t)$ as

$$c(t) = \alpha(t)e^{j\theta(t)} \qquad (7.21)$$

where

$$\alpha(t) = \sqrt{c_r^2(t) + c_i^2(t)} \quad \text{and} \quad \theta(t) = \tan^{-1}\left(\frac{c_i(t)}{c_r(t)}\right) \qquad (7.22)$$

⭐ SIMULATION **RayleighPDF:** To verify that the sum of phasors varying in amplitude $\alpha_n(t)$ and phase $\theta_n(t)$ with time to produce $c(t) = \sum_n \alpha_n(t)e^{-j\theta_n(t)}$ is equivalent to $c(t) = c_r(t) + jc_i(t)$, where $c_r(t)$ and $c_i(t)$ are zero-mean gaussian random variables by observing the probability distribution of $\left|\sum_n \text{rnd}(1)e^{-j\,\text{rnd}(2\pi)}\right|$ (where $\text{rnd}(x)$ generates a random number uniformly distributed between 0 and x) and comparing it with that of $|c_r(t) + jc_i(t)|$.

The *coherence bandwidth* B_c of a channel is the frequency range over which the spectral components of a signal that fall within this bandwidth experience the same amplitude fading. Its approximately given by $\left(\frac{1}{T_m}\right)$ Hz, where T_m is the *delay spread* of the channel, that is the time duration between the first (typically line-of-sight) and last (reflected) arriving components of the transmited signal $s(t)$. Thus, if the bandwidth B_s of the transmitted signal is less than B_c, then all spectral components fade simulatanoulsy and the channel is referred to as *frequency non-selective*. However, if $B_s > B_c$, then the channel is referred to as *frequency selective* and those spectral components outside the coherence bandwidth will fade differently.

7.4.1 BPSK over Rayleigh Fading Channel

For BPSK, assuming a frequency nonselective channel with the time-variations in $c(t)$ very slow compared with the bit interval $0 \le t \le T_b$, the model of this channel with AWGN is shown in Fig. 7.14, in which we may take $c(t)$ to be constant over a bit interval so that

$$c = \alpha e^{j\theta} \qquad (7.23)$$

where $\alpha = \sqrt{c_r^2 + c_i^2}$ has a Rayleigh probability distribution given by

$$f(\alpha) = \begin{cases} \frac{\alpha}{\sigma^2} e^{\frac{-\alpha^2}{2\sigma^2}} & \text{for} \quad \alpha \geq 0 \\ 0 & \text{otherwise} \end{cases} \qquad (7.24)$$

and $\theta = \tan^{-1}\left(\frac{c_i}{c_r}\right)$ has a uniform distribution.

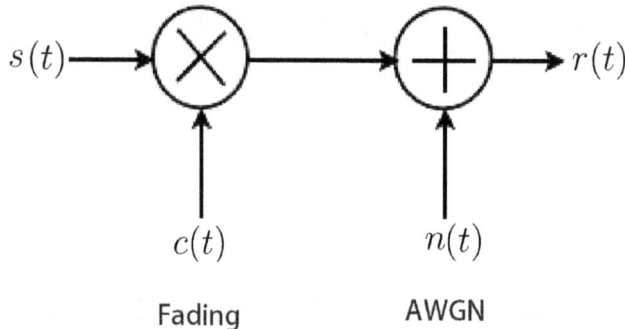

Figure 7.14: Frequency-nonselective channel model with AWGN.

⭐ SIMULATION **RayleighUniform:**   To verify that $\alpha = \sqrt{c_r^2 + c_i^2}$ and $\theta = \tan^{-1}\left(\frac{c_i}{c_r}\right)$ are characterized statistically by the Rayleigh and uniform probability distributions, respectively.

Making a further assumption that any phase offsets are perfectly estimated, we may set $\theta = 0$ and the transmitted BPSK signal $s_1(t) = \sqrt{E_b}\phi_1(t)$ (equation 7.8) will be received as $\alpha\sqrt{E_b}\phi_1(t)$ under a noiseless channel with the corresponding signal point coordinate given by

$$x_h = \int_0^{T_b} \left(\alpha\sqrt{E_b}\phi_1(t)\right)\phi_1(t)dt = \alpha\sqrt{E_b}\int_0^{T_b}\phi_1^2(t)dt = \alpha\sqrt{E_b} \qquad (7.25)$$

because the energy content $\int_0^{T_b} \phi_1^2(t)dt$ of the basis function $\phi_1(t)$ is 1 joule.

Thus for a given attenuation factor α, the signal point coordinates will be $\pm\alpha\sqrt{E_b}$ and similar to equation 7.13, the probability of an error in a binary digit will be given by

$$P_e(error|\alpha) = Q\left(\sqrt{\frac{(a_1 - a_2)^2}{2N_o}}\right) = Q\left(\sqrt{\frac{\left(2\alpha\sqrt{E_b}\right)^2}{2N_o}}\right) = Q\left(\sqrt{\frac{2\alpha^2 E_b}{N_o}}\right)$$

(7.26)

Now averaging over the probability distribution of α,

$$P_e = \int_0^\infty P_e(error|\alpha)\frac{\alpha}{\sigma^2}e^{\frac{-\alpha^2}{2\sigma^2}}d\alpha = \int_0^\infty Q\left(\sqrt{\frac{2\alpha^2 E_b}{N_o}}\right)\frac{\alpha}{\sigma^2}e^{\frac{-\alpha^2}{2\sigma^2}}d\alpha \qquad (7.27)$$

which can be shown (verified by simulation in this subsection) to be given by

$$P_e = \frac{1}{2}\left(1 - \sqrt{\frac{\gamma}{1+\gamma}}\right)$$

(7.28)

where γ is the mean value of the attenuated ratio $\frac{\alpha^2 E_b}{N_o}$, given by

$$\gamma = \frac{E_b}{N_o}E[\alpha^2] = \frac{E_b}{N_o}\left(2\sigma^2\right)$$

(7.29)

⭐ SIMULATION ChiSquared: To verify that α^2 has a chi-squared distribution $f(x) = \begin{cases} \frac{1}{2\sigma^2}e^{\frac{-x}{2\sigma^2}} & \text{for} \quad x \geq 0 \\ 0 & \text{otherwise} \end{cases}$ with $E[\alpha^2] = 2\sigma^2$ and

$$P_e = \int_0^\infty Q\left(\sqrt{\frac{2\alpha^2 E_b}{N_o}}\right)\frac{\alpha}{\sigma^2}e^{\frac{-\alpha^2}{2\sigma^2}}d\alpha = \frac{1}{2}\left(1 - \sqrt{\frac{\gamma}{1+\gamma}}\right).$$

7.4.2   Simulation : BPSK over Rayleigh Fading

⭐ SIMULATION **BPSKRayleighFading:** Figure 7.15 shows the P_e performance of uncoded BPSK over a Rayleigh Fading channel at various signal-to-noise ratios $SNR = 10\log_{10}\frac{E_b}{N_o}$, with $\sigma = \frac{1}{\sqrt{2}}$ in the simulation of the Rayleigh probability distribution of α to ensure $E[\alpha^2] = 1$ so that $\gamma = \frac{E_b}{N_o}$. Overlayed is the standard perormance of uncoded BPSK performance over an AWGN channel. Notice the dramatic loss in performance with fading.

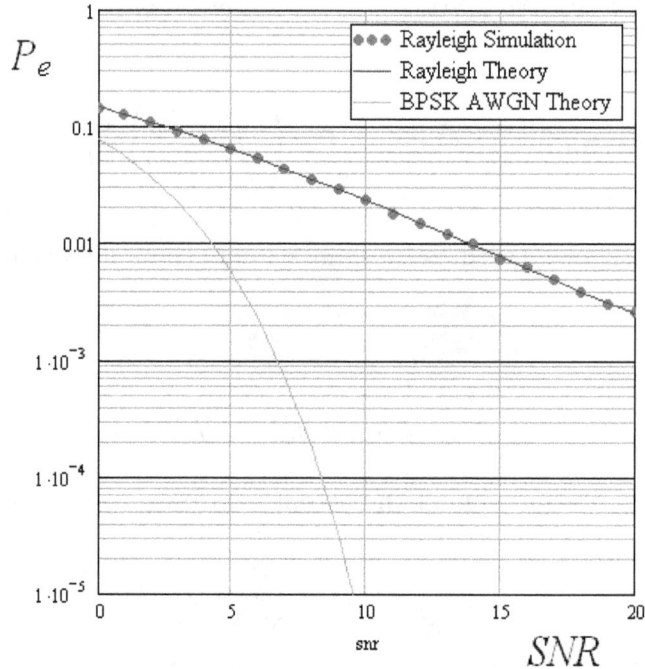

Figure 7.15: Probability of an error in a binary digit P_e versus the channel $SNR(dB)$ for uncoded BPSK over a Rayleigh fading channel.

7.5   Quadrature Phase Shift Keying (QPSK)

▶ **QPSK**

- Fundamentals

- Signal constellation diagram

- Comparison with BPSK

Recall that in the BPSK modulation scheme, only two possible signals are transmitted into the channel. The only difference between these two signals is a phase difference of 180 degrees. Using only two signals to communicate digital information across a given channel constrains the mapping of each signal to a single binary digit '0' or '1'. If a waveform was mapped to two binary digits, then four different signals would be required, because there are four different combinations of a pair of binary digits, namely 00, 01, 10, 11. The modulation scheme in this case is referred to as *quadrature phase-shift keying* (QPSK), in which the phase difference between each signal is now 90 degrees. Each pair of binary digits can be referred to as a *symbol*. Thus, each signal $s_i(t)$ is mapped to one of four possible symbols (00, 01, 10, 11) and is given by

$$s_i(t) = \sqrt{\frac{2E_s}{T_s}} \cos \left[2\pi f_c t + (2i - 1) \frac{\pi}{4} \right] \qquad (7.30)$$

where E_s is the signal energy per symbol, T_s is the time duration of each signal and $i = 1, 2, 3, 4$. Table 7.3 lists the four possible signals. Notice that the amplitude of each signal is given by $A = \sqrt{\frac{2E_s}{T_s}}$. Once again, to ensure an integer number of carrier cycles n_c within the time duration T_s, the frequency of the carrier $f_c = \frac{n_c}{T_s}$. To maintain the same information throughput as in BPSK, $T_s = 2T_b$, where T_b is the time-space occupied by a single binary digit. The amplitude of a QPSK signal is the same as a BPSK signal i.e. $A = \sqrt{\frac{2E_s}{T_s}} = \sqrt{\frac{2E_b}{T_b}}$, or equivalently $\frac{2E_s}{T_s} = \frac{2E_b}{T_b}$, from which we find $E_s = \frac{E_b T_s}{T_b}$. Given that the time duration of a QPSK signal is twice that of a BPSK signal, $T_s = 2T_b$ and therefore $E_s = \frac{2E_b T_b}{T_b} = 2E_b$.

▶ QPSK Signal Constellation Diagram

- Gray coding

- Comparison with BPSK

- Probability density function of a given signal point

To create the signal constellation diagram for the QPSK signal set, we require the orthonormal basis functions

$$\phi_1(t) = \sqrt{\frac{2}{T_s}} \cos\left(2\pi f_c t\right) \tag{7.31}$$

$$\phi_2(t) = \sqrt{\frac{2}{T_s}} \sin\left(2\pi f_c t\right) \tag{7.32}$$

where the energy of each basis function $E_{\phi_1} = E_{\phi_2} = 1$ joule.

- ⭐ SIMULATION **Ortho:** Verifies that the energy of each basis function is 1 joule and that they are orthonormal.

The horizontal coordinate x_h and the vertical coordinate x_v on the signal constellation (or space) diagram are respectively given by

$s_1(t) = \sqrt{\frac{2E_s}{T_s}} \cos\left[2\pi f_c t + \frac{\pi}{4}\right]$	$s_2(t) = \sqrt{\frac{2E_s}{T_s}} \cos\left[2\pi f_c t + \frac{3\pi}{4}\right]$
$s_3(t) = \sqrt{\frac{2E_s}{T_s}} \cos\left[2\pi f_c t + \frac{5\pi}{4}\right]$	$s_4(t) = \sqrt{\frac{2E_s}{T_s}} \cos\left[2\pi f_c t + \frac{7\pi}{4}\right]$

Table 7.3: QPSK signals.

$$x_h = \int_0^{T_s} r(t)\phi_1(t)dt \tag{7.33}$$

and

$$x_v = \int_0^{T_s} r(t)\phi_2(t)dt \tag{7.34}$$

where $r(t)$ is the received signal. For a noiseless channel, i.e. $r(t) = s_i(t)$, it is easily shown that

$$x_h = \sqrt{E_s}\cos\left[(2i-1)\frac{\pi}{4}\right] \text{ and } x_v = -\sqrt{E_s}\sin\left[(2i-1)\frac{\pi}{4}\right] \tag{7.35}$$

For example, for $i = 1$

$$s_1(t) = \sqrt{\frac{2E_s}{T_s}}\cos\left[2\pi f_c t + \frac{\pi}{4}\right] \tag{7.36}$$

$$= \sqrt{\frac{2E_s}{T_s}}\left\{\cos\left(2\pi f_c t\right)\cos\left(\frac{\pi}{4}\right) - \sin\left(2\pi f_c t\right)\sin\left(\frac{\pi}{4}\right)\right\} \tag{7.37}$$

$$= \sqrt{\frac{E_s}{T_s}}\left\{\cos\left(2\pi f_c t\right) - \sin\left(2\pi f_c t\right)\right\} \tag{7.38}$$

Thus

$$x_h = \int_0^{T_s} s_1(t)\phi_1(t)dt \tag{7.39}$$

$$= \sqrt{\frac{2}{T_s}}\sqrt{\frac{E_s}{T_s}}\left\{\int_0^{T_s} \cos^2\left(2\pi f_c t\right) - \sin\left(2\pi f_c t\right)\cos\left(2\pi f_c t\right)dt\right\} \tag{7.40}$$

Now making use of the standard trigonometry relationships, $\cos^2 \theta = \frac{1 + \cos 2\theta}{2}$ and $\sin 2\theta = 2 \sin \theta \cos \theta$,

$$x_h = \frac{\sqrt{2E_s}}{T_s} \int_0^{T_s} \left(\frac{1}{2} + \frac{\cos (4\pi f_c t)}{2} \right) dt - \frac{\sqrt{2E_s}}{T_s} \int_0^{T_s} \left(\frac{\sin (4\pi f_c t)}{2} \right) dt \quad (7.41)$$

$$= \frac{\sqrt{2E_s}}{T_s} \left(\frac{T_s}{2} \right) = \sqrt{\frac{E_s}{2}} \quad (7.42)$$

Similarly

$$x_v = \int_0^{T_s} s_1(t) \phi_2(t) dt \quad (7.43)$$

$$= \sqrt{\frac{2}{T_s}} \sqrt{\frac{E_s}{T_s}} \left\{ \int_0^{T_s} \sin (2\pi f_c t) \cos (2\pi f_c t) - \sin^2 (2\pi f_c t) \, dt \right\} \quad (7.44)$$

$$= \frac{\sqrt{2E_s}}{T_s} \left(-\frac{T_s}{2} \right) = -\sqrt{\frac{E_s}{2}} \quad (7.45)$$

To check these results, recall

$$x_h = \sqrt{E_s} \cos \left[(2i - 1) \frac{\pi}{4} \right] \text{ and } x_v = -\sqrt{E_s} \sin \left[(2i - 1) \frac{\pi}{4} \right] \quad (7.46)$$

For $i = 1$, we find that $x_v = -\sqrt{\frac{E_s}{2}}$ and $x_h = \sqrt{\frac{E_s}{2}}$ as expected. Plotting the point (x_h, x_v) for $i = 1, 2, 3, 4$, the QPSK signal constellation diagram is shown in Fig. 7.16, where each signal point corresponds to a symbol or equivalently, a pair of binary digits. The mapping between a signal point and a pair of binary digits has been carefully chosen so that neighboring signal points differ in only a single binary digit. For example, the signal point 11 and 10 differ in only a single binary digit position. This type of mapping is

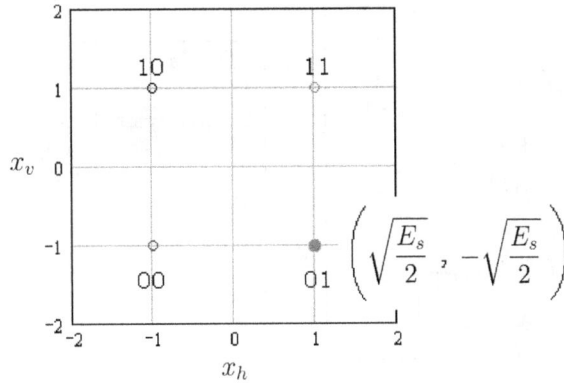

Figure 7.16: QPSK signal constellation.

called *Gray coding*. Refer to problems section in chapter 3 for further details. Based on the constellation diagram, the QPSK demodulator is shown in Fig. 7.17.

- ⭐ SIMULATION **QpskSignalSpace:** The QPSK signal constellation diagram is determined using equation 7.33. Add a slight phase shift to one of the QPSK signals to see its corresponding signal point moves. Consider also what happens if the basis functions are slightly out of phase.

Example : Signal Space Radius

For QPSK, show that the signal points lie on a circle of radius $\sqrt{E_s}$.

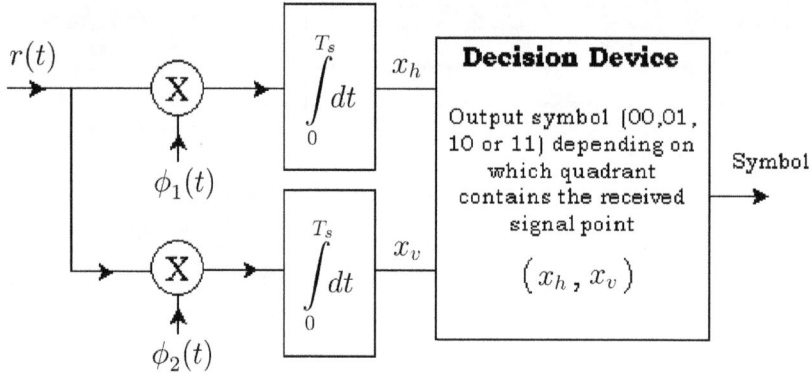

Figure 7.17: QPSK coherent demodulator.

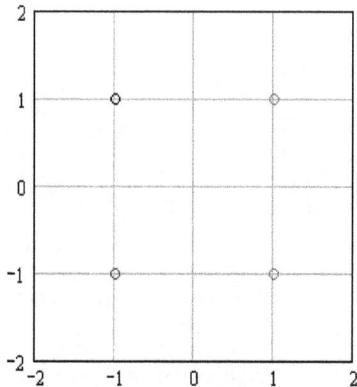

Solution

For QPSK, $x_h = \sqrt{E_s} \cos\left[(2i-1)\frac{\pi}{4}\right]$, $x_v = -\sqrt{E_s} \sin\left[(2i-1)\frac{\pi}{4}\right]$. For $i = 1$, we find that $x_v = -\sqrt{\frac{E_s}{2}}$ and $x_h = \sqrt{\frac{E_s}{2}}$. Radius of the circle r is given by $r^2 = x_h^2 + x_v^2 = \frac{E_s}{2} + \frac{E_s}{2}$. Therefore $r = \sqrt{E_s}$.

7.5.1   QPSK Performance

▷ **QPSK Performance**

- To determine an expression for the probability of a binary digit error of QPSK over an AWGN channel

- Probability of a symbol and binary digit error versus the channel signal-to-noise ratio over an AWGN channel.

Suppose the symbol 11 is repeatedly transmitted over an AWGN channel. Figure 7.18 shows a plot of the received signal points under a SNR of 4 dB.

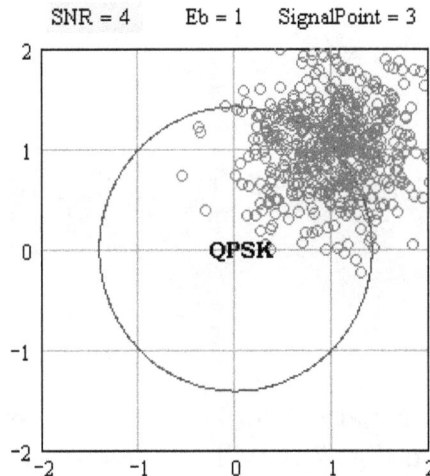

Figure 7.18: QPSK signal point (11) sent repeatedly under $SNR = 4$ dB.

- ⭐ SIMULATION **NoiseOnQpskSignalPoint:** Simulation corresponding to Fig. 7.18. Observe how the distribution of the transmitted signal point changes as the channel SNR varies from 0 to 20 dB.

If symbol 11 is sent under a noiseless channel, $x_h = \sqrt{\frac{E_s}{2}}$ and $x_v = \sqrt{\frac{E_s}{2}}$. Since $E_s = 2E_b$, we find that $\sqrt{\frac{E_s}{2}} = \sqrt{\frac{2E_b}{2}} = \sqrt{E_b}$ and therefore $x_h = x_v = \sqrt{E_b}$. The interesting feature is that $x_h = \sqrt{E_b}$ if a binary '1' is sent using BPSK under a noiseless channel. The implication is that to send the QPSK symbol 11 through the channel is equivalent to sending two consecutive binary digits '1' using BPSK. A visual representation of this process is shown in Fig. 7.20, where x_h and x_v correspond to the transmission of a binary digit via BPSK. A three dimensional illustration of the PDF which governs the movement of the signal point 11 on the constellation diagram in Fig. 7.18 is presented in Fig. 7.19.

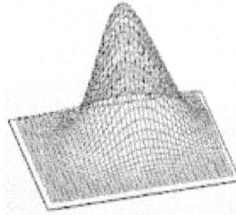

Figure 7.19: PDF of a QPSK signal point.

Using the QPSK demodulator shown in Fig. 7.17 over an AWGN channel, the PDFs of x_h and x_v in Fig. 7.20 will be gaussian with a standard deviation of $\sigma = \sqrt{\frac{N_o}{2}}$. The probability $P(x_h > 0) = P(x_v > 0) = (1 - \alpha)$, where α is the BSC crossover probability under BPSK, given by $\alpha = Q\left(\sqrt{\frac{2E_b}{N_o}}\right)$ (equation 7.13). Hence, the probability with which the QPSK demodulator will correctly decide that symbol 11 was transmitted is equal to the probability with which $P(x_h > 0)$ and $P(x_v > 0)$. Since these are statistically independent events, the probability of a correct decision P_c is given by

$$P_c = P(x_h > 0)P(x_v > 0) = (1 - \alpha)(1 - \alpha) \qquad (7.47)$$

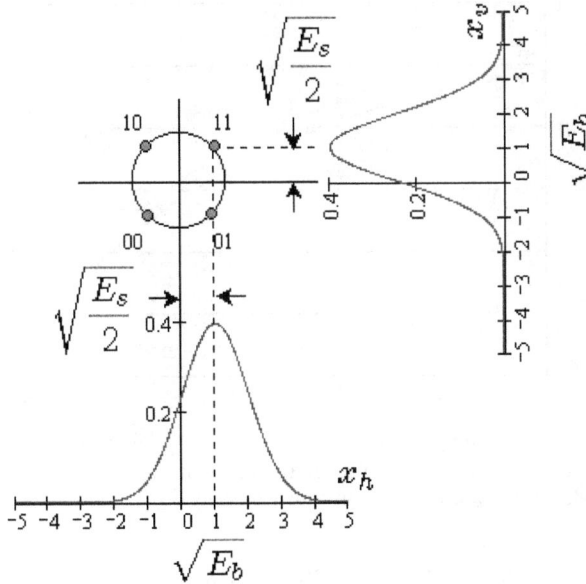

Figure 7.20: Sending a QPSK symbol is equivalent to sending two BPSK symbols.

which by symmetry should be true for any symbol. The probability of an error in a symbol $P_s = 1 - P_c$ and therefore

$$P_s = 1 - (1 - \alpha)^2 = 2\alpha - \alpha^2 \qquad (7.48)$$

which for high SNRs where $\alpha \ll 1$, $P_s \simeq 2\alpha$. By symmetry, it is reasonable to expect the symbol 11 to be most often incorrectly demodulated as one of the neighboring symbols 10 or 01 and therefore the QPSK probability of an error in a binary digit $P_e \simeq \alpha$. However, with Gray coding, the P_e of QPSK is *exactly* the same as that of BPSK. A comparison of the P_e and P_s performance curves of QPSK with BPSK over a range of $SNRs$ are presented in Fig. 7.21. It will be shown in section 7.15, that the channel bandwidth required for the transmission of a QPSK signal is half that of a BPSK signal, thereby making QPSK very popular. The only disadvantage is that the carrier phase synchronization must be maintained. This penalty is overshadowed by the fact that we have the performance of BPSK in half the bandwidth!

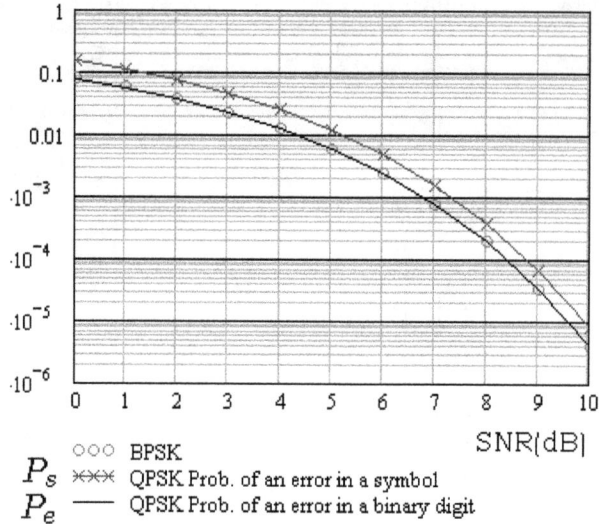

Figure 7.21: QPSK performance.

- ⭐SIMULATION **QpskTheory:** Simulation corresponding to Fig. 7.21. Use this code to verify your own ability to use the $Q(.)$ function tables and a calculator.

7.5.2   Simulation : QPSK

⭐SIMULATION **QPSKRotated:** The DCS shown in Fig. 7.22 is modeled to determine the probability of an error in a binary digit at the binary sink. Simulation curves are compared with the performance presented in Fig. 7.23.

- **Experiment:** Alter the simulation to observe the QPSK performance without the use of Gray coding.

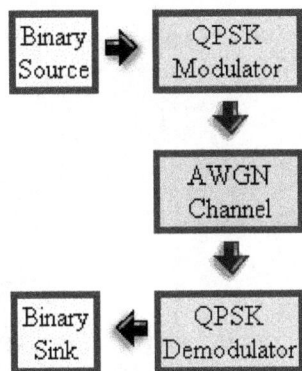

Figure 7.22: Simulated DCS : QPSK over an AWGN channel.

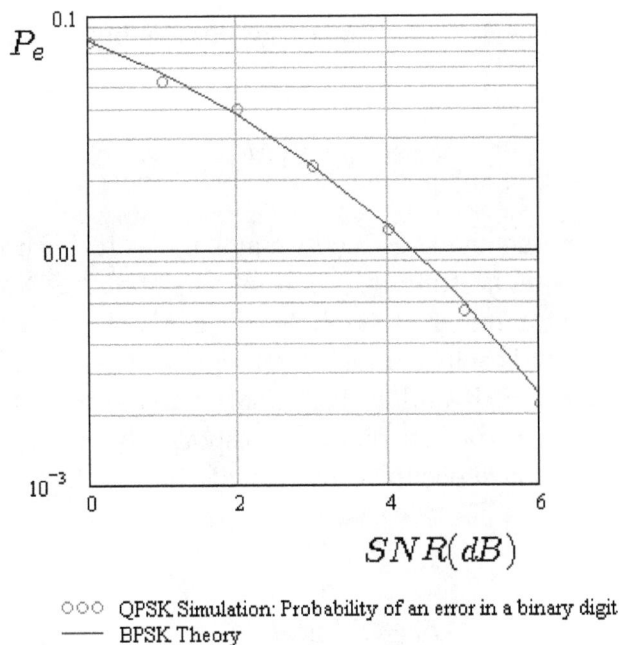

○○○   QPSK Simulation: Probability of an error in a binary digit
——   BPSK Theory

Figure 7.23: Performance of uncoded QPSK over an AWGN channel.

7.6   M-ary PSK Modulation (MPSK)

▶ **MPSK**

- Fundamentals

- Signal constellation diagram

In M-ary PSK, the phase of the carrier is set to one of M possible values. Each signal $s_i(t)$ of duration T_s is given by

$$s_i(t) = \sqrt{\frac{2E_s}{T_s}} \cos\left[2\pi f_c t + \frac{2\pi i}{M}\right] \qquad (7.49)$$

where $i = 0, 1, \cdots M - 1$, E_s is the energy per symbol/signal given by

$$E_s = E_b \log_2 M \qquad (7.50)$$

and the carrier frequency f_c is again equal to $\frac{n_c}{T_s}$ to ensure an integer number of cycles within the time duration of a symbol T_s. Notice that M also corresponds to the number of symbols or signals. For example, $M = 2$ corresponds to BPSK and each symbol/signal represents a single binary digit, $M = 4$ corresponds to QPSK and each symbol/signal represents two binary digits. In general, $\log_2 M$ binary digits correspond to each symbol/signal. Thus, if T_b is the time-space occupied by a single binary digit, then

$$T_s = T_b \log_2 M \qquad (7.51)$$

i.e. no reduction in information throughput. For example, for BPSK $T_s = T_b$ and for QPSK $T_s = 2T_b$.

7.6.1 Signal Constellation Diagram

The signal constellation diagram is determined using the same two basis functions as for QPSK, namely

$$\phi_1(t) = \sqrt{\frac{2}{T_s}} \cos\left(2\pi f_c t\right) \ \text{and} \ \phi_2(t) = \sqrt{\frac{2}{T_s}} \sin\left(2\pi f_c t\right) \qquad (7.52)$$

The horizontal coordinate x_h and the vertical coordinate x_v on the signal constellation (or space) diagram are respectively given by

$$x_h = \int_0^{T_s} r(t)\phi_1(t)dt \ \text{and} \ x_v = \int_0^{T_s} r(t)\phi_2(t)dt \qquad (7.53)$$

where $r(t)$ is the received signal. Hence, the demodulator for MPSK is the same as that for QPSK. For a noiseless channel, the MPSK signal points lie on a circle of radius $\sqrt{E_s}$, where each signal point is mapped to $\log_2 M$ binary digits. Under a noiseless environment, the horizontal coordinate x_h on the signal constellation diagram is given by

$$x_h = \int_0^{T_s} s_i(t)\phi_1(t)dt \ = \ \frac{2\sqrt{E_s}}{T_s} \int_0^{T_s} \cos\left(2\pi f_c t + \frac{2\pi i}{M}\right) \cos\left(2\pi f_c t\right) dt \qquad (7.54)$$

Using $\cos(A + B) = \cos A \cos B - \sin A \sin B$,

$$x_h = \frac{2\sqrt{E_s}}{T_s} \int_0^{T_s} \left[\cos^2\left(2\pi f_c t\right) \cos\left(\frac{2\pi i}{M}\right) \ - \ \sin\left(2\pi f_c t\right) \cos\left(2\pi f_c t\right) \sin\left(\frac{2\pi i}{M}\right) \right] \qquad (7.55)$$

Using $\cos^2 \theta = \frac{1+\cos 2\theta}{2}$ and $\sin 2\theta = 2 \sin \theta \cos \theta$

$$x_h = \frac{2\sqrt{E_s}}{T_s} \int_0^{T_s} \left[\left(\frac{1}{2} + \frac{\cos(4\pi f_c t)}{2} \right) \cos\left(\frac{2\pi i}{M}\right) - \frac{\sin(4\pi f_c t)}{2} \sin\left(\frac{2\pi i}{M}\right) \right]$$

(7.56)

$$= \frac{2\sqrt{E_s}}{T_s} \cos\left(\frac{2\pi i}{M}\right) \left[\frac{t}{2} + \frac{\sin(4\pi f_c t)}{8\pi f_c} \right]_0^{T_s} - \frac{2\sqrt{E_s}}{T_s} \sin\left(\frac{2\pi i}{M}\right) \left[-\frac{\cos(4\pi f_c t)}{8\pi f_c} \right]_0^{T_s}$$

(7.57)

Since there are an integer number n of carrier cycles within the time duration of a signal, $\sin(4\pi f_c T_s) = \sin(4\pi n) = 0$ and $\cos(4\pi f_c T_s) = \cos(4\pi n) = 1$. Hence

$$x_h = \sqrt{E_s} \cos\left(\frac{2\pi i}{M}\right)$$

(7.58)

Similarly, it is easily shown that $x_v = -\sqrt{E_s} \sin\left(\frac{2\pi i}{M}\right)$. The signal points lie on a circle of radius r, given by $r^2 = x_h^2 + x_v^2$. Thus

$$r^2 = E_s \left[\cos^2\left(\frac{2\pi i}{M}\right) + \sin^2\left(\frac{2\pi i}{M}\right) \right] = E_s$$

(7.59)

and therefore

$$r = \sqrt{E_s}$$

(7.60)

The signal constellation diagrams for $M = 2, 4, 8$ and 16 are presented in Figs. 7.24, 7.25, 7.26 and 7.27.

- ⭐ SIMULATION **Mpsk:** The signal constellation diagram for M-ary PSK is determined. Verify the signal points using $x_h = \sqrt{E_s} \cos\left(\frac{2\pi i}{M}\right)$ and $x_v = -\sqrt{E_s} \sin\left(\frac{2\pi i}{M}\right)$. Alter the code to determine what happens if the basis functions are not orthonormal.

Eb = 1

Radius of the circle = $\sqrt{Es}$ = 1

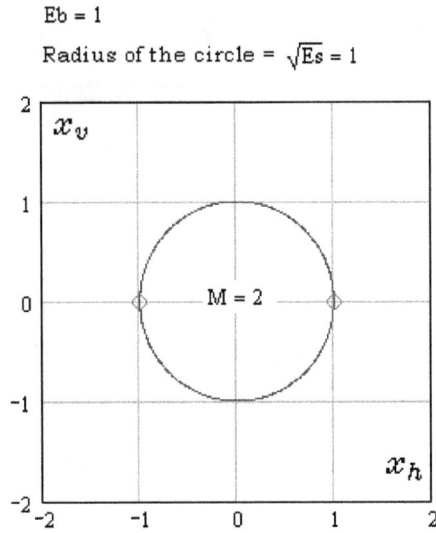

Figure 7.24: $M = 2$.

Eb = 1

Radius of the circle = $\sqrt{Es}$ = 1.414

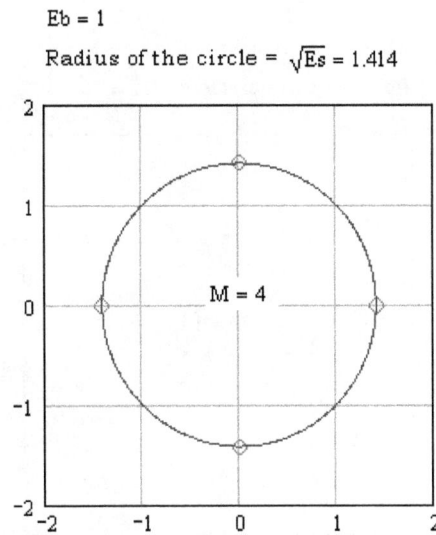

Figure 7.25: $M = 4$.

Eb = 1

Radius of the circle = $\sqrt{Es}$ = 1.732

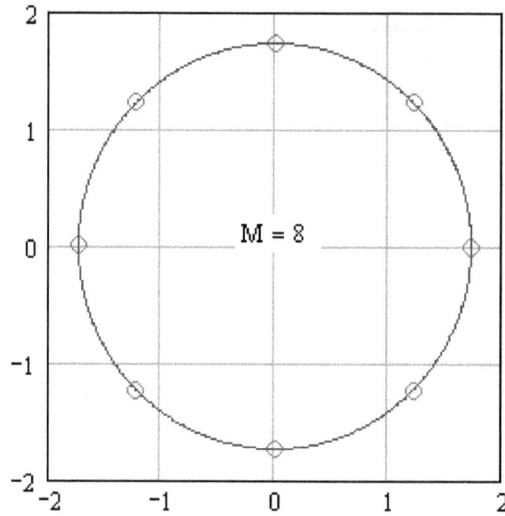

Figure 7.26: $M = 8$.

Eb = 1

Radius of the circle = $\sqrt{Es}$ = 2

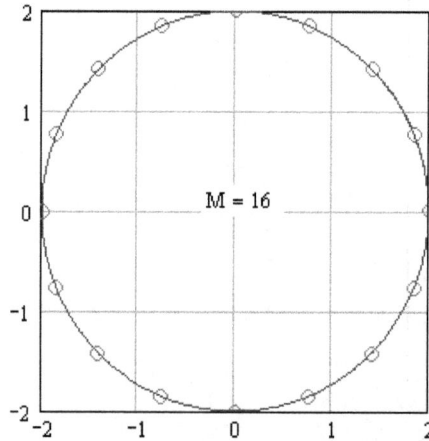

Figure 7.27: $M = 16$.

▷ **8-PSK**

- Example of MPSK with $M = 8$

By examining the spacing between signal points on a given circle, it is clear that as M is increased, we can expect the probability of an error in a symbol to degrade rapidly. For $M = 8$, Figs. 7.28, 7.29, 7.30 and 7.31 show the received signal points under various SNRs. As visually evident from these figures, as the channel SNR is reduced, the likelihood of symbol errors is increased. Notice the use of Gray coding to map three binary digits to each signal point.

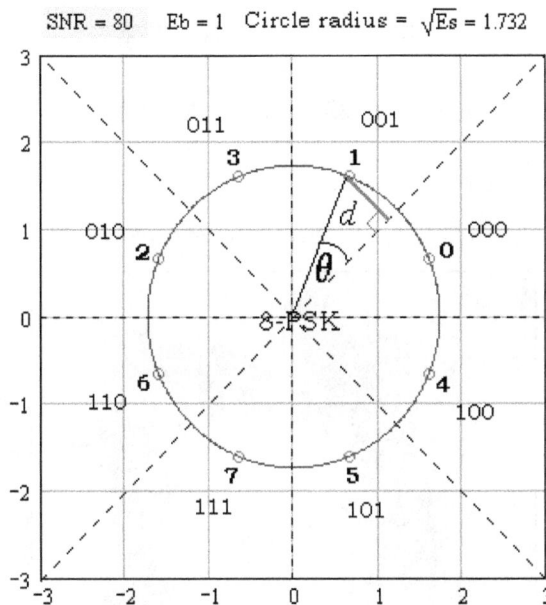

Figure 7.28: SNR = 80 dB.

7.6.2 MPSK Performance

▷ **8PSK Performance**

SNR = 20   Eb = 1   Circle radius = $\sqrt{Es}$ = 1.732

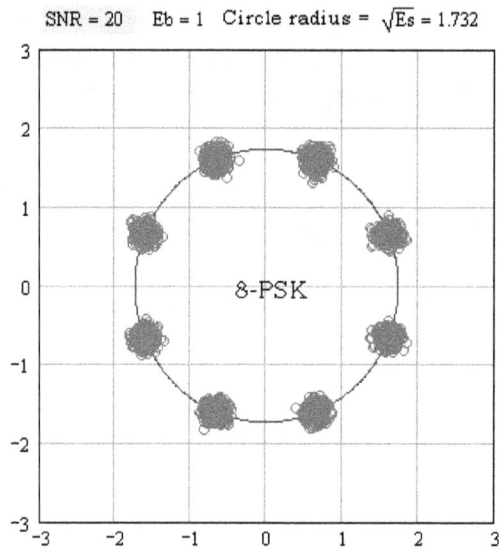

Figure 7.29: SNR = 20 dB.

SNR = 10   Eb = 1   Circle radius = $\sqrt{Es}$ = 1.732

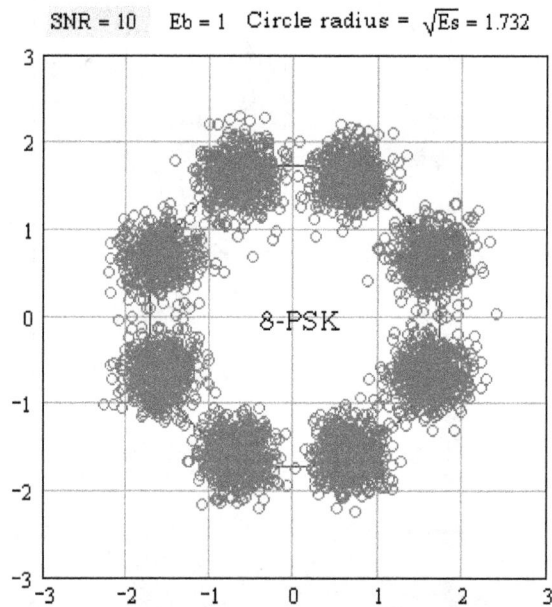

Figure 7.30: SNR = 10 dB.

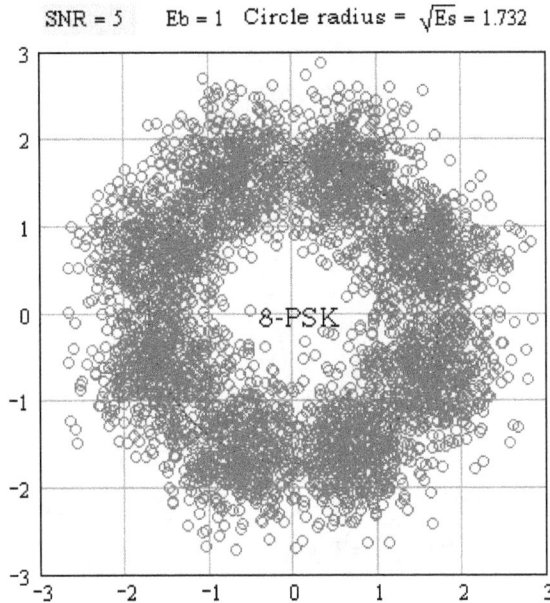

Figure 7.31: SNR = 5 dB.

- Insights on the performance of 8PSK

Referring to Figs. 7.28, the perpendicular distance d of signal point 1, which is also an angle θ, from the boundary is given by $d = \sqrt{E_s} \sin\left(\frac{\pi}{M}\right)$, because $\theta = \frac{\pi}{M}$ in general. Notice how the other boundary is also the same perpendicular distance away from this signal point by virtue of symmetry. Thus, the probability of a symbol error P_s must upper bounded by the probability that the received signal point is moved a distance d or greater perpendicularly towards either boundary, namely,

$$P_s \leq 2P\left(N_\perp > \sqrt{E_s} \sin\left(\frac{\pi}{M}\right)\right) \tag{7.61}$$

where $N_\perp$ is the noise component that moves the signal point perpendicularly towards a given boundary. A factor of 2 is required because a given signal point could be moved towards either boundary to create a symbol error. Given that the variance of the noise component is $\frac{N_o}{2}$, the probability $P\left(N_\perp > \sqrt{E_s}\sin\left(\frac{\pi}{M}\right)\right)$ is simply the area under a zero mean gaussian density function of standard deviation $\sigma = \sqrt{\frac{N_o}{2}}$, from $\sqrt{E_s}\sin\left(\frac{\pi}{M}\right)$ to ∞, given by

$$P\left(N_\perp > \sqrt{E_s}\sin\left(\frac{\pi}{M}\right)\right) = \int\limits_{\sqrt{E_s}\sin\left(\frac{\pi}{M}\right)}^{\infty} \frac{1}{\sigma\sqrt{2\pi}}\exp\left(\frac{-x^2}{2\sigma^2}\right)dx \qquad (7.62)$$

Integrating via substitution, let $u = \frac{x}{\sigma}$ for which $\frac{du}{dx} = \frac{1}{\sigma}$ and the lower integration limit changes to $\frac{\sqrt{E_s}}{\sigma}\sin\left(\frac{\pi}{M}\right) = \sqrt{\frac{2E_s}{N_o}}\sin\left(\frac{\pi}{M}\right)$. Thus

$$P\left(N_\perp > \sqrt{E_s}\sin\left(\frac{\pi}{M}\right)\right) = \int\limits_{\sqrt{\frac{2E_s}{N_o}}\sin\left(\frac{\pi}{M}\right)}^{\infty} \frac{1}{\sqrt{2\pi}}\exp\left(\frac{-z^2}{2}\right)du \qquad (7.63)$$

$$= Q\left(\sqrt{\frac{2E_s}{N_o}}\sin\left(\frac{\pi}{M}\right)\right) \qquad (7.64)$$

so that the probability of a symbol error for MPSK, denoted by $P_s(M)$, is given by

$$P_s(M) \leq 2Q\left(\sqrt{\frac{2E_s}{N_o}}\sin\left(\frac{\pi}{M}\right)\right) \quad \text{for } M \geqslant 4 \qquad (7.65)$$

where with the use of Gray coding, the probability of an error in a binary digit

$$P_e = \frac{P_s}{\log_2 M} \qquad (7.66)$$

Figure 7.32 illustrates the use of this formulae for $M = 4$ and 8, together with simulation results for 8-PSK using the signal constellation in Fig. 7.28. The BPSK and QPSK expressions for P_e derived previously are overlaid on this graph, where $\alpha = Q\left(\sqrt{\frac{2E_b}{N_o}}\right)$. The graphs in Fig. 7.32 confirm that the general formula for $P_s(M)$ is accurate.

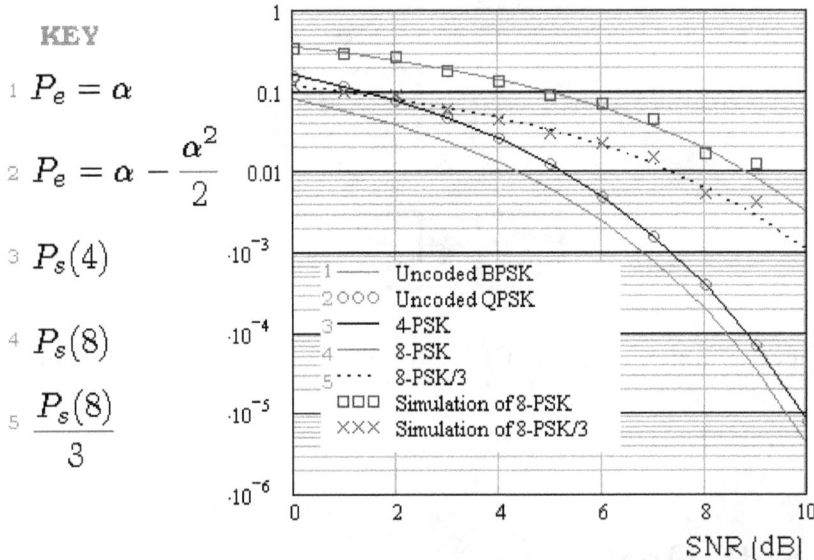

KEY

1 $P_e = \alpha$

2 $P_e = \alpha - \dfrac{\alpha^2}{2}$

3 $P_s(4)$

4 $P_s(8)$

5 $\dfrac{P_s(8)}{3}$

1 ——	Uncoded BPSK
2 ○○○	Uncoded QPSK
3 ——	4-PSK
4 ——	8-PSK
5 ····	8-PSK/3
□□□	Simulation of 8-PSK
×××	Simulation of 8-PSK/3

Figure 7.32: Uncoded M-ary PSK performance.

Since P_e for 8-PSK is worse than for both BPSK and QPSK, it may seem we should restrict ourselves to the use of either QPSK or BPSK. However, in section 8.9, we shall consider a combined modulation and error-control coding scheme with 8-PSK whose performance is better than QPSK at high $SNRs$ with the same channel bandwidth requirement as QPSK!

7.6.3 Simulation : 8-PSK

SIMULATION **EightPSKRotated:** The DCS shown in Fig. 7.33 is modeled. The corresponding simulation results are presented in Fig. 7.32.

- **Experiment:** Alter the simulation code to determine the performance of 8-PSK without the use of Gray coding.

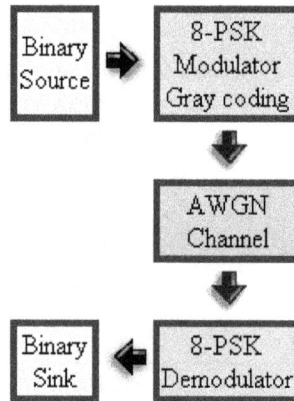

Figure 7.33: Simulated DCS : 8-PSK over an AWGN channel.

7.7   Soft-Decision

Hard-Decision

- The fundamentals of hard-decision decoding

  based on the probability density functions of

  BPSK signal points

In section 7.2, the BPSK demodulator output was restricted to either a binary digit 1 or 0 as summarized in Table 7.4 i.e. the demodulator is said to have made a *hard-decision* on whether the input signal represented a binary digit 1 or 0. In this case, the BPSK modulator, AWGN channel and the demodulator can be modeled as a BSC with forward transition probabilities as listed in Table 7.4 and illustrated in Fig. 7.34.

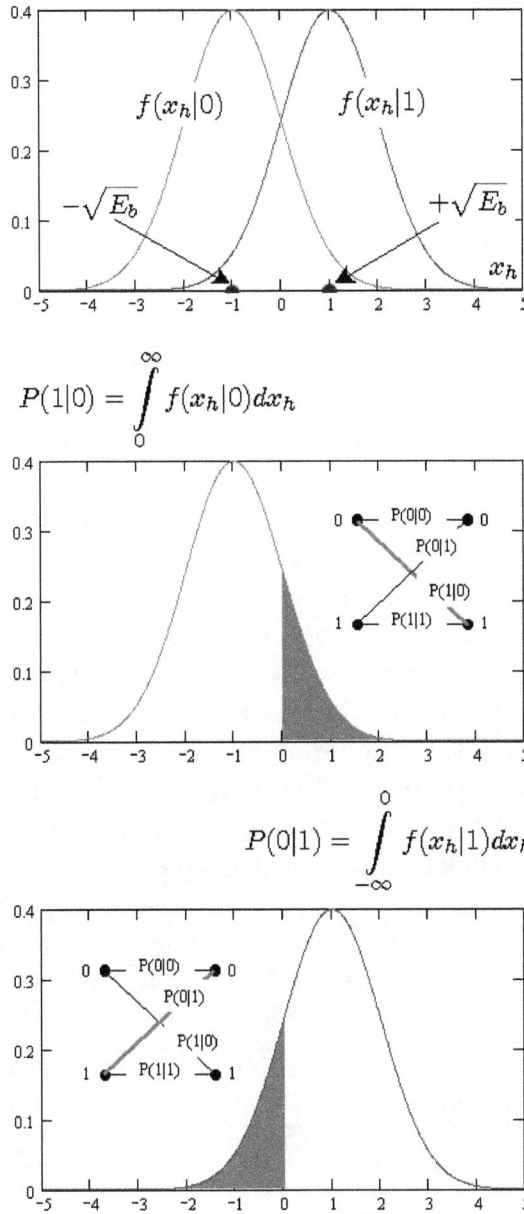

$$P(1|0) = \int\limits_{0}^{\infty} f(x_h|0)dx_h$$

$$P(0|1) = \int\limits_{-\infty}^{0} f(x_h|1)dx_h$$

Figure 7.34: BPSK hard-decision demodulation.

▷ Soft-Decision

- The fundamentals of soft-decision decoding
  based on the probability density functions
  of BPSK signal points

The alternative is to make a *soft-decision* by splitting the range of x_h into Q_{level} regions, where Q_{level} is the quantization level and output the corresponding *soft-symbol*. For example, for $Q_{level} = 8$, the range of x_h could be partitioned as shown in Fig. 7.35, where a is a parameter used to alter the spacing between the partitions. The corresponding soft symbols are presented in Table 7.5. For example, if x_h lies in the range $0 < x_h \leq a$, the demodulator will output the symbol 4. The symbol 4 indicates that the confidence of the decision made by the demodulator is a "weak binary digit 1". This is because the number 4 indicates that the value of x_h is close to the threshold, implying that the input signal is noisy and the corresponding hard-decision of a binary digit 1 is a weak decision. This extra or "soft" information can be utilized by the channel decoder to improve its error-correcting capability, as will be evident in the next chapter.

Since the input to the BPSK modulator is either a binary 0 or 1 and the output of the BPSK demodulator is a symbol from the set {0, 1, 2,

Binary digit transmitted	Output symbol	Condition	Forward transition probability
0	1	$x_h > 0$	$P(1\|0) = \int_0^\infty f(x_h\|0)dx_h$
1	0	$x_h \leq 0$	$P(0\|1) = \int_{-\infty}^0 f(x_h\|1)dx_h$
BSC crossover probability $\alpha = P(1\|0) = P(0\|1)$			

Table 7.4: BPSK forward transition probabilities.

3, 4, 5, 6, 7}, the communication channel can be modeled as a DMC. For example, the mathematical expression and corresponding shaded area under the appropriate PDF for the DMC transition probability $P(4|1)$ is highlighted in Fig. 7.35.

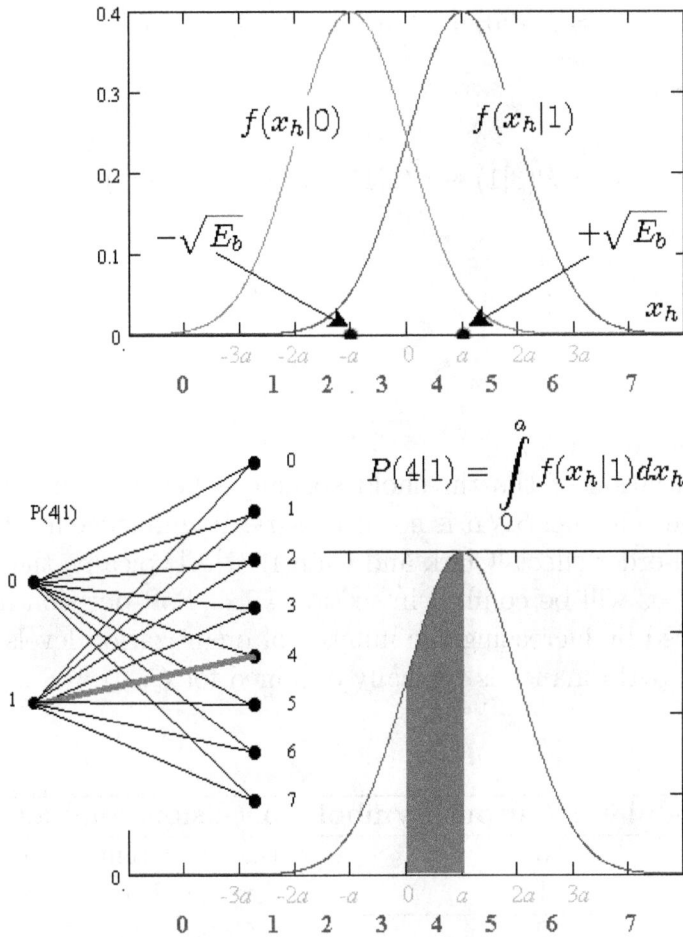

Figure 7.35: BPSK soft-decision demodulation.

Table 7.6 lists the remaining transition probabilities. By symmetry, $P(7|0) = P(0|1)$, $P(6|0) = P(1|1)$, $P(5|0) = P(2|1)$, $P(4|0) = P(3|1)$. Since the hard-decision of symbols 4, 5, 6, 7 is a binary digit 1, the BSC crossover probability

$$\alpha = P(4|0) + P(5|0) + P(6|0) + P(7|0) \tag{7.67}$$

or equivalently

$$\alpha = P(3|1) + P(2|1) + P(1|1) + P(0|1) \tag{7.68}$$

The optimization of the threshold spacing a between the quantization levels for a given channel SNR is not necessary. Its influence has been shown to be a second-order effect [Clark and Cain,1981]. Typically, the value for a is fixed at 2/7 as will be confirm in section 7.7.4. Soft-decision information can be improved by increasing the number of quantization levels. However, near optimum performance is typically obtained for $Q_{level} = 8$.

Demodulator Output Symbol	Decision confidence
0	Excellent binary digit 0
1	Very good binary digit 0
2	Good binary digit 0
3	Weak binary digit 0
4	Weak binary digit 1
5	Good binary digit 1
6	Very good binary digit 1
7	Excellent binary digit 1

Table 7.5: Soft symbols.

7.7.1 Example : Soft-Decision Transition Probability

Forward Transition Probability

- A detailed mathematical derivation of a forward transition probability via the area under the BPSK signal point PDF

The forward transition probability $P(4|0)$ is given by

$$P(4|0) = \int_0^a f(x_h|0)dx_h \tag{7.69}$$

where

$$f(x_h|0) = \frac{1}{\sigma\sqrt{2\pi}} \exp\left[-\frac{\left(x_h + \sqrt{E_b}\right)^2}{2\sigma^2}\right] \tag{7.70}$$

To integrate by substitution, let $z = \frac{\left(x_h+\sqrt{E_b}\right)}{\sigma}$. Thus $\frac{dz}{dx_h} = \frac{1}{\sigma}$ and for $x_h = 0$, $z = \frac{\sqrt{E_b}}{\sigma}$ and for $x_h = a$, $z = \frac{a+\sqrt{E_b}}{\sigma}$. Hence

$$P(4|0) = \int_{\frac{\sqrt{E_b}}{\sigma}}^{\frac{a+\sqrt{E_b}}{\sigma}} \frac{1}{\sqrt{2\pi}} \exp\left[-\frac{z^2}{2}\right] dz \tag{7.71}$$

and since $Q(u) = \int_u^\infty \frac{1}{\sqrt{2\pi}} \exp\left(-\frac{z^2}{2}\right) dz$,

$$P(4|0) = Q\left(\frac{\sqrt{E_b}}{\sigma}\right) - Q\left(\frac{a + \sqrt{E_b}}{\sigma}\right) \tag{7.72}$$

which is illustrated in Fig. 7.36.

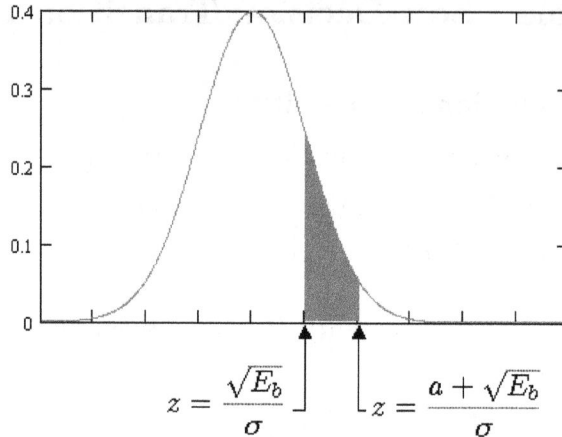

Figure 7.36: Q-function difference.

7.7.2   M-ary PSK with Soft-decision

▶ **M-ary Soft-Decision Decoding**

- Recap of hard and soft-decision for BPSK and its extension to infinite quantization level

- M-ary soft decision explained using the $M = 4$ QPSK signal constellation diagram

Incorporating soft-decision within M-ary PSK is simpler. Instead of partitioning the signal-space, Q_{level} is set to infinity i.e. the exact coordinates of the received signal point is output by the demodulator. This will be demonstrated in section 8.9, when we consider topic of the trellis-coded modulation (TCM).

7.7.3   Simulation I: Transition Probabilities

⭐ SIMULATION **SoftDecision:** The DCS shown in Fig. 7.37 is modeled in which the binary source outputs an all-zero sequence. The forward transition probabilities $P(0|0)$ to $P(7|0)$ are determined and compared with the theoretical values for a given channel *SNR* as shown in Fig. 7.38.

- **Experiment:** Change the threshold spacing a to observe its influence on these transition probabilities.

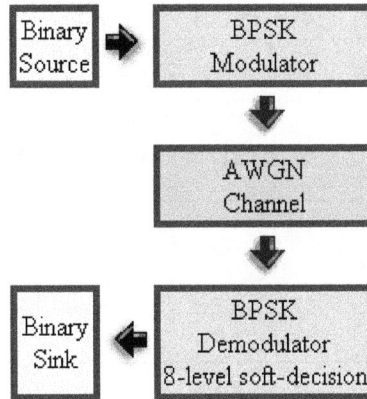

Figure 7.37: Simulated DCS : Focus on transition probabilities.

7.7.4 Simulation II: Optimum Spacing

▶ Optimum Threshold Spacing

- The influence of the quantization threshold spacing on the forward transition probabilities

- Analysis to determine the near optimum

- threshold spacing over low SNRs

⭐ SIMULATION **OptimumSpacing:** The expression $y(SNR, a) = \sum_{n=0}^{7} \sqrt{P(n|1)P(n|0)}$ [Clark and Cain,1981] is used to verify that $a = \frac{2}{7}$ provides near optimum performance over a range of $SNRs$. Figure 7.39 shows

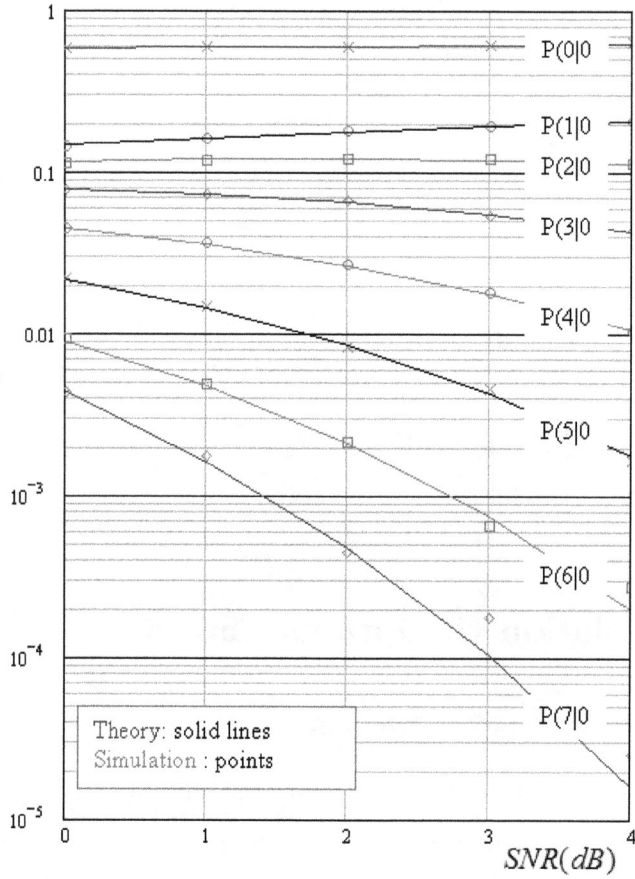

Figure 7.38: Forward transition probabilities for $a = \frac{2}{7}$.

the variation of $y(SNR, a)$ over the $SNRs$ 0 to 4 dB. The value of a for which $y(SNR, a)$ is minimized for a given SNR is shown in Fig. 7.40 together $y(SNR, a = 2/7)$. Clearly $a = 2/7 = 0.286$ is a reasonable value for optimum threshold spacing.

- **Experiment:** At high SNRs (e.g. 10 dB), what is the optimum value for a ?

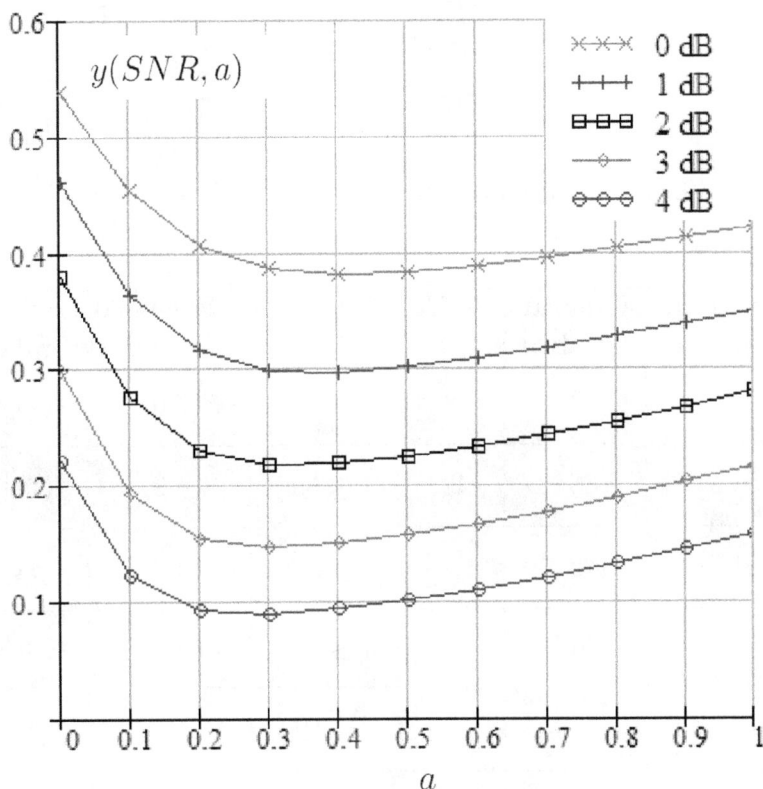

Figure 7.39: The variation of $y(SNR, a)$ with a for a range of $SNRs$.

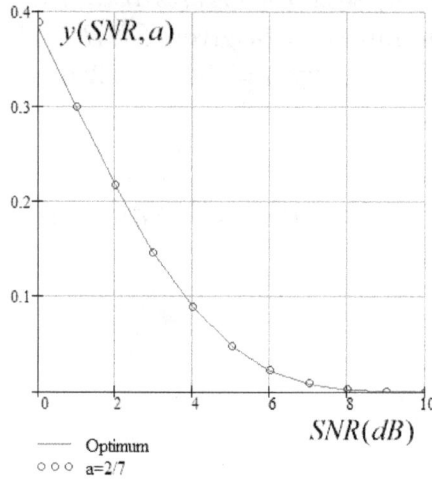

Figure 7.40: The minimum value of $y(SNR, a)$ versus $y(SNR, 2/7)$.

Binary digit transmitted	Output symbol	Condition	Forward transition probability
0	7	$x_h > 3a$	$P(7|0) = \int\limits_{3a}^{\infty} f(x_h|0)dx_h$
0	6	$2a < x_h \leq 3a$	$P(6|0) = \int\limits_{2a}^{3a} f(x_h|0)dx_h$
0	5	$a < x_h \leq 2a$	$P(5|0) = \int\limits_{a}^{2a} f(x_h|0)dx_h$
0	4	$0 < x_h \leq a$	$P(4|0) = \int\limits_{0}^{a} f(x_h|0)dx_h$
1	3	$-a < x_h \leq 0$	$P(3|1) = \int\limits_{-a}^{0} f(x_h|1)dx_h$
1	2	$-2a < x_h \leq -a$	$P(2|1) = \int\limits_{-2a}^{-a} f(x_h|1)dx_h$
1	1	$-3a < x_h \leq -2a$	$P(1|1) = \int\limits_{-3a}^{-2a} f(x_h|1)dx_h$
1	0	$x_h \leq -3a$	$P(0|1) = \int\limits_{-\infty}^{-3a} f(x_h|1)dx_h$

Table 7.6: BPSK with 8-level soft-decision.

7.8 Information Throughput

⊳ **Information Throughput**

- Analysis to show how the time duration of a
  MPSK signal and its energy depend on the code
  rate and the size of M for the constraint of no
  reduction in the information throughput

If error-control coding is incorporated within a DCS as illustrated in Fig. 7.41, then the binary digits which are transferred across the channel are code digits and **not** information digits. To ensure the information throughput is not reduced, we require the time space occupied by an MPSK signal which represent the n code digits, to be the same as that occupied by the k input digits to the channel encoder. Thus, if T_b is the time-space occupied by a binary digit and T_c is the time-space occupied by a code digit, so that $r_b = \frac{1}{T_b}$ and $r_c = \frac{1}{T_c}$, then we require

$$T_b k = n T_c \tag{7.73}$$

or equivalently,

$$T_c = \frac{k}{n} T_b = R_{code} T_b \tag{7.74}$$

The number of code digits per MPSK signal is equal to $\log_2 M$. Hence, the time-duration T_s of an MPSK signal is given by

$$T_s = T_c \log_2 M = R_{code} T_b \log_2 M \tag{7.75}$$

If the time duration of the carrier is reduced by a factor of R_{code}, whilst maintaining the same amplitude A and carrier frequency f_c, the energy of the signal must be reduced by a factor of R_{code} so that the energy per MPSK symbol is given by

$$E_s = R_{code} E_b \log_2 M \qquad\qquad (7.76)$$

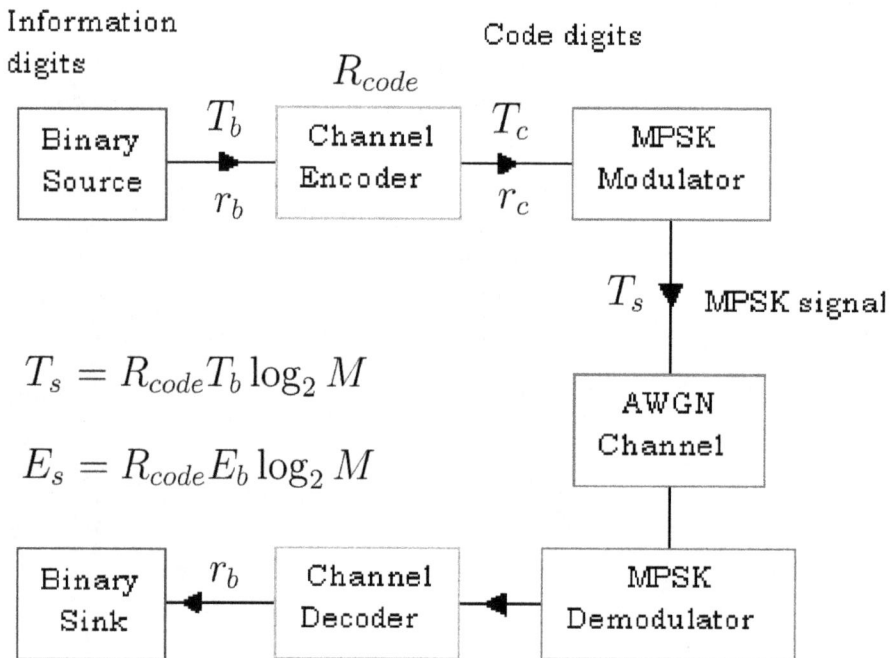

Information
digits   Code digits

R_{code}

| Binary Source | T_b | Channel Encoder | T_c | MPSK Modulator |

r_b r_c

T_s ▼ MPSK signal

AWGN
Channel

$$T_s = R_{code} T_b \log_2 M$$

$$E_s = R_{code} E_b \log_2 M$$

| Binary Sink | r_b | Channel Decoder | | MPSK Demodulator |

Figure 7.41: DCS with error-control coding.

▶ **Throughput Example**

- Time-space of three information binary digits entering a rate 1/2 channel encoder

- Time-space of the corresponding code digits

- Time-space of the signals transmitted via an 8PSK modulator

In section 7.6, it was shown that the MPSK signal constellation points lie on a circle of radius $\sqrt{E_s}$. However, with error-control coding, the signal constellation diagrams for $M = 2$ and 8 are presented in Figs. 7.42, 7.43, 7.44, 7.45 for $R_{code} = 1$ and 2/3. Notice how the signal points come closer together, indicating a reduction in signal energy, because the time duration of an MPSK signal has been reduced to maintain the information throughput. An interesting feature evident from Fig. 7.42 and 7.45, is that the energy of a uncoded QPSK signal is the same as an 8-PSK signal with $R_{code} = 2/3$, which also implies that the time duration of these two signals is the same. We shall refer to this important feature in section 8.10, when we consider trellis-coded modulation.

Eb = 1

Radius of the circle = $\sqrt{Es}$ = 1.414

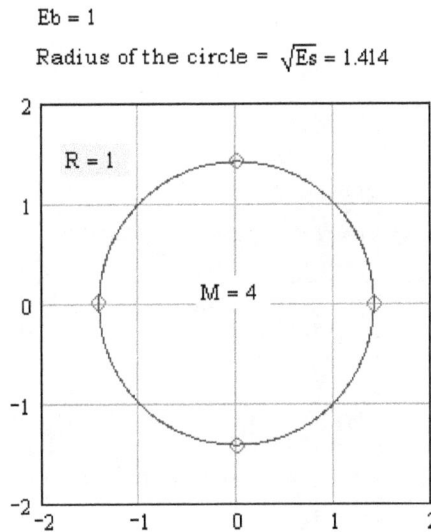

Figure 7.42: $M = 4$ and $R_{code} = 1$.

- ⭐ SIMULATION **MpskErrorControl:** To determine the M-ary PSK signal constellation diagram with error-control coding.

Eb = 1

Radius of the circle = $\sqrt{Es}$ = 1.155

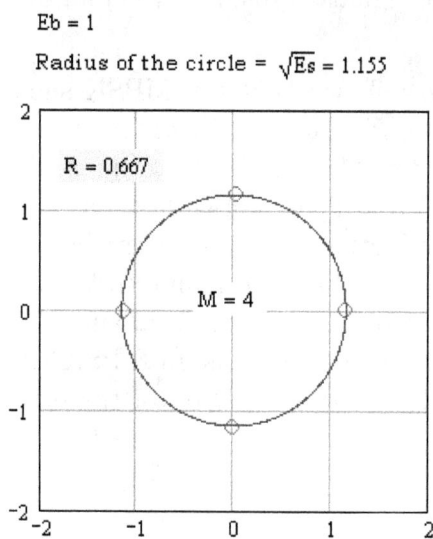

R = 0.667

M = 4

Figure 7.43: $M = 4$ and $R_{code} = 2/3$.

Eb = 1

Radius of the circle = $\sqrt{Es}$ = 1.732

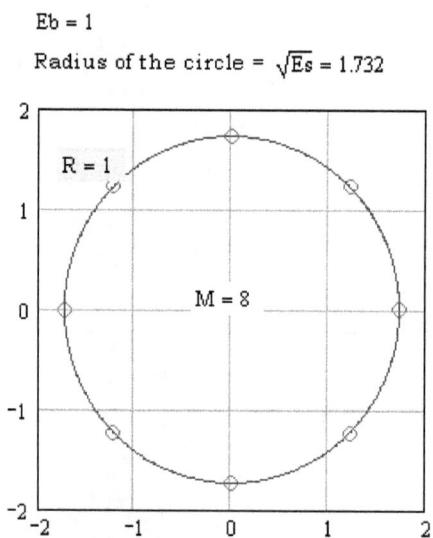

R = 1

M = 8

Figure 7.44: $M = 8$ and $R_{code} = 1$.

Eb = 1

Radius of the circle = $\sqrt{Es}$ = 1.414

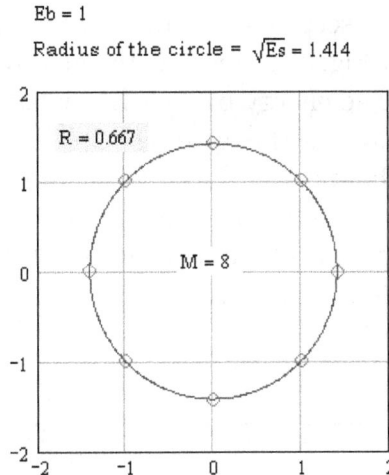

Figure 7.45: $M = 8$ and $R_{code} = 2/3$.

7.8.1 BPSK with Error-Control Coding

BPSK with Error-Control Coding

- The influence of error-control coding on the
  BPSK performance is demonstrated via the
  BPSK signal constellation diagram

Using BPSK to transfer code digits across an AWGN channel with no reduction in the information throughput, the energy per symbol from equation 7.76 $E_s = R_{code} E_b \log_2 2 = R_{code} E_b$. The signal points under a noiseless environment are now located at $\pm\sqrt{E_s} = \sqrt{R_{code} E_b}$, so that the probability of an error P_e at the binary sink is now given by

$$P_e = Q\left(\sqrt{\frac{2E_s}{N_o}}\right) = Q\left(\sqrt{\frac{2E_b R_{code}}{N_o}}\right) \tag{7.77}$$

As expected, this expression reverts back to the uncoded BPSK performance equation for $R_{code} = 1$. An interesting consequence of using error-control coding is that if the energy of a BPSK signal is reduced below E_b, then the carrier is more susceptible to noise within the AWGN channel because the signal points in Fig. 7.4, now located at $\sqrt{R_{code}E_b}$, will move closer together for $R_{code} < 1$. The energy of each BPSK signal $E_s = R_{code}E_b$ is reduced by a factor of R_{code}. In this case, the BPSK demodulator will make more errors for a given channel SNR as evident from Fig. 7.46. This is not a point of concern if the combination of the channel encoder and decoder can reduce the probability of an error in a binary digit P_e at the output of the decoder below the uncoded BPSK P_e performance.

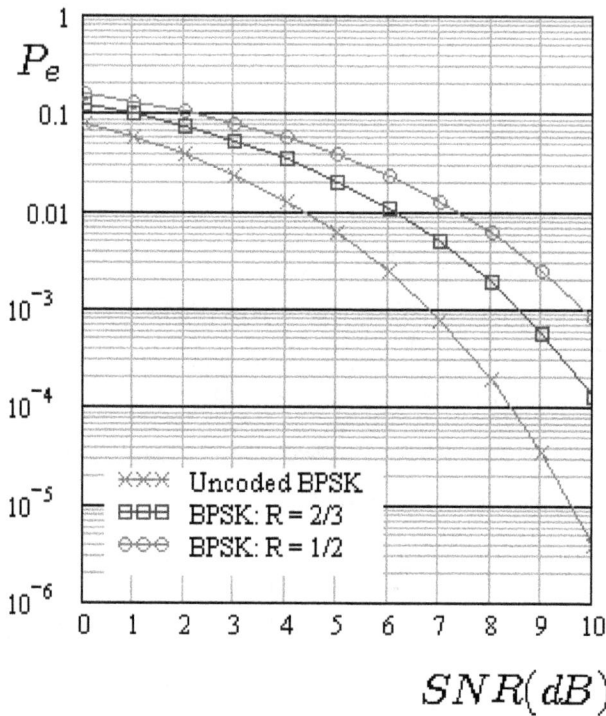

Figure 7.46: Influence of error-control coding on the BPSK performance.

- ⭐ SIMULATION **BpskErrorControl:** Simulation corresponding to Fig. 7.46.

7.8.2   Simulation : Repetition Code II

⭐ SIMULATION **RepetitionCodeII**: The simulation undertaken in section 3.3 is repeated for a rate 1/3 repetition code but now with the information throughout taken into account i.e. the information throughput is not allowed to be reduced with the use of error-control coding. The simulation results are presented in Fig. 7.47. In the soft-decision case, the coordinate $x_h = \int_0^{T_b} r(t)\phi_1(t)dt$ (from section 7.2) for each code digit is added. If the sum is greater than zero, than the decoder outputs a binary digit 1, otherwise a binary digit 0. Notice that the repetition code is not able to improve the performance below uncoded BPSK even with the use of soft-decision. This is not surprising because with no reduction in the information throughput, uncoded BPSK and soft-decision repetition code decoding are equivalent i.e. with soft-decision decoding, the code rate can be set to $1/n$, where n is 1, 2, 3, 4, etc with no improvement over BPSK. Refer to the simulation code for the hard and soft-decision performance results for the scenario where the information throughput is allowed to be reduced.

- **Experiment:** Modify the simulation code to allow the reduction of information throughput.

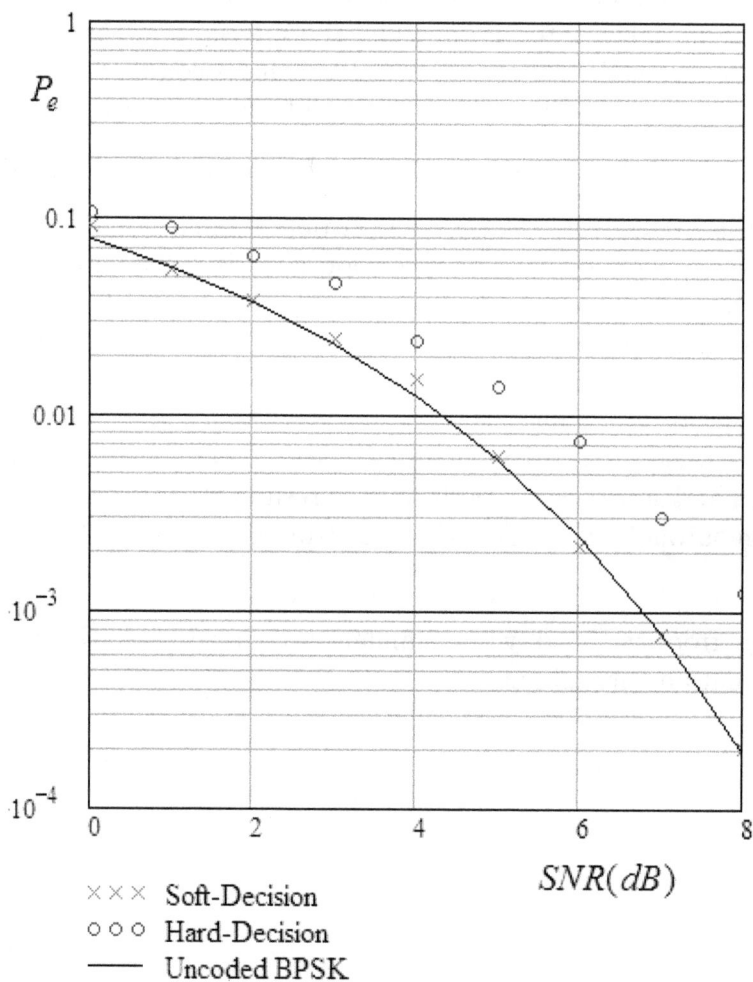

Figure 7.47: Rate 1/3 hard and soft-decision repetition code performance with no reduction in information throughput.

7.9 Quadrature Amplitude Modulation (QAM)

7.9.1 Signal Constellation

▶ **QAM**

- Fundamentals

- 16-QAM signal constellation

Recall that in M-ary PSK, the signal points are distributed on a circle of radius $\sqrt{E_s}$, where E_s is the energy per symbol given by $E_s = E_b \log_2 M$. In *quadrature amplitude modulation* (QAM), the signal points are no longer constrained to lie on a circle. Each signal $s_i(t)$ of duration $T_s = T_b \log_2 M$ is given by

$$s_i(t) = \sqrt{\frac{2}{T_s}} A \left[a_i \cos(2\pi f_c t) + b_i \sin(2\pi f_c t) \right] \qquad (7.78)$$

where $i = 0, 1, \cdots M - 1$, f_c is the carrier frequency and $A = \sqrt{\frac{3E_b \log_2 M}{2(M-1)}}$ in which E_b is the **average** energy per binary digit. The amplitudes a_i and b_i are coordinates of the signal points selected to create for example, a *square* constellation by ensuring that both a_i and b_i are from the set $\in \left\{ \pm 1, \pm 3, \cdots, \pm \left(\sqrt{M} - 1 \right) \right\}$.

▶ **QAM and QPSK**

- Analysis of 4-QAM to show that its equivalent to QPSK

Notice that for $M = 4$, QAM and 4-PSK (i.e. QPSK) are equivalent. Indeed, the basis functions $\phi_1(t)$ and $\phi_2(t)$ are the same as for M-ary PSK i.e.

$$\phi_1(t) = \sqrt{\frac{2}{T_s}} \cos\left(2\pi f_c t\right) \text{ and } \phi_2(t) = \sqrt{\frac{2}{T_s}} \sin\left(2\pi f_c t\right) \qquad (7.79)$$

and the signal $s_i(t)$ is equivalently given by

$$s_i(t) = Aa_i\phi_1(t) + Ab_i\phi_2(t) \tag{7.80}$$

▶ QAM Signal Energy

- The energy of a QAM signal is related to
  its constellation diagram

The energy of a QAM signal E_s is given by

$$E_s = \int_0^{T_s} s_i^2(t)dt \tag{7.81}$$

$$= \int_0^{T_s} \left(Aa_i\phi_1(t) + Ab_i\phi_2(t)\right) \left(Aa_i\phi_1(t) + Ab_i\phi_2(t)\right) dt \tag{7.82}$$

$$= (Aa_i)^2 \int_0^{T_s} \phi_1^2(t)dt + 2A^2 a_i b_i \int_0^{T_s} \phi_1(t)\phi_2(t)dt + (Ab_i)^2 \int_0^{T_s} \phi_2^2(t)dt \tag{7.83}$$

Given that the basis functions have unit energy so that $\int_0^{T_s} \phi_1^2(t)dt =$ $\int_0^{T_s} \phi_2^2(t)dt = 1$ and they are orthogonal $\int_0^{T_s} \phi_1(t)\phi_2(t)dt = 0$, we find

$$E_s = A^2 \left(a_i^2 + b_i^2\right) \tag{7.84}$$

Notice that unlike MPSK, not all the signals have the same energy. The average energy per symbol $E_{av} = E_b \log_2 M$ is given by

$$E_{av} = \frac{1}{M} \sum_{i=0}^{M-1} A^2 \left(a_i^2 + b_i^2 \right) = \frac{A^2 2 \left(M - 1 \right)}{3} \tag{7.85}$$

As before, the horizontal coordinate x_h and the vertical coordinate x_v on the signal constellation (or space) diagram are respectively given by

$$x_h = \int_0^{T_s} r(t) \phi_1(t) dt \tag{7.86}$$

and

$$x_v = \int_0^{T_s} r(t) \phi_2(t) dt \tag{7.87}$$

where $r(t)$ is the received signal. For a noiseless channel,

$$x_h = \int_0^{T_s} s_i(t) \phi_1(t) dt = A a_i \int_0^{T_s} \phi_1^2(t) dt + A b_i \int_0^{T_s} \phi_1(t) \phi_2(t) dt \tag{7.88}$$

$$x_v = \int_0^{T_s} s_i(t) \phi_2(t) dt = A a_i \int_0^{T_s} \phi_1(t) \phi_2(t) dt + A b_i \int_0^{T_s} \phi_2^2(t) dt \tag{7.89}$$

Since $\phi_1(t)$ and $\phi_2(t)$ are orthogonal and the energy content of these basis functions is 1 joule,

$$x_h = A a_i \tag{7.90}$$

$$x_v = Ab_i \tag{7.91}$$

For example, for $M = 4$ for which $i = 0, 1, 2, 3$ and $a = \begin{bmatrix} -1 \\ -1 \\ 1 \\ 1 \end{bmatrix}$, $b =$

$\begin{bmatrix} -1 \\ 1 \\ -1 \\ 1 \end{bmatrix}$, $A = \sqrt{\frac{3E_b \log_2 4}{2(4-1)}} = \sqrt{E_b}$ and for any signal i, the energy per symbol

$$E_s = A^2 \left(a_i^2 + b_i^2 \right) = E_b \left(1 + 1 \right) = 2E_b \tag{7.92}$$

as expected from equation 7.50. However for 16-QAM with $A = 1$, the amplitudes a_i and b_i corresponding to the signal constellation diagram in Fig. 7.48 are given by

$$a = \begin{bmatrix} -3 \\ -3 \\ -3 \\ -3 \\ -1 \\ -1 \\ -1 \\ -1 \\ 3 \\ 3 \\ 3 \\ 3 \\ 1 \\ 1 \\ 1 \\ 1 \end{bmatrix} \quad b = \begin{bmatrix} -3 \\ -1 \\ 3 \\ 1 \\ -3 \\ -1 \\ 3 \\ 1 \\ -3 \\ -1 \\ 3 \\ 1 \\ -3 \\ -1 \\ 3 \\ 1 \end{bmatrix} \tag{7.93}$$

where $(a_0, b_0) = (-3, -3)$, $(a_1, b_1) = (-3, -1)$, $\cdots$, $(a_{15}, b_{15}) = (1, 1)$. The signal points labeled I, II and III in Fig. 7.48 have the coordinates (A, A), $(3A, A)$ and $(3A, 3A)$, respectively. The average energy per symbol $E_{av} = 10$ Joules and since each symbol represents $\log_2 M = 4$ binary digits, the average energy per binary digit E_b is thus $\frac{10}{4} = \frac{5}{2}$ Joules.

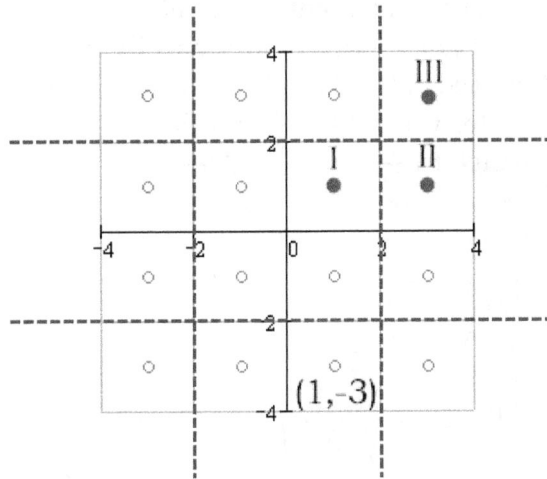

Figure 7.48: QAM signal constellation for $M = 16$ and $E_b = 15/6$ J.

- ⭐ SIMULATION **QAMConstellation:** To determine the signal constellation diagram in Fig. 7.48 and to view the corresponding signals. Experiment with different values of E_b and observe the influence on the signal point spacing.

To determine an expression for the probability of a symbol error, consider the signal point labeled I in Fig. 7.48. Notice there are three other signal points of this type with coordinates $(-A, A)$, $(-A, -A)$ and $(A, -A)$. Given that the probability density function of the noise component on a given signal point is a zero mean gaussian function $\frac{1}{\sigma\sqrt{2\pi}} \exp\left(\frac{-x^2}{2\sigma^2}\right)$ with standard deviation $\sigma = \sqrt{\frac{N_o}{2}}$, the probability that a signal point of type I will be correctly decoded P_I is given by

$$P_I = \left[\int_{-A}^{A} \frac{1}{\sigma\sqrt{2\pi}} \exp\left(\frac{-x^2}{2\sigma^2}\right) dx\right] \left[\int_{-A}^{A} \frac{1}{\sigma\sqrt{2\pi}} \exp\left(\frac{-x^2}{2\sigma^2}\right) dx\right] \qquad (7.94)$$

because we require the joint event of the noise component to lie within the range $-A$ to A both horizontally and vertically, to ensure a signal point of this type is not moved into a neighboring quadrant. Integrating via substitution, let $u = \frac{x}{\sigma}$ for which $\frac{du}{dx} = \frac{1}{\sigma}$ and the lower and upper integration limits respectively change to $\frac{-A}{\sigma}$ and $\frac{A}{\sigma}$. Thus

$$\begin{aligned} P_I &= \left[Q\left(\frac{-A}{\sigma}\right) - Q\left(\frac{A}{\sigma}\right)\right]^2 = \left[1 - 2Q\left(\frac{A}{\sigma}\right)\right]^2 \qquad (7.95) \\ &= \left[1 - 2Q\left(\sqrt{\frac{2A^2}{N_o}}\right)\right]^2 \qquad (7.96) \end{aligned}$$

Similarly, the probability that a signal point of type II will be correctly decoded P_{II} is given by

$$P_{II} = \left[\int_{-A}^{A} \frac{1}{\sigma\sqrt{2\pi}} \exp\left(\frac{-x^2}{2\sigma^2}\right) dx\right] \left[\int_{-A}^{\infty} \frac{1}{\sigma\sqrt{2\pi}} \exp\left(\frac{-x^2}{2\sigma^2}\right) dx\right] \qquad (7.97)$$

$$= \left[1 - 2Q\left(\sqrt{\frac{2A^2}{N_o}}\right)\right]\left[1 - Q\left(\sqrt{\frac{2A^2}{N_o}}\right)\right] \qquad (7.98)$$

Finally, the probability P_{III} is given by

$$P_{III} = \left[\int_{-A}^{\infty} \frac{1}{\sigma\sqrt{2\pi}} \exp\left(\frac{-x^2}{2\sigma^2}\right) dx\right] \left[\int_{-A}^{\infty} \frac{1}{\sigma\sqrt{2\pi}} \exp\left(\frac{-x^2}{2\sigma^2}\right) dx\right] \qquad (7.99)$$

$$= \left[1 - Q\left(\sqrt{\frac{2A^2}{N_o}}\right)\right]^2 \qquad (7.100)$$

For 16-QAM, given that there are 4 signal points of type I, 4 of type III and 8 of type II, the average probability of a symbol error P_s is given by

$$P_s = 1 - \frac{1}{16}\left[4P_I + 8P_{II} + 4P_{III}\right] \qquad (7.101)$$

In general, there will be $\left(\sqrt{M} - 2\right)^2$ signal points of type I, $4\left(\sqrt{M} - 2\right)$ of type II and 4 type III signal points. In this case, the probability of a symbol error P_s is given by

$$P_s = 1 - \frac{1}{M}\left[\left(\sqrt{M} - 2\right)^2 P_I + 4\left(\sqrt{M} - 2\right)P_{II} + 4P_{III}\right] \qquad (7.102)$$

At high SNRs, it is easily shown that this approximates to

$$P_s \approx 4\left(1 - \frac{1}{\sqrt{M}}\right)Q\left(\sqrt{\frac{2A^2}{N_o}}\right) \qquad (7.103)$$

With Gray coding, the probability of an error in a binary digit $P_e = \frac{P_s}{\log_2 M}$ because there is only a single error in a binary digit if a signal point is incorrectly demodulated as one of its neighbors.

▷ QAM Performance

- The probability of a symbol error versus the channel SNR ratio is analyzed for M-ary QAM

- 16-QAM compared with 16-PSK

- Coding gain of QAM over MPSK

Using equation 7.103 for P_s, Fig. 7.49 shows the uncoded performance we can expect for various values of M over an AWGN channel. With reference to DSL mentioned in section 6.1, the ADSL variation G.DMT utilizes the bandwidth above the 4 kHz voice band to provide data rates of 6 Mb/s (download) and 640 kb/s (upload) using 256 carriers with 15 binary digits per carrier via 32768 QAM [Couch, 2001].

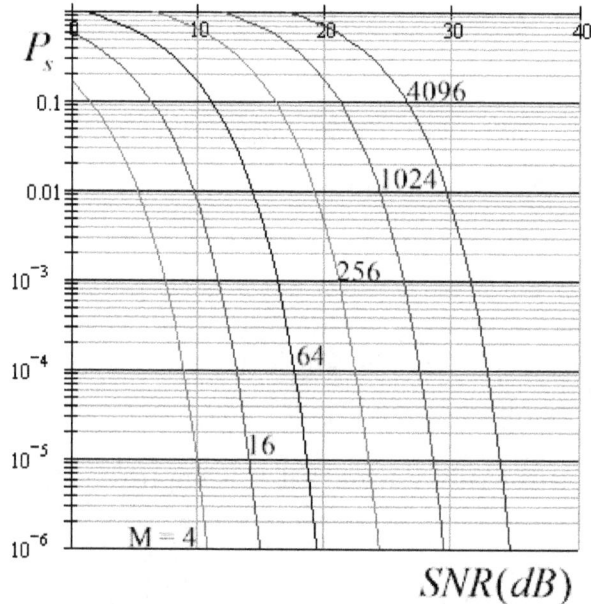

Figure 7.49: Uncoded QAM P_s performance over an AWGN channel.

A comparison of the PSK and QAM performance for $M = 16$ is shown in Fig. 7.50. Notice that 16-QAM outperforms 16-PSK with a coding gain of approximately 4 dB at $P_s = 10^{-6}$. For a given channel SNR, a comparison of the distance between neighboring signal points for QAM and MPSK for $M > 4$ reveals that the MPSK points are closer together (refer to the problems section). Thus not surprisingly, the performance of QAM is better than MPSK over an ideal AWGN channel. However, unlike a PSK signal, a QAM signal envelope is not constant and therefore would not be used over a non-linear channel.

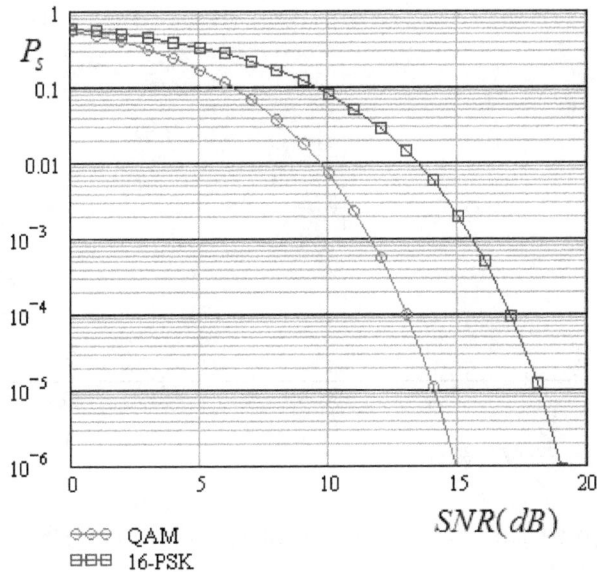

Figure 7.50: $M = 16$ PSK and QAM P_s performance comparison.

Over a range of M values, Fig. 7.51 shows the coding gain with respect to both the symbol and binary digit error probability of QAM over MPSK. As expected for $M = 4$, QAM and 4-PSK (i.e. QPSK) are equivalent and the coding gain is zero.

- ⭐ SIMULATION **QAMPSKComparison:** Simulation corresponding to Figs. 7.50 and 7.51.

7.9.2 Simulation : QAM

⭐ SIMULATION **QAM:** The $M = 16$ QAM modulation scheme presented in Fig. 7.48 is simulated over an AWGN channel. The corresponding simulation results are presented in Fig. 7.52.

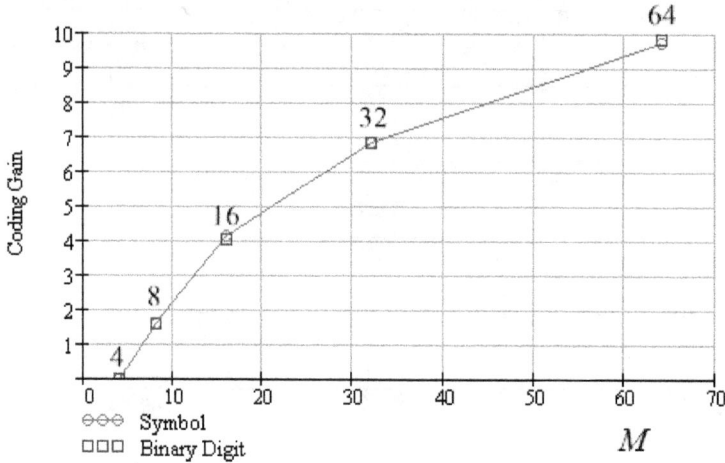

Figure 7.51: Coding gain of QAM over MPSK.

⭐ SIMULATION **FourQAM:** The performance of both 4-QAM and QPSK (or equivalently BPSK) are shown to be the same as shown in Fig. 7.53.

- **Experiment:** Modify the simulation code to investigate the performance without the use of Gray coding. Implement $M = 64$ QAM with Gray coding.

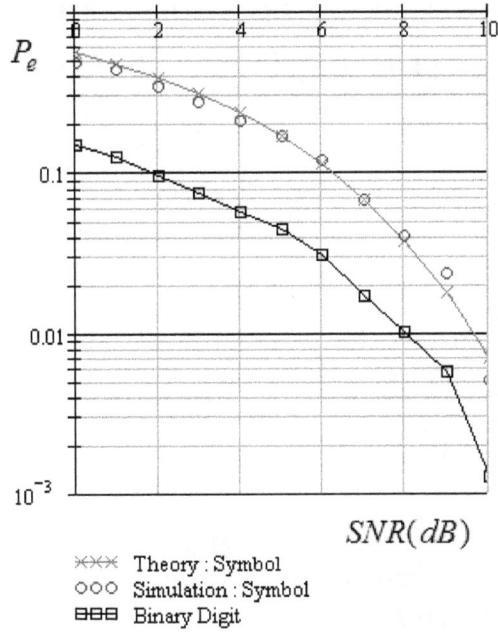

Figure 7.52: Symbol and binary digit error performance for $M = 16$ QAM.

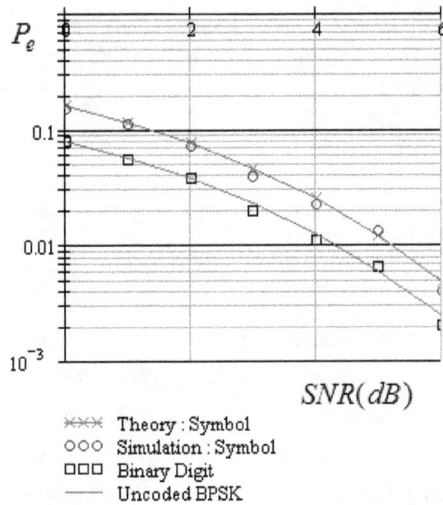

Figure 7.53: QAM Symbol and binary digit error performance for $M = 4$ compared with uncoded BPSK.

7.10   Orthogonal Frequency Division Multiplexing (OFDM)

▷ OFDM Fundamentals

- Fundamentals

- OFDM concept

- Complex representation of signal points

- OFDM signal

We shall now consider OFDM because it makes use of QAM, even though this modulation scheme falls under the category of M-ary Frequency Shift Keying (MFSK) to be covered later in section 7.12. Up to this point, we have considered the use of a single modulated carrier of frequency f_c to transfer binary digits over an AWGN channel. We shall now consider the use of multiple carriers to transfer binary digits simultaneously over a given channel as mentioned in section 5.4.2 in light of ADSL, which currently provides broadband internet access into most homes over the last copper mile. Specifically *orthogonal frequency division multiplexing* (OFDM), in which the available channel bandwidth W is split into K subchannels, such that width of each subchannel Δf is given by

$$\Delta f = \frac{W}{K} \tag{7.104}$$

Now suppose we assign a subcarrier $\cos(2\pi f_k t)$ to each subchannel, where $k = 0, 1, \cdots K-1$. Then we could transfer **simultaneously**, K signal points over the channel by modulating each subcarrier. These K subcarriers could then be combined to create an OFDM *multicarrier* signal $v(t)$. For example, if M-ary quadrature amplitude modulation (QAM) or M-ary phase shift keying (MPSK) is used for each subcarrier, then each subcarrier would have to be correlated with two basis functions $\phi_1(t) = \sqrt{\frac{2}{T_s}}\cos(2\pi f_k t)$ and $\phi_2(t) = \sqrt{\frac{2}{T_s}}\sin(2\pi f_k t)$ to demodulate the associated signal point. Unfortunately, this is very difficult to implement in practice for a large value of K. It was only recently [Bingham, 1990] that a very clever and simple solution was found to reduce the complexity of the modulator and demodulator as follows. The first step is to ensure that adjacent subcarrier frequencies are separated by $\Delta f = \frac{1}{T}$, where $T = KT_s$ is the OFDM signal duration. Notice that the time duration of the subcarrier signal is increased from T_s to T. In this case, the subcarriers will be orthogonal to each other over the symbol interval T, namely

$$\int_0^T \cos\left(2\pi f_k t + \phi_k\right) \cos\left(2\pi f_j t + \phi_j\right) dt = 0 \qquad (7.105)$$

where f_k is the mid-frequency in the kth subchannel and $f_k - f_j = \frac{n}{T}$, $n = 1, 2, ...$ By ensuring the subcarriers are orthogonal, the subcarriers can overlap, which reduces the overall OFDM system bandwidth and removes the need for sharp filters.

- ⭐ SIMULATION **OrthoSubcarriers:** You may select any two subcarriers and verify that they are orthogonal, irrespective of their phase.

▶ **OFDM Example**

- Four QPSK signal points are transmitted simultaneously to create a complex OFDM signal

- Samples of the OFDM signal are determined via the inverse discrete Fourier transform

- The four signal points are demodulated via the discrete Fourier transform

The next step is to represent the signal point associated with each sub-carrier in complex form. For simplicity, consider the parallel transmission of K signal points via QPSK without the use of error-control coding with $E_b = 1$ joule as summarized in Table 7.7. The four signal points lie on a circle of radius $\sqrt{E_s} = \sqrt{2E_b} = \sqrt{2}$. The *complex signal point* $X = re^{j\theta}$ associated with each signal point is also listed in this table. Comparing this table with the QPSK signal constellation diagram in Fig. 7.16, we note that X is simply the coordinate of the signal point in which we take the horizontal axis to be the real axis and the vertical axis to be the imaginary axis, where $j = \sqrt{-1}$.

For the parallel transmission of K signal points, the corresponding complex OFDM *baseband* multicarrier signal $g(t)$ is given by

$$g(t) = \sum_{k=0}^{K-1} X_k e^{j\frac{2\pi kt}{T}} = \sum_{k=0}^{K-1} r_k e^{j\left(\frac{2\pi kt}{T} + \theta_k\right)} \tag{7.106}$$

and the corresponding physical baseband and bandpass OFDM multicarrier signals are given by

$$v_{baseband}(t) = \operatorname{Re}\left(g(t)\right) = \sum_{k=0}^{K-1} r_k \cos\left(\frac{2\pi kt}{T} + \theta_k\right) \tag{7.107}$$

$$v_{bandpass}(t) = \operatorname{Re}\left(g(t)e^{j2\pi f_{shift}t}\right) = \sum_{k=0}^{K-1} r_k \cos\left[2\pi\left(\frac{k}{T} + f_{shift}\right)t + \theta_k\right] \tag{7.108}$$

Signal point		r	θ (deg)	$X = re^{j\theta}$
0	00	$\sqrt{2}$	225	$\sqrt{2}e^{j\frac{225\pi}{180}} = -1 - j$
1	01	$\sqrt{2}$	315	$\sqrt{2}e^{j\frac{315\pi}{180}} = 1 - j$
2	10	$\sqrt{2}$	135	$\sqrt{2}e^{j\frac{135\pi}{180}} = -1 + j$
3	11	$\sqrt{2}$	45	$\sqrt{2}e^{j\frac{45\pi}{180}} = 1 + j$

Table 7.7: QPSK complex signal points.

where r_k and θ_k will correspond to the signal point sent on a given sub-carrier and the frequency f_{shift} in equation 7.108 is the OFDM carrier frequency shift of the OFDM baseband signal and **not** the subcarrier frequency $\frac{k}{T}$. For example, if the signal points 11220, corresponding to the binary stream 0101101000, are to be sent in parallel, then $K = 5$ and the complex OFDM baseband multicarrier signal $g(t)$ is given by

$$g(t) = \sum_{k=0}^{K-1} X_k e^{j\frac{2\pi kt}{T}} = X_0 + X_1 e^{j\frac{2\pi t}{T}} + X_2 e^{j\frac{4\pi t}{T}} + X_3 e^{j\frac{6\pi t}{T}} + X_4 e^{j\frac{8\pi t}{T}} \quad (7.109)$$

Note that there are now $K = 5$ identical QPSK signal constellations (each operating at a different subcarrier frequency given by $\frac{k}{T}$) and in this case, X_0 and X_1 correspond to the signal point 1, X_2 and X_3 correspond to the signal point 2 and X_4 represents the signal point 1. Thus, $X_0 = X_1 = 1 - j$, $\theta_0 = \theta_1 = \frac{315\pi}{180}$, $X_2 = X_3 = -1 + j$, $\theta_2 = \theta_3 = \frac{135\pi}{180}$, $X_4 = -1 - j$, $\theta_4 = \frac{225\pi}{180}$. We could find the Re $(g(t))$ to determine $v_{baseband}(t)$, or alternatively, using equation 7.107

$$v_{baseband}(t) = \text{Re}\,(g(t)) = \sqrt{2}\cos\left(\frac{2\pi(0)t}{T} + \frac{315\pi}{180}\right) + \sqrt{2}\cos\left(\frac{2\pi(1)t}{T} + \frac{315\pi}{180}\right)$$

$$+\sqrt{2}\cos\left(\frac{2\pi(2)t}{T} + \frac{135\pi}{180}\right) + \sqrt{2}\cos\left(\frac{2\pi(3)t}{T} + \frac{135\pi}{180}\right) + \sqrt{2}\cos\left(\frac{2\pi(4)t}{T} + \frac{225\pi}{180}\right)$$

$$(7.110)$$

To generate the OFDM multicarrier signal without actually implementing equation 7.107, the clever solution is to make use of the inverse discrete Fourier Transform (IDFT), equation 1.71, that

$$x_n = \frac{1}{\sqrt{N}} \sum_{k=0}^{N-1} X_k e^{j\frac{(2\pi nk)}{N}} \quad \text{for } n = 0, 1, \cdots N - 1 \quad (7.111)$$

where $\frac{1}{\sqrt{N}}$ is a scale factor.   Comparing equation 7.107 with equation 7.111, we find that for $t = \frac{nT}{N}$ and $N = K$

$$g(t = \frac{nT}{N}) = \sum_{k=0}^{K-1} X_k e^{j\frac{2\pi knT}{T(N)}} = x_n \sqrt{N} \qquad (7.112)$$

Notice that the $\text{Re}(x_n\sqrt{N})$ for $n = 0, 1, \cdots N-1$ are the *samples* of the physical waveform $v_{baseband}(t)$ and that the sequence $\{x_n\}$ to be transferred over the channel is **complex**. Before we consider how to overcome this problem, assume that $\{x_n\}$ is transferred over the channel and let $\{y_n\}$ denote the received sample sequence. Then the received complex signal point sequence $\{Y_k\}$ can be recovered from $\{y_n\}$ via the discrete Fourier Transform (DFT), equation 1.70, as follows

$$Y_k = \frac{1}{\sqrt{N}} \sum_{n=0}^{N-1} y_n e^{-j\frac{(2\pi nk)}{N}} \qquad (7.113)$$

- ⭐ SIMULATION **OFDMqpsk:** The foregoing analysis is illustrated for the specific example in which the OFDM signal represents the simultaneous transfer of $K = 5$ signal points 11220.

▶ OFDM

- Method to create a real sampled sequence

  is explained by sending four QPSK signal points

- The OFDM baseband signal corresponding to the

  real samples is created and confirmed to be equivalent

  to adding the individual four QPSK signals

To make the sequence $\{x_n\}$ real, we begin by creating a new sequence $\{X'_k\}$ with $N = 2\,(K+1)$ values as illustrated below for $K = 5$.

Index	k =	0	1	2	3	4	5	6	7	8	9	10	11
Value	X'_k =	0	X_0	X_1	X_2	X_3	X_4	0	X_4^*	X_3^*	X_2^*	X_1^*	X_0^*

The symmetry of $\{X'_k\}$ with

$$X'_{N-1-k} = X_k^* \text{ for } k = 0, 1, \cdots, K-1 \tag{7.114}$$

and

$$X'_0 = X'_{K+1} = 0 \tag{7.115}$$

is necessary to ensure that the samples x_n are real values and that we can recover all of the K signal points. Notice that we have now doubled the numbers of samples of the physical waveform $v_{baseband}(t)$. Consider the IDFT of this new sequence $\{X'_k\}$ given by

$$x_n = \frac{1}{\sqrt{N}} \sum_{k=0}^{N-1} X'_k e^{j\frac{(2\pi nk)}{N}} \tag{7.116}$$

$$
\begin{aligned}
x_n\sqrt{N} &= X'_0 + X'_1 e^{j\frac{(2\pi n(1))}{N}} + \cdots + X'_{N-1} e^{j\frac{(2\pi n(N-1))}{N}} \tag{7.117} \\
&= \sum_{k=0}^{K-1} X_k e^{j\frac{(2\pi n(k+1))}{N}} + \sum_{k=0}^{K-1} X_k^* e^{-j\frac{(2\pi n(k+1))}{N}} \tag{7.118} \\
&= \sum_{k=0}^{K-1} r_k \left[e^{j\left[\frac{(2\pi n(k+1))}{N}+\theta_k\right]} + e^{-j\left[\frac{(2\pi n(k+1))}{N}+\theta_k\right]} \right] \tag{7.119} \\
&= \sum_{k=0}^{K-1} 2r_k \cos\left(\frac{(2\pi n(k+1))}{N} + \theta_k \right) \tag{7.120}
\end{aligned}
$$

Thus

$$x_n \frac{\sqrt{N}}{2} = \sum_{k=0}^{K-1} r_k \cos\left(\frac{(2\pi n(k+1))}{N} + \theta_k\right) \tag{7.121}$$

$$v_{baseband}\left(t = \frac{nT}{N}\right) = \text{Re}\left(g'(t)\right) = x_n \frac{\sqrt{N}}{2} \tag{7.122}$$

for

$$g'(t) = \sum_{k=0}^{K-1} X_k e^{j\frac{2\pi(k+1)t}{T}} \tag{7.123}$$

and the OFDM signal of duration T now has subcarriers of frequency $\frac{(k+1)}{T}$Hz. The real sequence $\{x_n\}$ can be transferred over the channel using the combination of a digital-to-analog (D/A) convertor at the transmitter and an analog-to-digital (A/D) converter at the receiver to recover the samples $\{y_n\}$. Finally, the received complex signal point sequence $\{Y_k\}$ is determined via the DFT of $\{y_n\}$ via

$$Y_k = \frac{1}{\sqrt{N}} \sum_{n=0}^{N-1} y_n e^{-j\frac{(2\pi nk)}{N}} \tag{7.124}$$

The slight difference is that now $k = 0, 1, \cdots, K$ and the demodulated complex points are within $Y_1, Y_2, \cdots, Y_K$ with $Y_0 = 0$.

- ⭐ SIMULATION **OFDMqpsk2:** The new sequence $\{X'_k\}$ is used to determine the real valued samples of $v_{baseband}(t)$. Notice that the frequency of the subcarriers are now $\frac{1}{T}, \frac{2}{T}, \frac{3}{T}, \frac{4}{T}, \frac{5}{T}$ and therefore, the shape of $v_{baseband}(t)$ will be slightly different to that in ⭐ SIMULATION **OFDMqpsk.**

7.10.1   OFDM Based on QAM

In practical applications such as ADSL and wireless LANs, M-ary QAM (refer to section 7.9) is used for each subcarrier signal $s_i(t)$ given by

$$s_i(t) = \sqrt{\frac{2}{T_s}} A \left[a_i \cos(2\pi f_k t) + b_i \sin(2\pi f_k t) \right] \qquad (7.125)$$

where $i = 0, 1, \cdots M-1$, f_k is the subcarrier frequency and the amplitudes a_i and b_i are coordinates of the signal points.   Note here that i represents a signal point. Ensuring $A = 1$, the complex representation X_k of a signal point is once again given by $X_k = r_k e^{j\theta_k}$, where k is an index variable ranging from 0 to $K - 1$, r_k is the distance of the signal point i from the origin given by $\sqrt{a_i^2 + b_i^2} = \sqrt{E_s}$ at the angle θ_k given by angle(a_i, b_i) where

$$\text{angle}(x, y) = \begin{cases} \tan^{-1}\left(\frac{y}{x}\right) & \text{if } x > 0 \text{ and } y \geq 0 \\ \tan^{-1}\left(\frac{y}{x}\right) + 2\pi & \text{if } x > 0 \text{ and } y < 0 \\ \tan^{-1}\left(\frac{y}{x}\right) + \pi & \text{if } x < 0 \\ \frac{\pi}{2} & \text{if } x = 0 \text{ and } y > 0 \\ \frac{3\pi}{2} & \text{if } x = 0 \text{ and } y < 0 \\ 0 & \text{if } x = 0 \text{ and } y = 0 \end{cases} \qquad (7.126)$$

For example for $K = 6$ using the 16-QAM signal constellation shown in Fig. 7.54, the signal points to be transmitted simultaneously could be 0, 3, 15, 4, 2, 14 in which case $X_0 = r_0 e^{j\theta_0}$, $X_1 = r_1 e^{j\theta_1}$, $\cdots$, $X_5 = r_5 e^{j\theta_5}$. For example, $r_0 = \sqrt{3^2 + 3^2} = 3\sqrt{2}$ with $\theta_0 = \frac{5\pi}{4}$, so that $X_0 = r_0 e^{j\theta_0} = r_0 \cos(\theta_0) + jr_0 \sin(\theta_0) = 3\sqrt{2} \left[\cos\left(\frac{5\pi}{4}\right) + j \sin\left(\frac{5\pi}{4}\right) \right] = 3\sqrt{2} \left[-\frac{1}{\sqrt{2}} - j\frac{1}{\sqrt{2}} \right] = -3 - 3j$. Hence

$$X_k = \begin{Bmatrix} 0 \\ 3 \\ 15 \\ 4 \\ 2 \\ 14 \end{Bmatrix} = \begin{Bmatrix} -3 - 3j \\ -3 + 3j \\ 3 + 3j \\ -1 - 3j \\ -3 + j \\ 3 + j \end{Bmatrix} \qquad (7.127)$$

and

$$X'_k = \left\{ \begin{array}{c} 0 \\ -3 - 3j \\ -3 + 3j \\ 3 + 3j \\ -1 - 3j \\ -3 + j \\ 3 + j \\ 0 \\ 3 - j \\ -3 - j \\ -1 + 3j \\ 3 - 3j \\ -3 - 3j \\ -3 + 3j \end{array} \right\} \qquad (7.128)$$

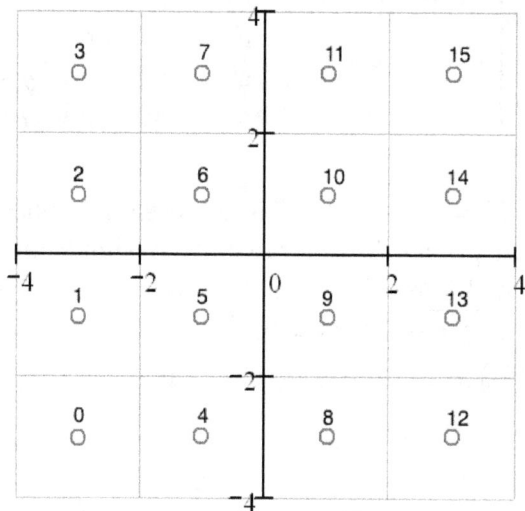

Figure 7.54: 16-QAM signal constellation.

- ⭐ SIMULATION **ComplexQAM**: To verify that for QAM $s_i(t) = \sqrt{\frac{2}{T_s}} A \left[a_i \cos(2\pi f_k t) + b_i \sin(2\pi f_k t) \right] = \mathrm{Re} \left[\sqrt{\frac{2}{T_s}} \left(\sqrt{a_i^2 + b_i^2} \right) e^{j\theta_i} e^{-j(2\pi f_k t)} \right]$ for a given QAM subcarrier signal of frequency f_k for $A = 1$.

Each subcarrier can have a different value for M. Thus K subcarriers will represent K independent complex signal points $X_0, X_1, X_2, \cdots, X_{K-1}$ which could potentially be from K different sized signal constellations. The *baseband* OFDM multicarrier signal $v_{baseband}(t)$ is given by

$$v_{baseband}(t) = \mathrm{Re}\left(g'(t) \right) \tag{7.129}$$

where

$$g'(t) = \sum_{k=0}^{K-1} X_k e^{j \frac{2\pi(k+1)t}{T}} \tag{7.130}$$

and the *bandpass* OFDM multicarrier signal of duration $T = KT_s$ is given by

$$v_{bandpass}(t) = \mathrm{Re}\left(g'(t) e^{j2\pi f_{shift} t} \right) \tag{7.131}$$

Notice how the $e^{j2\pi f_{shift} t}$ term equivalently shifts the frequency of each subcarrier up by f_{shift} Hz because $g'(t) e^{j2\pi f_{shift} t} \equiv \sum_{k=0}^{K-1} X_k e^{j \frac{2\pi(k+1+f_{shift})t}{T}}$. The DFT of the received sequence $\{y_n\}$ is used to determine the complex signal point sequence $\{Y_k\}$. Finally, the coordinates (a_k', b_k') of the received signal points are given by

$$a_k' = \mathrm{Re}\left(Y_k \right) \tag{7.132}$$

and

$$b_k' = \mathrm{Im}\left(Y_k \right) \tag{7.133}$$

7.10.2   OFDM Signal Bandwidth

▷ **OFDM Signal Bandwidth**

- MPSK signal bandwidth

- OFDM null bandwidth

Given that K signal points will be transferred simultaneously per OFDM signal, recall that the time duration T of the multicarrier signal $v(t)$ is increased by a factor of K compared to the time duration T_s of a single carrier of frequency f_k used transfer only one signal point over the channel i.e. $T = KT_s$. Thus, the data rate on a given subchannel is less than the total data rate by a factor of K. Given that the spacing between each subcarriers is $\Delta f = \frac{1}{T}$ Hz, the null bandwidth B_{null} of the OFDM signal is given by

$$B_{null} = \frac{K+1}{T} \qquad\qquad (7.134)$$

To illustrate this null bandwidth, consider once again the previous example in which the symbols 0, 3, 15, 4, 2, 14 are transmitted using the 16-QAM signal constellation shown of Fig. 7.54. The corresponding OFDM baseband waveform and its spectrum are shown in Fig. 7.55 in which the highest frequency is $\frac{K}{T} = \frac{6}{12} = 0.5$ Hz and $B_{null} = \frac{7}{12}$ Hz.

⭐ SIMULATION **OFDM16QAMExample**: Simulation corresponding to Fig. 7.55. Notice that we can use the samples $x_n \frac{\sqrt{N}}{2}$ to determine its FFT without the need to sample $v_{baseband}(t)$. In this case, the sampling frequency is given by $\frac{N}{T}$ and the FFT spectrum is determined in Mathcad simply via $W = CFFT\left(x\frac{\sqrt{N}}{2}\right)$.

- **Experiment**: For MATLAB users, verify the results in Fig. 7.55. Refer to the Mathcad simulation for the solution.

Since $T = KT_s$, $B_{null} = \frac{K+1}{KT_s} \approx \frac{1}{T_s}$ for a large value of K i.e. multicarrier OFDM does not significantly change the data rate or effect the signal bandwidth when compared to the single-carrier system.

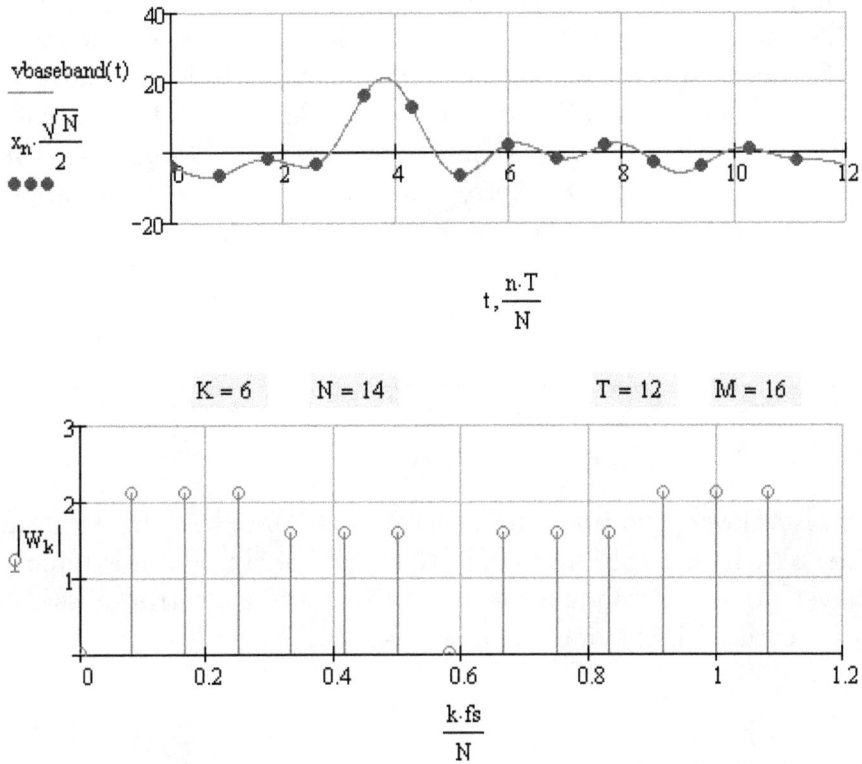

Figure 7.55: OFDM waveform signal and its spectrum for transmitted symbols 0, 3, 15, 4, 2, 14.

IEEE 802.11a wireless LAN standard

In the IEEE 802.11a wireless LAN standard, $\Delta f = 312.5$ kHz, $T_s = 0.05\ \mu$s, $m = 16$, $K' = 48$ from a total of $K = 64$ subcarriers, R_{code} can be either 1/2, 2/3 or 3/4 and the modulation on the subchannels can be either BPSK, QPSK, 16-QAM or 64-QAM depending on the channel SNR. The symbol duration $T = \frac{1}{\Delta f}$ can be designed to be greater than the duration of the channel impulse response (or the multipath delay) by selecting a suitable large value for K to ensure that each subchannel experiences little intersymbol interference (ISI) degradation i.e. subchannels experience relatively flat fading.   Furthermore, the ISI between OFDM symbols can be eliminated by increasing the length of the $\{x_n\}$ sequence by m samples as follows. The values $x_{N-m}, x_{N-m+1}, \cdots, x_{N-1}$, referred to as the *cyclic prefix*, are appended to the beginning of the sequence $\{x_n\}$ e.g. for $K = 4$, $N = 10$ and $m = 3$, the new $\{x_n\}$ sequence, which represents one OFDM symbol as illustrated below.

-3	-2	-1	0	1	2	3	4	5	6	7	8	9
x_7	x_8	x_9	x_0	x_1	x_2	x_3	x_4	x_5	x_6	x_7	x_8	x_9

There is a now cyclic prefix between each OFDM symbol. The value of m is chosen such that ISI affects only the cyclic prefix, which is removed at the receiver prior to demodulation. Note that the total transmission time T_{OFDM} for each OFDM symbol is now given by

$$T_{OFDM} = T + mT_s = KT_s + mT_s = T_s\,(K+m) \qquad (7.135)$$

If the same error-control code and modulation scheme is used for each subcarrier and K' is the number of subcarriers actually used from within the total available of K carriers, then each OFDM signal of duration T_{OFDM} seconds corresponds to K' symbols. If an encoder with code rate R_{code} is employed, then each of these symbols corresponds to $\log_2(M)$ code digits, or equivalently $R_{code}\log_2(M)$ information binary digits. Thus, the data rate of the OFDM system is given by $\left(\frac{R_{code} K' \log_2(M)}{T_{OFDM}}\right)$ binary digits/sec. Thus, the maximum data rate is $\left(\frac{\frac{3}{4}(48)\log_2(64)}{4\mu s}\right) = 54$ Mbps.

7.10.3 Simulation I : OFDM

⭐ SIMULATION **OFDM:** The multicarrier signal $v(t)$ is generated using K subcarriers modulated with $M = 16$ QAM. Specifically, four binary digits are generated randomly and the decimal representation of this binary word is used to select the associated complex signal point X_k. This process is repeated a further $(K - 1)$ times and the sequence $\{X_k\}$ is used to create $\{X_k'\}$, from which $\{x_n\}$ is determined to show that $v_{baseband}\left(t = \frac{nT}{N}\right) = x_n \frac{\sqrt{N}}{2}$ and $v_{bandpass}(t) = \mathrm{Re}\left(g'(t)e^{j2\pi f_{shift}t}\right)$. For no channel noise ($\{y_n\} = \{x_n\}$), the demodulated signal points are shown to correspond exactly to the transmitted signal points.

- **Experiment:** Try a large value for K. Modify the simulation code to add noise within the OFDM system to determine the performance of P_e versus E_b/N_o (dB) on each subchannel over an AWGN channel. In practice, one must take into account the loss in E_b/N_o (dB) due to the use of a cyclic prefix.

7.10.4 Simulation II : OFDM M-ary

⭐ SIMULATION **OFDMary:** The DFT of a baseband multicarrier signal $v(t)$, generated using K subcarriers modulated for M-ary QAM is determined. The bandwidth of the OFDM signal is shown to correspond to $B_{null} = \frac{K+1}{KT_s} \approx \frac{1}{T_s}$. Finally, the OFDM signal is reconstructed using the DFT spectrum.

- **Experiment:** For MATLAB users, insert the code necessary to determine the DFT of the OFDM signal. Refer to the Mathcad simulation for the solution.

7.11 Frequency Shift Keying (FSK)

▶ **FSK**

- Fundamentals

- Binary FSK

- Minimum shift keying

In *frequency shift keying* (FSK), the information is embedded within the frequency of the carrier. For binary FSK, each signal $s_i(t)$ of duration T_b is given by

$$s_i(t) = \sqrt{\frac{2E_b}{T_b}} \cos(2\pi f_i t) \qquad (7.136)$$

where $i = 1, 2$ and E_b is the transmitted signal energy per binary digit. To illustrate the FSK modulation scheme, suppose the signals $s_1(t)$ and $s_2(t)$ represent a binary digit 1 and 0, respectively. These two signals for $f_1 = 2$ Hz, $f_2 = 5$ Hz, $T_b = 1$ second and $E_b = 1$ joule are shown in Fig. 7.56. Thus, the FSK signal corresponding to the binary digit sequence 11010001 is shown in Fig. 7.57. Note that a FSK signal with no phase discontinuity between symbols is referred to as a *continuous phase shift keying* (CPFSK) signal. Binary FSK with a frequency separation of $\frac{1}{T_b}$ Hz and no pulse shaping is referred to as Sunde's FSK and binary CPFSK with a frequency separation of $\frac{1}{2T_b}$ Hz and no pulse shaping is referred to as *minimum shift keying* (MSK).

- ⭐ SIMULATION **FSKsignal:** You may view the FSK signal corresponding to a binary stream of your choice. What is shape of the FSK signal waveform for an all-zero or all-one sequence ?

A key feature of the signals $s_1(t)$ and $s_2(t)$ is that they are orthogonal provided the difference $|f_2 - f_1| = \frac{n}{2T_b}$ where n is an integer greater than or equal to 1. For example, suppose $f_1 = \frac{n_c}{T_b}$ and $f_2 = \left(\frac{n_c}{T_b} + f\right)$ where n_c is a

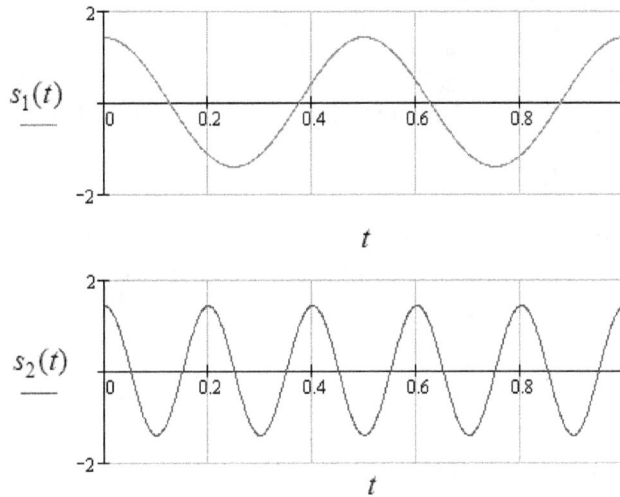

Figure 7.56: Example of FSK signals $s_1(t)$ and $s_2(t)$.

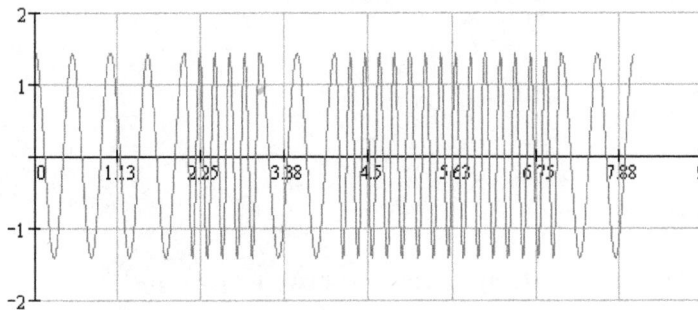

Figure 7.57: FSK signal representing the binary sequence 11010001.

fixed integer and f is the frequency difference $(f_2 - f_1)$. Figure 7.58 shows the variation of $\int\limits_{0}^{T_b} s_1(t)s_2(t)dt$ with f for $E_b = 1$ joule and $T_b = 1$ second with

$$s_1(t) = \sqrt{\frac{2E_b}{T_b}}\cos\left(2\pi\frac{n_c}{T_b}t\right) \text{ and } s_2(t) = \sqrt{\frac{2E_b}{T_b}}\cos\left[2\pi\left(\frac{n_c}{T_b}+f\right)t\right]$$

$$(7.137)$$

As expected, $s_1(t)$ and $s_2(t)$ are orthogonal for $f = \frac{n}{2T_b}$ with $n = 1, 2, 3$,etc. The key point is that the minimum frequency separation required to ensure $s_1(t)$ and $s_2(t)$ are orthogonal is $\frac{1}{2T_b}$ Hz.

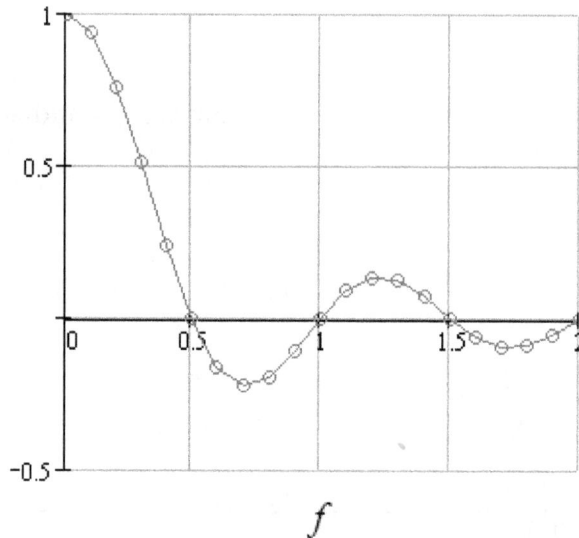

Figure 7.58: Variation of the integral $\int\limits_{0}^{T_b} s_1(t)s_2(t)dt$ with f.

- ⭐ SIMULATION **MinSpacing:** Simulation corresponding to Fig. 7.58. What happens if $|f_2 - f_1| \neq \frac{n}{2T_b}$?

▷ **BFSK**

- Signal constellation diagram

- The effect of noise on a signal point via animation

- BPSK demodulator and performance

The binary FSK basis functions are given by

$$\phi_1(t) = \sqrt{\frac{2}{T_b}} \cos{(2\pi f_1 t)} \qquad (7.138)$$

$$\phi_2(t) = \sqrt{\frac{2}{T_b}} \cos{(2\pi f_2 t)} \qquad (7.139)$$

Thus, the signal $s_i(t)$ can be expressed as

$$s_i(t) = \sqrt{E_b}\phi_i(t) \qquad (7.140)$$

Unlike BPSK, the signal constellation diagram of binary FSK is two dimensional. The horizontal coordinate x_h and the vertical coordinate x_v on the signal constellation (or space) diagram are respectively given by

$$x_h = \int_0^{T_b} r(t)\phi_1(t)dt \qquad (7.141)$$

$$x_v = \int_0^{T_b} r(t)\phi_2(t)dt \qquad (7.142)$$

where $r(t)$ is the received signal. For a noiseless channel,

$$x_h = \int_0^{T_b} s_i(t)\phi_1(t)dt = \sqrt{E_b} \int_0^{T_b} \phi_i(t)\phi_1(t)dt \qquad (7.143)$$

$$x_v = \int_0^{T_s} s_i(t)\phi_2(t)dt = \sqrt{E_b} \int_0^{T_b} \phi_i(t)\phi_2(t)dt \qquad (7.144)$$

For $i = 1$, the coordinates $(x_h, x_v) = (\sqrt{E_b}, 0)$ and for $i = 2$, $(x_h, x_v) = (0, \sqrt{E_b})$ as illustrated in Fig. 7.59.

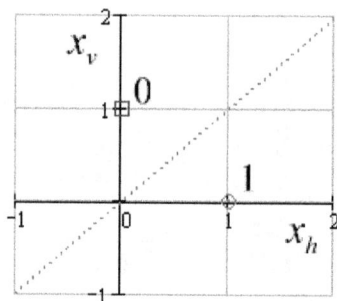

Figure 7.59: Binary FSK signal constellation.

- ⭐ SIMULATION **FSKConstellation:** The signal constellation diagram of Fig. 7.59 is determined. Futhermore, the signal corresponding to a binary digit 1 is sent through an AWGN channel 500 times. The distribution of the received signal points are shown on a single graph. Experiment with different values of E_b and the channel SNR.

7.11.1 Probability of an Error

The coherent binary FSK demodulator is shown in Fig. 7.60, where the coordinates of the output signal point is given by (x_h, x_v). Let $s_1(t)$ and $s_2(t)$ correspond to a binary digit one and zero, respectively. Figure 7.61 shows the signal point distribution of 500 binary ones sent over an AWGN with $SNR = 5$ dB via binary FSK. By symmetry, the line at angle of 45 degrees to the horizontal is the optimum decision threshold to decide in favour of a binary digit 1 or 0. Clearly if $(x_v > x_h)$ then the demodulator would make an error. Alternatively by symmetry, the probability of a binary digit error $P_e = P(x_h > x_v | 0)$ is given by (refer to problems section)

$$P_e = Q\left(\sqrt{\frac{E_b}{N_o}}\right) \qquad (7.145)$$

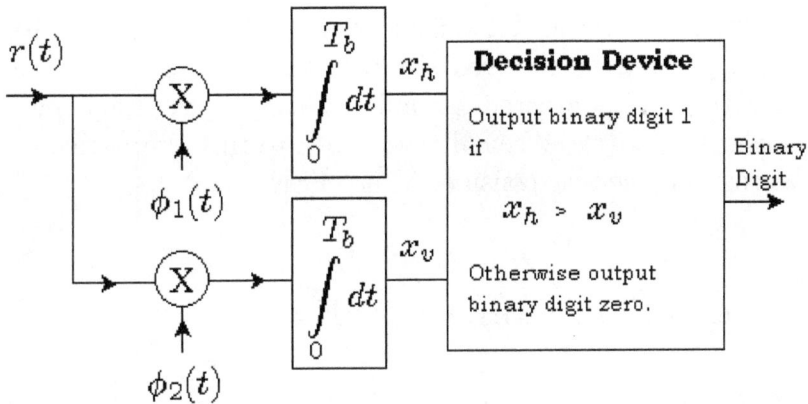

Figure 7.60: Coherent binary FSK demodulator.

An alternative way of thinking about the demodulation process is to consider that for a given transmitted signal $s_i(t)$, the coordinate x_i along $\phi_i(t)$ is given by

$$x_i = \int_0^{T_b} s_i(t)\phi_i(t)dt \qquad (7.146)$$

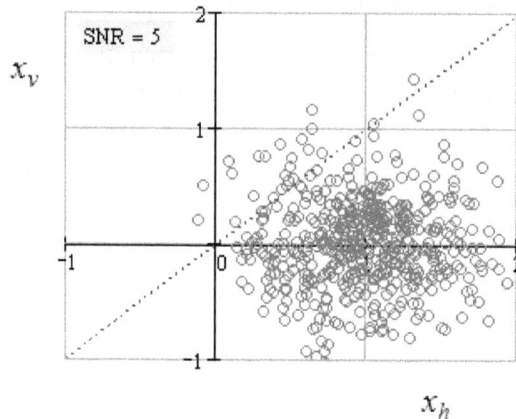

Figure 7.61: Binary digit 1 transmitted 500 times via binary FSK over a 5 dB AWGN channel.

For binary FSK, a given transmitted signal $s_i(t)$ corresponds to x_1 and x_2. If $x_1 > x_2$, then the received signal is taken to be $s_1(t)$ because the correlation with $\phi_1(t)$ is greater. Similarly, if $x_2 > x_1$, then the received signal is taken to be $s_2(t)$. We shall come back to this alternative viewpoint in the next section when we consider M-ary FSK.

7.11.2   Simulation : Binary FSK

SIMULATION **FSK:** The DCS shown in Fig. 7.62 is modeled. The corresponding simulation results are presented in Fig. 7.63 and compared with the expected theoretical results and uncoded BPSK. Clearly, BPSK outperforms Binary FSK with a coding gain of approximately 3dB. However, a comparison between modulation schemes using only the performance measure P_e for a given SNR is not fair. We have to take into account the required channel bandwidth, deviation of the channel from the assumed ideal AWGN channel, demodulator complexity, etc. Such issues will be further explored in section 7.15, where most of the modulation schemes presented in this chapter are compared.

- **Experiment:** Change the values of f_1 and f_2. In particular, determine the performance of BFSK using a frequency separation of $\frac{n}{2T_b}$ and compare it with the $\frac{n}{T_b}$ frequency separation performance for different values of the integer n.

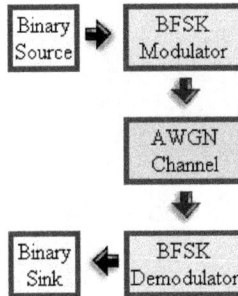

Figure 7.62: Simulated DCS : Binary FSK over an AWGN channel.

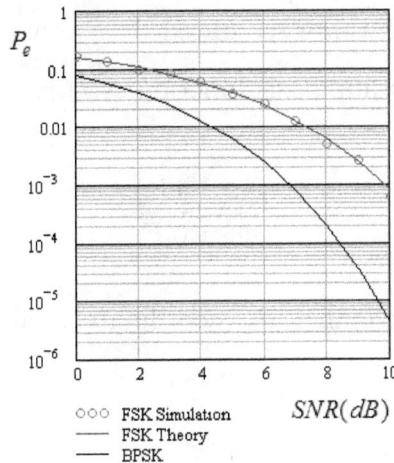

Figure 7.63: Uncoded Binary FSK and PSK performance comparison.

7.12   M-ary FSK

▶ **MFSK**

- 4 FSK signals

- Demodulation of MPSK

- Probability of a symbol error

- Performance curves

In M-ary FSK, a signal $s_i(t)$ is given by

$$s_i(t) = \sqrt{\frac{2E_s}{Ts}} \cos\left[2\pi\left(f_0 + \frac{i}{2Ts}\right)t\right] \qquad (7.147)$$

where $i = 0, 1, \cdots, M-1$, $T_s = T_b \log_2(M)$ is the symbol duration, $E_s = E_b \log_2(M)$ is the energy per symbol and f_0 is a carrier frequency. Since the signal frequencies are separated by $\frac{1}{2T_s}$, we find $\int_0^{T_s} s_i(t)s_j(t)dt = 0$ for $i \neq j$. The corresponding basis function $\phi_i(t)$ is given by

$$\phi_i(t) = \frac{s_i(t)}{\sqrt{E_s}} \qquad (7.148)$$

Thus for a noiseless channel, the coordinate x_i along the $\phi_i(t)$ direction is given by

$$x_i = \int_0^{T_s} s_i(t)\phi_i(t)dt = \frac{1}{\sqrt{E_s}}\int_0^{T_s} s_i^2(t)dt = \frac{E_s}{\sqrt{E_s}} = \sqrt{E_s} \qquad (7.149)$$

For example for $M = 4$, the four FSK signals are

$$s_0(t) = \sqrt{\frac{2E_s}{Ts}} \cos\left[2\pi \left(f_0\right) t\right] \tag{7.150}$$

$$s_1(t) = \sqrt{\frac{2E_s}{Ts}} \cos\left[2\pi \left(f_0 + \frac{1}{2Ts}\right) t\right] \tag{7.151}$$

$$s_2(t) = \sqrt{\frac{2E_s}{Ts}} \cos\left[2\pi \left(f_0 + \frac{2}{2Ts}\right) t\right] \tag{7.152}$$

$$s_3(t) = \sqrt{\frac{2E_s}{Ts}} \cos\left[2\pi \left(f_0 + \frac{3}{2Ts}\right) t\right] \tag{7.153}$$

Unlike MPSK or QAM, the signal space diagram is now M - dimensional and not constrained to two dimensions. In the previous section, it was highlighted that the correlation of a signal $s_i(t)$ along a given basis function ϕ_i can be used to demodulate a given signal point. In general for M basis functions, x_i is calculated for all i i.e. the set $\{x_0, x_1, \cdots, x_{M-1}\}$ is first determined and then the index of the maximum x_i value is taken to correspond to the transmitted signal index. For example, for $M = 4$, if x_0 is the maximum value of the set $\{x_0, x_1, x_2, x_3\}$, then the transmitted signal is assumed to be $s_0(t)$ and the symbol or binary digits corresponding to $s_0(t)$ are output by the demodulator.

If indeed signal $s_0(t)$ is transmitted, then it will be decoded correctly if x_0 is greater than x_1, x_2, and x_3. Taking a closer look at this specific example, the PDF of x_0 under a AWGN channel is given by $\frac{1}{\sigma\sqrt{2\pi}}\exp\left[-\frac{\left(z-\sqrt{E_s}\right)^2}{2\sigma^2}\right]$ with variance $\sigma^2 = \frac{N_o}{2}$ and the other three ($M - 1$ in general) are governed by a zero-mean Gaussian PDF, namely $\frac{1}{\sigma\sqrt{2\pi}}\exp\left[-\frac{z^2}{2\sigma^2}\right]$. Given that x_1, x_2, and x_3 are mutually statistically independent Gaussian random variables, the joint probability that x_1, x_2, and x_3 are less than x_0 for a given x_0, denoted by $P(x_1 < x_0, x_2 < x_0, x_3 < x_0 | x_0)$, is given by

$$P(x_1 < x_0, x_2 < x_0, x_3 < x_0 | x_0) = \left(\int_{-\infty}^{x_0} \frac{1}{\sigma\sqrt{2\pi}}\exp\left[-\frac{z^2}{2\sigma^2}\right] dz\right)^3 \tag{7.154}$$

Equivalently, under a zero-mean unit variance distribution using a change of variable $x = \frac{z}{\sigma}$, we have

$$P(x_1 < x_0, x_2 < x_0, x_3 < x_0 | x_0) = \left(\int_{-\infty}^{\frac{x_0}{\sigma}} \frac{1}{\sqrt{2\pi}} \exp\left[-\frac{x^2}{2} \right] dx \right)^3 \qquad (7.155)$$

Thus the probability of a correct decision P_c for any value of M averaged over all x_0 is given by

$$P_c = \int_{-\infty}^{\infty} \left(\int_{-\infty}^{\frac{x_0}{\sigma}} \frac{1}{\sqrt{2\pi}} \exp\left[-\frac{x^2}{2} \right] dx \right)^{M-1} \frac{1}{\sigma\sqrt{2\pi}} \exp\left[-\frac{(x_0 - \sqrt{E_s})^2}{2\sigma^2} \right] dx_0$$

$$(7.156)$$

Let $y = \frac{x_0}{\sigma}$, then $\frac{dy}{dx_0} = \frac{1}{\sigma}$ and substituting $\sigma = \sqrt{\frac{N_o}{2}}$, we have

$$P_c = \frac{1}{\sqrt{2\pi}} \int_{-\infty}^{\infty} \left(\frac{1}{\sqrt{2\pi}} \int_{-\infty}^{y} \exp\left[-\frac{x^2}{2} \right] dx \right)^{M-1} \exp\left[\frac{-1}{2} \left(y - \sqrt{\frac{2E_s}{N_o}} \right)^2 \right] dy$$

$$(7.157)$$

Hence, the symbol error probability $P_s = 1 - P_c$ is given by

$$P_s = \frac{1}{\sqrt{2\pi}} \int_{-\infty}^{\infty} \left[1 - \left(\frac{1}{\sqrt{2\pi}} \int_{-\infty}^{y} e^{-\frac{x^2}{2}} dx \right)^{M-1} \right] e^{\left[\frac{-1}{2} \left(y - \sqrt{\frac{2E_s}{N_o}} \right)^2 \right]} dy \qquad (7.158)$$

which is upper bounded by

$$P_s \le (M-1) Q \left(\sqrt{\frac{E_s}{N_o}} \right) \tag{7.159}$$

Recall that in MPSK, a signal point is most likely to be incorrectly decoded as a neighboring signal point, so that with Gray coding, $(P_e)_{MPSK} = \frac{P_s}{\log_2 M}$. However in MFSK, a given signal point can be incorrectly decoded with equal probability as any one of the other $(M-1)$ signal points. It is easily shown (refer to problems section) that the corresponding probability of an error in a binary digit $P_e = \frac{M}{2(M-1)} P_s$.

In *Gaussian minimum shift keying* (GMSK), a pulse shaping filter known as the Gaussian filter is used prior to the CPFSK modulator to minimize the spectral occupancy outside the main FSK lobe. This method is employed within the second generation GSM cell phones.

7.12.1   Simulation : M-ary FSK

SIMULATION **MFSK:** The DCS shown in Fig. 7.64 is modeled. The corresponding simulation results are presented in Fig. 7.65 and compared with theory. As expected, the formula for P_s is an upper bound to the actual symbol error probability.

- **Experiment:** Alter the signal frequency separation and other variables (T_b, E_b, etc.) to re-enforce your understanding. Is Gray coding necessary ?

SIMULATION **MFSKSymbol:** The code in simulation **MFSK** is modified to show that each of the remaining $(M-1)$ symbols are equally likely to be output in the event of an error. For example, Fig. 7.66 shows the probability of symbol 01, 10 and 11 being decoded given that the symbol 00 is repeatedly transmitted 10,000 times. As expected, all three curves essentially lie on top of each other over a range of SNRs.

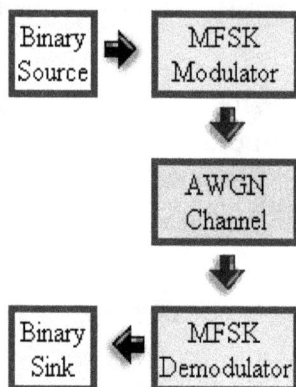

Figure 7.64: Simulated DCS : MFSK over an AWGN channel.

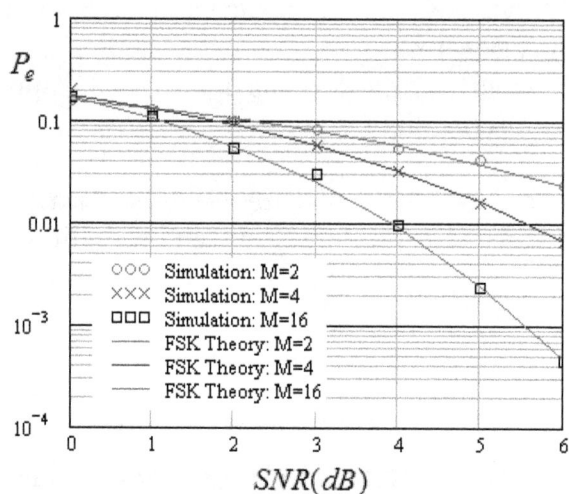

Figure 7.65: MFSK simulation and theoretical performance over an AWGN channel.

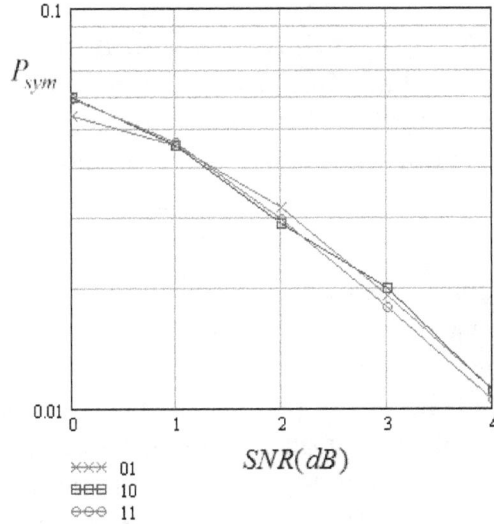

Figure 7.66: Probability of symbol 01, 10 and 11 being decoded given that only the signal corresponding to symbol 00 is transmitted.

7.13 Differential Phase Shift Keying (DPSK)

Differential phase shift keying (DPSK) is a modulation scheme that does not require coherent reference signals to demodulate the received signal because the information is conveyed by phase shifts relative to the previous signal interval. Such a scheme is ideally suited for channels or specific applications (e.g. spread-spectrum systems) where it is practically not feasible to coherently demodulate the incoming signal.

Consider first the differential encoding of a binary sequence. The differentially encoded binary digit time k, denoted by v_k, is given by

$$v_k = \begin{cases} 1 & \text{if} \quad v_{k-1} = m_k \\ 0 & \text{otherwise} \end{cases} \tag{7.160}$$

where m_k is the input information binary digit at time k. The output v_0 can be set arbitrarily to either 0 or 1. To decode, simply output a binary 1 if $v_k = v_{k-1}$, otherwise a binary digit 0.

- ⭐ SIMULATION **DifferentialEncoding:** A random binary sequence is differentially encoded and decoded. The inversion of encoded digits is shown to have no effect on the decoder output. This is an inherent advantage of differential encoding. Set $v_0 = 0$ and then 1 to observe its influence.

Over an AWGN channel, it is reasonable to assume that the phase shift of the received signal over two symbol times $2T_s$ varies slowly. In this case, the phase difference between two successive received signals can be taken to be essentially independent of the carrier phase shift incurred due to transmission delays. In this case to make use of differential encoding, we simply advance the phase of the signal at time k by $\frac{2\pi i}{M}$ radians over the signal at time $k - 1$. For example, for binary DSPK, the signal phase ϕ_k at time k is given by

$$\phi_k = \text{mod}\left[\left(\pi i + \phi_{k-1}\right), 2\pi\right] \qquad (7.161)$$

where i is 0 or 1 depending on whether the current information binary digit is 0 or 1, respectively i.e. the phase of the signal is advanced by π if the input binary digit is 1, otherwise 0. To extract the information bit embedded within the phase difference, the phase difference $\text{mod}\left[\left(\phi_k - \phi_{k-1}\right), 2\pi\right]$ is computed and the binary digit corresponding to this angle on the BPSK signal constellation diagram is output.

- ⭐ SIMULATION **DifferentialEncoding2:** A random binary sequence is differentially encoded via $\phi_k = \text{mod}\left[\left(\pi i + \phi_{k-1}\right), 2\pi\right]$ and appropriately decoded.

The M-ary DPSK demodulator is shown in Fig. 7.67. The received signal $r(t)$ is correlated with the reference signals $\sqrt{\frac{2}{T_s}} \cos\left(2\pi f_c t\right)$ and $\sqrt{\frac{2}{T_s}} \sin\left(2\pi f_c t\right)$, which are not phase-locked to $r(t)$, to produce a point on a signal constellation diagram. Let θ_k represent the angle of this received signal vector and let θ_{k-1} represent the angle of the previous signal vector. Then as in simulation **DifferentialEncoding2**, the information binary digits are extracted via $\text{mod}\left[\left(\theta_k - \theta_{k-1}\right), 2\pi\right]$.

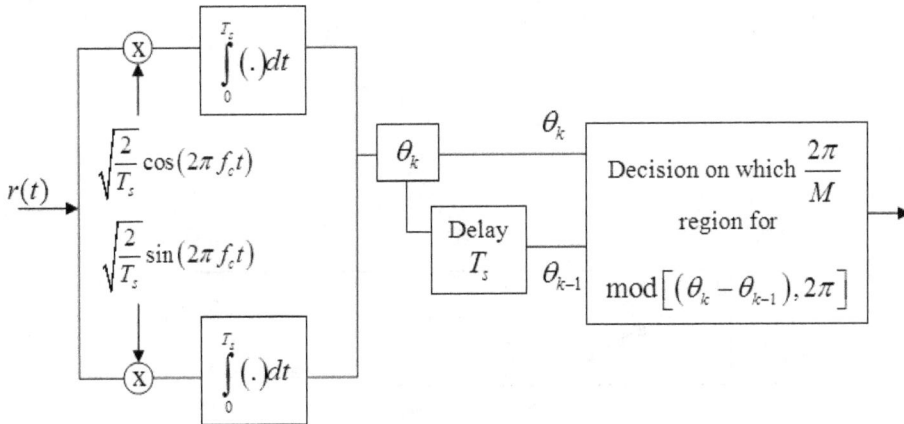

Figure 7.67: DPSK demodulator.

It can be shown [Pawula, 1998] that the probability of a symbol error P_s for M-ary DSPK over an AWGN channel is given by

$$P_s = \frac{1}{\pi} \int_0^{\pi\left(1-\frac{1}{M}\right)} \exp\left[-\frac{\log_2(M)\frac{E_b}{N_o}\sin\left(\frac{\pi}{M}\right)^2}{1+\cos\left(\frac{\pi}{M}\right)\cos\phi}\right] d\phi \qquad (7.162)$$

Using this formula, Fig. 7.68 shows the performance of DPSK for $M = 2, 4, 8, 16$. For $M = 2$, this expression reduces to $\frac{1}{2}e^{\left[-\frac{E_b}{N_o}\right]}$. A comparison of binary DPSK with BPSK is shown in Fig. 7.69. Notice that the performance of DPSK is worse than BPSK because the received signal is not coherently demodulated.

- ⭐ SIMULATION **MDPSK:** Simulation corresponding to Figs. 7.68 and 7.69. Use this code to compare the performance of MPSK and MDPSK for any value of M not simply $M = 2$ as in Fig. 7.69.

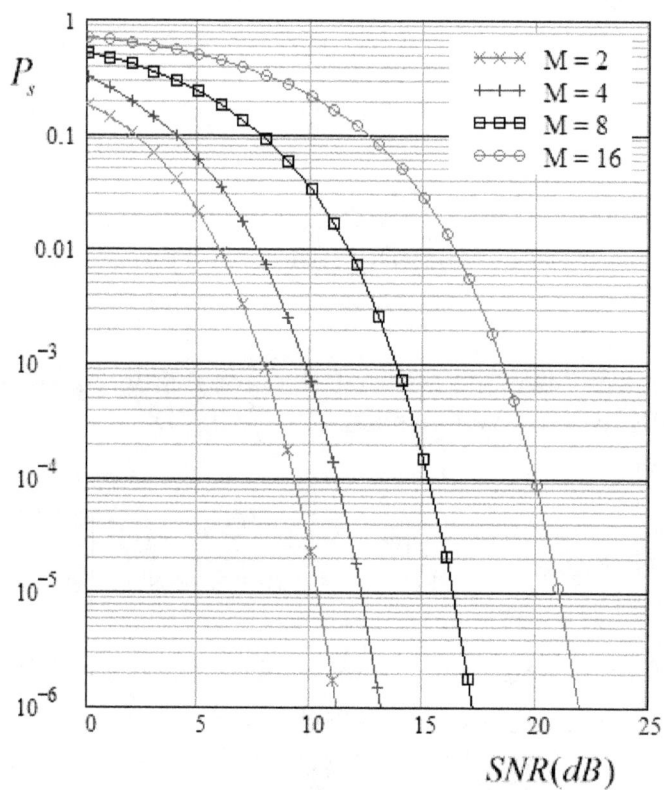

Figure 7.68: Probability of symbol error for M-ary DPSK.

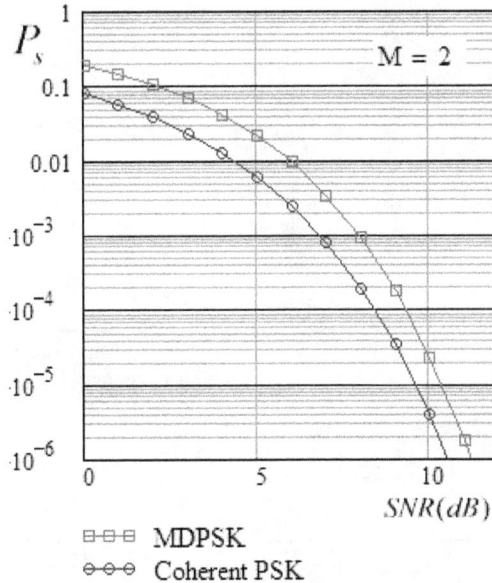

Figure 7.69: Comparison of binary DPSK with BPSK.

7.13.1 Simulation I: Noncoherent and coherent Binary DPSK

⭐ SIMULATION **BinaryDPSK:** The binary DPSK modulation scheme is simulated over an AWGN channel. The simulation results overlay the theoretical as shown in Fig. 7.70. Compared with the performance of coherent BPSK, we note a loss in coding gain of approximately 0.5 dB.

7.13.2 Simulation II: Coherent Binary DPSK

⭐ SIMULATION **CoherentBDSPK:** Simulation of *coherent* binary DPSK.

⭐ SIMULATION **CoherentBDSPK2:** Another approach which is an extension of the BPSK code in section 7.3. The corresponding simulation results are presented in Fig. 7.71. It can be shown [Lindsey and Simon,

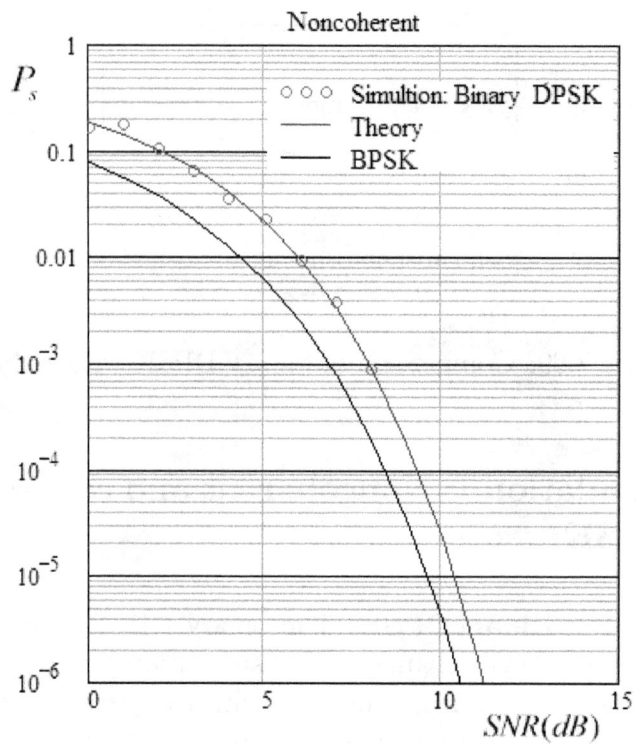

Figure 7.70: Binary DPSK.

1973] that the probability of an error in a binary digit for coherently detected binary DSPK is given by

$$P_e = 2Q\left(\sqrt{\frac{2E_b}{N_o}}\right)\left[1 - Q\left(\sqrt{\frac{2E_b}{N_o}}\right)\right] \qquad (7.163)$$

The performance is slightly worse (0.25 dB) than coherent BPSK because an error will typically propagate between pairs of binary digit decisions.

- **Experiment:** Verify the elimination of phase ambiguity.

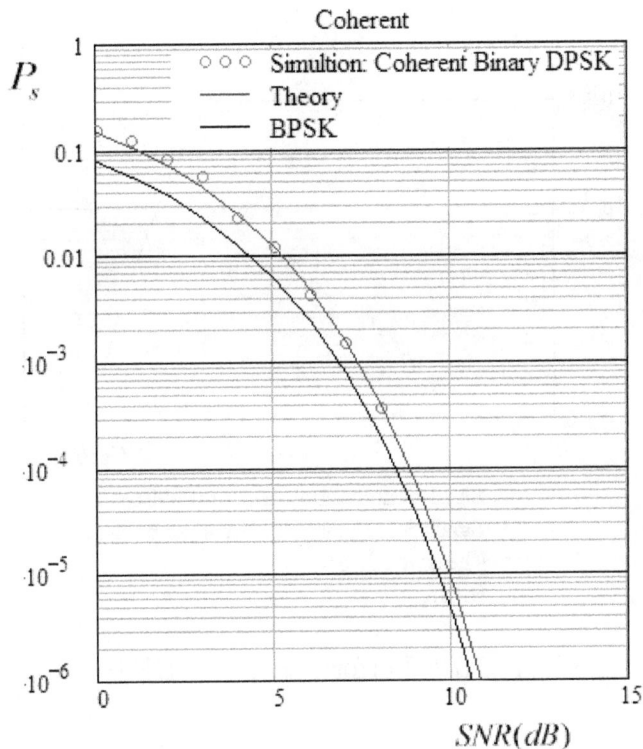

Figure 7.71: Coherent binary DPSK.

7.13.3 Simulation III: 4-DPSK

⭐ SIMULATION **4DSPK:** 4-ary DPSK is simulated over an AWGN channel. The simulation results for the probability of a symbol error and a binary digit error, together with the corresponding theoretical performance curve are shown in Fig. 7.72.

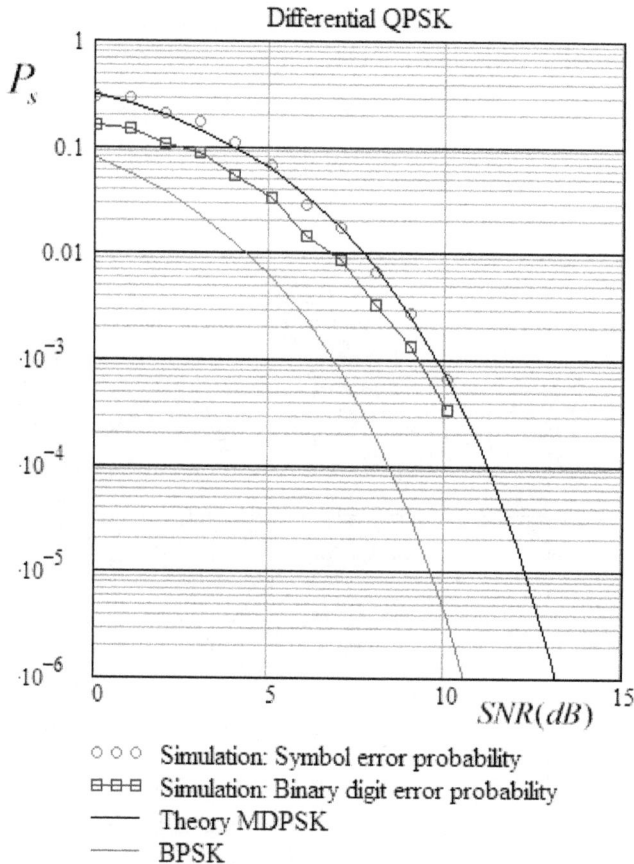

Figure 7.72: Performance of 4-DPSK.

- **Experiment:** Modify the code to implement *8DPSK*. Compare your simulation results with theory.

7.14 Noncoherent FSK

In practice, noncoherent MFSK is preferred over coherent MFSK because it is very difficult to maintain the phase coherence of M signals. For binary noncoherent FSK,

$$s_1(t) = \sqrt{\frac{2E_b}{T_b}} \cos\left(2\pi f_1 t + \theta\right) \tag{7.164}$$

and

$$s_2(t) = \sqrt{\frac{2E_b}{T_b}} \cos\left(2\pi f_2 t + \theta\right) \tag{7.165}$$

with $f_1 = \frac{n_c}{T_b}$, $f_2 = \frac{n_c+n}{T_b}$, where n_c and n are an integers i.e. we have ensured that $f_2 - f_2 = \frac{n}{T_b}$ is an integer multiple of $\frac{1}{T_b}$. To demodulate the received signal $r(t)$ with unknown phase, we simply correlate it with the reference signals

$$\phi_1(t) = \sqrt{\frac{2}{T_b}} \cos\left(2\pi f_1 t\right), \quad \phi_2(t) = \sqrt{\frac{2}{T_b}} \sin\left(2\pi f_1 t\right) \tag{7.166}$$

$$\phi_3(t) = \sqrt{\frac{2}{T_b}} \cos\left(2\pi f_2 t\right), \quad \phi_4(t) = \sqrt{\frac{2}{T_b}} \sin\left(2\pi f_2 t\right) \tag{7.167}$$

as shown in Fig. 7.73.

Under a noiseless channel, Fig. 7.74 shows the variation of

$$z_1 = \int_0^{T_b} r(t)\phi_1(t)dt, \quad z_2 = \int_0^{T_b} r(t)\phi_2(t)dt \tag{7.168}$$

$$z_3 = \int_0^{T_b} r(t)\phi_3(t)dt, \quad z_4 = \int_0^{T_b} r(t)\phi_4(t)dt \tag{7.169}$$

for $r(t) = s_1(t)$ and $s_2(t)$ with the phase difference θ over the range 0 to 2π radians. Notice that when $r(t) = s_1(t)$, $z_3 = z_4 = 0$ i.e. correlation with $\phi_1(t)$

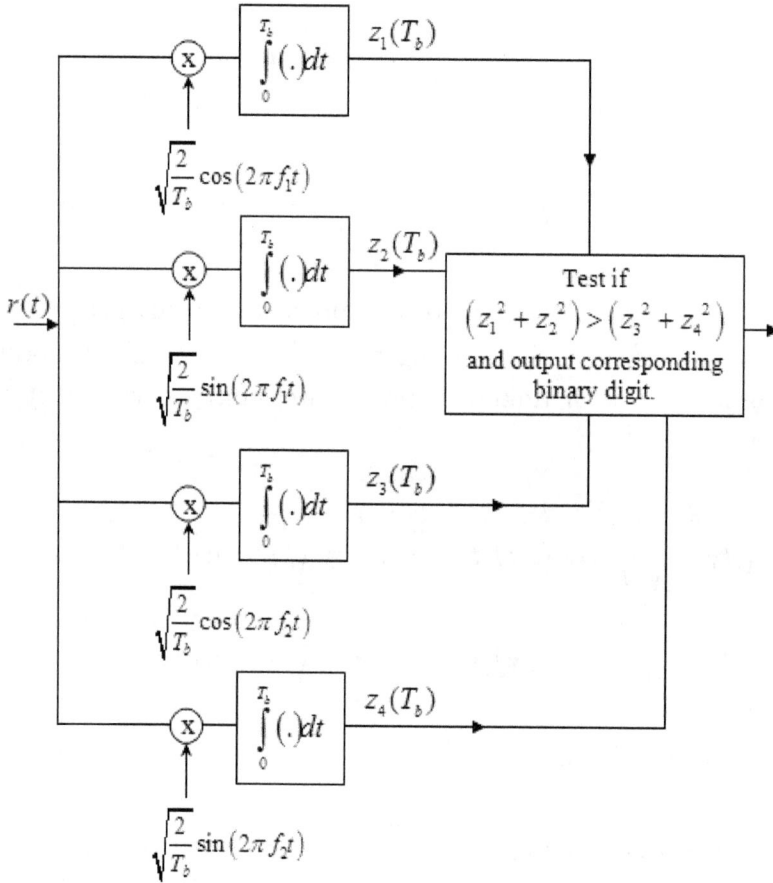

Figure 7.73: Binary Noncoherent FSK demodulator.

and $\phi_2(t)$ is used to detect $s_1(t)$. Similarly, if $r(t) = s_2(t)$, then $z_1 = z_2 = 0$. The variation of z_i^2, $(z_1^2 + z_2^2)$ and $(z_3^2 + z_4^2)$ are presented in Figs. 7.75 and 7.76. Notice that either $(z_1^2 + z_2^2)$ or $(z_3^2 + z_4^2)$ will be equal to E_b, depending on whether the received signal is either $s_1(t)$ or $s_2(t)$, respectively. Thus the binary digit corresponding to whether $(z_1^2 + z_2^2) > (z_3^2 + z_4^2)$ is output accordingly.

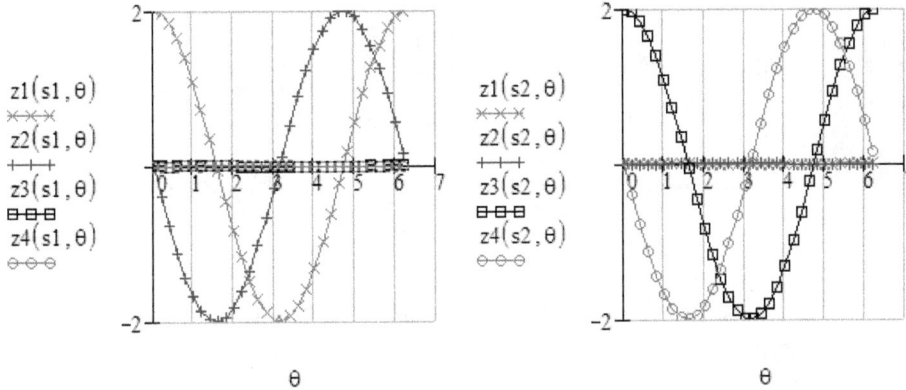

Figure 7.74: Correlator outputs under ideal conditions for the received signals $s_1(t)$ and $s_2(t)$.

• ⭐ SIMULATION **OrthogonalSignaling:** Simulation corresponding to Figs. 7.75 and 7.76. Alter the code to investigate what happens under noisy conditions.

In general, it can be shown [Ziemer and Peterson, 2001] that the probability of a symbol error P_s for M-ary noncoherent FSK over an AWGN channel is given by

$$P_s = \sum_{k=1}^{M-1} \binom{M-1}{k} \frac{(-1)^{k+1}}{k+1} \exp\left[-\frac{k \log_2(M)}{(k+1)}\left(\frac{E_b}{N_o}\right)\right] \qquad (7.170)$$

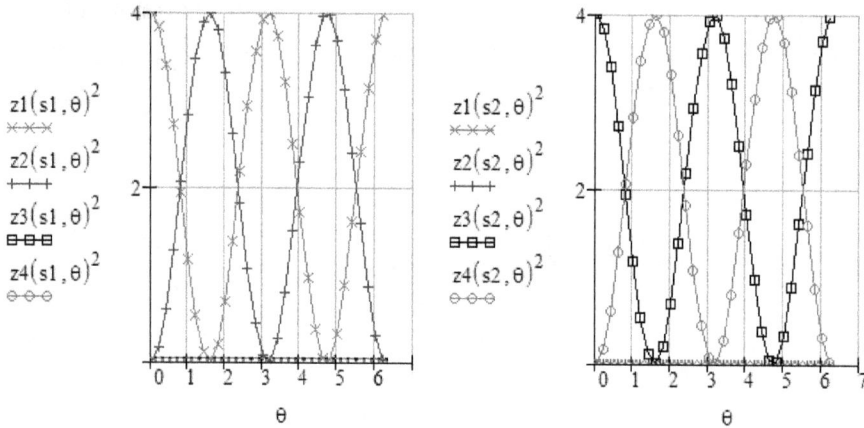

Figure 7.75: Correlator outputs squared.

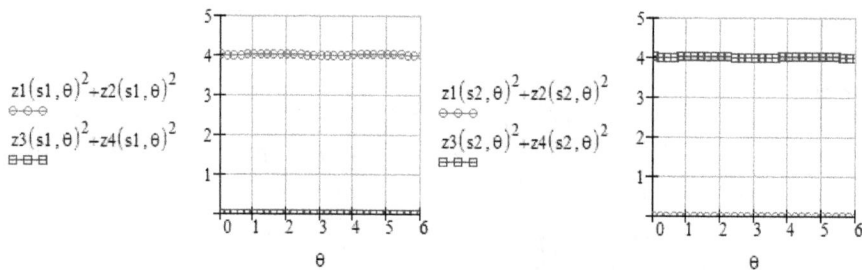

Figure 7.76: Variation of $(z_1^2 + z_2^2)$ and $(z_3^2 + z_4^2)$ with θ for $r(t) = s_1(t)$ and $s_2(t)$.

which for $M = 2$, reduces to $\frac{1}{2}\exp\left(-\frac{E_b}{2N_o}\right)$. Using this formula, Fig. 7.77 shows the performance of noncoherent MPSK for $M = 2, 4, 8, 16$. A comparison of noncoherent 4FSK with coherent 4FSK is shown in Fig. 7.78. As expected, the noncoherent performance is worse. For $\dot{M} > 2$, the probability of an error in a binary digit P_e for noncoherent MFSK is given by

$$P_e = \frac{P_s}{2}\left(\frac{M}{M-1}\right) \tag{7.171}$$

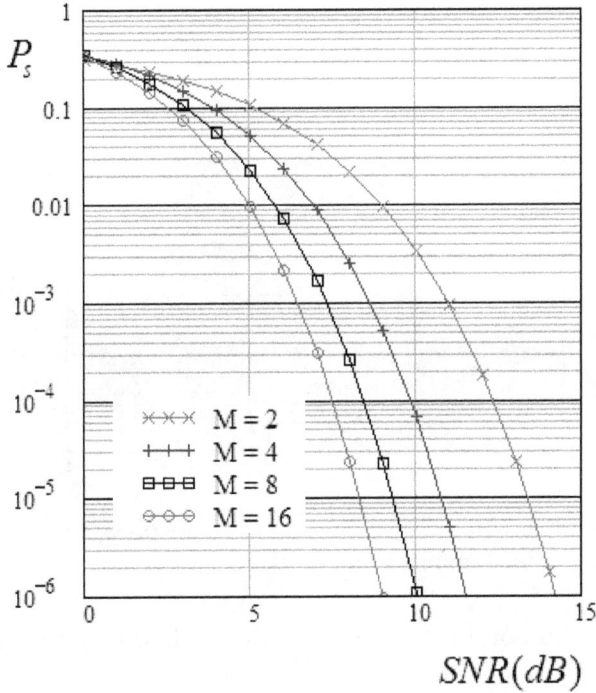

Figure 7.77: Probability of symbol error for noncoherent MFSK.

- ⭐ SIMULATION **MNFSK:** Simulation corresponding to Figs. 7.77 and 7.78. Try other values for M.

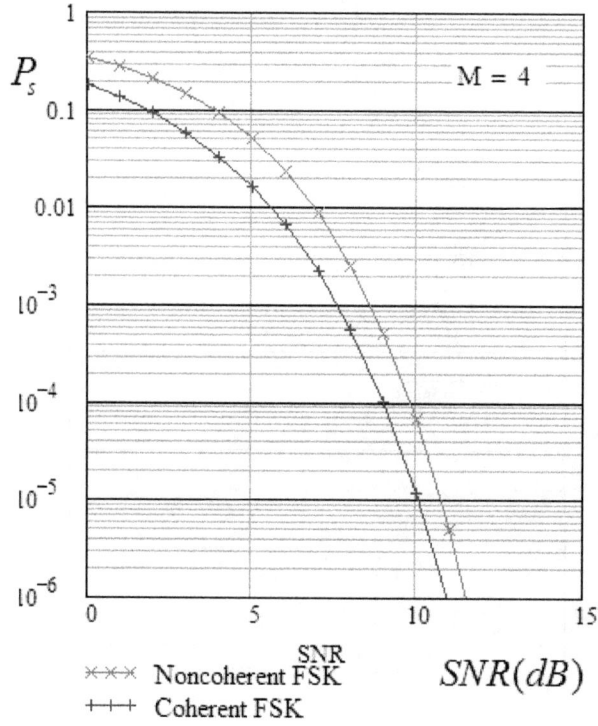

Figure 7.78: Comparison of noncoherent and coherent FSK for $M = 4$.

7.14.1   Simulation : Noncoherent FSK

⭐ SIMULATION **NonCoherentFSK:** The binary noncoherent FSK modulation scheme is simulated over an AWGN channel. The simulation results lie along the theoretical curve as shown in Fig. 7.79. Compared with the performance of coherent FSK, we note a loss in coding gain of approximately 0.7 dB. Typically, this penalty is worth paying to minimize the complexity of the demodulator.

- **Experiment:** (a) What happens if $f_2 - f_2 \neq \frac{n}{T_b}$? (b) If n binary digits are sent over an AWGN via binary noncoherent FSK, the probability of an error in this word of n digits is simply $P_{word} = 1 - (1 - P_e)^n$, where $P_e = \frac{1}{2} \exp\left(-\frac{E_b}{2N_o}\right)$. Compare the performance of P_{word} with P_s for M-ary noncoherent FSK.

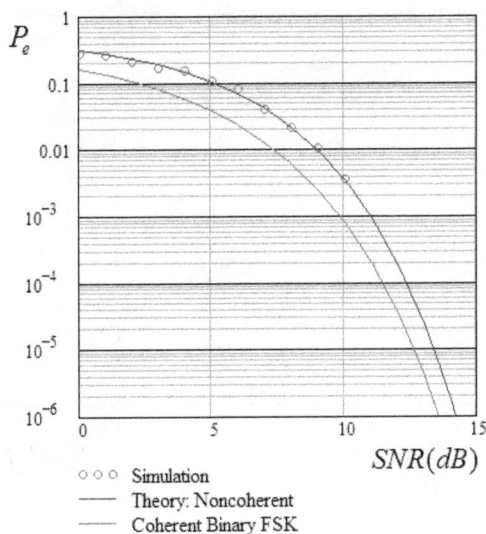

Figure 7.79: Noncoherent binary FSK.

7.15   Comparison of Modulation Schemes

▶ **Performance Comparison**

- Performance of BPSK, BFSK and 4-QAM

- Comparison of M-ary PSK, M-ary FSK
  and M-ary QAM with a detailed look at $M = 16$.

In this section, we shall compare the coherently demodulated schemes presented in the previous sections. Figures. 7.80 to 7.85 present the probability of a symbol and binary digit error for M-ary PSK, QAM and FSK over a range of SNRs. For $M = 16$, a comparison between the symbol and binary digit error probability over a range of SNRs for PSK, QAM and FSK are presented in Figs. 7.87 to 7.89.

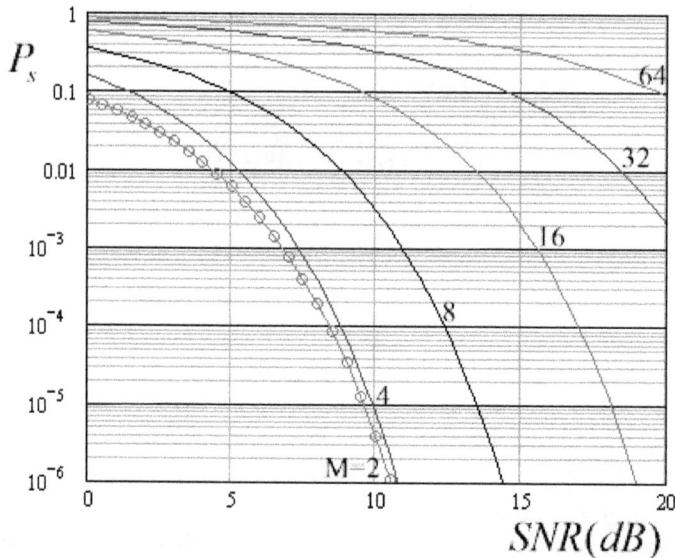

Figure 7.80: M-ary PSK symbol error probability.

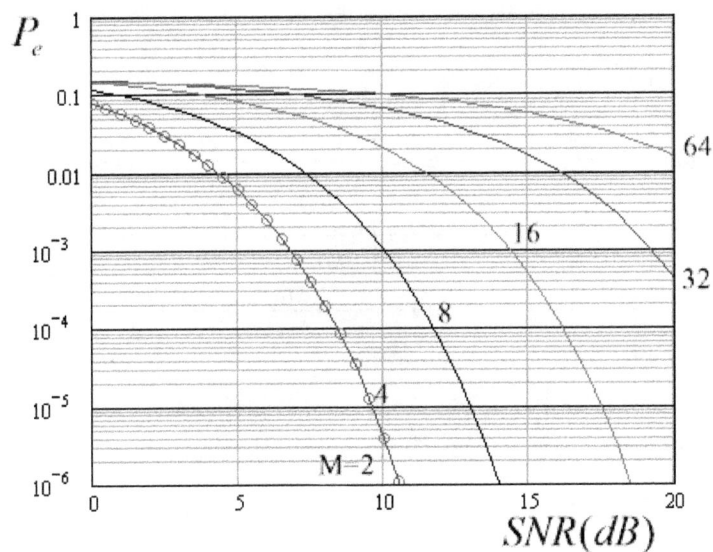

Figure 7.81: M-ary PSK binary digit error probability.

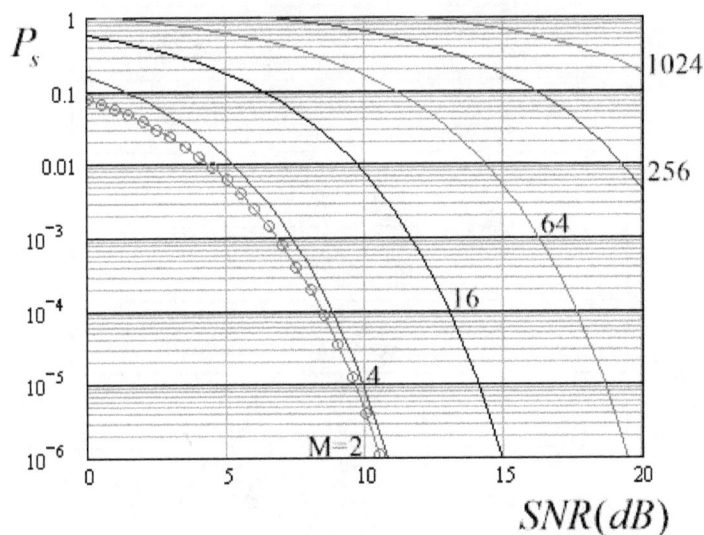

Figure 7.82: M-ary QAM symbol error probability.

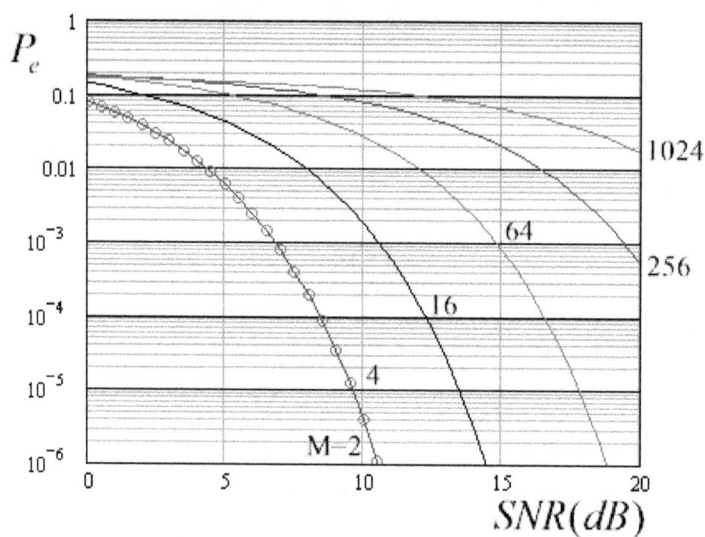

Figure 7.83: M-ary QAM binary digit error probability.

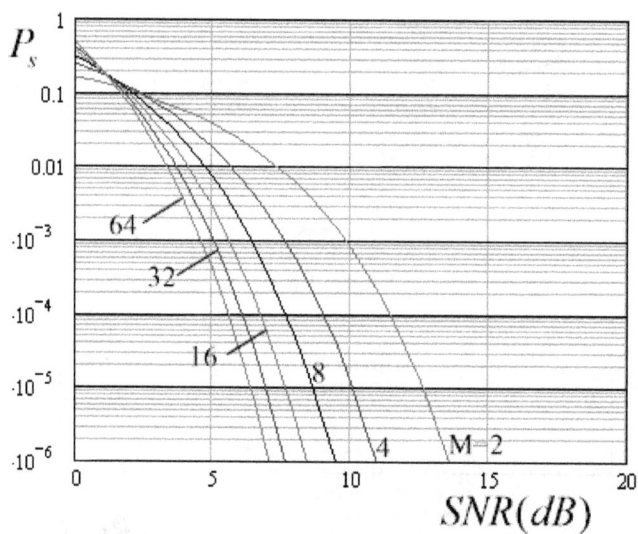

Figure 7.84: M-ary FSK symbol error probability.

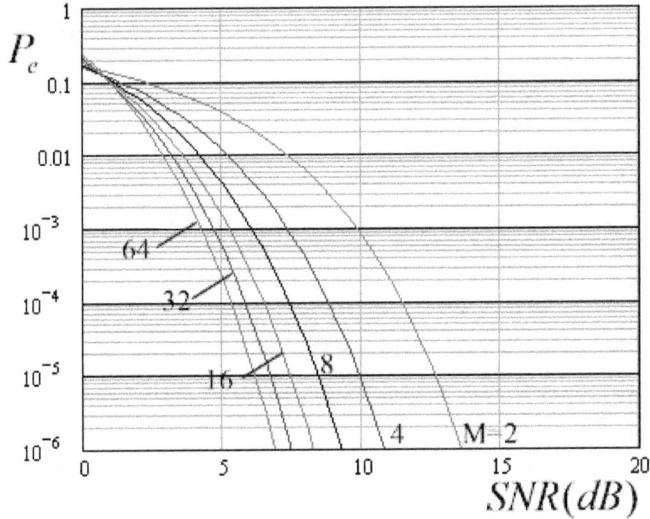

Figure 7.85: M-ary FSK binary digit error probability.

- ⭐ SIMULATION **Comparison:** Details of all the comparison graphs. Similar to Fig. 7.86, generate graphs to compare M-ary PSK, FSK and QAM for any value of M.

For both M-ary PSK and M-ary QAM, as M is increased, the performance is degraded. However, for M-ary FSK, the performance is improved for a larger value of M. The penalty incurred for this improved performance is not evident from Fig. 7.84. Similarly, the advantage of using M-ary PSK or QAM for a large value of M is not evident from Figs. 7.80 or 7.82. This is because we have *not* taken into account the bandwidth requirement for a given modulation scheme. In the next subsection, we shall investigate the bandwidth requirement for a given modulation scheme to enable a fair comparison of all the modulation schemes.

7.15.1 Bandwidth Efficiency

▶ **Bandwidth Efficiency**

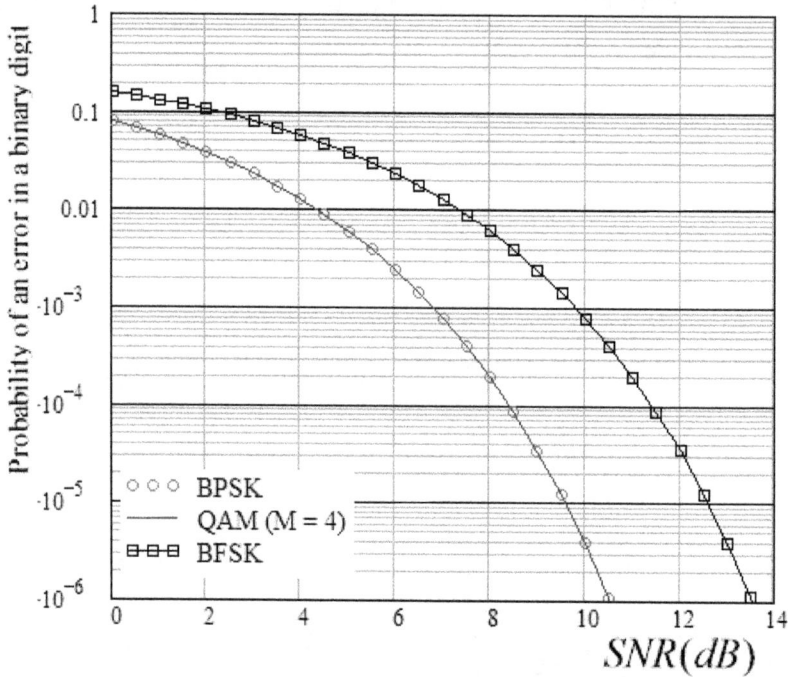

Figure 7.86: Comparison of BPSK, QAM ($M = 4$) and BFSK.

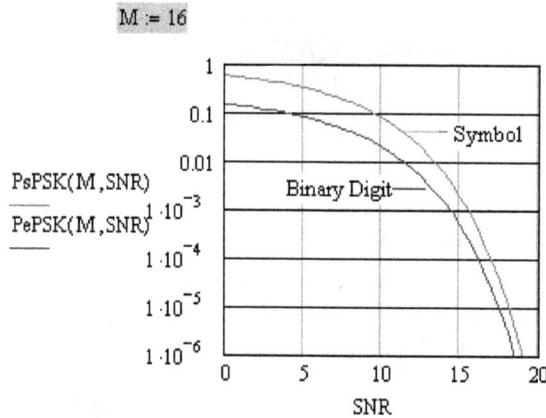

Figure 7.87: Symbol and binary digit error probability comparison for 16-PSK.

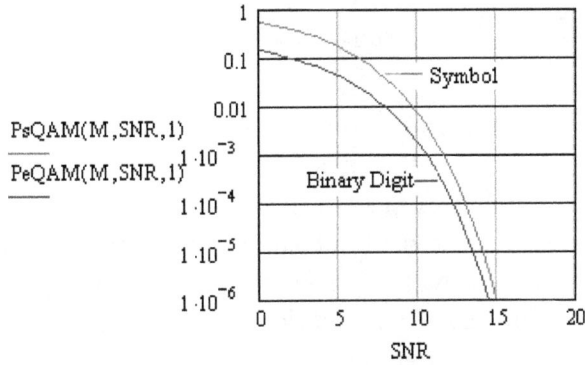

Figure 7.88: Symbol and binary digit error probability comparison for 16-QAM.

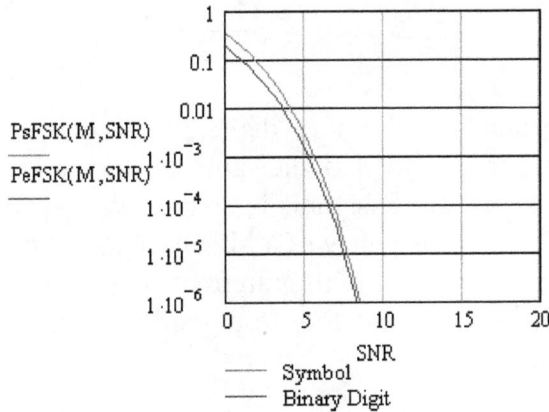

Figure 7.89: Symbol and binary digit error probability comparison for 16-FSK.

- Bandwidth efficiency of MPSK and MFSK

- Comparison of MPSK, QAM and MFSK
  on Shannon's channel capacity curve

To determine the bandwidth requirement of an M-ary PSK signal, let us first consider the bandwidth of a BPSK signal. If a baseband signal is multiplied by a carrier signal $\cos(2\pi f_c t)$, the spectrum of the baseband signal is shifted to be centered about the carrier frequency f_c (section 1.5). For a rectangular pulse of width τ seconds, the frequency range from 0 Hz to the first null is equal to $\frac{1}{\tau}$ Hz (section 1.4). If this spectrum is shifted to be centered over f_c Hz, then the first *null-to-null* bandwidth is equal to $\frac{2}{\tau}$. Since $\tau = T_b$ is the time duration of a binary digit, the first null-to-null bandwidth is given by $\frac{2}{T_b}$. In general for MPSK, the time duration of a signal $T_s = T_b \log_2 M$. Thus, the width of the main spectral lobe of a MPSK signal, or the first null-to-null bandwidth, is given by

$$B_{MPSK} = \frac{2}{T_s} = \frac{2}{T_b \log_2 M} \qquad (7.172)$$

or equivalently,

$$B_{MPSK} = \frac{2r_b}{\log_2 M} \qquad (7.173)$$

where r_b is the number of binary digits per second equal to $\frac{1}{T_b}$. The bandwidth requirement for QAM is the same as that for MPSK. However, the spectral analysis for MFSK signals is nontrivial [Haykin, 2001]. Recall from section 7.12 that the frequency of a MFSK signal is given by $\left(f_0 + \frac{i}{2T_s}\right)$ where $i = 0, 1, \cdots, M-1$. The minimum frequency is f_0 and the maximum frequency is $\left(f_0 + \frac{M-1}{2T_s}\right)$ and therefore to a good approximation, the MFSK bandwidth is simply

$$B_{MFSK} = \frac{M}{2T_s} = \frac{M}{2T_b \log_2 M} = \frac{Mr_b}{2\log_2 M} \qquad (7.174)$$

Thus the *bandwidth efficiency* ρ, defined by the ratio $\frac{r_b}{B}$ (binary digits/sec/Hz), for MPSK and MFSK is given by

$$\rho_{PSK} = \rho_{QAM} = \frac{r_b}{B_{MPSK}} = \frac{\log_2 M}{2} \qquad (7.175)$$

$$\rho_{FSK} = \frac{r_b}{B_{MFSK}} = \frac{2 \log_2 M}{M} \qquad (7.176)$$

For example, if a bandwidth of 4 kHz is available to transmit 8000 binary digits/sec, the required bandwidth efficiency is $8000/4000 = 2$ binary digits/second/Hz. To double the rate at which binary digits are sent over this bandwidth, we would require a bandwidth efficiency of $16000/4000 = 4$ binary digits/second/Hz. We may now compare all the modulation schemes on a "level playing field". Fixing the required probability of an error in a binary digit P_e for a given modulation scheme to be 10^{-5}, Fig. 7.90 shows a plot of the bandwidth efficiency versus the $\frac{E_b}{N_o}$ in decibels required to achieve $P_e = 10^{-5}$ for the M-ary PSK, QAM and M-ary FSK. The key to the points labeled with the letters a, b, c ....q is presented in Table 7.8. For example, point "a" corresponds to BPSK which requires $\frac{E_b}{N_o} = 9.6$ dB to achieve $P_e = 10^{-5}$ and has a bandwidth efficiency of 0.5 binary digits/sec/Hz. As another example, point "j" corresponds to 64-QAM which requires 22.6 dB to achieve $P_e = 10^{-5}$ and has a bandwidth efficiency of 4 binary digits/sec/Hz. In Fig. 7.90, the curve labeled "Shannon Capacity Curve" will be formally introduced in section 8.11.

Some of the key features evident from Fig. 7.90 are as follows:

- For MPSK, as M is increased, the bandwidth efficiency improves. The penalty incurred is that $\frac{E_b}{N_o}(dB)$ has to be increased to maintain the required P_e. This is because as M increases the signal points, which lie on a circle of radius $\sqrt{E_s}$, come closer together resulting in an increase in the probability of a symbol error. Thus, MPSK is suitable for communication channels that are band-limited (e.g. telephone), where we desire $\frac{r_b}{B} > 1$.

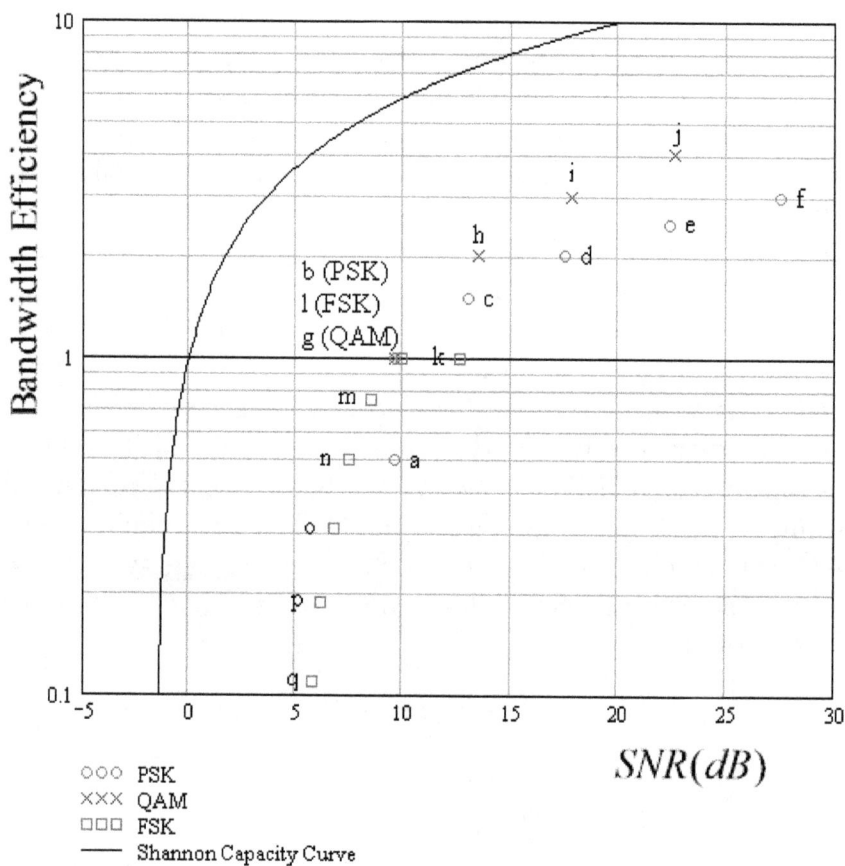

Figure 7.90: The bandwidth efficiency of various modulation schemes over a range of SNRs for a binary digit error probability of 10^{-5}.

- QAM is better than MPSK, because for a given bandwidth efficiency, the required increase in $\frac{E_b}{N_o}(dB)$ to maintain P_e is less. This is because the QAM signal space is rectangular and for a given signal power, the QAM signal points are further apart than the corresponding MPSK signal points.

- For MFSK, as M is increased, the bandwidth efficiency decreases. However, the advantage is a reduction in the required $\frac{E_b}{N_o}(dB)$ for a given P_e. This is because the signal space is M-dimensional. Thus, MFSK is suitable for commuication channels that are power-limited.

- QPSK has twice the bandwidth efficiency of BPSK and requires the same $\frac{E_b}{N_o}(dB)$ for a given P_e assuming a perfect coherent demodulator.

- 4-FSK has the same bandwidth efficiency as BFSK and requires a less $\frac{E_b}{N_o}(dB)$ for a given P_e.

- We can trade bandwidth efficiency with signal power for a given required level of performance.

- Any given modulation scheme is far from the Shannon channel capacity curve. However, with the use of error-control coding, the operating point can be moved closer to this curve.

7.15.2 Simulation I: Periodic BPSK Signal Bandwidth

SIMULATION **BPSKbandwidth:** A periodic BPSK signal is created by multiplying a periodic polar NRZ signal with $\cos(2\pi f_c t)$, where $f_c = 6$ Hz is the carrier frequency, as shown in Figs. 7.91 and 7.92. The time duration of a binary digit $T_b = 1$ sec. The spectral components are determined via the exponential Fourier series using both the double-sided and single-sided spectrum representations. The first bandpass signal null-to-null bandwidth is shown to be $\frac{2}{T_b}$ Hz. Allowing only the components within this bandwidth to filter through, the original and filtered baseband and bandpass amplitude spectra are shown in Figs. 7.93 and 7.94, respectively. The corresponding reconstructed baseband and bandpass signals overlaid are shown in Fig. 7.95.

Notice that the effect of the phase shifts in the carrier is to expand the bandwidth occupied by the BPSK signal.

- **Experiment:** Change the carrier frequency, amplitude, time duration and the bandwidth of the baseband and band-pass filters to observe the effect on the reconstructed signals. Modify the simulation code to first shape the polar NRZ baseband signal through a root raised cosine filter before multiplying it with $\cos(2\pi f_c t)$. Verify that the bandwidth of the transmitted signal is reduced.

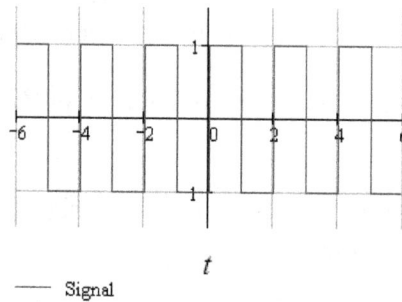

Figure 7.91: Baseband signal.

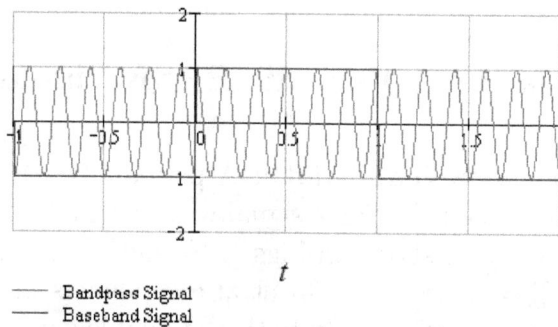

Figure 7.92: Bandpass and baseband signals.

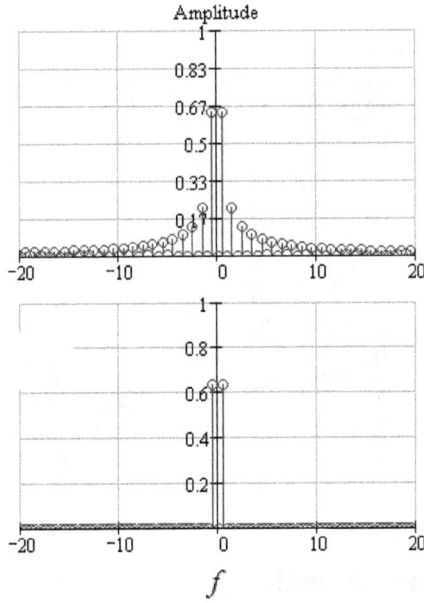

Figure 7.93: Original and filtered baseband amplitude spectrum.

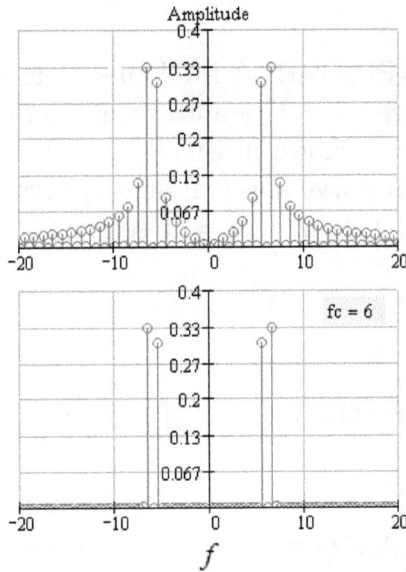

Figure 7.94: Original and filtered bandpass amplitude spectrum.

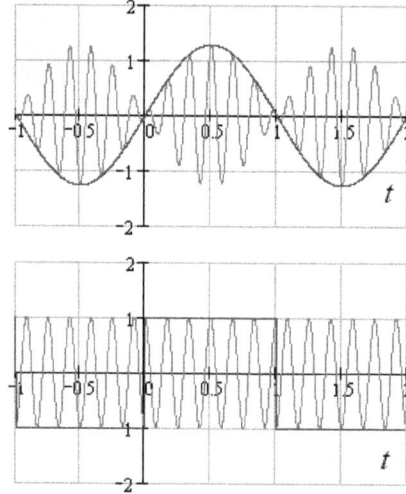

Figure 7.95: Reconstructed and original baseband and bandpass signals overlay.

7.15.3   Simulation II: Random BPSK Signal PSD

SIMULATION **BPSKpsd:** A random stream of binary digits are used to generate a BPSK signal. The autocorrelation function of this signal is compared with theory as shown in Fig. 7.96 and then its Fourier transform to determine the signal power spectral density (PSD) as shown in Fig. 7.97 for $f_c = 5$ Hz, $A = 1$ volt and $T_b = 1$ secs. The theoretical PSD from section 7.3.3 is given by

$$PSD_{BP}(f) = \frac{A^2 T_b}{4} \operatorname{sinc}^2 \left[T_b(f - f_c) \right] + \frac{A^2 T_b}{4} \operatorname{sinc}^2 \left[T_b(f + f_c) \right] \qquad (7.177)$$

- **Experiment:** Modify the simulation code to determine the PSD of a random QPSK signal.

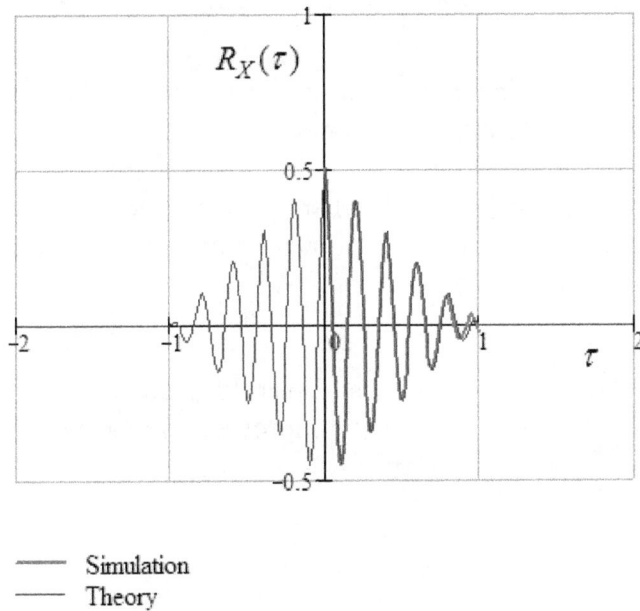

Figure 7.96: Autocorrelation function of a random BPSK signal.

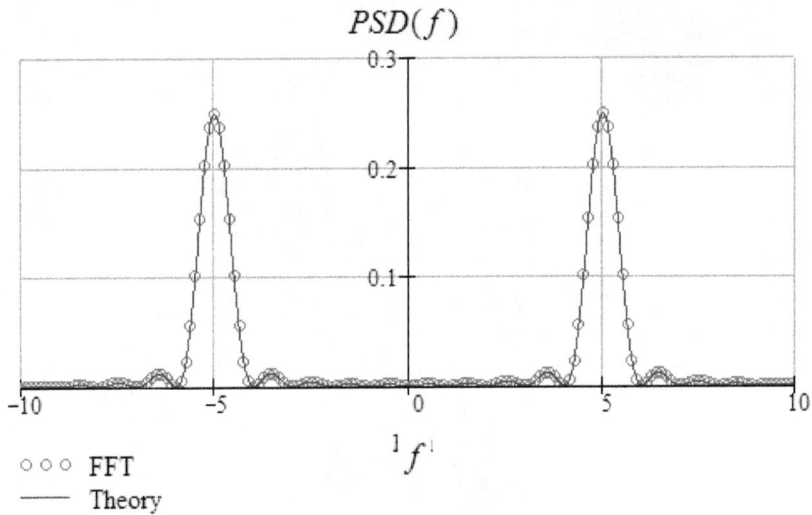

Figure 7.97: Random BPSK signal power spectral density (PSD).

7.15.4 Simulation III: BFSK Signal Power

SIMULATION **BFSK:** The baseband signal in Fig. 7.91 is used to create a periodic BFSK signal. Experiment with the frequency (f_1, f_2) of each signal. Using a band-pass filter which allows the spectral components in the range $\left(f_1 - \frac{r_b}{2}\right)$ to $\left(f_2 + \frac{r_b}{2}\right)$, where $f_2 > f_1$, the signal is reconstructed from the spectral components and shown to contain approximately 90% of the original signal power.

- **Experiment:** Increase the band-pass filter range and observe how the reconstructed FSK signal shape improves. Experiment with the signal frequencies. For example try $(f_2 - f_1) = r_b$.

M	PSK	QAM	FSK
	Letter (SNR(dB), Bandwidth Efficiency)		
2	a $(9.6, 0.5)$		k $(12.6, 1)$
4	b $(9.6, 1)$	g $(9.6, 1)$	l $(9.9, 1)$
8	c $(13, 1.5)$		m $(8.5, 0.75)$
16	d $(17.5, 2)$	h $(13.5, 2)$	n $(7.5, 0.5)$
32	e $(22.4, 2.5)$		o $(6.8, 0.313)$
64	f $(27.5, 3)$	i $(17.8, 3)$	p $(6.2, 0.188)$
128			q $(5.8, 0.109)$
256		j $(22.6, 4)$	

Table 7.8: Key for Fig. 6.68.

7.16   Problems

1. The binary digit sequence 1001 is transferred over an AWGN channel. If the modulation scheme is BPSK, what is the probability of correctly receiving this sequence of digits if $\dfrac{E_b}{N_o} = 0$ dB ?

2. If a transmitted BPSK signal $s_1(t) = \cos(20\pi t)$ is of duration 0.5 seconds, determine:

(a)   Amplitude A of the signal and its carrier frequency ?
(b)   Energy per binary digit E_b ?
(c)   Coordinates of the signal points at the receiver under a noiseless channel ?

3. Binary digits are transferred at a rate of 90000 binary digits per second over an AWGN channel using the BPSK modulation scheme. The amplitude of a given BPSK signal is 5 mV. If the single-sided noise power spectral density is 10^{-11} W/Hz, determine the

(a)   Probability of an error in a binary digit if hard-decision coherent demodulation is used at the receiver.
(b)   Number of errors made per hour.
(c)   Average signal power.

4. A DCS consists of a binary source, BPSK modulator, AWGN channel, BPSK demodulator and a binary sink. If the channel SNR is 4 dB

(a)   What is the BSC crossover probability ?
(b)   If a rate $\frac{1}{2}$ code is introduced within the DCS, what the new BSC crossover probability ?

5. Consider a DCS in which BPSK is used over an AWGN channel. If the reference signal within the BPSK demodulator is given by

$$\phi_1(t) = \sqrt{\frac{2}{T_b}} \cos(2\pi f_c t + \theta).$$

(a) Derive an expression for the probability an error in a binary digit. Confirm that your expression is correct setting $\theta = 0$.

(b) Under what conditions will the BSC crossover probability be equal to $\dfrac{1}{2}$.

(c) What value of the phase error θ would increase the probability of an error in a binary digit from 2×10^{-4} for $\theta = 0$ to 1.2×10^{-3}.

6. Consider the use of a BPSK modulation scheme in which the transmitted signals are given by $s_1(t) = \sqrt{\dfrac{2E_b}{T_b}} \cos(2\pi f_c t)$ and $s_2(t) = -\sqrt{\dfrac{E_b}{2T_b}} \cos(2\pi f_c t)$. Using the basis function $\phi_1(t) = \sqrt{\dfrac{2}{T_b}} \cos(2\pi f_c t)$, determine the optimum threshold T for hard-decision demodulation.

7. Using uncoded coherent BPSK over an AWGN, we find that the probability of an error in a binary digit is given by $P_e = Q\left(\sqrt{\dfrac{2E_b}{N_o}}\right)$ if the hard-decision threshold T is set at zero. Derive a similar expression but now for any value of T, where $0 < T < \sqrt{E_b}$. Check your answer by setting $T = 0$.

8. If Gray coding is used with M-ary PSK in which the symbol error probability is P_s, what is average probability of an error in a binary digit? Express your answer in terms of M and P_s.

9. Consider the standard BPSK modulation scheme in which $s_1(t) = \sqrt{\dfrac{2E_b}{T_b}} \cos(2\pi f_c t)$, $s_2(t) = -\sqrt{\dfrac{2E_b}{T_b}} \cos(2\pi f_c t)$ and $\phi_1(t) = \sqrt{\dfrac{2}{T_b}} \cos(2\pi f_c t)$.

At the receiver, the signal point coordinate $x_h = \displaystyle\int_0^{T_b} r(t)\phi_1(t)\,dt$, where the received signal $r(t)$ is either $s_1(t)$ or $s_2(t)$ under a noiseless channel. Now suppose there is a synchronization error in the timing such that $x_h = \displaystyle\int_{kT_b}^{T_b+kT_b} r(t)\phi_1(t)\,dt$, where k is a constant. Determine x_h if:

(a) Signal $s_1(t)$ is followed by $s_2(t)$.
(b) Signal $s_2(t)$ is followed by $s_1(t)$.
(c) Signal $s_1(t)$ is followed by $s_1(t)$.
(d) Signal $s_2(t)$ is followed by $s_2(t)$.
(e) Hence determine an expression for the probability of an error in a binary digit.

10. Consider the BPSK demodulator shown below in which $\phi_1(t) = k \cos(2\pi f_c t + \theta)$. The low-pass filter will not allow high frequency components to pass through to the output. If $s_1(t) = \cos(2\pi f_c t)$ and $s_2(t) = -\cos(2\pi f_c t)$,

(a)   Determine an expression for the output of the low-pass filter.

(b)   Explain what happens as θ ranges from 0 to π.

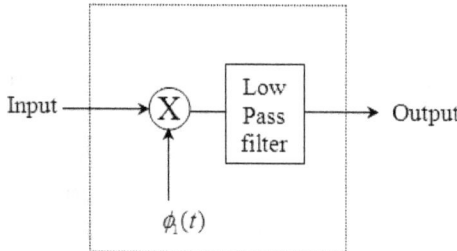

11. In section 7.7, 8-level soft-decision BPSK demodulation was considered. Now suppose only 4 levels of quantization are used as illustrated below.

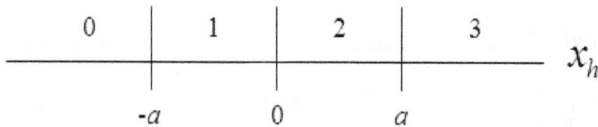

(a)   Draw a diagram indicating the transition probabilities which represent the discrete memoryless channel (DMC).

(b)   Determine an expression for forward transition probabilities $P(0|0)$ and $P(1|0)$.

(c)   Using your answer to part (b), verify that the equivalent BSC crossover probability α is given by

$$\alpha = Q\left(\sqrt{\frac{2E_b}{N_o}}\right).$$

12. The probability of an error in a symbol P_s for M-ary PSK is approximately given by $P_s = 2Q\left(\sqrt{\frac{2E_s}{N_o}} \sin\left(\frac{\pi}{M}\right)\right)$, where $Q(.)$ is the function. For $M = 8$ with Gray coding and $r_b = 98000$ binary digits/sec, the average carrier amplitude at the receiver is measured to be 7 mV. Assuming the single sided noise power spectral density $N_o = 2 \times 10^{-11}$ W/Hz, determine the average number of symbol and binary digit errors made per second by the demodulator.

13. A digital communication system consists of a binary source, a channel encoder of code rate R_{code}, a BPSK modulator, AWGN channel, demodulator, decoder and a sink. It can be shown that if the channel encoder is a rate $1/3$ repetition code, then the probability of an error P_e in a binary digit at the output of the decoder is given by $P_e = 3\alpha^2(1-\alpha) + \alpha^3$, where α is the BSC crossover probability.

(a) Determine the value of P_e for a channel SNR of 0 dB if the information throughput is allowed to be reduced.

(b) Repeat part (a), but now for no reduction in the information through-put.

(c) What do you conclude about the error-control capability of the repe-tition code.

14. Consider an M-ary phase shift keying modulation scheme in which the signal transmitted is given by $s_i(t) = \sqrt{\dfrac{2E_s}{T_s}}\cos\left(2\pi f_c t + \dfrac{2\pi i}{M}\right)$, where E_s is the energy per signal, T_s is the time duration, f_c is the carrier frequency and $i = 0, 1, \ldots, (M-1)$.

(a) With the use of a channel encoder of code rate R_{code} and no reduction in the information throughput, show that $E_s = R_{code} E_b \log_2 M$, where E_b is the energy per binary digit.

(b) If an M-ary communication system transmits at a rate of 5000 symbols per second, what is the equivalent rate in binary digits per second for $M = 2$ and $M = 4$?

(c) What is the required transmission bandwidth to achieve a rate of 100000 binary digits per second using 16-PSK if error-control coding is not used ? With error-control coding, is the required bandwidth increased or decreased ? Explain your answer.

15. Suppose the probability density function (PDF) of a given BPSK signal point is rectangular as shown below. Determine an expression for the probability of an error in a binary digit. State any necessary assumptions.

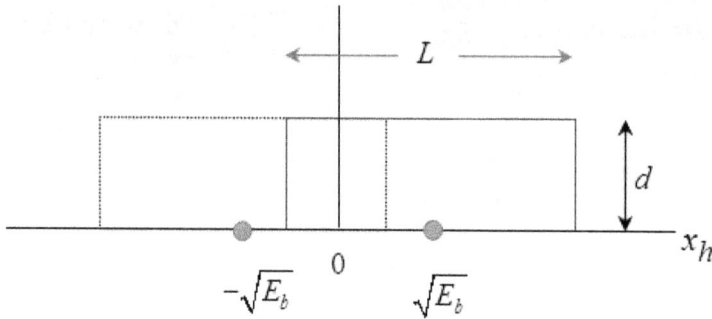

16. The front end of a DCS consists of a DMS which outputs one of four possible symbols, a source encoder and an MPSK modulator. If modulator outputs 4000 symbols per second,

(a)   What is the rate at which binary digits enter the modulator for $M = 2, 4, 8, 16, 32$ and 64.

(b)   If each symbol in the DMS is equally likely to be output, what is the rate at which symbols are output from the DMS for each value of M in part (a) ?

17. What is the required transmission bandwidth to ensure a date rate of 4000 binary digits per second using a rate $\frac{1}{2}$ FEC code together with MPSK for $M = 2, 4, 8, 16, 32$ and 64.

18.   Consider an M-ary digital communication system, in which the $M = 2^N$ equally likely signal points lie at the corners of a N-dimensional hypercube, centered at the origin.

(a) For $N = 3$, draw a diagram of the 3-dimensional hypercube signal and assign binary digits to each signal point using Gray coding.

(b) Show that the probability of an error in a binary digit P_e, for an AWGN channel is given by $P_e = \dfrac{1}{N}\left[1 - \left(1 - Q\left(\sqrt{\dfrac{2E_b}{N_o}}\right)\right)^N\right]$ where E_b is the energy per binary digit, N_o is the single sided noise power spectral density and $Q(.)$ is the Q-function.

19. Determine the autocorrelation function corresponding to the PSD of a random BPSK signal given by
$$PSD(f) = \frac{A^2 T_b}{4}\,\text{sinc}^2\left[T_b(f - f_c)\right] + \frac{A^2 T_b}{4}\,\text{sinc}^2\left[T_b(f + f_c)\right] \text{ for } A = 1$$
volt, $T_b = 1$ s and $f_c = 5$ Hz and the corresponding signal power.

20. Verify that for a given channel SNR, the distance between neighboring signal points for MPSK points are closer together then QAM signal points for $M > 4$.

21. (a) Using rectangular data pulses, what is the first and second null-to-null bandwidth of MPSK and QAM ?

(b) For $M = 4$, 16 and 64 compare the bandwidths for a data rate of 9600 bits/sec and 14400 bits/sec.

(c) The side-lobes can be reduced by employing raised cosine pulse shaping. In this case, the transmission bandwidth of a QAM or MPSK signal is given by $B_{shape} = \left(\dfrac{1+r}{T_s} \right)$, where r is the rolloff factor. Plot a graph of B_{shape} versus r (0, 0.25, 0.5, 0.75, 1) for the date rates of 9600 bits/sec and 14400 bits/sec.

22. Using coherent binary frequency shift keying (BFSK) with $r_b = 98000$ binary digits/sec, determine the average carrier power required to ensure that the probability of an error in a binary digit equal to 10^{-6}, 10^{-5}, 10^{-4} and 10^{-3}. Plot your results and comment. You may assume the function $Q(u) \approx \dfrac{1}{2\sqrt{\pi}} \exp\left(-\dfrac{u^2}{2} \right)$ and that the single sided noise power spectral density $N_o = 2 \times 10^{-11}$ W/Hz.

23. Show that the probability of an error in a binary digit for coherent binary FSK is given by $P_e = Q\left(\sqrt{\dfrac{E_b}{N_o}} \right)$.

24. Consider a coherent orthogonal 64-ary FSK system in which the symbol time $T_s = 0.5$ ms.

(a) If the probability of a symbol error is found to be 10^{-4}, what is the average probability of an error in a binary digit with the use of Gray coding ?

(b) What is the carrier amplitude A ? You may assume the function $Q(u) \approx \dfrac{1}{2\sqrt{\pi}} \exp\left(-\dfrac{u^2}{2} \right)$ and that the single sided noise power spectral density $N_o = 2 \times 10^{-11}$ W/Hz. How does A depend on T_s to maintain a given P_s ?

25. Consider a 4-ary FSK modulation scheme. Let the symbols A, B, C, D correspond to the four FSK signals and let the mapping scheme between a symbol and the pair of binary digits be as A (00), B (01), C (10), D (11). Each symbol is equally likely to be transmitted. If an all-zero sequence is transmitted, determine an expression for the probability of an error in a binary digit P_e in terms of the probability of a symbol error P_s. Make use of the fact that the symbol A is equally likely to be received as symbol B, C, or D. Generalize your analysis to show that for MFSK, $P_e = \dfrac{M P_s}{2(M-1)}$.

26. The horizontal signal space coordinate x_h for BPSK is given by

$$x_h = \int_0^{T_b} r(t)\phi_1(t)dt = \int_0^{T_b} s_i(t)\phi_1(t)dt + \int_0^{T_b} n(t)\phi_1(t)dt$$

$$= \left\{ \begin{array}{ll} \sqrt{E_b} + n_1 & if \quad s_i(t) = s_1(t) \\ -\sqrt{E_b} + n_2 & if \quad s_i(t) = s_2(t) \end{array} \right\}$$

Prove that for an AWGN channel, the Gaussian random variables n_1 and n_2 have zero mean and standard deviation $\sigma = \sqrt{\frac{N_o}{2}}$.

27. (a) If the QPSK signal points 0, 1, 2, 3 correspond to the complex coordinates $(-1 - i)$, $(1 - i)$, $(-1 + i)$, $(1 + i)$, respectively, draw the QPSK signal constellation diagram. Use this diagram to determine the energy per QPSK signal E_s and the energy per binary digit E_b.

(b) If $X_k = \begin{bmatrix} 1 - i \\ -1 + i \\ 1 - i \\ 1 + i \end{bmatrix}$, which signal points have been transmit-

ted. Using the function $Euler(n, k) = e^{j\frac{\pi n k}{2}}$, which for $k = 0 \cdots 3$ yields

$Euler(2, k) = \begin{bmatrix} 1 \\ -1 \\ 1 \\ -1 \end{bmatrix}$, verify that the real values of first sample value

$(n = 0)$ and the third sample value $(n = 2)$ both lie on the OFDM baseband signal $V_{baseband}(t)$ at the value of 2 volts. (Do not transform X_k into X_k').

28. The signal space for BPSK and the OFDM baseband signal $v_{bandband}(t)$ determined from the real part of $g'(t) = \sum_{k=0}^{K-1} X_k e^{j\frac{2\pi(k+1)t}{T}}$ are shown below. This OFDM baseband signal of duration $T = 1$ second corresponds to the transmission of the two signal points 01.

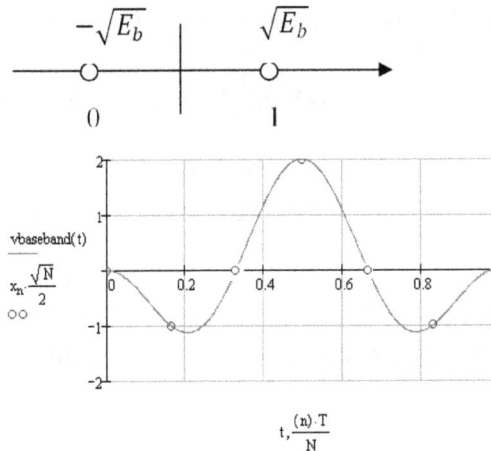

(a) Show that the complex signal point sequence $X_k = \begin{bmatrix} -1 \\ 1 \end{bmatrix}$ if we take E_b to be 1 joule and hence prove that $v_{bandband}(t) = -\cos\left(\frac{2\pi t}{T}\right) + \cos\left(\frac{4\pi t}{T}\right)$. Indicate which cos(.) component corresponds to the signal point 0 and which correspond to the signal point 1. Verify that these two component frequencies are separated by $\frac{1}{T}$ Hz.

(b) Given that the real samples x_n are given by the inverse discrete Fourier transform (IDFT) $x_n = \frac{1}{\sqrt{N}} \sum_{k=0}^{N-1} X'_k e^{j\frac{2\pi nk}{N}}$ where X'_k is determined from the complex signal point sequence X_k, verify that $x_1 \frac{\sqrt{N}}{2} = -1$ for the second sample point.

(c) How do we recover the transmitted signal points using these real samples of the OFDM baseband signal?

(d) Explain what happens as K is increased and derive an expression for the bandwidth of the OFDM baseband signal in terms of T_b, the time duration of a BPSK signal. What is the implication?

29. Signal points are transmitted simultaneously using orthogonal frequency division multiplexing (OFDM), in which each subcarrier is a QAM signal given by $s_i(t) = \sqrt{\frac{2}{T_s}} A \left[a_i \cos(2\pi f_c t) + b_i \sin(2\pi f_c t)\right]$, where $i = 0, 1, \cdots M-1$, $A = \sqrt{\frac{3E_b log_2 M}{2(M-1)}}$ with a_i and b_i from the set $\in \left\{\pm 1, \pm 3, \cdots, \left(\sqrt{M}-1\right)\right\}$. The basis functions are $\phi_1(t) = \sqrt{\frac{2}{T_s}} \cos(2\pi f_c t)$ and $\phi_2(t) = \sqrt{\frac{2}{T_s}} \sin(2\pi f_c t)$.

(a) Sketch the single-sided amplitude spectrum if $K = 10$ signal points are transmitted using OFDM with each subcarrier using a 4-QAM constellation with $T_s = 2$ seconds and $A = 1$.

(b) What is the value of the highest frequency spectral component and bandwidth of the OFDM signal?

(c) How can the bandwidth of the OFDM signal be reduced?

(d) Explain what happens to the amplitude spectrum as K is increased.

30. The figure below shows the signal space for 16-QAM, in which for example, the signal point 4 corresponds to the coordinates $(-1, +3i)$. Using Orthogonal Frequency Division Multiplexing (OFDM), the signal points $(0,7,7,10)$ are sent in parallel.

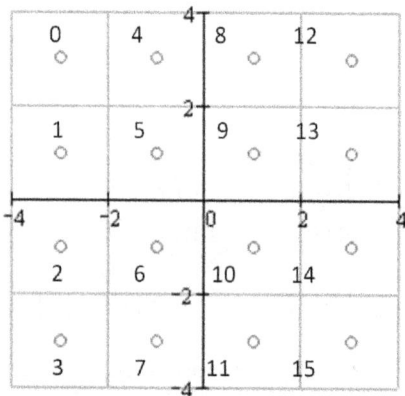

(a) Determine the corresponding X_k and X'_k complex sequences.

(b) Explain how X'_k is used to determine the real samples of the OFDM signal.

(c) For $K = 10$, the amplitude spectrum of an OFDM signal is shown below.

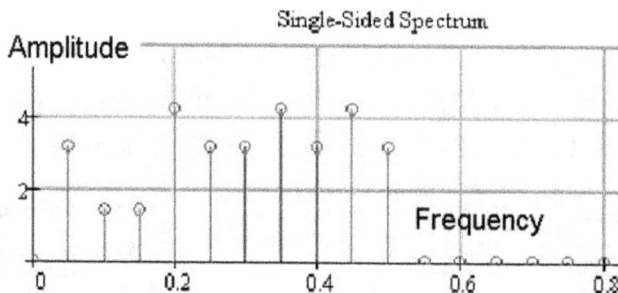

(i) What is the frequency resolution of the FFT used to determine this spectrum? (ii) What is the time duration T of the corresponding OFDM signal? (iii) How many real samples of the OFDM signal are available? (iv) What is the time duration T_s of a QAM signal? (v) What is the highest frequency within this OFDM signal? (vi) Determine the bandwidth of this OFDM signal as K is increased?

31. Consider the graph shown in Fig. 7.90 which compares various modulation schemes on Shannon's channel capacity curve for an AWGN channel for a probability of an error in a binary digit P_e of 10^{-5}. The key to the symbols are listed in Table 7.8.

(a) Comment on how the bandwidth efficiency of MPSK depends on M. What is the penalty incurred for an improved MPSK bandwidth efficiency and explain why this penalty occurs. For what type of channel is MPSK suitable?

(b) Which is better, QAM or MPSK and explain why.

(c) Comment on how the bandwidth efficiency of MFSK depends on M. What is the advantage of a reduction in MFSK bandwidth efficiency and explain why this occurs. For what type of channel is MFSK suitable?

(d) It can be shown that for coherent orthogonal MFSK, the probability of a symbol error $P_s \leq (M-1)\,Q\left(\sqrt{\frac{E_s}{N_o}}\right)$. For a 64-ary FSK system in which the symbol time $T_s = 0.1$ ms, (i) If the probability of a symbol error P_s is found to be 10^{-5}, what is the average probability of an error in a binary digit with the use of Gray coding ? (ii) What is the carrier amplitude A? You may assume $Q(u) \approx \frac{1}{2\sqrt{\pi}}\exp\left(-\frac{u^2}{2}\right)$ and that the single sided noise power spectral density $N_o = 10^{-11}$ W/Hz.

32. If the bandwidth available on a telephone line is B Hz, compare the data rate that can be supported over this bandwidth using QPSK only versus OFDM with QPSK subcarriers.

If you can't explain it to a six year old, you don't understand it yourself.
— *Albert Einstein*

Chapter 8

Block and Convolutional Codes

8.1 Introduction

8.1.1 FEC Codes

Block Codes

- Recap of repetition codes

- Concept of block codes

- Number of code words

The fundamentals of error control coding was established in section 3.3 in which we considered the performance of a simple repetition code. The importance of ensuring that the information throughput is not effected with the use of error-control coding was highlighted in section 3.3 with further details in section 7.8. The performance of a given error-control coding scheme is typically analyzed within the DCS illustrated in Fig. 8.1, with the combination of the modulator, AWGN channel and demodulator being modeled as a DMC.

Forward error correction (FEC) codes may be divided into two types, namely, *block* codes and *convolutional* codes. In block codes, the encoder segments the incoming binary stream into blocks of k digits and processes

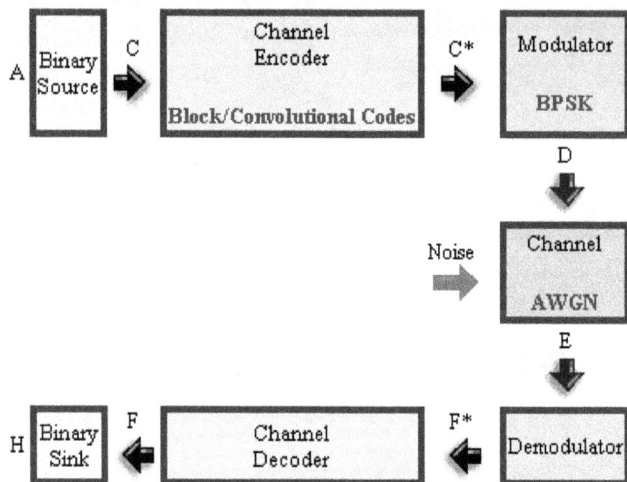

Figure 8.1: Chapter 8 DCS.

each block individually by adding redundancy (extra digits) according to a predefined algorithm (e.g. repetition code). However in convolutional codes, previous input blocks of k digits influence the output code word. In both cases, the output of the encoder is a code word of n code digits, where $n > k$. Block and covolutional codes are used extensivly in data, image, audio and video processing communication systems to protect the binary data. We shall begin with a definition of coding gain and apply Shannon's channel coding theorem before presenting the fundamentals of block and convolutional coding.

8.1.2   Coding Gain

▷ Coding Gain

- Concept of coding gain

- Advantage and disadvantage of error-control coding

- Asymptotic coding gain

- Throughput not reduced with the use of error-control coding

The objective of an error-control coding scheme is to provide a coding gain. For example, suppose Fig. 8.2 represents the performance of the DCS in Fig. 8.1. The curve labeled *uncoded* refers to the performance without the use of a channel encoder and decoder. The other curve is with the use of a channel encoder and decoder and with no reduction in the information throughput. For given binary digit error probability P_e at the binary sink, the difference between the required SNR for an uncoded and coded DCS is referred to as the *coding gain*. For example, the coding gain at $P_e=$ 0.00001 is approximately 1.7 dB in Fig. 8.2. The implication is that by using error-control coding, less signal energy per information binary digit is required to achieve a given P_e. Recall the energy of a BPSK signal depends on the amplitude and the time duration of the signal. Thus either of these two quantities may be reduced by an appropriate amount without sacrificing performance by compensating with error control coding. The *asymptotic coding gain* is the gain corresponding to a vanishing small value of P_e or equivalently, at a very high SNR.

8.1.3   Application of the Channel Coding Theorem

▶ **Application Coding Theorem**

- Channel capacity of a BSC using BPSK

- Minimum SNR required for essentially error-free communication using a rate 1/2 code with BPSK and hard-decision decoding

- Channel capacity with 8-level soft-decision

- Channel capacity with infinite soft-decision levels

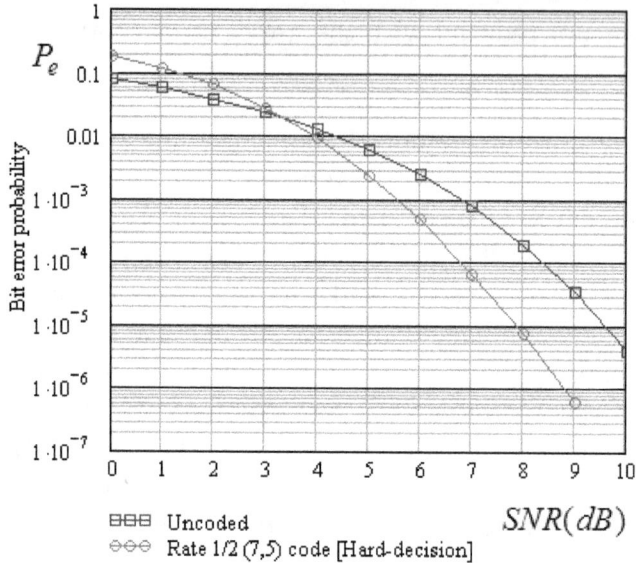

Figure 8.2: Example of how error-control coding improves performance.

- Minimum SNR required for a range of code rates and quantization levels

- Coding gain of soft-decision over hard-decision

Shannon's channel coding theorem was introduced in section 4.6. Using BPSK over an AWGN channel, we can use this theorem to determine the minimum channel SNR (dB) for virtually error-free communication simply from its code rate as follows. If the probability of a binary digit 0 and 1 output from the binary source in Fig. 8.1 is 0.5, then Shannon's channel coding theorem promises the existence of an error-control coding scheme which is capable of achieving an arbitrarily low probability of error provided $R_{code} \leq C_s$, or equivalently, $C_s \geq R_{code}$. If the BPSK demodulator makes hard-decisions, then the DMC is a BSC. In this case, recall from section 4.5.4 that the channel capacity is given by $C_s = 1 - \Omega(\alpha)$ bits/symbol, where the BSC crossover probability $\alpha = Q\left(\sqrt{\frac{2E_b R_{code}}{N_o}}\right)$ for BPSK. For $R_{code} = 1/2$, Fig. 8.3 shows the variation of the channel capacity C_s with the channel SNR. For $\left(\frac{E_b}{N_o}\right)$dB $= 1.77$ dB, we find that $R_{code} = C_s$. The implication of this result is that there is an error-control coding scheme which utilizes a rate 1/2 code that can achieve virtually error-free communication over an AWGN using BPSK with hard-decision decoding provided the channel SNR is greater than or equal to 1.77 dB.

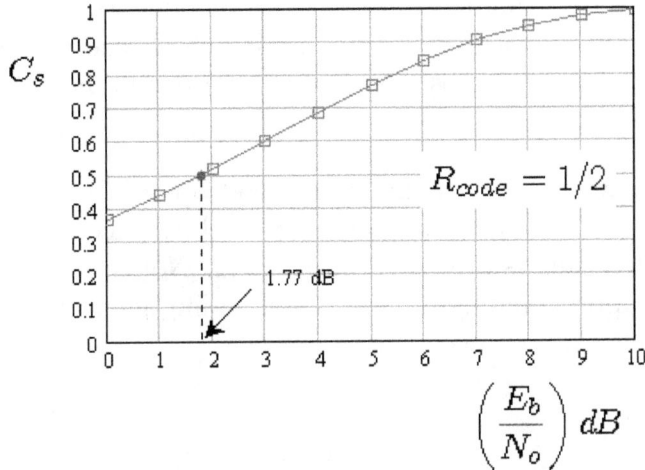

Figure 8.3: Channel capacity of an AWGN channel using a rate 1/2 code with BPSK and hard-decision decoding.

With soft-decision decoding, it can be shown [Odenwalder, 1985] that in general for a DMC with Q_{level} soft-decision demodulation, the channel capacity C_s is given by

$$C_s = \sum_{j=0}^{Q_{level}-1} P(j|0) \log_2 \left[\frac{2P(j|0)}{P(j|0) + P(j|1)} \right] \text{ bits/symbol} \qquad (8.1)$$

where a given forward transition probability $P(j|0)$ is easily determined as shown in section 7.7. If the quantization level is increased to infinity, then it can be shown [Odenwalder, 1985] that the channel capacity is given by

$$C_s = \frac{1}{2} \log_2 \left(1 + 2R_{code} \frac{E_b}{N_o} \right) \text{ bits/symbol} \qquad (8.2)$$

Figure 8.4 presents the curves for the minimum channel SNR in dB versus the code rate necessary for essentially error-free communication using hard-decision, 8-level and infinite soft-decision decoding. Table 8.1 summarizes some of the key results. Rate 1/2 turbo codes [Berrou, Glavieux and Thitimajshima, 1993] and low density parity check codes [Gallager, 1962] can provide a performance close (within 0.7 dB) to the theoretical minimum SNR of 0 dB.

Figure 8.5 shows the coding gain over uncoded BPSK operating at a SNR of 9.6 dB. The additional coding gain of 8-level soft-decision decoding over hard-decision decoding is presented in Fig. 8.6. Notice for a rate 1/2 code, we can expect approximately an additional coding gain of 1.5 dB with the use of 8-level soft-decision decoding.

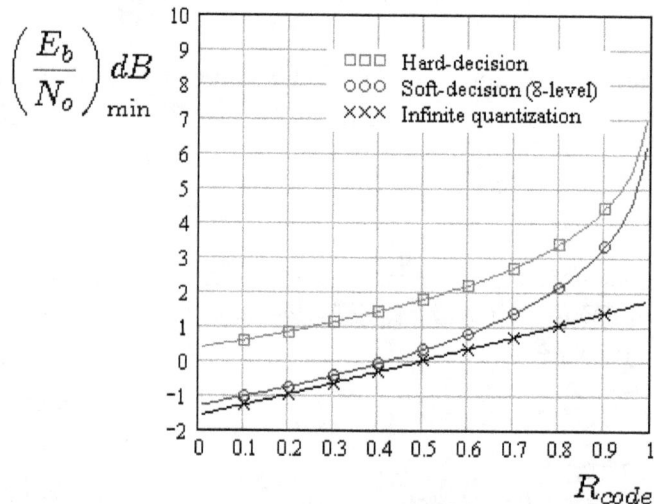

Figure 8.4: Minimum SNR for essentially error-free communication and the expected coding gain.

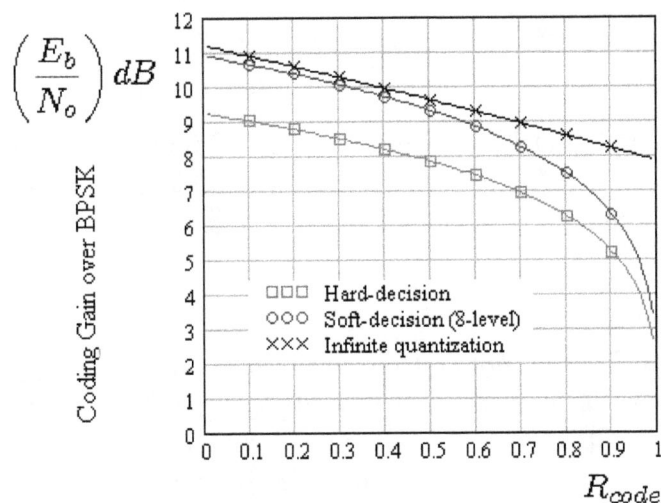

Figure 8.5: Coding gain over BPSK at a SNR of 9.6 dB over an AWGN channel.

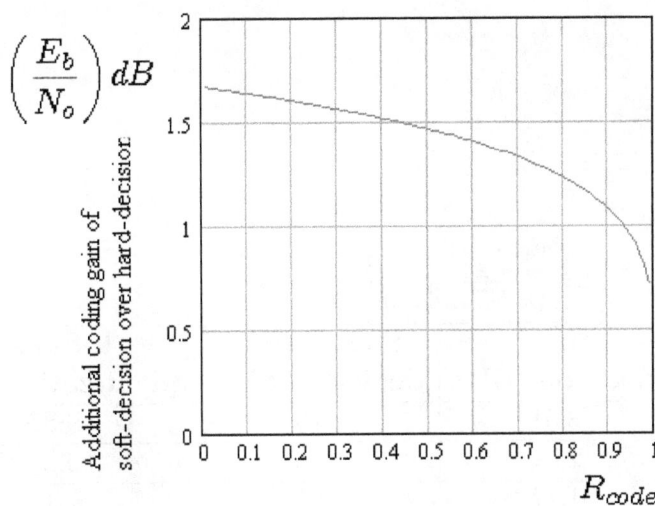

Figure 8.6: Additional coding gain of soft-decision over hard-decision.

- ⭐ SIMULATION **ChannelCoding:** Simulation corresponding to Figs. 8.2, 8.4, 8.5 and 8.6.   Alter the soft-decision quantization level spacing from its set value of $\frac{2}{7}$ to observe its influence on the results in Table 8.1.

Referring to Fig. 8.4, notice that as the code rate R_{code} approaches zero, the minimum SNR required with infinite quantization level approaches - 1.6 dB. This result agrees with Shannon's channel capacity theorem to be presented in section 8.11, which defines the fundamental limit on the maximum rate at which a communication system can operate error-free for a band-limited, power-limited Gaussian channel.

Code Rate	Hard decision (dB)	8-level soft decision (dB)	Infinite quantization (dB)
3/4	3.01	1.73	0.86
2/3	2.51	1.16	0.57
1/2	1.77	0.31	0
1/4	0.97	-0.61	-0.82

Table 8.1: Minimum channel SNR in dB necessary for virtually error-free communication using BPSK over an AWGN channel.

8.2 Linear Block Codes

8.2.1 Fundamentals

Linear Block Codes

- Generator matrix

- Modulo-2 addition

- Multiplication of two matrices

- Systematic block code

- Linear nature of block codes

- Hamming distance and weight

- Error-correcting capability of a simple block code

The code rate R_{code} of a block code is given by $R_{code} = \frac{k}{n}$, where k information binary digits are mapped onto a code word of n digits i.e. for every k binary digits input to the block code encoder, n code digits are output. A block code is conveniently referred to as a (n, k) block code. In general, a code word $\mathbf{c}$ is given by

$$\mathbf{c} = \mathbf{mG} \tag{8.3}$$

where $\mathbf{m}$ is the *message* vector, or equivalently the *information* vector and $\mathbf{G}$ is a matrix with n columns and k rows. For example, consider the following generator matrix:

$$\mathbf{G} = \begin{bmatrix} 1 & 0 & 0 & 0 & 1 & 1 & 0 \\ 0 & 1 & 0 & 0 & 1 & 0 & 1 \\ 0 & 0 & 1 & 0 & 0 & 1 & 1 \\ 0 & 0 & 0 & 1 & 1 & 1 & 1 \end{bmatrix} \tag{8.4}$$

Given that there are 7 columns and 4 rows, this is a $(7, 4)$ block code. For $\mathbf{m} = \begin{bmatrix} 1 & 0 & 1 & 0 \end{bmatrix}$, the corresponding code word is given by

$$\mathbf{c} = \begin{bmatrix} 1 & 0 & 1 & 0 \end{bmatrix} \begin{bmatrix} 1 & 0 & 0 & 0 & 1 & 1 & 0 \\ 0 & 1 & 0 & 0 & 1 & 0 & 1 \\ 0 & 0 & 1 & 0 & 0 & 1 & 1 \\ 0 & 0 & 0 & 1 & 1 & 1 & 1 \end{bmatrix} \tag{8.5}$$

$$= \begin{bmatrix} 1 & 0 & 1 & 0 & 1 & 0 & 1 \end{bmatrix} \tag{8.6}$$

where the addition is modulo-2 as illustrated in the table below.

Modulo-2 addition		
A	B	A $\oplus$ B
0	0	0
0	1	1
1	0	1
1	1	0

(8.7)

An interesting feature of this $(7, 4)$ block code is that its *systematic*, which means that a code word contains the message or information vector unaltered as part of the code word. In this case, the first four digits are the input k digits and $(7 - 4) = 3$ code digits are appended to form a code word. For all possible input combinations of k digits, namely 2^k, the corresponding code words are listed in Table 8.2. Out of the entire 2^n possible n digit words, only 2^k words are utilized as code words. A decoding error will occur if the noise within the channel is such that the received word is interpreted as a code word other than the transmitted code word.

- ⭐ SIMULATION **Generator74:** Simulation to determine the code words listed in Table 8.2.

A *linear* block code has the unique feature that the modulo-2 addition of any two code words produces another code word. Selecting two code words at random from Table 8.2 for the (7, 4) code,

$$
\begin{array}{llccccccc}
\mathbf{c1} & & 1 & 0 & 1 & 0 & 1 & 0 & 1 \\
\mathbf{c2} & + & 1 & 1 & 1 & 0 & 0 & 0 & 0 \\
\hline
\mathbf{c3} & & 0 & 1 & 0 & 0 & 1 & 0 & 1
\end{array}
\tag{8.8}
$$

where **c3** is also a code word. The *hamming distance* between any two binary words of the same length is equal to be the total number of differing binary digits. For example, the hamming distance between $\mathbf{c1} = 1010101$ and $\mathbf{c2} = 1110000$ is 3. The total number of ones in a code word is referred to as its *Hamming weight*. Notice that the hamming weight of $\mathbf{c3} = \mathbf{c2} + \mathbf{c1}$ is equal to the hamming distance between **c1** and **c2**. Thus, we need only count the number of ones in each code word listed in Table 8.2 to establish the distance of each code word from the all zero code word (000000). This is because the hamming weight of a code word is equal to its hamming distance from the all-zero code word. Furthermore, referring to Table 8.2, the linearity feature ensures that since there are 7 code words a distance 4 from the all-zero code word, then there will also be 7 code words a distance 4 from any other non-zero code word. Hence, we need only consider the all-zero code word to establish the distance properties of a linear block code.

- ⭐ SIMULATION **HammingDistance74:** The hamming distance between each code word and the all-zero code word is determined for the $(7, 4)$ block code.

For the systematic $(7, 4)$ block code in Table 8.2, we find that the *minimum distance* $d_{\min}$ between any two code words is 3. The implication is that if the received word from the channel has a single digit in error, it will still be closer to the transmitted code word. However, if there are two or more code digit errors, then the received word may be closer to one of the other code words. In general, a linear block code is able to correct all up to t errors if and only if

$$t \le \left\lfloor \frac{(d_{\min} - 1)}{2} \right\rfloor \tag{8.9}$$

where $\lfloor x \rfloor$ denotes the largest integer less than or equal to x. For this $(7, 4)$ block code, $t \le \lfloor 1 \rfloor$.

8.2.2   Systematic Block Codes

▶ **Systematic Block Codes**

- Generator matrix of a systematic block code

  is shown to be the combination of an identity and

  parity-check matrix

The generator matrix of a **systematic** block code consists of a parity matrix $\mathbf{P}$ appended to an identity matrix $\mathbf{I}_k$ of size k such that $\mathbf{G} = \begin{bmatrix} \mathbf{I}_k & \mathbf{P} \end{bmatrix}$. For example, referring to the $(7, 4)$ systematic block code

$$\mathbf{I}_4 = \begin{bmatrix} 1 & 0 & 0 & 0 \\ 0 & 1 & 0 & 0 \\ 0 & 0 & 1 & 0 \\ 0 & 0 & 0 & 1 \end{bmatrix} \text{ and } \mathbf{P} = \begin{bmatrix} 1 & 1 & 0 \\ 1 & 0 & 1 \\ 0 & 1 & 1 \\ 1 & 1 & 1 \end{bmatrix} \tag{8.10}$$

so that

$$\mathbf{G} = \begin{bmatrix} 1 & 0 & 0 & 0 & 1 & 1 & 0 \\ 0 & 1 & 0 & 0 & 1 & 0 & 1 \\ 0 & 0 & 1 & 0 & 0 & 1 & 1 \\ 0 & 0 & 0 & 1 & 1 & 1 & 1 \end{bmatrix} \tag{8.11}$$

The parity matrix $\mathbf{P}$ is selected to ensure that the k rows of the generator matrix are linearly independent of each other i.e. it is not possible to express any row of the generator matrix, which is also a code word, as a linear combination of the remaining rows.

8.2.3 Parity-Check Matrix and the Syndrome

▶ **Parity-Check Matrix**

- Construction of the **H** matrix

- Difference between the transmitted code word
  and the received noisy word

- Syndrome concept

- Using the syndrome to correct channel errors

- Undetected channel errors

To decode a received code word, we may utilize the syndrome **s** defined by

$$\mathbf{s} = \mathbf{r}\mathbf{H}^T \tag{8.12}$$

where **r** is the received word and **H** is a *parity-check matrix*

$$\mathbf{H} = \begin{bmatrix} \mathbf{P}^T & \mathbf{I}_{(n-k)} \end{bmatrix} \tag{8.13}$$

created from the generator matrix

$$\mathbf{G} = \begin{bmatrix} \mathbf{I}_k & \mathbf{P} \end{bmatrix} \tag{8.14}$$

The structure of the matrix **H** has is designed to ensure that

$$\mathbf{G}\mathbf{H}^T = \mathbf{0} \tag{8.15}$$

where $\mathbf{0}$ is a k by $(n - k)$ all-zero matrix. If $\mathbf{r}$ is a valid code word, then

$$\mathbf{s} = \mathbf{r}\mathbf{H}^T = \mathbf{m}\left(\mathbf{G}\mathbf{H}^T\right) = 0 \qquad (8.16)$$

For example, for the $(7, 4)$ systematic block code, the transpose of the parity matrix $\mathbf{P}$ is given by

$$\mathbf{P}^T = \begin{bmatrix} 1 & 1 & 0 & 1 \\ 1 & 0 & 1 & 1 \\ 0 & 1 & 1 & 1 \end{bmatrix} \qquad (8.17)$$

and since

$$\mathbf{I}_{n-k} = \mathbf{I}_{7-4} = \mathbf{I}_3 = \begin{bmatrix} 1 & 0 & 0 \\ 0 & 1 & 0 \\ 0 & 0 & 1 \end{bmatrix} \qquad (8.18)$$

we find

$$\mathbf{H} = \begin{bmatrix} \mathbf{P}^T & \mathbf{I}_{(n-k)} \end{bmatrix} = \begin{bmatrix} 1 & 1 & 0 & 1 & 1 & 0 & 0 \\ 1 & 0 & 1 & 1 & 0 & 1 & 0 \\ 0 & 1 & 1 & 1 & 0 & 0 & 1 \end{bmatrix} \qquad (8.19)$$

to confirm that

$$\mathbf{G}\mathbf{H}^T = \begin{bmatrix} 1 & 0 & 0 & 0 & 1 & 1 & 0 \\ 0 & 1 & 0 & 0 & 1 & 0 & 1 \\ 0 & 0 & 1 & 0 & 0 & 1 & 1 \\ 0 & 0 & 0 & 1 & 1 & 1 & 1 \end{bmatrix} \begin{bmatrix} 1 & 1 & 0 \\ 1 & 0 & 1 \\ 0 & 1 & 1 \\ 1 & 1 & 1 \\ 1 & 0 & 0 \\ 0 & 1 & 0 \\ 0 & 0 & 1 \end{bmatrix} = \begin{bmatrix} 0 & 0 & 0 \\ 0 & 0 & 0 \\ 0 & 0 & 0 \\ 0 & 0 & 0 \end{bmatrix} = \mathbf{0} \quad (8.20)$$

- ⭐ SIMULATION **Parity74:** Verification that $\mathbf{GH}^T = 0$. Alter the generator transfer function matrix to see what happens.

Let $\mathbf{e}$ represent the error vector such that $\mathbf{r} = \mathbf{c} + \mathbf{e}$, where $\mathbf{c}$ is the transmitted code word. For example, if $\mathbf{c} = 1010101$ and $\mathbf{e} = 0000010$ then $\mathbf{r} = 1010111$. In this case

$$\mathbf{s} = \mathbf{rH}^T = (\mathbf{c} + \mathbf{e})\,\mathbf{H}^T = \mathbf{cH}^T + \mathbf{eH}^T. \tag{8.21}$$

But since $\mathbf{c} = \mathbf{mG}$, we find $\mathbf{s} = \mathbf{mGH}^T + \mathbf{eH}^T$. However $\mathbf{GH}^T = 0$, and therefore the syndrome is simply given by

$$\mathbf{s} = \mathbf{eH}^T \tag{8.22}$$

Since the minimum distance of this block code is 3, we can correct all single digit errors in the receive word $\mathbf{r}$ can be corrected by evaluating the syndrome $\mathbf{s} = \mathbf{rH}^T$ and using a look-up table to find the error vector that produces the same syndrome. For example, this look-up table, which is referred to as the *decoding table*, for the (7,4) code is presented in Table 8.3. The estimated code word $\widehat{\mathbf{c}}$ is then given by $(\mathbf{r} + \mathbf{e})$. For example, if $\mathbf{r} = 1010111$, then the syndrome is given by

$$\mathbf{s} = \mathbf{rH}^T = \begin{bmatrix} 1 & 0 & 1 & 0 & 1 & 1 & 1 \end{bmatrix} \begin{bmatrix} 1 & 1 & 0 \\ 1 & 0 & 1 \\ 0 & 1 & 1 \\ 1 & 1 & 1 \\ 1 & 0 & 0 \\ 0 & 1 & 0 \\ 0 & 0 & 1 \end{bmatrix}$$

$$= \begin{bmatrix} (1+1) & (1+1+1) & (1+1) \end{bmatrix} = \begin{bmatrix} 0 & 1 & 0 \end{bmatrix} \tag{8.23}$$

But $\mathbf{s} = 010$ corresponds to the error vector $\mathbf{e} = 0000010$. Thus the estimated code word $\widehat{\mathbf{c}}$ is given by

$$
\begin{array}{rccccccc}
\mathbf{r} & 1 & 0 & 1 & 0 & 1 & 1 & 1 \\
\mathbf{e} \ + & 0 & 0 & 0 & 0 & 0 & 1 & 0 \\
\hline
\widehat{\mathbf{c}} & 1 & 0 & 1 & 0 & 1 & 0 & 1
\end{array}
\qquad (8.24)
$$

The error detection and correction capability of this (7, 4) block code is very poor. For example, if $\mathbf{e} = 0010110$, then the transmitted code word $\mathbf{c} = 1010101$ would be received as $\mathbf{r} = 1000110$ which is a valid code word. In this case, the syndrome is zero and the channel errors are undetected by the decoder.

- ⭐ SIMULATION **Syndrome74:** To determine the entries in Table 8.3 and verify the foregoing analysis. Experiment with other error vectors e.g. with more than one error.

8.2.4 Hamming Codes

▶ **Hamming Codes**

- Construction of Hamming codes

- (7, 4) systematic Hamming block code

- (15,11) systematic Hamming block code

A (n, k) block code is referred to as a *Hamming* code if

$$
n = 2^m - 1 \qquad (8.25)
$$

$$
k = 2^m - 1 - m \qquad (8.26)
$$

where m is any positive integer and the columns of the corresponding parity-check matrix $\mathbf{H}$ consist of all possible m-digit patterns of ones and zeros except the all-zero pattern. For example, the $(7, 4)$ block code considered in this section is a Hamming code, for which $m = 3$. Referring to its parity-check matrix (rewritten here for convenience)

$$\mathbf{H} = \begin{bmatrix} 1 & 1 & 0 & 1 & 1 & 0 & 0 \\ 1 & 0 & 1 & 1 & 0 & 1 & 0 \\ 0 & 1 & 1 & 1 & 0 & 0 & 1 \end{bmatrix} \tag{8.27}$$

notice that all possible 3-digit patterns of ones and zeros are contained within the columns and that the column vector $\begin{bmatrix} 0 \\ 0 \\ 0 \end{bmatrix}$ is not included. Based on this property of a Hamming code, we can simply write down the parity-check matrix for say a $(15, 11)$ Hamming code with $m = 4$ as follows:

$$\mathbf{H} = \begin{bmatrix} 0 & 1 & 0 & 1 & 0 & 1 & 0 & 1 & 0 & 1 & 0 & 1 & 1 & 0 & 1 \\ 0 & 0 & 1 & 1 & 0 & 0 & 1 & 1 & 0 & 0 & 1 & 1 & 0 & 1 & 1 \\ 1 & 1 & 1 & 1 & 0 & 0 & 0 & 0 & 1 & 1 & 1 & 1 & 0 & 0 & 1 \\ 1 & 1 & 1 & 1 & 1 & 1 & 1 & 1 & 0 & 0 & 0 & 0 & 0 & 0 & 1 \end{bmatrix} \tag{8.28}$$

The order of the columns does not matter, but to ensure the systematic feature of the block code, the columns are rearranged so that the last $(n - k)$ columns correspond to a unit matrix $\mathbf{I}_{(n-k)}$, so that $\mathbf{H}$ and the generator matrix $\mathbf{G}$ are of the form

$$\mathbf{H} = \begin{bmatrix} \mathbf{P}^T & \mathbf{I}_{(n-k)} \end{bmatrix} \tag{8.29}$$

$$\mathbf{G} = \begin{bmatrix} \mathbf{I}_k & \mathbf{P} \end{bmatrix} \tag{8.30}$$

- SIMULATION **Switch74:** The parity matrix $\mathbf{P}$ columns are switched within $\mathbf{G} = \begin{bmatrix} \mathbf{I}_k & \mathbf{P} \end{bmatrix}$.

8.2.5 Simulation : (7, 4) Block Code

SIMULATION **Block74:** The decoder for the $(7, 4)$ block code is shown to correct only single code digit errors.

- **Experiment:** Consider multiple channel errors. Implement the DCS shown in Fig. 8.1 to determine the coding gain of the $(15, 11)$ block code.

Input	Code word
0000	0000000
0001	0001111
0010	0010011
0011	0011100
0100	0100101
0101	0101010
0110	0110110
0111	0111001
1000	1000110
1001	1001001
1010	1010101
1011	1011010
1100	1100011
1101	1101100
1110	1110000
1111	1111111

Table 8.2: (7,4) systematic block code.

Syndrome $\mathbf{s}$			Error vector $\mathbf{e}$						
0	0	0	0	0	0	0	0	0	0
0	0	1	0	0	0	0	0	0	1
0	1	0	0	0	0	0	0	1	0
1	0	0	0	0	0	0	1	0	0
1	1	1	0	0	0	1	0	0	0
0	1	1	0	0	1	0	0	0	0
1	0	1	0	1	0	0	0	0	0
1	1	0	1	0	0	0	0	0	0

Table 8.3: (7,4) Decoding table.

8.3   Cyclic Codes

8.3.1   Fundamentals

▶ **Cyclic Codes**

- Introduction to cyclic codes

- Code polynomial

- Cyclic shift of a code word

Cyclic codes are a subset of linear block codes whose structural properties inherently provide good protection against a burst of channel errors. The linearity property ensures that the sum of any two code words is also a code word, but now with the following additional feature. If $(c_{n-1}, c_{n-2}, \cdots, c_1, c_0)$ denotes a code word of length n, then a cyclic shift of a code word **c** to the **left** is also a code word i.e. $\mathbf{c} = (c_{n-2}, \cdots, c_1, c_0, c_{n-1})$ is a code word. For example, if $(c_{n-1}, c_{n-2}, \cdots, c_1, c_0) = (1010011)$, then the following cyclic shifts are also code words,

$$
\begin{array}{c}
1010011 \\
0100111 \\
1001110 \\
0011101 \\
0111010 \\
1110100 \\
1101001
\end{array}
\tag{8.31}
$$

Not all code words can be produced by shifting a single code word. We have to make use of the linearity feature (addition of two code words to produce another code word) together with cyclic shifts to generate all the code words. This process is easily illustrated by a polynomial representation of a code word. A *code polynomial* is given by

$$c(X) = c_{n-1}X^{n-1} + \cdots + c_2X^2 + c_1X + c_0 \qquad (8.32)$$

where for a binary code, the coefficients $c_{n-1}, \ldots, c_1, c_0$ are 0's and 1's and the additions are modulo 2. For example, the code word (1010011) is represented by the polynomial $c(X) = X^6 + X^4 + X + 1$. If we multiply $c(X)$ by X, then $c(X)$ will shift to the left so that

$$Xc(X) = X\left(X^6 + X^4 + X + 1\right) = X^7 + X^5 + X^2 + X$$

which is equivalent to 10100110. However, to produce a **cyclic** shift, that is to shift (1010011) to (0100111), we need to add $(X^7 + 1)$ to $(X^7 + X^5 + X^2 + X)$ to yield

$$X^7 + X^5 + X^2 + X + X^7 + 1 = X^5 + X^2 + X + 1 \qquad (8.33)$$

which is equivalent to the required cyclic shifted code word (0100111). The cyclic shift process is illustrated below.

code word	1010011
shift to the left	10100110
add $(X^7 + 1)$	10000001
Answer	00100111

where the addition of $(X^7 + 1)$ is only required if a shift to the left produces a X^n term.

8.3.2   Generator Polynomial

▶ Systematic Cyclic Codes

- Introduction

- Generator polynomial

- Parity polynomial from the message polynomial

- (7,4) systematic cyclic block code

- Polynomial division

- Code word from the message and parity polynomial

- Valid code word check

Let $g(X)$ represent a polynomial of degree $(n-k)$ which generates a (n,k) cyclic block code. Then a code word polynomial $c(X)$ is given by

$$c(X) = a(X)g(X) \qquad (8.34)$$

where $a(X)$ is a quotient polynomial with degree $(k-1)$. For example, consider a (n,k) *systematic* cyclic block code. Recall that for a systematic code word, the information sequence is contained within the code word. Let $m(X)$ represent the information sequence of k digits and let $p(X)$ represent the parity sequence of $(n-k)$ digits. Then a systematic code word is given by

$$c(X) = X^{n-k}m(X) + p(X) \qquad (8.35)$$

where the multiplication by X^{n-k} shifts $m(X)$ by $(n-k)$ places to the left. If we now add $p(X)$ to both sides of the equation,

$$c(X) + p(X) = X^{n-k}m(X) \qquad (8.36)$$

or equivalently

$$X^{n-k}m(X) = a(X)g(X) + p(X) \qquad (8.37)$$

Now dividing by $g(X)$,

$$\frac{X^{n-k}m(X)}{g(X)} = a(X) + \frac{p(X)}{g(X)} \tag{8.38}$$

implying that if we take modulo $g(X)$ of the shifted message sequence $X^{n-k}m(X)$ (i.e. divide by $g(X)$) then the **remainder** is equal to $p(X)$ because the degree of $g(X)$ is larger than that of $p(X)$. To illustrate the encoding process by an example, consider the generator polynomial

$$g(X) = X^3 + X + 1 \tag{8.39}$$

which generates a $(7,4)$ systematic cyclic code word.   To encode the message sequence 1010, we undertake the following steps

$m(X) = X^3 + X \equiv 1010$
$X^{n-k}m(X) = X^3 m(X) = X^6 + X^4 \equiv 1010000$
Now we divide $X^{n-k}m(X) \equiv 1010000$ by $g(X) \equiv 1011$ as follows.

$$
\begin{array}{r}
\ \ 1\ \ 0\ \ 0\ \ 1 \quad a(X) \\
1\ \ 0\ \ 1\ \ 1\,\overline{|\,1\ \ 0\ \ 1\ \ 0\ \ 0\ \ 0\ \ 0} \\
1\ \ 0\ \ 1\ \ 1 \\
\overline{\ 1\ \ 0\ \ 0\ \ 0} \\
1\ \ 0\ \ 1\ \ 1 \\
\overline{\ 0\ \ 1\ \ 1 \quad p(X)}
\end{array}
\tag{8.40}
$$

The remainder 011 are the parity digits and therefore the code word is given by

$$
\begin{array}{r}
1\ \ 0\ \ 1\ \ 0\ \ 0\ \ 0\ \ 0 \quad X^{n-k}m(X) \\
+ \quad 0\ \ 1\ \ 1 \quad p(X) \\
\hline
1\ \ 0\ \ 1\ \ 0\ \ 0\ \ 1\ \ 1 \quad c(X)
\end{array}
\tag{8.41}
$$

or equivalently, since $p(X) = X + 1$ and $X^{n-k}m(X) = X^6 + X^4$,

$$c(X) = X^{n-k}m(X) + p(X) = X^6 + X^4 + X + 1 \equiv 1010011 \qquad (8.42)$$

with the quotient $a(X) = (X^3 + 1)$. To check that the long division was carried out correctly, we can confirm that $X^{n-k}m(X) = a(X)g(X) + p(X)$ as follows:

$$a(X)g(X) + p(X) = (X^3 + 1)(X^3 + X + 1) + (X + 1)$$

$$= X^6 + X^4 + X^3 + X^3 + X + 1 + X + 1 = X^6 + X^4 \equiv 1010000 \qquad (8.43)$$

as expected. Furthermore, as an additional check

$$c(X) = a(X)g(X) = (X^3 + 1)(X^3 + X + 1)$$

$$= X^6 + X^4 + X^3 + X^3 + X + 1 = X^6 + X^4 + X + 1 \equiv 1010011 \qquad (8.44)$$

as expected. To confirm that $c(X)$ is a code word generated by $g(X)$, we can simply divide $c(X)$ by $g(X)$ and confirm that the remainder is zero i.e.

$$
\begin{array}{r}
\phantom{1\ 0\ 1\ 1\,\overline{)}}1\ \ 0\ \ 0\ \ 1\quad a(X) \\
1\ 0\ 1\ 1\,\overline{)\,1\ \ 0\ \ 1\ \ 0\ \ 0\ \ 1\ \ 1} \\
\underline{1\ \ 0\ \ 1\ \ 1} \\
1\ \ 0\ \ 1\ \ 1 \\
\underline{1\ \ 0\ \ 1\ \ 1} \\
0\ \ 0\ \ 0
\end{array}
\qquad (8.45)
$$

Based on the analysis presented in the previous subsection, we can expect that the generator polynomial of a (n, k) cyclic code must be a factor of

$(X^n + 1)$. In terms of our ongoing example, $(X^7 + 1)$ divided by $g(X) = X^3 + X + 1$ leaves a remainder of zero i.e.

$$
\begin{array}{r}
1 \;\; 0 \;\; 1 \;\; 1 \;\; 1 \\
1 \;\; 0 \;\; 1 \;\; 1 \;\overline{)\; 1 \;\; 0 \;\; 0 \;\; 0 \;\; 0 \;\; 0 \;\; 0 \;\; 1} \\
1 \;\; 0 \;\; 1 \;\; 1 \\
\hline
1 \;\; 1 \;\; 0 \;\; 0 \;\; 0 \;\; 1 \\
1 \;\; 0 \;\; 1 \;\; 1 \\
\hline
1 \;\; 1 \;\; 1 \;\; 0 \;\; 1 \\
1 \;\; 0 \;\; 1 \;\; 1 \\
\hline
1 \;\; 0 \;\; 1 \;\; 1 \\
1 \;\; 0 \;\; 1 \;\; 1 \\
\hline
0 \;\; 0 \;\; 0
\end{array}
\tag{8.46}
$$

8.3.3 Cyclic Code Encoder

▶ **Cyclic Code Encoder**

- Type I encoder

- Type II encoder

- General encoder

The encoder for the $(7,4)$ cyclic code with the generator polynomial $g(X) = X^3 + X + 1 \equiv 1011$ is presented in Fig. 8.7. We shall refer to the configuration of the encoder as type I. To clarify the operation of the encoder, the state of the shift register before and after the input of the binary digit x is presented in Fig. 8.7. The message sequence to be encoded is 0110. The table in this figure shows how the contents of the shift register contain the parity bits once the message sequence has been inserted and flushed with $(n-k)$ zeros. This type of configuration simulates the division of $X^{n-k}m(X)$ by $g(X)$, with the remainder being the contents of the shift registers after flushing with $(n-k)$ zeros.

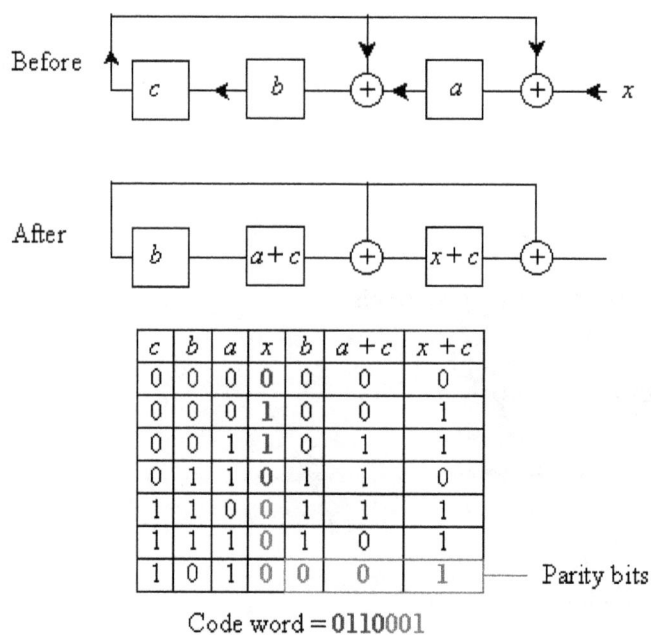

c	b	a	x	b	$a+c$	$x+c$
0	0	0	0	0	0	0
0	0	0	1	0	0	1
0	0	1	1	0	1	1
0	1	1	0	1	1	0
1	1	0	0	1	1	1
1	1	1	0	1	0	1
1	0	1	0	0	0	1

—— Parity bits

Code word $= \mathbf{0110001}$

Figure 8.7: Type I $(7, 4)$ cyclic code encoder.

To confirm the code word in Fig. 8.7,

$$
\begin{array}{r}
 1\ \ 1\ \ 1\ \ \ a(X)\\
1\ 0\ 1\ 1 \overline{| 0\ \ 1\ \ 1\ \ 0\ \ 0\ \ 0\ \ 0}\\
1\ \ 0\ \ 1\ \ 1\\
\overline{1\ \ 1\ \ 1\ \ 0\ \ 0}\\
1\ \ 0\ \ 1\ \ 1\\
\overline{1\ \ 0\ \ 1\ \ 0}\\
1\ \ 0\ \ 1\ \ 1\\
\overline{0\ \ 0\ \ 1}\ \ p(X)
\end{array}
$$

(8.47)

from which the code word is given by

$$
\begin{array}{r}
0\ \ 1\ \ 1\ \ 0\ \ 0\ \ 0\ \ 0\ \ \ X^{n-k}m(X)\\
+\ \ \ \ \ \ \ \ \ \ \ \ \ \ \ \ \ \ 0\ \ 0\ \ 1\ \ \ p(X)\\
\overline{0\ \ 1\ \ 1\ \ 0\ \ 0\ \ 0\ \ 1\ \ \ c(X)}
\end{array}
$$

(8.48)

To avoid flushing the encoder with zeros, a more efficient encoder configuration, which we shall refer to as type II, is presented in Fig. 8.8. Once again the same information sequence is encoded to illustrate the encoder operation. Using this encoder, the code words corresponding to all possible input combinations of 4 digits are listed in Table 8.4.

A general form of a cyclic code encoder is presented in Fig. 8.9, where the tap connections depend on the generator polynomial

$$
g(X) = g_{n-k}X^{n-k} + \cdots + g_2X^2 + g_1X + g_0
$$

(8.49)

For example, for the $(7, 4)$ cyclic code with $g(X) = X^3 + X + 1$, we have $g_{n-k} = g_3 = 1$, $g_2 = 0$, $g_1 = 1$, $g_0 = 1$. Notice that the g_{n-k} and g_0 tap connections are always connected so that

$$
g_{n-k} = g_0 = 1
$$

(8.50)

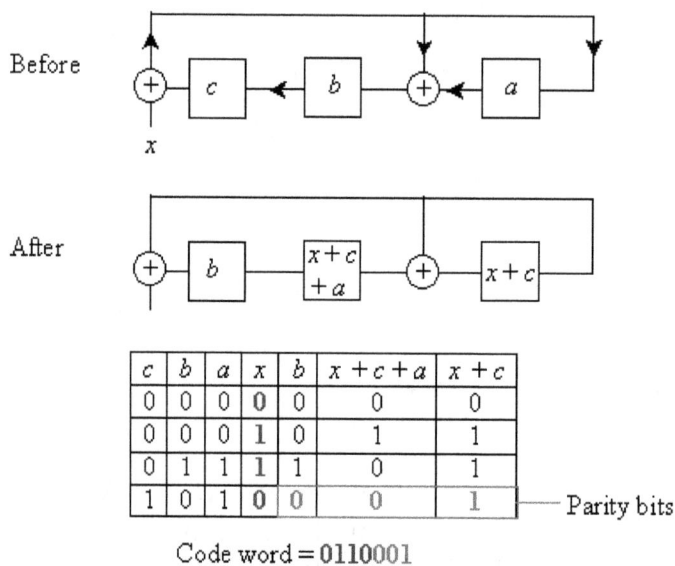

c	b	a	x	b	$x+c+a$	$x+c$	
0	0	0	0	0	0	0	
0	0	0	1	0	1	1	
0	1	1	1	1	0	1	
1	0	1	0	0	0	1	Parity bits

Code word $= 0110001$

Figure 8.8: Type II $(7,4)$ cyclic code encoder.

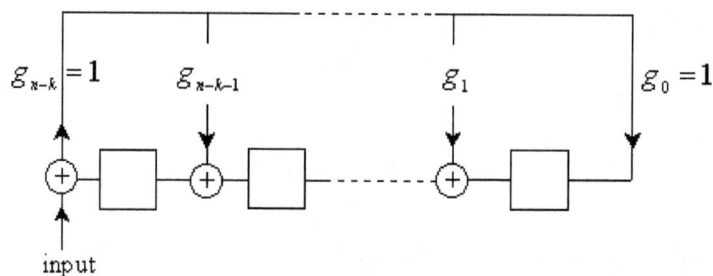

Figure 8.9: General cyclic code encoder.

8.3.4 Syndrome

▶ **Syndrome**

- Error polynomial

- Syndrome definition

- (7,4) systematic code example

- All-zero syndrome for a received valid code word

- Correcting a single code digit error via the syndrome

Let $r(X)$ denote the received word polynomial given by

$$r(X) = r_{n-1}X^{n-1} + \cdots + r_2X^2 + r_1X + r_0 \qquad (8.51)$$

where the coefficients $(r_{n-1}, r_{n-2}, \cdots, r_1, r_0)$ are 0s and 1s and correspond to the noisy version of the transmitted code word $(c_{n-1}, \ldots, c_1, c_0)$. Let $e(X)$ represent the error polynomial such that $r(X) = c(X) + e(X)$. Equivalently,

$$e(X) = r(X) + c(X) \qquad (8.52)$$

For example, if the code word 1010011 is transmitted and is received as 1000011, then we find

$$
\begin{array}{cccccccl}
  & 1 & 0 & 0 & 0 & 0 & 1 & 1 & r(X) \\
+ & 1 & 0 & 1 & 0 & 0 & 1 & 1 & c(X) \\
\hline
  & 0 & 0 & 1 & 0 & 0 & 0 & 0 & e(X)
\end{array}
\qquad (8.53)
$$

Alternatively, using the polynomial representation, we have

$$r(X) = X^6 + X + 1, c(X) = X^6 + X^4 + X + 1 \qquad (8.54)$$

and therefore

$$e(X) = r(X) + c(X) = X^6 + X + 1 + X^6 + X^4 + X + 1 = X^4 \qquad (8.55)$$

as expected. If we divide $r(X)$ by the generator polynomial $g(X)$, then we may express $r(X)$ as

$$r(X) = q(X)g(X) + s(X) \qquad (8.56)$$

where $q(X)$ is the quotient and the remainder $s(X)$ is referred to as the *syndrome polynomial*.   Thus

$$e(X) = r(X) + c(X) = q(X)g(X) + s(X) + c(X). \qquad (8.57)$$

where $c(X) = a(X)g(X)$, so that

$$e(X) = q(X)g(X) + s(X) + a(X)g(X) = [q(X) + a(X)]\, g(X) + s(X) \quad (8.58)$$

$$= u(X)g(X) + s(X) \qquad (8.59)$$

where $u(X) = q(X) + a(X)$ is the quotient of dividing $e(X)$ by $g(X)$ and $s(X)$ is the remainder. The implication is that the syndrome of the received noisy code word $r(X)$ is the same as the syndrome of the corresponding error polynomial.   For example, consider once again the $(7,4)$ systematic cyclic code for which $g(X) = X^3 + X + 1$. If the code word 1010011 is transmitted and is received as 1000011, then the syndrome determined from the received noisy code word is given by

$$
\begin{array}{r}
1\ \ 0\ \ 1\ \ 1\quad q(X) \\
1\ \ 0\ \ 1\ \ 1\ \overline{\big)\ 1\ \ 0\ \ 0\ \ 0\ \ 0\ \ 1\ \ 1} \\
1\ \ 0\ \ 1\ \ 1 \\
\overline{1\ \ 1\ \ 0\ \ 1\ \ 1} \\
1\ \ 0\ \ 1\ \ 1 \\
\overline{1\ \ 1\ \ 0\ \ 1} \\
1\ \ 0\ \ 1\ \ 1 \\
\overline{1\ \ 1\ \ 0\quad s(X)}
\end{array}
\tag{8.60}
$$

To confirm the long division, $q(X) = X^3 + X + 1$ and $s(X) = X^2 + X$ and therefore

$$
r(X) = q(X)g(X) + s(X) = \left(X^3 + X + 1\right)\left(X^3 + X + 1\right) + X^2 + X \tag{8.61}
$$

$$
= X^6 + X^4 + X^3 + X^4 + X^2 + X + X^3 + X + 1 + X^2 + X \tag{8.62}
$$

$$
= X^6 + X + 1 \equiv 1000011 \tag{8.63}
$$

as expected. Now using the error polynomial $e(X) = X^4$ to determine the syndrome, we find

$$
\begin{array}{r}
1\ \ 0\quad u(X) \\
1\ \ 0\ \ 1\ \ 1\ \overline{\big)\ 0\ \ 0\ \ 1\ \ 0\ \ 0\ \ 0\ \ 0} \\
1\ \ 0\ \ 1\ \ 1 \\
\overline{1\ \ 1\ \ 0\quad s(X)}
\end{array}
\tag{8.64}
$$

In both cases, we find the syndrome digits to be 110. As a final check, recall that $u(X) = q(X) + a(X)$. In the previous section, it was shown that $a(X) = (X^3 + 1)$ and since $q(X) = X^3 + X + 1$, we have

$$
u(X) = X^3 + X + 1 + X^3 + 1 = X \tag{8.65}
$$

or equivalently, the binary sequence 10 as expected from the carrying out the long division of $e(X)$ by $g(X)$. If there are no channel errors so that $e(X) = 0$, then dividing $r(X)$ by $g(X)$ must produce the quotient $q(X) = a(X)$ with $s(X) = 0$, and

$$e(X) = [q(X) + a(X)] \, g(X) + s(X) \qquad (8.66)$$

$$= [a(X) + a(X)] \, g(X) + s(X) = s(X) = 0 \qquad (8.67)$$

A key feaure is that the syndrome polynomial is zero if the received word is a valid code word and not necessarily error free. For example, if the code word 1010011 is transmitted and received as 0100111, the syndrome is given by

```
                              1   0   1
        1  0  1  1 | 0   1   0   0   1   1   1
                     1   0   1   1
                     ─────────────
                             1   0   1   1
                             1   0   1   1
                             ─────────────
                             0   0   0   s(X)
```

For all possible single error positions within the received noisy code word, Table 8.5 lists the corresponding syndromes for the $(7, 4)$ systematic cyclic code with $g(X) = X^3 + X + 1$. Hence, this $(7, 4)$ systematic cyclic code can be used to correct single code digit errors by evaluating $\widehat{c}(X) = r(X) + e(X)$, where $\widehat{c}(X)$ is the code word estimate and the error polynomial $e(X)$ is identified from the syndrome in Table 8.5. Of course if multiple errors occurred within the channel, the code word estimate would be incorrect.

8.3.5 Syndrome Former

▶ **Syndrome Former**

- Using the encoder as a syndrome former (mathematical proof)

- Type I and Type II syndrome formers

- Syndrome table

- How to estimate the transmistted code word from the syndrome

Recall that to encode $m(X)$, the information sequence of length k digits is shifted to the left by $(n-k)$ places to produce $X^{n-k}m(X)$, which is a sequence of length n. This sequence is then divided by $g(X)$ to determine the parity digits $p(X)$. Similarly, to determine the syndrome, the received word of length n digits is divided by $g(X)$ to determine the syndrome $s(X)$. The implication is that the encoder of type I (Fig. 8.7) can also be used to determine the syndrome. Indeed, referring to the table presented in Fig. 8.7, if we assume that the input sequence 0110000 is the received word $\mathbf{r}$, then the corresponding syndrome is 001. From Table 8.5, this syndrome corresponds to the error pattern 0000001, in which case

$$
\begin{array}{r}
\mathbf{r} \quad\;\; 0\;\;1\;\;1\;\;0\;\;0\;\;0\;\;0 \\
\mathbf{e} \;+\; 0\;\;0\;\;0\;\;0\;\;0\;\;0\;\;1 \\
\hline
\widehat{c} \quad\;\; 0\;\;1\;\;1\;\;0\;\;0\;\;0\;\;1
\end{array}
\qquad (8.68)
$$

Alternatively, the syndrome corresponding to all possible single code digit errors using the encoder of type II (Fig. 8.8) are presented in Table 8.6. Notice that these are different to those presented in Table 8.5. For example, if the sequence 0110000 is input to the encoder of type II, then the corresponding syndrome is 011 as illustrated in Fig. 8.10. From Table 8.6, this syndrome 011 once again corresponds to the error pattern 0000001, thereby giving $\widehat{\mathbf{c}} = 0110001$ as before.

c	b	a	x	b	x + c + a	x + c
0	0	0	0	0	0	0
0	0	0	1	0	1	1
0	1	1	1	1	0	1
1	0	1	0	0	0	1
0	0	1	0	0	1	0
0	1	0	0	1	0	0
1	0	0	0	0	1	1

—— Syndrome

Figure 8.10: Syndrome formed using the type II encoder.

At first glance, forming the syndrome using an encoder of type II may seem unnecessary. On closer examination, we find that if we flush this syndrome former with zero's until the syndrome is equal to 1 followed by $(n - k - 1)$ zeros, then the number of zeros inserted will correspond to the single code digit error position, starting from the nth digit position r_{n-1} towards r_0. For the example presented in Fig. 8.10, where the received noisy code word $(r_6, r_5, r_4, r_3, r_2, r_1, r_0)$ is 0110000, we found the syndrome using an encoder of type II to be 011. Figure 8.11 illustrates that a further six zeros have to be inserted before the syndrome becomes 100, which implies that the error pattern corresponding to the syndrome 011 is 0000001. The convenient feature of this approach is that a syndrome look-up table is not required.

8.3.6   Other Block Codes

In this subsection, specific block codes that have become well known over the years are briefly mentioned. For further details, please refer to several well known text books which specialize in error-control coding [e.g. Peterson and Weldon (1972)]. We start with the so called "perfect codes", in which every received code word is at most a distance $t = \left\lfloor \frac{(d_{min} - 1)}{2} \right\rfloor$ from one of

c	b	a	x	b	x + c + a	x + c
0	0	0	0	0	0	0
0	0	0	1	0	1	1
0	1	1	1	1	0	1
1	0	1	0	0	0	1
0	0	1	0	0	1	0
0	1	0	0	1	0	0
1	0	0	0	0	1	1
0	1	1	0	1	1	0
1	1	0	0	1	1	1
1	1	1	0	1	0	1
1	0	1	0	0	0	1
0	0	1	0	0	1	0
0	1	0	0	1	0	0

Received word = 0110000

Inserted zeros = 000000

6 places

Error pattern = 0000001

Figure 8.11: Error pattern extraction.

the possible transmitted code words. A well known example is the (23, 12) cyclic Golay code with $d_{\min} = 7$, for which the generator polynomial is either

$$g(X) = X^{11} + X^{10} + X^6 + X^5 + X^4 + X^2 + 1 \qquad (8.69)$$

or

$$g(X) = X^{11} + X^9 + X^7 + X^6 + X^5 + X + 1 \qquad (8.70)$$

A cyclic code used specifically for error *detection* is referred to as a *cyclic redundancy check* (CRC) code. A (n, k) CRC code can detect all error bursts of length $(n - k)$ or less, all combinations of $(d_{\min} - 1)$ errors or less, all error patterns with an odd number of errors if $g(X)$ has an even number of coefficients and a fraction of error bursts of length $(n - k + 1)$ or more. Some of the most popular CRC codes are listed in Table 8.7. For example, the CRC-32 is typically used with data blocks on a CD-ROM.

Up to this point, we have only considered code words which consist of binary digits i.e. the code word symbols are from an alphabet of two symbols $\{0, 1\}$. In general, we can make use of q symbols $\{0, 1, \ldots, q-1\}$ using the concept of a *finite field*, or more commonly referred to as *Galois field* $\mathrm{GF}(q)$. For example, modulo-2 addition is an operation defined within $\mathrm{GF}(2)$. A class of popular nonbinary $\mathrm{GF}(q > 2)$ cyclic codes are *Bose Chauduri Hocquenghem* (BCH) codes, which can detect and correct up to t random errors per code word and *Reed-Solomon* (RS) codes, which are a sub-class of nonbinary BCH cyclic codes. Table 8.8 lists the performance of some well known relatively powerful block codes, including trellis based codes to be covered later, to provide an overall feel for the performance capability of blocks codes. Each scheme is compared on the basis of the signal-to-noise ratio $\frac{E_b}{N_o}(dB)$ required to achieve a probability of an error in a binary digit $P_e = 10^{-5}$. Although this is not entirely fair, because the bandwidth efficiency has not been taken into account, we still gain a reasonable indication of the error-control capability of block codes.

Block codes are well suited for applications where a burst of channel errors are likely to occur, for example a scratch on a CD-ROM, or where the use of a complex decoding algorithm with soft-decision decoding would become a bottleneck on the required data rate. For example, an optical fibre network, in which high rate block codes such as the (255, 239) RS code are utilized to also minimize the increase in the bandwidth requirement with the use of error-control coding.

A block and a convolutional code, together with an interleaver, may be combined to create a *concatenated* error-control coding scheme which improves performance. The combination of a high rate RS code with a convolutional code is a popular concatenated scheme. Namely, the code words output by a RS encoder are interleaved and re-encoded by a convolutional code encoder before being transmitted over a given channel. At the receiver, the output of the Viterbi decoder is likely to contain a burst of errors. These digits are then de-interleaved, thereby spreading the error bursts and then processed by the RS decoder. For example, the Galileo space mission used a (255, 233) RS code, together with a rate $1/4$, $K = 14$ convolutional code, to achieve a P_e of 10^{-5} at $\frac{E_b}{N_o} = 0.95$ dB. With the advent of turbo codes and low density parity check codes, the use such concatenated schemes has become almost obsolete.

8.3.7 Simulation : (7,4) Cyclic Code

SIMULATION **Code7-4:** All the features presented and discussed in this section are demonstrated for the $(7, 4)$ cyclic code.

- **Experiment:** Consider multiple channel errors. Implement other cyclic codes.

Input	Code word
0000	0000000
0001	0001011
0010	0010110
0011	0011101
0100	0100111
0101	0101100
0110	0110001
0111	0111010
1000	1000101
1001	1001110
1010	1010011
1011	1011000
1100	1100010
1101	1101001
1110	1110100
1111	1111111

Table 8.4: (7,4) Cyclic code.

Error pattern	Syndrome
0000001	001
0000010	010
0000100	100
0001000	011
0010000	110
0100000	111
1000000	101

Table 8.5: Type I (7,4) Syndrome table.

Error pattern	Syndrome
0000001	011
0000010	110
0000100	111
0001000	101
0010000	001
0100000	010
1000000	100

Table 8.6: Type II (7,4) Syndrome table.

Name	Generator polynomial
CRC-12	$g(X) = X^{12} + X^{11} + X^3 + X^2 + X + 1$
CRC-ANSI	$g(X) = X^{16} + X^{15} + X^2 + 1$
CRC-CCITT	$g(X) = X^{16} + X^{12} + X^5 + 1$
CRC-32	$g(X) = (X^{16} + X^{15} + X^2 + 1)(X^{16} + X^2 + X + 1)$

Table 8.7: Popular CRC codes.

To achieve $P_e = 10^{-5}$		$\frac{E_b}{N_o}(dB)$
Uncoded	BPSK	9.6
Block codes	(23,12) Golay code	7.4
	(255,239) RS code	7
	(255,123) BCH code	5.7
Trellis codes	$R_{code} = 1/2$, $K = 7$ NASA convolutional code with soft-decision Viterbi decoding	5.1
	$R_{code} = 1/2$ (37, 57, 256*256) turbo code	0.7

Table 8.8: Performance of well known block codes.

8.4　Convolutional Codes

The error correction capability of a **simple** block code is generally poor because a given code word uniquely corresponds to k information digits. If any given code word is severely corrupted, then the corresponding information digits are incorrectly decoded. An alternative strategy is to map the information digits onto a *sequence* of code words and enforce an interdependency between code words. This is the underlying concept behind convolutional codes.

8.4.1　Encoder

▷ **Convolutional Codes**

- Convolutional code encoder

- State diagram

- State table

A simple rate $\frac{1}{2}$ convolutional code encoder is shown in Fig. 8.12. A rectangular box represents an element of a serial shift register and the symbol $\oplus$ represents the modulo-2 operation. A parameter which describes the influence of the shift register on the output code word is the *constraint length* K, defined to be the number of shifts over which a single input digit can influence the encoder output. For a rate $1/n$ code, K is simply the number of elements in the serial shift register i.e. for the convolutional code encoder shown in Fig. 8.12, $K = 3$. For a rate $1/n$ code, the *state* of the encoder is defined to be the contents of the $K - 1$ shift registers. For example, if the contents of the three element shift-register is 110, then the state of the encoder is 11. If a binary digit zero is then input to the encoder, the contents of the shift-register becomes 011 and the state switches from [11] to [01]. Of course, the encoder at any given time can only be in any one of four possible states [00] , [01], [10], [11] corresponding to the $K - 1 = 2$ binary digit tuple. Table 8.9 summarizes the encoding of the input binary information sequence is 10011, where the first input digit is 1, the next is 0, etc. Prior to encoding an input binary stream, the encoder shift register is always initialized with binary digit zeros i.e. the encoder state is initially [00].

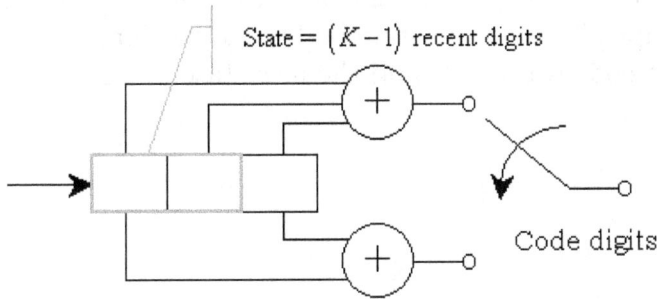

Figure 8.12: Rate 1/2 convolutional code encoder.

- ⭐ SIMULATION **ConvoEncoder:** An implementation of the convolutional encoder in Fig. 8.12.

Referring to Fig. 8.12, the code digits output by the encoder are multiplexed into a serial stream of binary digits. For every binary digit that enters the encoder, two code digits are output. Hence the code rate R_{code} is 1/2. In general, the code rate $R_{code} = k/n$, where k is the number of binary digits input to the encoder and n is the corresponding output code word.

Input	1	0	0	1	1
State	10	01	00	10	11
Output	11	10	11	11	01

Table 8.9: Encoder operation.

8.4.2   State Diagram

The *state diagram* for the rate $1/2$, $K = 3$ convolutional code is shown in Fig. 8.13. Each state is connected by a branch, shown as a solid or a dotted line, corresponding to an input binary digit 0 or 1, respectively. Each branch is labeled with the input binary digit and the corresponding output code word. For example, if the initial state is [11] and the input binary digit is 0, then the output code word is 01 and the new state is [01].

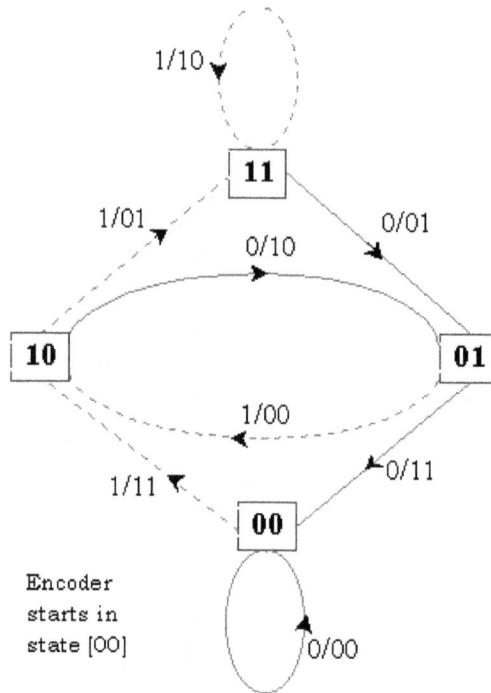

Figure 8.13: State diagram.

8.4.3   State Table

The easiest way to draw the state diagram is to first determine the *state table* as shown in Table 8.10 by simply writing down all possible inputs to all possible initial states. Then determine the final state and the corresponding output code word in each case. From the state table, it is relatively easy to draw the corresponding state diagram.

8.4.4 Generator Polynomials

Generator Polynomial

- Generator polynomials construction

- Code word via generator polynomials

Generator polynomials can be used to describe the tap connections to the shift-register in the encoder as shown in Fig. 8.14. The generator polynomial for the upper branch is given by

$$G^{(1)}(D) = 1 + D + D^2 \tag{8.71}$$

and the polynomial for the lower branch is given by

Input	Initial State	Final State	Output Code Word
0	00	00	00
1	00	10	11
0	01	00	11
1	01	10	00
0	10	01	10
1	10	11	01
0	11	01	01
1	11	11	10

Table 8.10: State table.

$$G^{(2)}(D) = 1 + D^2 \qquad\qquad (8.72)$$

where D denotes an element delay.

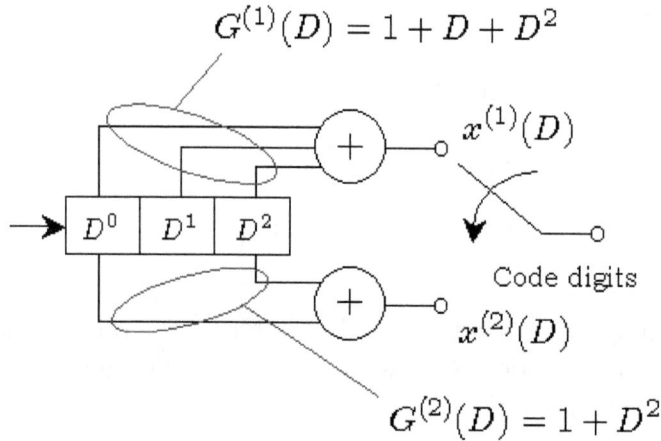

Figure 8.14: Encoder generator polynomials.

In general, for a rate $1/2$ code with a constraint length K, the generator polynomials for the upper and lower branch are given by

$$G^{(1)}(D) = g_0^{(1)} + g_1^{(1)}D + \cdots + g_{K-1}^{(1)}D^{K-1} \qquad (8.73)$$

$$G^{(2)}(D) = g_0^{(2)} + g_1^{(2)}D + \cdots + g_{K-1}^{(2)}D^{K-1} \qquad (8.74)$$

where the coefficients of the polynomials $g_i^{(j)}$ are either 0 or 1 depending on the shift register connections. Note that the plus signs in these equations and in all the equations to follow in this section, do **not** represent algebraic addition (i.e. $3 + 2 = 5$), but modulo-2 addition $\oplus$. i.e. $D^x + D^x = 0$ where x is an integer. The circle around the plus sign is normally left out for convenience. The polynomials representing the sequence of binary digits from the upper and lower arm of the encoder, denoted by $x^{(1)}(D)$ and $x^{(2)}(D)$ respectively as shown in Fig. 8.14 are given by

$$x^{(1)}(D) = m(D)G^{(1)}(D) \tag{8.75}$$

$$x^{(2)}(D) = m(D)G^{(2)}(D) \tag{8.76}$$

where $m(D)$ is the *message polynomial* that represents the input binary sequence, given by

$$m(D) = m_0 + m_1 D + m_2 D^2 + \cdots \tag{8.77}$$

where $m_i \in \{0,1\}$. For example, the input binary sequence (10011) is represented by the message polynomial

$$m(D) = m_0 D^0 + m_1 D^1 + m_2 D^2 + m_3 D^3 + m_4 D^4 \tag{8.78}$$

$$= \left(1 * D^0\right) + \left(0 * D^1\right) + \left(0 * D^2\right) + \left(1 * D^3\right) + \left(1 * D^4\right) \tag{8.79}$$

$$= 1 + D^3 + D^4 \tag{8.80}$$

for which

$$x^{(1)}(D) = \left(1 + D + D^2\right)\left(1 + D^3 + D^4\right) \tag{8.81}$$

$$= 1 + D^3 + D^4 + D + D^4 + D^5 + D^2 + D^5 + D^6 \tag{8.82}$$

$$= 1 + D + D^2 + D^3 + \left(D^4 + D^4\right) + \left(D^5 + D^5\right) + D^6 \tag{8.83}$$

$$= 1 + D + D^2 + D^3 + D^6 \qquad (8.84)$$

i.e. the output sequence from upper arm of encoder is 1111001. Similarly

$$x^{(2)}(D) = \left(1 + D^3 + D^4\right)\left(1 + D^2\right) = 1 + D^2 + D^3 + D^4 + D^5 + D^6 \quad (8.85)$$

i.e. the output sequence from lower arm of encoder is 1011111. Finally the two output sequences [1111001] and [1011111] are multiplexed as shown in Fig. 8.15 and summarized in Table 8.11.

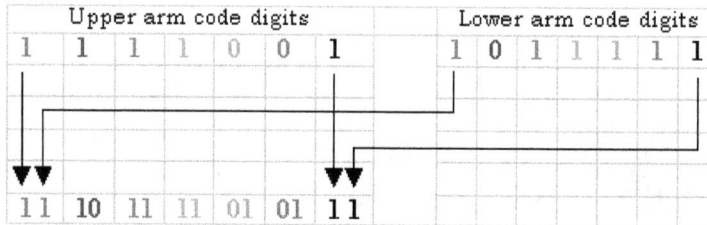

Figure 8.15: Multiplex of the upper and lower arm binary sequences.

As expected, the code word sequence in Table 8.11 is the same as in Table 8.9. An interesting feature of using the generator polynomials to determine the output code word sequence is that the output of the encoder corresponds to an input sequence 1001100, even though the two additional zeros (highlighted in bold) were not present in the message polynomial. These two zeros simply flush the encoder back to the all-zero state [00].

Input	1	0	0	1	1	0	0
Output	11	10	11	11	01	01	11

Table 8.11: Multiplexed output.

8.4.5   Generator Transfer-Function Matrix

▶ **Generator Matrix**

- Generator matrix of the simple rate 1/2 (7,5) code

- How to draw the convolutional code encoder

  from the generator matrix

- Vector and octal representation of convolutional codes

In general, the encoder for a rate $R_{code} = k/n$ convolutional code can be described by a $(k \times n)$ generator transfer-function matrix $G(D)$, given by

$$
G(D) = \begin{bmatrix} G_1^{(1)}(D) & G_1^{(2)}(D) & \cdots & G_1^{(n)}(D) \\ G_2^{(1)}(D) & & & G_2^{(n)}(D) \\ \vdots & & & \vdots \\ G_k^{(1)}(D) & G_k^{(2)}(D) & \cdots & G_k^{(n)}(D) \end{bmatrix} \tag{8.86}
$$

where the generator polynomial in the ith row and the jth column of the matrix $G(D)$ is given by

$$
G_i^{(j)}(D) = g_{i0}^{(j)} + g_{i1}^{(j)} D + \cdots g_{im}^{(j)} D^m \tag{8.87}
$$

where m is the *memory order* of the code defined as [Lin and Costello, 1983]

$$
m = \max_{\substack{1 \leq j \leq n \\ 1 \leq i \leq k}} \left[\deg G_i^{(j)} \right] \tag{8.88}
$$

For example, the $G(D)$ for the rate 1/2 code shown in Fig. 8.12 is given by

$$G(D) = \begin{bmatrix} G_1^{(1)}(D) & G_1^{(2)}(D) \end{bmatrix} = \begin{bmatrix} 1 + D + D^2 & 1 + D^2 \end{bmatrix} \qquad (8.89)$$

In this case, the memory order $m = 2$. If the input sequence to the encoder is 10011, then the output of the encoder is given by

$$m(D)G(D) = \begin{bmatrix} 1 + D + D^2 \end{bmatrix} \begin{bmatrix} (1 + D + D^2) & (1 + D^2) \end{bmatrix}$$

$$= \begin{bmatrix} x^{(1)}(D) & x^{(2)}(D) \end{bmatrix} \qquad (8.90)$$

As another example, suppose

$$G(D) = \begin{bmatrix} 1 + D & 1 + D & 1 \\ D & 0 & 1 + D \end{bmatrix} \qquad (8.91)$$

The corresponding convolutional code is shown in Fig. 8.16, where $k = 2$, $n = 3$ and the memory order $m = 1$.

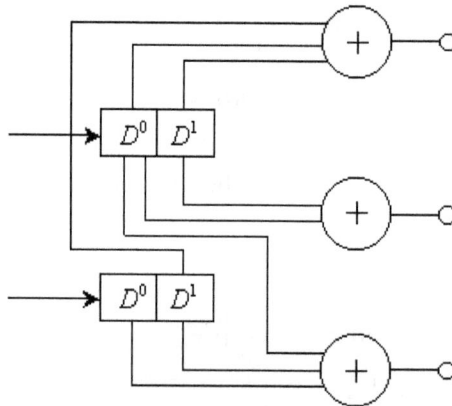

Figure 8.16: Rate 2/3 convolutional code encoder.

- ⭐ SIMULATION **R2-3Encoder:** An implementation of the convolutional encoder shown in Fig. 8.16.

To specify the generator transfer-function matrix for a given convolutional code using the polynomial representation is too cumbersome. Instead, the $G(D)$ is typically specified in either octal or vector form as illustrated in Table 8.12 i.e. the binary digits represent the tap connections or equivalently, $g_{im}^{(j)}$ for a given shift register. We shall use the octal representation to describe a given convolutional code. For example, the code in Fig. 8.12 would be referred to as a rate 1/2 (7, 5) code or equivalently, a rate 1/2 (111,101) code.

Generator transfer-function matrix	Code Vectors	Octal form (right-justified)
$G(D) = \begin{bmatrix} 1 + D + D^2 & 1 + D^2 \end{bmatrix}$	111, 101	7,5
$G(D) = \begin{bmatrix} 1 + D & 1 + D & 1 \\ D & 0 & 1 + D \end{bmatrix}$	011, 011, 010	3, 3, 2
	001, 000, 011	1, 0, 3

Table 8.12: Vector or octal representation of the generator transfer-function matrix.

8.5   Code Trellis

8.5.1   Fundamentals

▷ Code Trellis

- Code trellis from the state diagram

- Concatenation of code trellis diagrams

A code trellis is essentially another way of drawing the state diagram. For example, the state diagram for the rate 1/2 (7, 5) convolutional code and its corresponding code trellis diagram are shown in Figs. 8.17 and 8.18, respectively. On the code trellis diagram, the four possible states [00], [01], [10], [11] are labeled using the decimal equivalent representation 0, 1, 2, 3 for convenience.

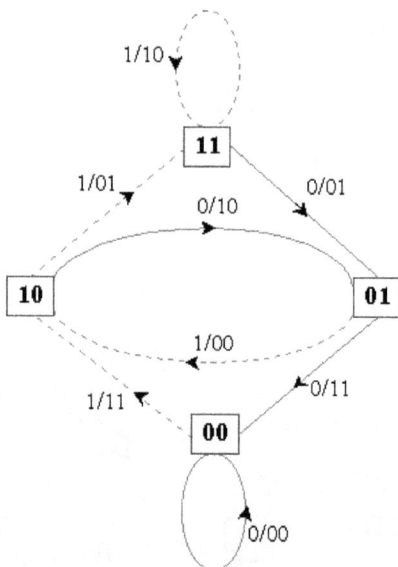

Figure 8.17: Rate 1/2 (7, 5) state diagram.

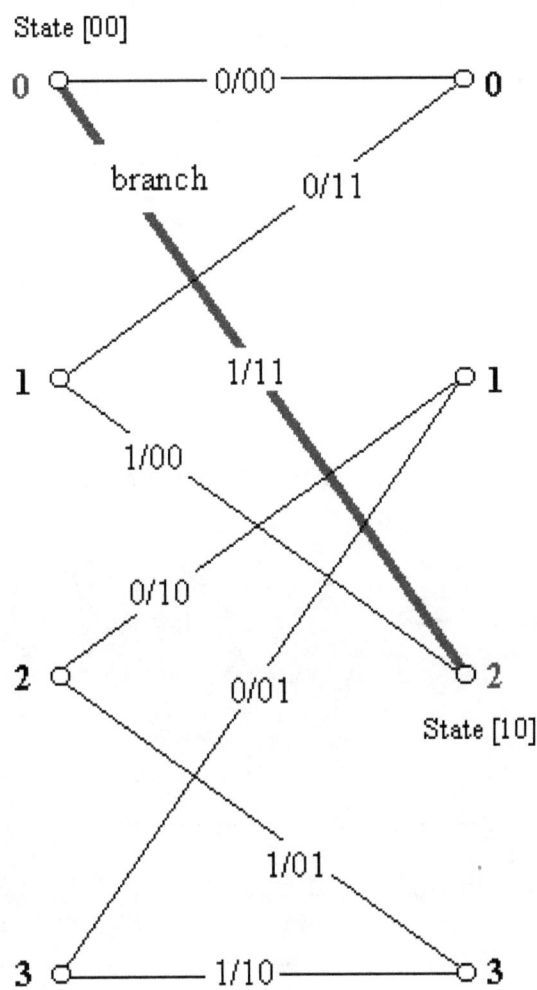

Figure 8.18: Rate 1/2 (7, 5) code trellis.

To illustrate the benefit of a code trellis, suppose the encoder was in the state [00], or equivalently the decimal state 0. If a binary digit 1 is input to the encoder, then from the code trellis diagram, the output code word is 11 and the new state is [10], or equivalently 2, as we trace along the *branch* starting from the left and ending on the right-hand-side of the code trellis, as highlighted in Fig. 8.18. This visual representation of a branch from left to right allows us to determine the state of the encoder at any given time using many code trellis diagrams next to each other as shown in Fig. 8.19. This serial concatenation of several code trellis diagrams is referred to as a *trellis*. The number of code trellis diagrams used to create the trellis is referred to as the *trellis depth*. For example, in Fig. 8.19, the trellis depth is 7.

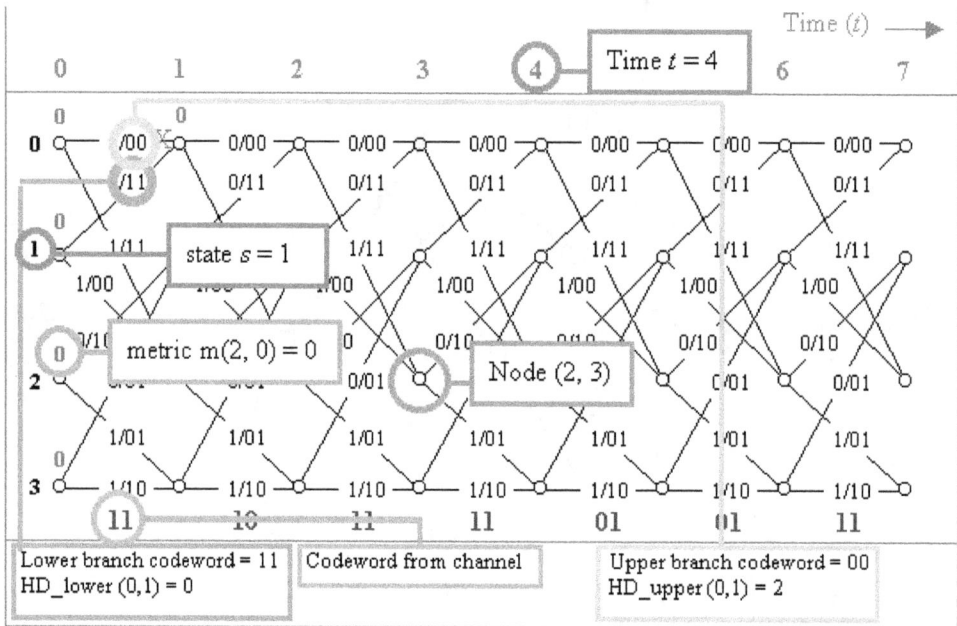

Figure 8.19: Trellis diagram.

At any given state within the trellis, there are two branches entering the state from the left-hand-side as highlighted in Fig. 8.19. These are referred to as the *upper* and *lower* branch i.e. the state [00] has an upper branch which comes from the state [00] and a lower branch which comes from state

[01]. The *branch code word* is the code word associated with a branch. For example, the upper branch entering state [01] has the branch code word 10. It is labeled 0/10, which means that a binary digit 0 input to the encoder in state [10], will output the code word 10 and move to the state [01].

8.5.2   Viterbi Algorithm

▶ **Viterbi Algorithm**

- Decoding a noiseless code sequence of the rate 1/2

  (7,5) code using the Viterbi algorithm

A trellis is used to decode the binary digits received from the DMC using the *Viterbi algorithm*. In this case, the channel decoder is commonly referred to as a *Viterbi decoder*. The Viterbi algorithm is best explained by example and is easy to understand if you refer to the corresponding video clip. It will save you having to fight your way through the explanation to follow :-).

Referring once again to Fig. 8.19, notice how the binary digits received from the channel are written underneath the trellis. These binary digits are in fact the code words shown in Table 8.11. i.e. we have assumed that the information sequence 10011 was encoded and the code digits 11101111010111 were sent through a noiseless channel. Let node(s, t) represent any given state in the trellis at time t. For example, the node$(0, 1)$ represents the state $s = 0$ at time $t = 1$ and the node$(3, 5)$ represents the state $s = 3$ at time $t = 5$. Referring to Fig. 8.19, the upper branch code word entering node$(0, 1)$ is 00 and the lower branch code word is 11. The corresponding word received from the channel written under this code trellis is 11. Given that we are considering a hard-decision demodulator that outputs binary digits, we may use the *hamming distance* to quantify the difference between a given branch code word and a received word from the channel. Let the hamming distance for the upper branch entering a state s at time t be represented by HD_upper(s, t) and the hamming distance for the lower branch by HD_lower(s, t). Thus, for this example, HD_upper$(0, 1) = 2$ and HD_lower$(0,1) = 0$, as illustrated in Fig. 8.19. Finally, let the *metric* for a state s at time t be represented by met(s, t), which is the cumulative hamming distance between the branch code words that lead up to the node(s, t) and the corresponding received channel words. The Viterbi algorithm for a rate 1/2 convolutional code is as follows:

(a) At time $t = 0$, initialize all state metrics to zero i.e. met(0,0) = met(1,0) = met(2,0) = met(3,0) = 0 if the starting state of the encoder is unknown.

(b) At time t, for a given state s, compare the received word with each branch code word entering this state to calculate HD_upper(s, t) and HD_lower(s, t). For example, HD_upper(0,1) = 2 and HD_lower(0,1) = 0, as illustrated in Fig. 8.19.

(c) Calculate y_up = HD_upper(s, t) + met(s^*, $t-1$), where s is the state at time t and s^* is the pervious state at time $(t-1)$ for a given branch. For example, for the first state $s = 0$ at $t = 1$, y_up = HD_upper(0,1) + met(0,0) = 2 + 0 = 2.

(d) Calculate y_low = HD_lower(s, t) + met(s^*, $t - 1$) For example, for the first state $s = 0$ at $t = 1$, y_low = HD_lower(0,1) + m(1,0) = 0 + 0 = 0.

(e) Identify the *surviving* branch entering the state at time t as follows: Choose the upper branch as the survivor if y_up < y_low and let y_final = y_up. Otherwise choose the lower branch and let y_final = y_low. If y_up = y_low, then randomly select any branch as the survivor. For example, for the first state $s = 0$ at $t = 1$, y_final = y_low = 0.

(f) The branch which does not survive is marked with an "X". Only one surviving branch is allowed per node. For example, for the first state $s = 0$ at $t = 1$, the upper branch is marked with an "X". This means that this branch does not survive. Only the lower branch entering the state 00 survives.

(g) Set the state metric met(s, t) = y_final. For example, for the first state $s = 0$ at $t = 1$, met(0,1) = y_final = 0.

(h) Repeat steps [b] to [g] until you reach the end of the trellis at time $t = 7$.

(i) From all the final state metrics [met(0,7) met(1,7) met(2,7) met(3, 7)], choose the minimum metric and trace back the path from this state.

(j) Output the information binary digits which correspond to the branches on this trace back path. In practice a slightly different approach is adopted. This will be discussed later in this section.

- ⭐ SIMULATION **SimpleViterbiDecoder:** A small sequence of binary digits are encoded via the rate 1/2 (7, 5) convolutional code and decoded using the Viterbi algorithm over a noiseless channel.

8.5.3 Metrics

▶ **Metrics**

- What do the metrics mean?
- Trace back from a given state within the trellis
- Maximum likelihood decoding
- Number of branches entering a given node within the trellis
- Number of surviving branches at any given time
- Trellis depth

Referring to Fig. 8.20, if we trace back the path which starts at $s = 2$, $t = 5$, then the hamming distance between the code words on that trace-back path and the words received from the channel is 3, which corresponds to met(2, 5) = 3. i.e. a node metric of 3 means that the code word sequence on a path traced back from this state differs with the received word sequence in 3 positions. By choosing to trace back from the state with the minimum metric at time $t = 7$ (step [i]), the code word sequence within the trellis that is as close as possible (minimum hamming distance) to the received word sequence from the channel is identified. i.e. we have *maximum likelihood decoding*.

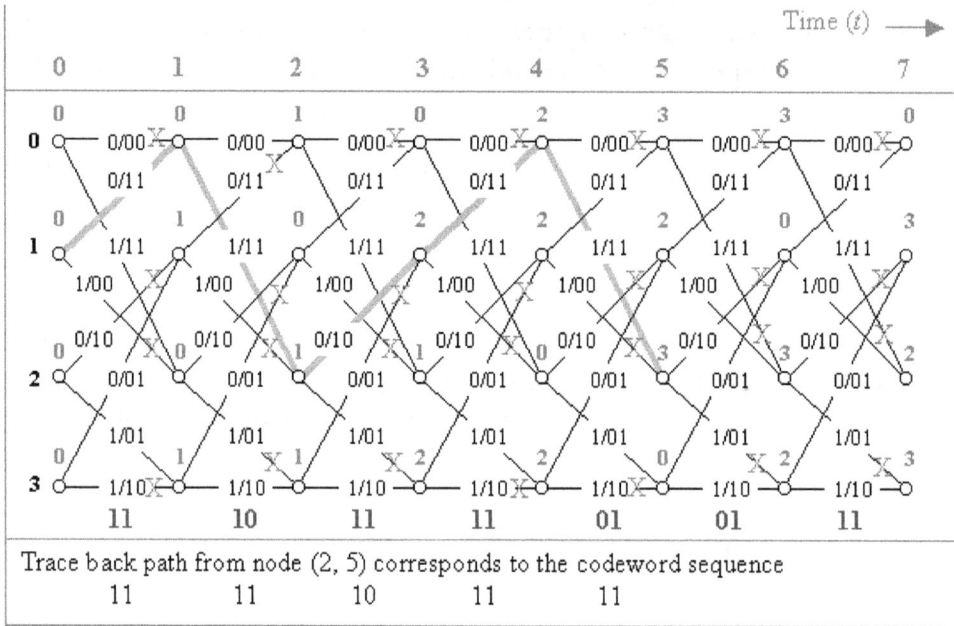

Figure 8.20: Cumulative metric within the trellis.

Step [a] can now be explained. By setting each state metric to zero, we are taking into account that the encoder may have started in any of the four possible states. This is typically the case because even though the encoder does in fact start in the all-zero state, the transmitted code word sequence may have been segmented and sent as a series of packets. In this case, the starting state of any given segment cannot be assumed to be the all-zero state [00]. If however, we know that the encoder started in the all-zero state, then we need only calculate the metrics which emanate from the state $s = 0$ at time $t = 0$. i.e. for our ongoing rate $1/2$ code example, we need only calculate the metrics met(0,1) and met(2,1).

In general for a $R_{code} = k/n$, 2^k branches will enter and exit from a given node on the trellis. The number of states will be 2^{km}, where m is the memory order of the code. The only slight modification of the Viterbi algorithm is as follows. In step [b], the metrics of the 2^k branches entering are compared and the survivor is the one with the smallest cumulative metric. All other branches are marked 'X'. Thus, at any given time, there are only 2^{km} surviving paths.

8.5.4 Practical Implementation of the Viterbi Decoder

In practice, thousands of binary digits are encoded and decoded. Of course the use of a very large trellis depth would be too memory intensive and impractical. Fortunately, for a convolutional code with memory order m, it has been shown [Heller and Jacobs, 1971] that if the trellis depth is $\geq 5m$ and only the oldest message bit within the trellis is decoded, then the error correcting capability of the Viterbi decoder would not be noticeably diminished. This is because at this depth and beyond, it is very likely that the 2^{km} surviving paths will merge into a single path as they are traced back from the end of the trellis. For the rate 1/2 (7, 5) code since $m = 2$, a trellis depth of at least 10 is required. The simulation in the next subsection will demonstrate this feature.

Once the path with the minimum cumulative metric is traced back and the corresponding information digit output, the trellis is then shifted to the left and the next word from the channel is processed. Once again, only the oldest information digit is output by the decoder. This procedure is repeated until all the information digits have been decoded, as illustrated by example in Figs. 8.21 to 8.26 using a trellis depth of only 3.

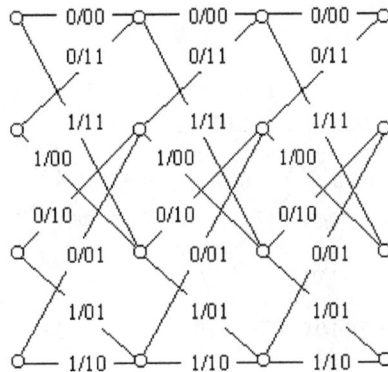

Figure 8.21: Trellis of depth 3.

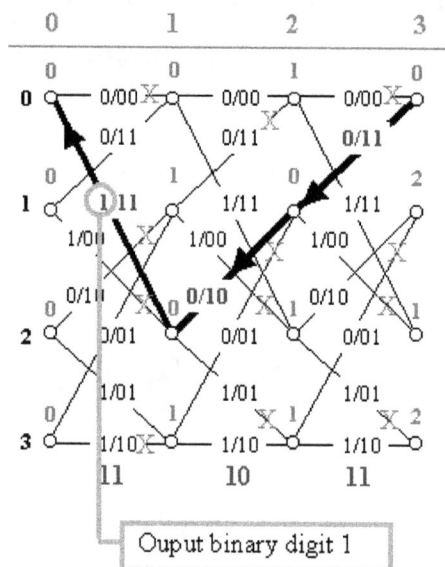

Figure 8.22: Decode first binary digit.

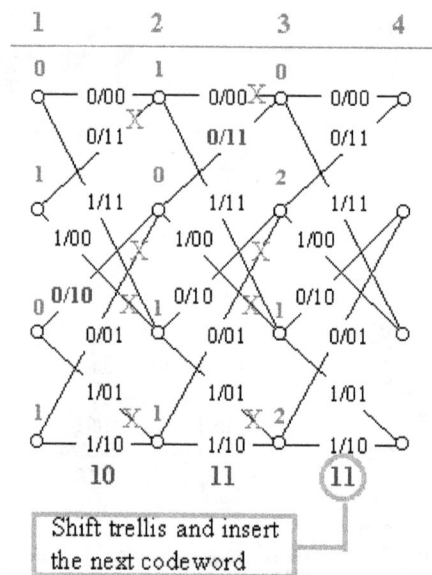

Figure 8.23: Shift trellis and insert the next received noisy code words.

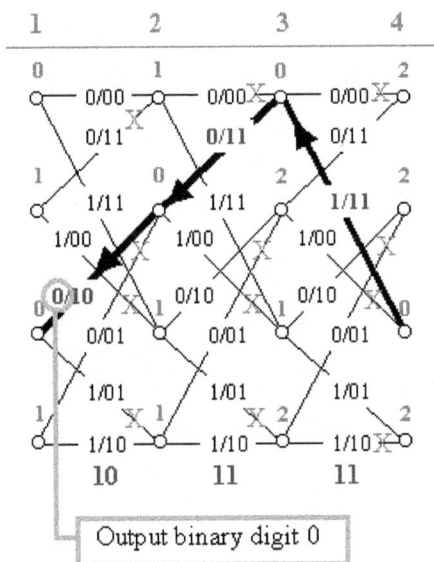

Figure 8.24: Decode binary digit.

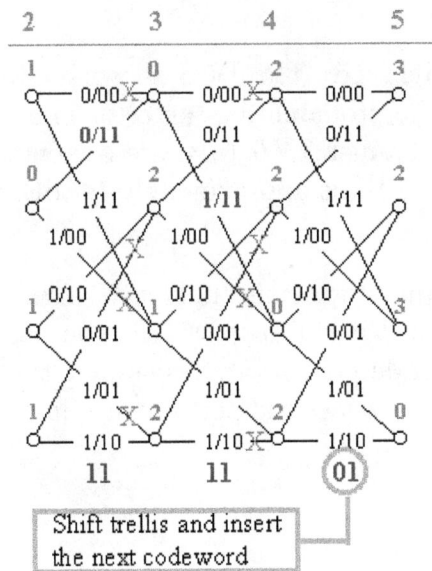

Figure 8.25: Shift trellis and insert the next received noisy code words.

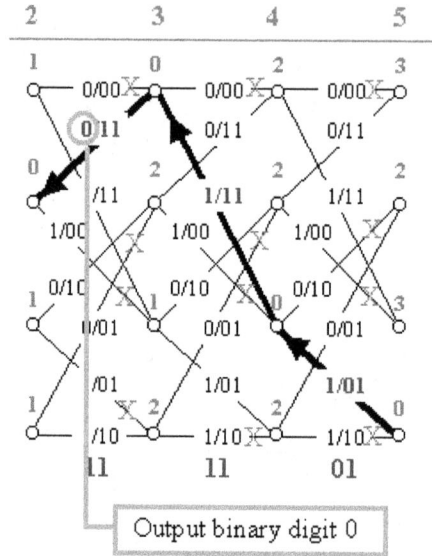

Figure 8.26: Decode third binary digit.

8.5.5 Simulation : Hard-Decision Viterbi Decoder

SIMULATION **VitHard:** The DCS shown in Fig. 8.27 is modeled. Simulation results for the probability of an error in a binary digit P_e at the binary sink versus the channel SNR (dB) are presented in Fig. 8.28, where the coding gain at $P_e = 10^{-5}$ is approximately 1.7 dB.

- **Experiment:** Investigate the influence of the trellis depth on the performance of the Viterbi decoder. If you are feeling adventurous, modify the Viterbi decoder to always assume that the encoder started in the all-zero state. Does the performance improve ?

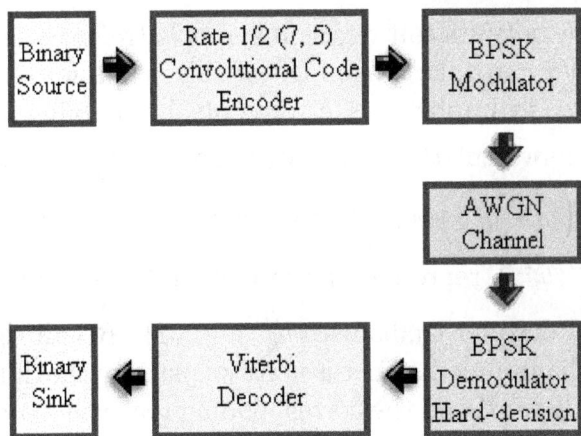

Figure 8.27: Simulated DCS : BPSK over an AWGN channel with hard-decision Viterbi decoding.

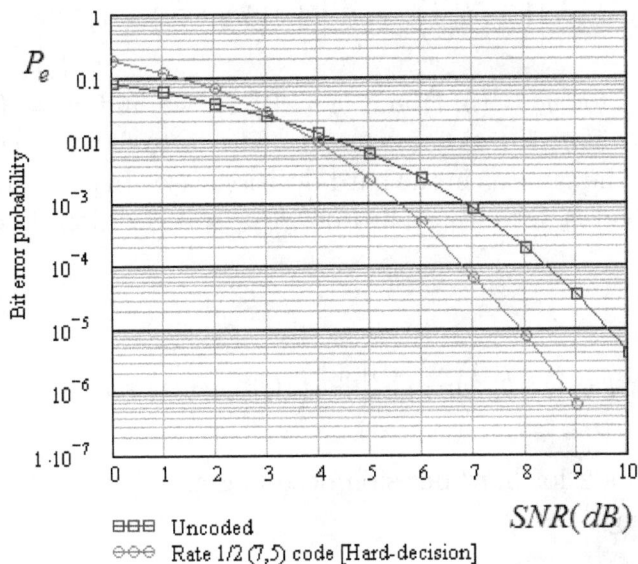

Figure 8.28: Hard-decision Viterbi decoder performance.

8.6   Soft-Decision Viterbi Decoding

If the demodulator outputs soft-decision symbols (refer to section 7.7), then the Viterbi decoder can take advantage of this soft-information within the metric calculations to improve its error-control capability as follows. Let $x_k = \left(x_k^{(1)}, x_k^{(2)}\right)$ represent the transmitted code word by the convolutional encoder, let $y_k = \left(y_k^{(1)}, y_k^{(2)}\right)$ represent the corresponding noisy received word and let $b_k = \left(b_k^{(1)}, b_k^{(2)}\right)$ represent the branch code word on a given branch within the Viterbi decoder trellis with $b_k^{(i)} \in \{0, 1\}$. Recall that in the hard-decision case, the hamming distance is used to quantify the difference between a given branch code word and the corresponding received word. For example, suppose $x_k = (1, 0)$ is transmitted using BPSK and demodulated as $y_k = (5, 4)$ using 8-level soft-decision. If $y_k = (5, 4)$ and is hard-decisioned to (11) and $b_k = (0, 1)$, then the hamming distance between y_k and b_k is 1. However, to utilize the 8-level soft-decision symbols, the distance d_s between the branch code word and the received word can be calculated using

$$d_s = \left|\left(7 * b_k^{(1)}\right) - y_k^{(1)}\right| + \left|\left(7 * b_k^{(2)}\right) - y_k^{(2)}\right| \qquad (8.92)$$

For example, for $y_k = (5, 4)$ and $b_k = (0, 1)$, we find $d_s = |(7 * 0) - 5| + |(7 * 1) - 4| = |-5| + |7 - 4| = 5 + 3 = 8$. The rest of the Viterbi algorithm remains exactly the same as in the hard-decision case. In general, the distance d_s between y_k and b_k for either hard or Q_{level} soft decision decoding can be written as

$$d_s = \left|\left((Q_{level} - 1) * b_k^{(1)}\right) - y_k^{(1)}\right| + \left|\left((Q_{level} - 1) * b_k^{(2)}\right) - y_k^{(2)}\right| \qquad (8.93)$$

where Q_{level} is 2 for hard-decision decoding and 4, 8 or 16, etc. for soft-decision decoding.

- ⭐ SIMULATION **SoftDistance:** An example using equation 8.93.

8.6.1   Simulation : Soft-Decision Viterbi Decoder

SIMULATION **VitSoft:** The DCS shown in Fig. 8.29 is modeled. Simulation results for the probability of error in a binary digit P_e at the binary sink versus the channel SNR (dB) are presented in Fig. 8.30.

- **Experiment:** Investigate the influence of the trellis depth and the separation between the quantization levels on the performance of the Viterbi decoder.

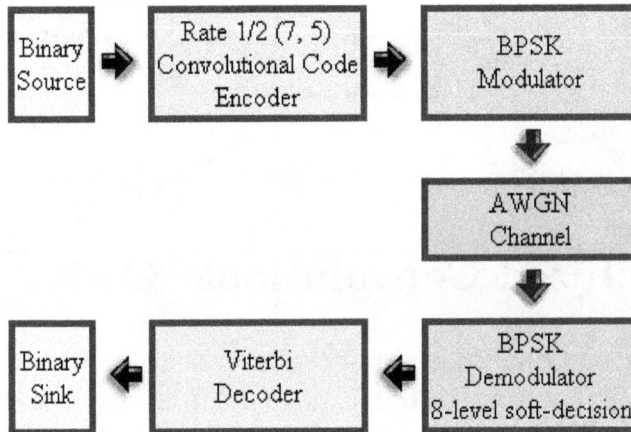

Figure 8.29: Simulated DCS : BPSK over an AWGN channel with soft-decision Viterbi decoding.

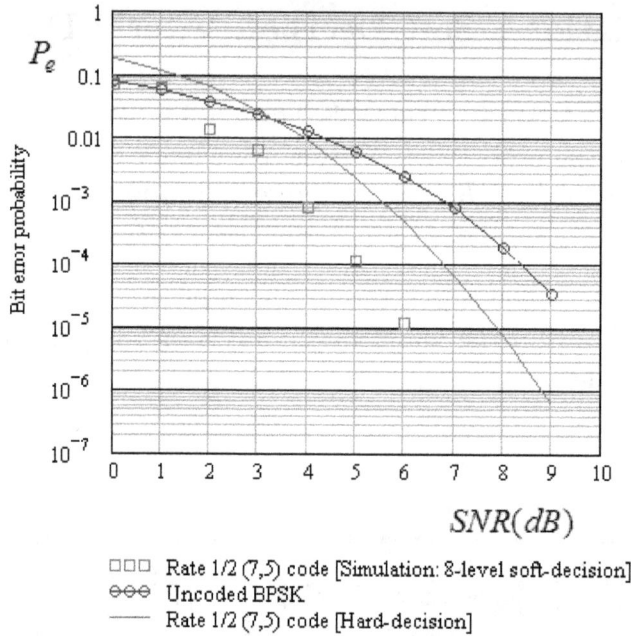

Figure 8.30: Soft-decision Viterbi decoder performance.

8.7   Punctured Convolutional Codes

A punctured convolutional code is a high-rate code that is created by periodically deleting or puncturing, certain code digits from the code digit sequence output by the encoder. A rate $R_{code} = k/n$ punctured convolutional code encoder is shown in Fig. 8.31. It consists of an original low rate $\frac{1}{n_o}$ encoder followed by a symbol selector that punctures from every kn_o code digits, $(kn_o - n)$ digits according to a chosen pattern that is usually described by a *perforation* or *puncturing* matrix P. This matrix P has n_o rows and k columns and its binary elements correspond to deleting '0' or transmitting '1' a code digit from every k information digits encoded by the original low rate encoder. By modifying the elements of the perforation matrix, both the punctured convolutional code structure and its rate can be varied.

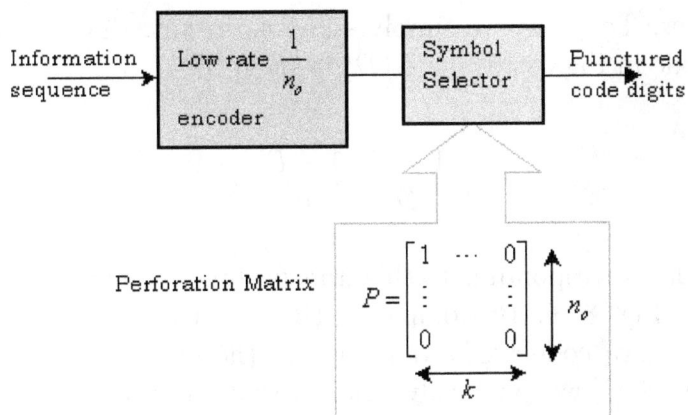

Figure 8.31: Punctured convolutional code.

8.7.1 Rate 2/3 Punctured Code

Using the rate 1/2 (7, 5) code as the original low rate code with the puncturing matrix $P = \begin{bmatrix} 1 & 1 \\ 1 & 0 \end{bmatrix}$ creates a rate 2/3 punctured convolutional code as shown in Fig. 8.32.

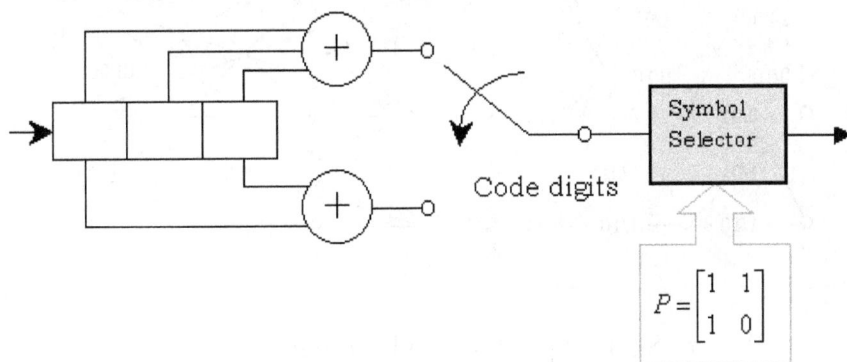

Figure 8.32: Rate 2/3 punctured convolutional code.

The method to determine the generator transfer-function matrix $G(D)$ of the punctured convolutional code will be addressed in a problem at the end of this chapter. For now, we simply state the result, that for the rate 2/3 punctured code in Fig. 8.32, the $G(D)$ is given by

$$G(D) = \begin{bmatrix} 1+D & 1+D & 1 \\ D & 0 & 1+D \end{bmatrix}. \tag{8.94}$$

The encoder corresponding to this transfer-function matrix was presented in section 8.4, Fig. 8.16. By comparing the code trellis diagram of this rate 2/3 code with two code trellis diagrams of the original rate 1/2 code, as shown in Fig. 8.33, we can easily confirm that the encoder in Fig. 8.16 is equivalent to the rate 2/3 punctured code in Fig. 8.32. For example, the highlighted branch in Fig. 8.33 corresponds to the input 11 to the rate 1/2 encoder starting in the state 1. The output code word sequence in this case is 0001, which when punctured becomes 000 and corresponds to the rate 2/3 code trellis branch 11/000 from state 1 to 3.

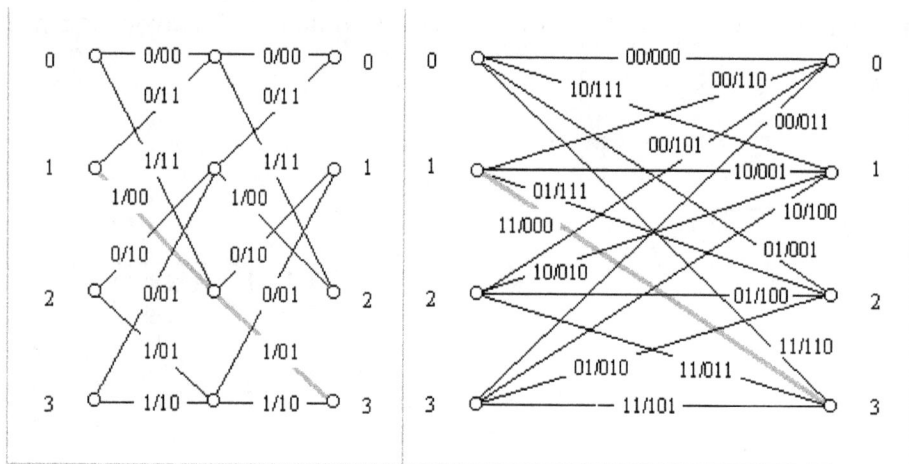

Figure 8.33: Punctured code trellis comparison.

- ⭐ SIMULATION **Rate2-3StateTable:** To determine the state table for the rate 2/3 code and confirm the analysis in Fig. 8.33.

8.7.2 Viterbi Decoder for Punctured Convolutional Codes

To decode the punctured code sequence, it is not necessary to create a Viterbi decoder for the high rate punctured convolutional code. A processing block based on the puncturing matrix P is inserted prior to the Viterbi decoder for the original low rate $R = \frac{1}{n_o}$, which inserts an arbitrary symbol e.g. 'X' within the sequence received from the channel in those positions that were punctured. Of course, correct synchronization is essential. For example, if the code word sequence output from the original low rate code is 00, 00, 00, 00 and the puncturing matrix $P = \begin{bmatrix} 1 & 1 \\ 1 & 0 \end{bmatrix}$, then the code word sequence transmitted into the channel is 000, 000. Prior to the Viterbi decoder, the unknown punctured code digits are replaced by "X". Under a noiseless channel with perfect synchronization, the input to the Viterbi decoder is now the sequence 00, 0X, 00, 0X. The only alteration of the Viterbi algorithm for the original low rate code is once again the manner in which the metrics are calculated. Using the notation in section 8.6, the distance d_p between a branch code word b_k and the received word y_k is now given by

$$
d_p = \begin{cases}
\left| (Q_{level} - 1)\, b_k^{(1)} - y_k^{(1)} \right| & \text{if } y_k^{(2)} = X \\[2mm]
\left| (Q_{level} - 1)\, b_k^{(2)} - y_k^{(2)} \right| & \text{if } y_k^{(1)} = X \\[4mm]
\left| (Q_{level} - 1)\, b_k^{(1)} - y_k^{(1)} \right| + \left| (Q_{level} - 1)\, b_k^{(2)} - y_k^{(2)} \right| & \text{otherwise}
\end{cases}
$$

$$(8.95)$$

Notice from the above formulae that the punctured code digit, which is replaced by the symbol "X", is assumed to be received without error. The penalty incurred is a reduction in the error-control capability when compared to the original low rate code. Nevertheless, there are a few advantages of the punctured coding technique as follows:

- It provides an implementation of high-rate decoders with little additional complexity over the decoder used for the original low rate code. An increase in the code rate lowers the channel bandwidth requirement, as will be evident in section 8.11.

- It allows the user to change the code rate as required, producing a variable rate system that allows for example, unequal error protection within a given data frame.

- Even though punctured coded tend not to include the best known convolutional codes for a given rate, the performance loss is fairly minimal.

8.7.3   Simulation : Rate 2/3 Punctured Code with Viterbi Decoding

SIMULATION **VitHardSoftPunctured:** The DCS shown in Fig. 8.34 is modeled.

- **Experiment:** Consider the influence of the trellis depth and the separation between the quantization levels on the performance of the Viterbi decoder. Try a different puncturing matrix. Be careful to ensure that the punctured code is not catastrophic.

Simulation results for the probability of error in a binary digit versus the channel SNR are presented in Fig. 8.35. The performance of the rate 1/2 (7, 5) code under both hard and soft-decision decoding is overlaid on this graph from which we note that soft-decision decoding provides an additional gain of approximately 1.5 dB as expected from the application of Shannon's channel coding theorem in section 8.1.3. Notice that the performance of the punctured code under both hard or soft-decision decoding is slightly worse than the original rate 1/2 code, because the puncturing process has weakened the code structure.

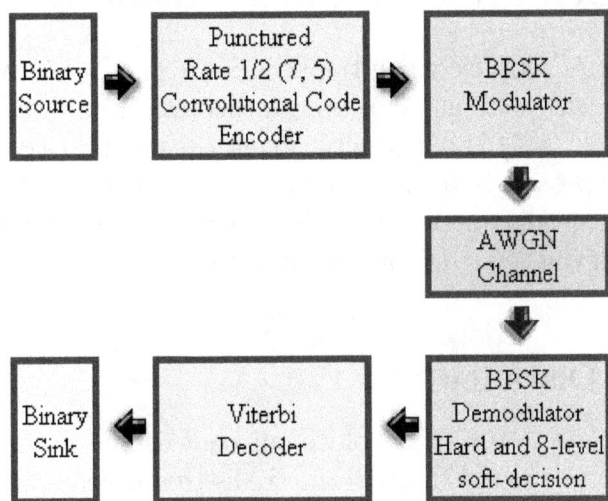

Figure 8.34: Simulated DCS : BPSK over an AWGN with punctured hard and soft-decision Viterbi decoding.

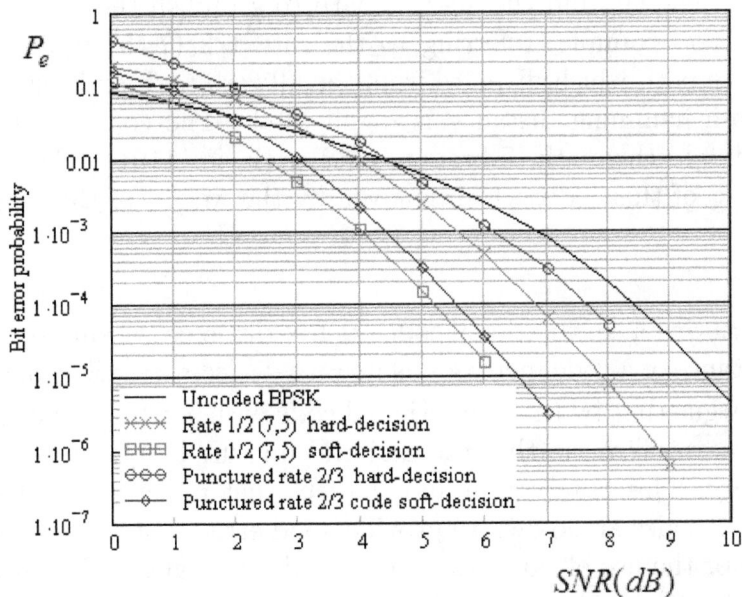

Figure 8.35: Comparison of simulation results with uncoded BPSK.

8.8 Performance of Convolutional Codes

The performance of simple convolutional codes (e.g. $K = 7$) can be predicted as we shall see in this section. However, a theoretical analysis for more powerful codes (e.g. $K = 7$) is not feasible given the complexity of the analysis. The standard approach is to use computer simulations [Odenwalder, 1976] to determine the generator polynomials that will yield the best coding gain for a given constraint length K and code rate R_{code}.

8.8.1 Free Distance

The *free distance* d_{free} of a convolutional code is the minimum hamming distance between two paths through the trellis that diverge from one another at a given node and remerge at another node. The larger the free distance of a code, the smaller is the probability of tracing back a valid but incorrect code word sequence through the trellis. Therefore, free distance is a good indicator of the error-control capability of a convolutional code. Since convolutional codes are linear, it is convenient to use all the paths which deviate and remerge from the all-zero path to establish the free distance of the code. For example, referring to the rate 1/2 (7, 5) trellis shown in Fig. 8.36, the path which diverges from the all-zero path and remerges with the minimum hamming distance is highlighted. The code word sequence corresponding to this path (3 branches in length) is 111011. Thus, the free distance d_{free} of this code, which is given by the weight of this sequence, is equal to 5.

By puncturing a convolutional code, its free distance is reduced. Although the free distance of a code increases as its rate $\frac{1}{n_o}$ decreases, Haccoon and Begin (1989) argue that by using an original code with a rate $\frac{1}{n_o}$ lower than 1/2 does not necessarily mean that the derived punctured codes will have larger free distances. This observation and the fact that punctured codes derived from rate 1/2 codes are known to provide good error-control properties [ref], tends to favor the use of rate 1/2 original codes to generate high-rate punctured convolutional codes. In particular, most practical high-rate punctured codes are of rate $\frac{(n-1)}{n}$, created by deleting $(k-1)$ code digits from every $2k$ code digits.

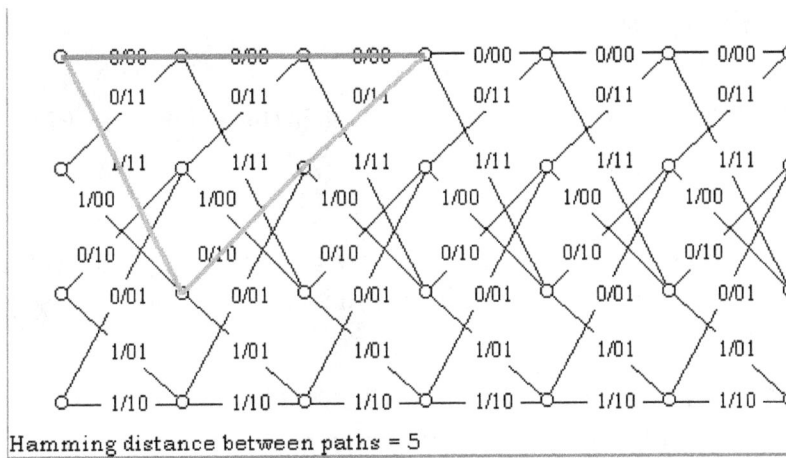

Figure 8.36: Free distance of the rate 1/2 (7, 5) convolutional code.

8.8.2 Transfer Function

In addition to free distance, the performance of a given convolutional code also depends on the number of paths of different weights and the weight of the corresponding input sequences. It is easily shown [Lin and Costello, 1983] that the *transfer function* $T(D, L, N)$ of the rate 1/2 (7, 5) code is given by

$$T(D, L, N) = \frac{D^5 L^3 N}{1 - DL\,(1+L)\,N} \tag{8.96}$$

$$= D^5 L^3 N + D^6 L^4 N^2 + D^6 L^5 N^2 + D^7 L^5 N^3 + D^7 L^6 N^3 + D^7 L^6 N^3 + D^7 L^7 N^3 + \cdots \tag{8.97}$$

where the exponent of

- D is the hamming distance from the all-zero path

- L is the number of branches or equivalently, the length of the path

- N is the number of binary one digits in the input sequence or equivalently, its weight.

so that the expression

- $D^5 L^3 N$, indicates there is only one path (highlighted in Fig. 8.36) of distance 5 of length 3 from the all-zero path and the weight of the input sequence to generate this path is 1.

- $D^6 L^4 N^2 + D^6 L^5 N^2$, indicates there are two paths of distance 6 of length 4 and 5 from the all-zero path and the weight of the input sequence is 2 in each case.

- $D^7 L^5 N^3 + D^7 L^6 N^3 + D^7 L^6 N^3 + D^7 L^7 N^3$, indicates there are four paths of distance 7 of length 5, 6, 6 and 7 from the all-zero path and the weight of the input sequence is 3 in each case.

It can be shown [Viterbi, 1971] that using coherent BPSK over an AWGN channel, the probability of an error in a binary digit at the output of a decoder is bounded by (with $L = 1$)

$$P_e^{bound} = Q\left(\sqrt{2d_{free}\frac{R_{code}E_b}{N_o}}\right) \exp\left(d_{free}\frac{R_{code}E_b}{N_o}\right) \frac{dT(D,N)}{dN}\Bigg|_{N=1,D=\exp\left(-\frac{R_{code}E_b}{N_o}\right)} \tag{8.98}$$

Hence, for the rate 1/2 (7, 5) code with $d_{free} = 5$ and $R_{code} = 1/2$, we find

$$T(D,N) = \frac{D^5 N}{1 - 2DN} \tag{8.99}$$

$$\frac{dT(D,N)}{dN}\Bigg|_{N=1} = \frac{D^5}{(1-2D)^2} \tag{8.100}$$

and therefore

$$P_e^{bound} = Q\left(\sqrt{\frac{5E_b}{N_o}}\right) \exp\left(\frac{5E_b}{2N_o}\right) \frac{\exp\left(-\frac{5E_b}{2N_o}\right)}{\left(1 - 2\exp\left(-\frac{E_b}{2N_o}\right)\right)^2} \tag{8.101}$$

As evident from Fig. 8.37, this bound is in good agreement with the simulation results of the rate 1/2 (7, 5) code via BPSK over an AWGN channel with 8-level soft-decision Viterbi decoding.

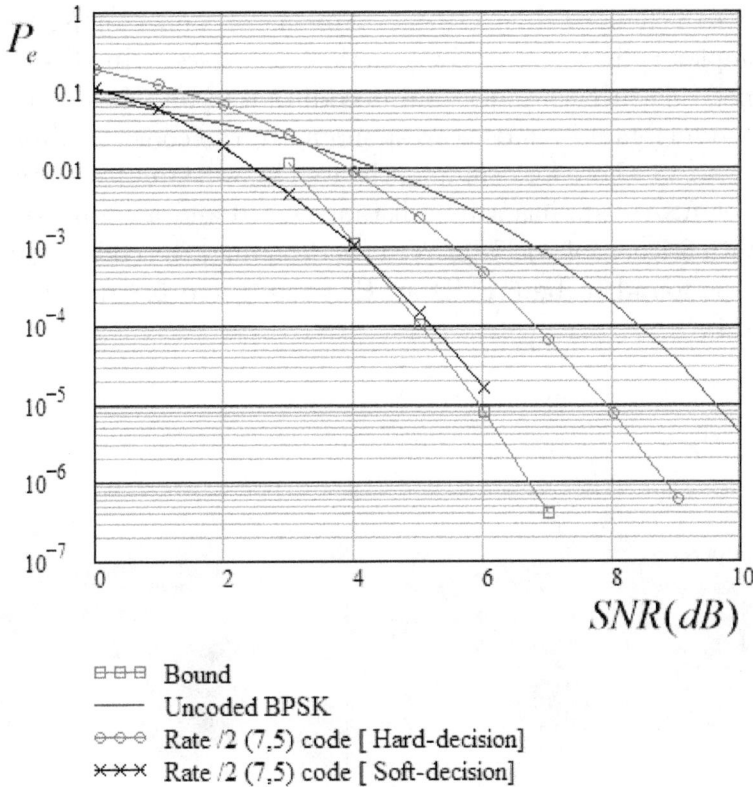

Figure 8.37: Rate 1/2 (7, 5) code theoretical performance bound.

• ⭐ SIMULATION **Bound:** Simulation corresponding to Fig. 8.37 and the variation of the bound coding gain with respect to uncoded BPSK.

Although the method to determine $T(D, L, N)$ is not complicated, its complexity increases exponentially with the constraint length of the code, making it a suitable means of analysis for simple codes only. In general, it can be shown [Jacobs, 1974] that the asymptotic coding gain in decibels, compared with coherent uncoded BPSK over an AWGN channel, is less than or equal to $10 \log_{10} \left(R_{code} d_{free} \right)$. Table 8.13 lists the maximum d_{free} rate 1/2 and 1/3 convolutional codes (nonsystematic) with a constraint length K in the range 3 to 9 [Odenwalder, 1976]. For larger values of K, the Viterbi decoder complexity is too high for practical purposes. In this case, other decoding methods such as sequential decoding [Lin and Costello, 1983] are used. The coding gain bounds corresponding to the codes in Table 8.13 are shown in Fig. 8.38 where for example, the well known standard rate 1/2 (171, 133) $K = 7$ NASA code has a free distance of 10 and a coding gain bound of 7 dB. With 8-level soft-decision Viterbi decoding, the coding gain for this NASA code is found to be 5.1 dB at $P_e = 10^{-5}$ and 5.8 dB at $P_e = 10^{-7}$ [Jacobs, 1974].

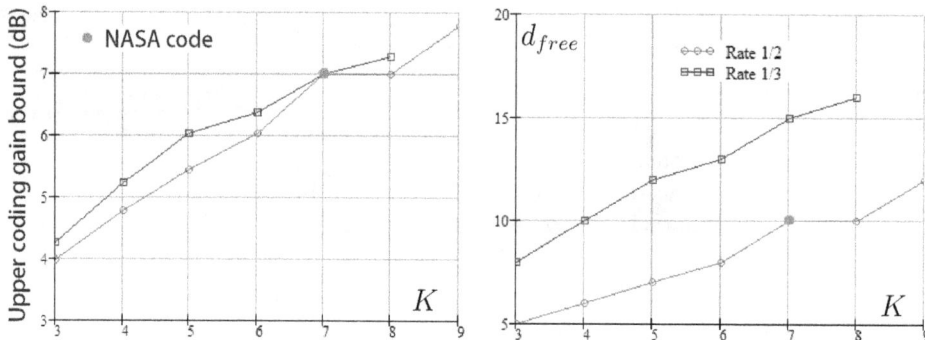

Figure 8.38: Coding gain bound and free distance of optimum rate 1/2 and 1/3 codes.

- ⭐ SIMULATION **CodingGain:** For further details on how to determine the values shown in Fig. 8.38.

$R_{code} = 1/2$		
K	d_{free}	Octal
3	5	7, 5
4	6	17, 15
5	7	35, 23
6	8	75, 53
7	10	171, 133
8	10	371, 247
9	12	753, 561
$R_{code} = 1/3$		
K	d_{free}	Octal
3	8	7, 7, 5
4	10	17, 15, 13
5	12	37, 33, 25
6	13	75, 53, 47
7	15	171, 145, 133
8	16	367, 331, 225

Table 8.13: Optimum Rate 1/2 and Rate 1/3 convolutional codes.

8.9   Trellis Coded Modulation (TCM)

8.9.1   Introduction

▷ **TCM Encoder**

- TCM encoder operation

- State table

- Code trellis of a rate 2/3 TCM code

In this section, we shall consider a combined error-control coding and modulation scheme as illustrated in Fig. 8.39, known as *trellis coded modulation* (TCM), which can provide a coding gain without increasing the channel bandwidth requirement.   TCM has been adopted in numerous telephone modem standards since its invention in 1980 by Gottfried Ungerboeck.   The most famous being the 36.6 bit/s V.34 ITU standard which provides a data rate close to the theoretical Shannon limit for narrowband phone lines that have a bandwidth from 300 to 3100 Hz.   For further information please refer back to section 5.4.2.   Of course, TCM has also been applied to more recent error-control codes to improve the bandwidth efficiency of modern communication systems e.g. Turbo TCM.

8.9.2   TCM Encoder

▷ **TCM**

- Introduction to TCM

- Link between output code word and MPSK

- Euclidean distance

- Soft-decision information within TCM

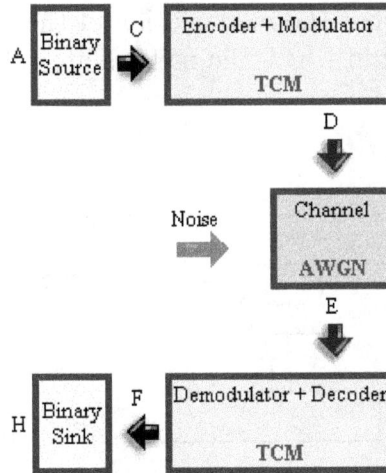

Figure 8.39: Combined error-control coding and modulation DCS.

Consider the rate 2/3 *systematic* convolutional code encoder shown in Fig. 8.40. The input to the encoder is the binary pair $m^{(1)}m^{(2)}$ and the corresponding output code word is $x^{(1)}x^{(2)}x^{(3)}$, which is mapped to the corresponding signal point of an 8-PSK signal constellation diagram. For example, as illustrated in Fig. 8.41, since the input digits 11 produce the code word 110, the signal corresponding to the signal point 6 is transmitted through the channel. From section 7.8, equation 7.76, the energy per signal E_s is given by $E_s = R_{code}E_b \log_2 M = \frac{2}{3} * 1 * \log_2 8 = 2$. Hence, the radius of the 8-PSK signal constellation circle is $\sqrt{E_s} = \sqrt{2}$.

- ⭐ SIMULATION **TCMEncoder:** An implementation of the encoder shown in Fig. 8.40. Alter the input and view the corresponding output.

At the receiver, unlike a conventional 8-PSK demodulator which outputs the code word corresponding to the received signal point, the *coordinates* of the received signal point are passed onto a modified Viterbi decoder. Just as in the case of soft-decision or punctured convolutional code decoding, we shall see in section 8.10 that the modification of the Viterbi decoder is once again only in the manner in which the metrics are calculated.

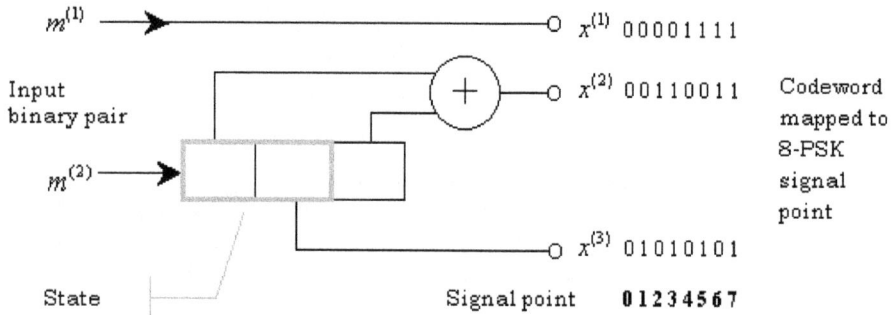

Figure 8.40: TCM encoder.

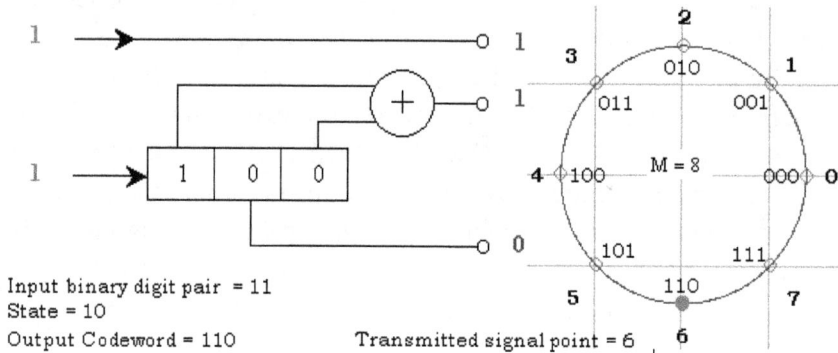

Figure 8.41: TCM encoder mapping scheme.

8.9.3 Code Trellis

The code trellis for the encoder in Fig. 8.40 is presented in Fig. 8.42. There are now two parallel branches between states, where each branch corresponds to $m^{(1)}m^{(2)}/x^{(1)}x^{(2)}x^{(3)}$. For example, the highlighted branch shown in Fig.

8.42 goes from state [01] to state [00] and corresponds to 10/110. For convenience in Fig. 8.42, each input and output pair is replaced by their decimal equivalent i.e. 2/6. Notice how the branch code word 110 corresponds to the signal point 6.

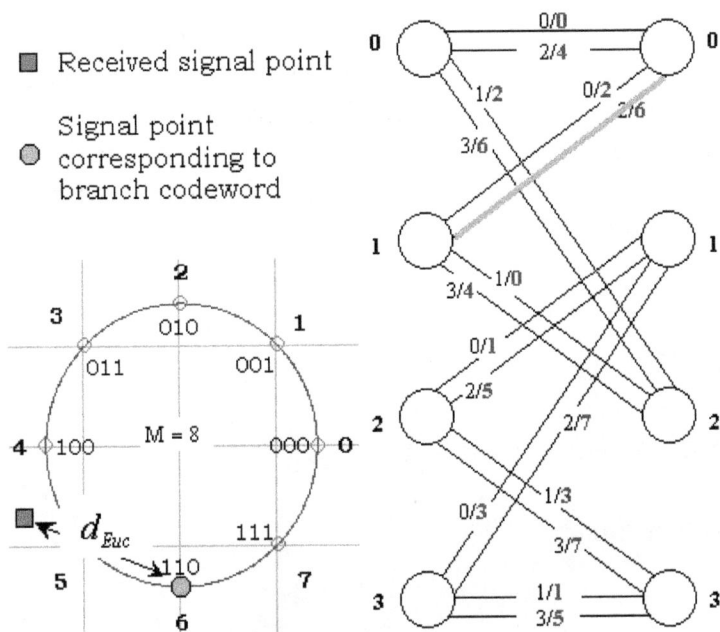

Figure 8.42: Rate 2/3 code trellis diagram.

• ⭐ SIMULATION **TCMStateTable:** To determine the encoder state table.

The code trellis metrics and the surviving branches are determined as follows. Suppose the code word 110 is output from the TCM encoder and the corresponding received signal point is as shown in Fig. 8.42. To determine the surviving branch entering the state [00], one could map the received signal point to its corresponding code word and then apply the Viterbi algorithm i.e. determine the hamming distance between the branch code word and the received word. However, to take full advantage of the available soft-decision information, the distance between the signal point corresponding to the branch code word and the received signal point is used to determine the metric for a given branch. The distance between any two signal points on a signal constellation diagram is referred to as the *Euclidean distance* d_{Euc} as illustrated in Fig. 8.42. From an energy perspective, since the Euclidean distance of each signal point is $\sqrt{E_s}$ from the center of the circle, the euclidean distance squared d_{Euc}^2 is an energy term between two signal points. Thus, for the purpose of metric calculations, it is the d_{Euc}^2 between the received signal point and the branch code word signal point that is used to determine the node metrics within the decoder trellis. Once again, only one surviving branch is allowed per node. The rest of the Viterbi algorithm remains the same as will be evident in the next section.

8.10 TCM Decoder

8.10.1 Noiseless Channel

▶ TCM Decoder

- Decoder trellis

- Using the trellis to determine the encoder output

- Detailed explanation of the decoding algorithm

- Decoding example for a noiseless channel

The Viterbi algorithm was explained in section 8.5, assuming a noiseless channel. Under the same assumption, a transmitted signal point will be received without error. In this case within the TCM decoder, all the euclidean distance calculations will only be between the signal points on the 8-PSK circle as shown in Fig. 8.43. If (x_b, y_b) and (x_r, y_r) represent the rectangular coordinates of the signal point corresponding to the branch code word and received signal point, respectively, then $d_{Euc}^2 = (x_b - x_r)^2 + (y_b - y_r)^2$.

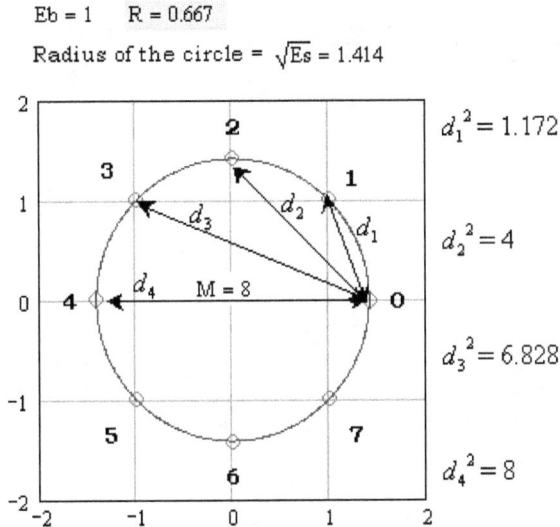

Figure 8.43: Euclidean distance squared between 8-PSK signal points.

- ⭐ SIMULATION **EuclideanDistance:** The euclidean distance between each signal point is determined.

- ⭐ SIMULATION **PolarToRectangular:** How to convert from polar to rectangular coordinates.

If the input sequence to the rate 2/3 TCM encoder is 11, 00, 00, 11, then the corresponding signal point sequence to be transmitted via 8-PSK is 6, 1, 2 and 6. Given that we are assuming a noiseless channel, these signal points are shown beneath the trellis diagram in Fig. 8.44, which is created using the code trellis diagram of Fig. 8.42. Using the euclidean distances shown in Fig. 8.43, Fig. 8.45 illustrates how to determine the surviving branch entering a given state. All the node metrics and the trace back path are shown in Fig. 8.46. Of course the information binary digits corresponding to the trace back path are then output by the decoder. Refer to the corresponding video lecture for a detailed explanation of the decoding process.

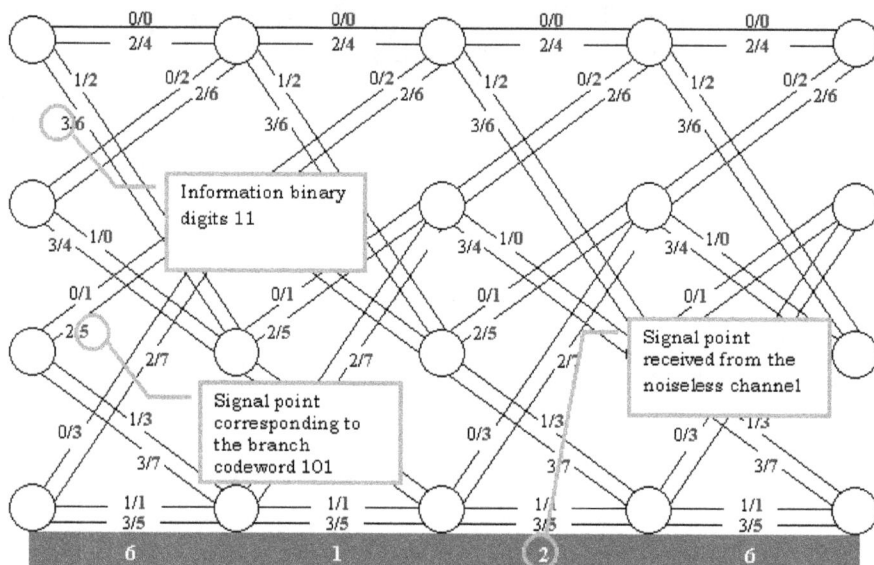

Figure 8.44: TCM decoder trellis.

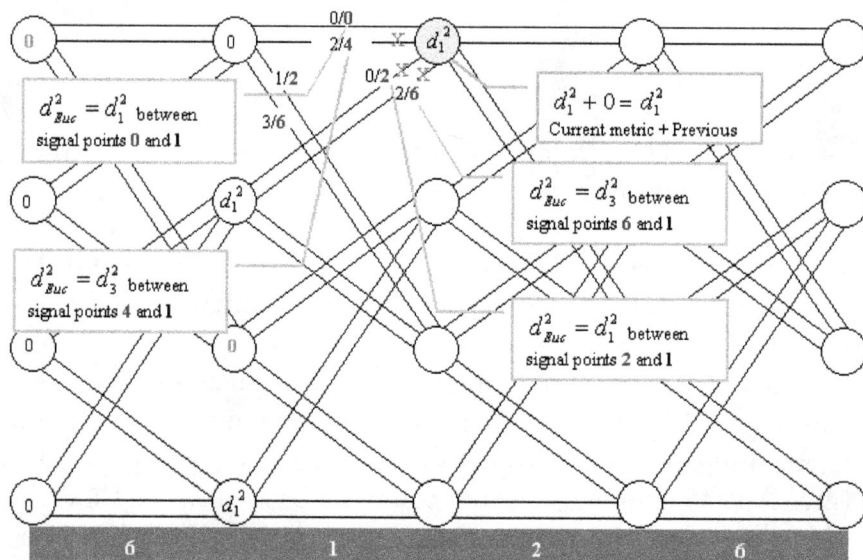

Figure 8.45: Metric calculations.

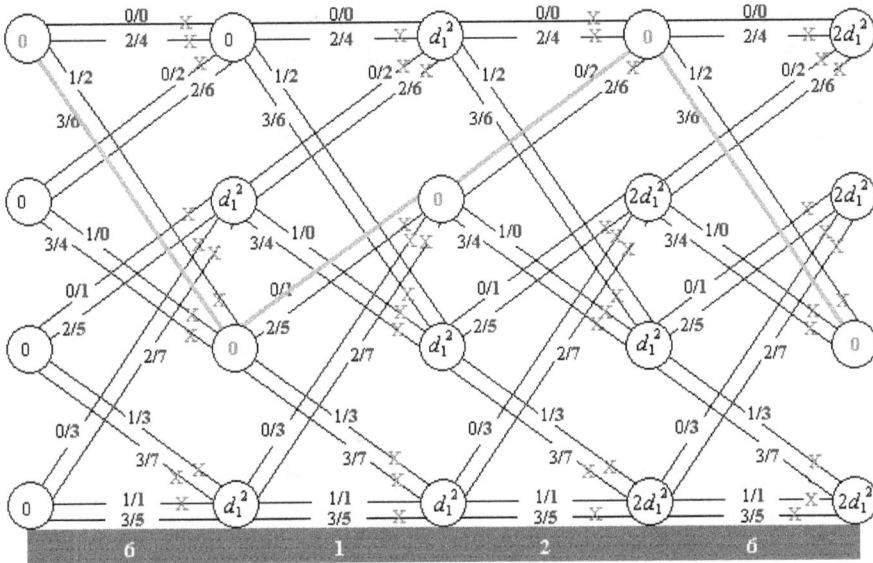

Figure 8.46: Node metrics and the trace back path.

8.10.2 Ungerboeck Codes

The encoder that was shown in Fig. 8.40 is in fact the best known rate $R_{code} = \frac{2}{3}$ 8-PSK trellis code in which the 8-PSK signal points are mapped *naturally* i.e. 0, 1, 2, 3, 4, 5, 6, 7 without the use of Gray coding [Ungerboeck, 1971]. Ungerboeck designed codes that provide a FEC capability without an increase in the required channel bandwidth by using an encoder structure of the type shown in Fig. 8.47. For every k input information binary digits, k^* are input to a rate $\frac{k^*}{k^*+1}$ convolutional encoder. The code rate R_{code} of the overall TCM encoder is

$$R_{code} = \frac{k}{k+1} \tag{8.102}$$

For example, for the Ungerboeck code in Fig. 8.47, $k^* = 1$ and $k = 2$. The key aspect of Ungerboeck's design is to use a technique, referred to as *set partitioning*, to map the k information binary digits onto 2^{k+1} signal points.

Figure 8.48 illustrates the set partitioning technique for the rate 2/3 encoder in Fig. 8.47. Namely, the $(k^* + 1)$ code digits are first used to select one of the $2^{(k^*+1)}$ partitions of the $2^{(k+1)}$ signal constellation at the $(k+1)$ *level,* where each level is labeled in Fig. 8.48. The remaining $(k - k^*)$ information binary digits are then used to select a particular signal point from the given partition.

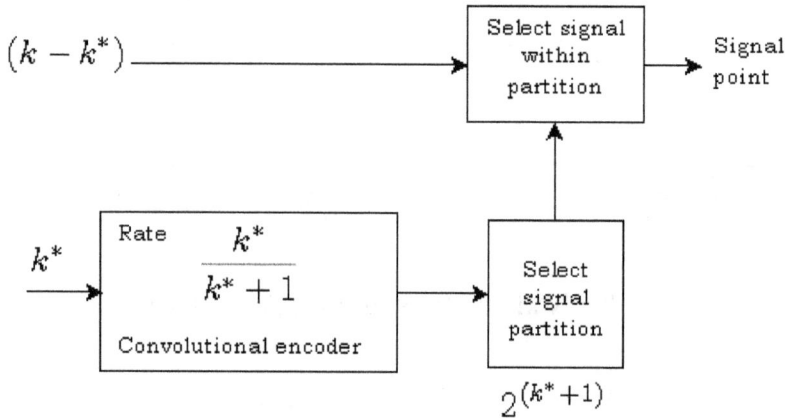

Figure 8.47: Ungerboeck encoder.

▶ TCM Coding Gain

- How to determine the coding gain of a
  rate 2/3 TCM code

It can be shown [Ungerboeck, 1971, 1987] that the asymptotic coding gain $G(dB)$ (i.e. the coding gain at a very high SNR) of an Ungerboeck code is given by

$$G(dB) = 10\log_{10}\left(\frac{d^2_{free}}{d^2_{ref}}\right) \tag{8.103}$$

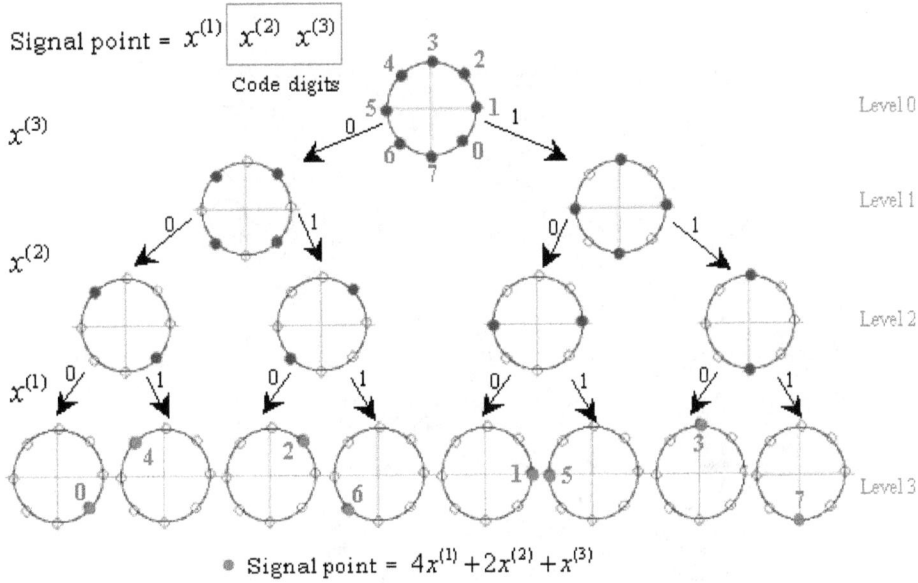

Figure 8.48: Set partitioning technique.

where d_{free} is the minimum euclidean distance between a pair of valid signal point sequences, referred to as the *free euclidean distance* of the code and d_{ref} is the minimum euclidean distance of the corresponding uncoded reference modulation scheme, operating with the same signal energy per binary digit. Using M-ary PSK with error-control coding, the energy per signal $E_s = R_{code} E_b \log_2 M$. Thus, for the rate 2/3 TCM code using 8-PSK, $E_s^{(TCM)} = \frac{2}{3} E_b \log_2 8 = 2E_b$. The signal energy for the uncoded QPSK case is given by $E_s^{(QPSK)} = (1)E_b \log_2 4 = 2E_b$. Without a loss of generality, we shall take $E_b = 1/2$, so that the signal points of both the coded and uncoded case lie conveniently on a circle of radius $\sqrt{E_s} = 1$. The free euclidean distance d_{free} is determined from the Viterbi decoder trellis of code using the all-zero signal point sequence as the first valid sequence and identifying the path with the minimum cumulative euclidean distance that deviates and returns to this path through the trellis, as illustrated in Fig. 8.49.

Since the signal point sequence 0400000... differs from the signal point sequence 000000... by only a single parallel branch transition as illustrated

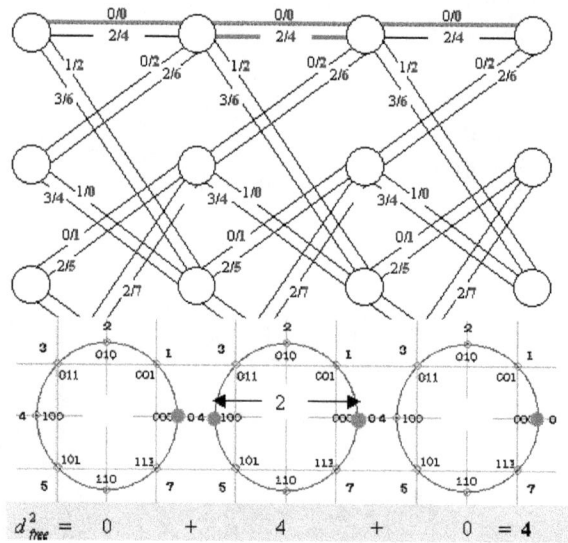

Figure 8.49: Free euclidean distance squared between valid signal point sequences.

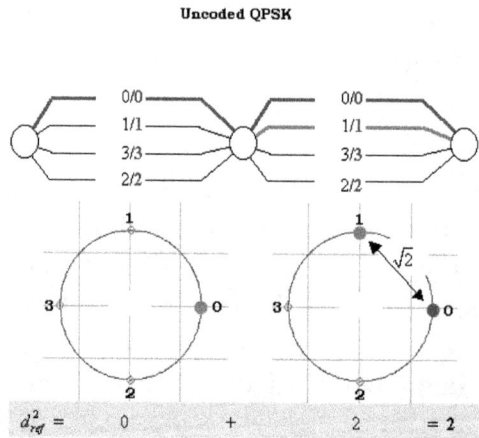

Figure 8.50: Uncoded QPSK trellis.

in Fig. 8.49, $d_{free} = 2$. The uncoded QPSK trellis is shown in Fig. 8.50. The minimum euclidean distance d_{ref} is simply the distance of $\sqrt{2}$ between the two adjacent signal points. Hence $G(dB) = 10 \log_{10} \left(\frac{4}{2} \right) = 3$ dB. The asymptotic coding gain of other Ungerboeck codes [Ungerboeck, 1971] based on 8-PSK with natural mapping, with respect to uncoded QPSK, are presented in Fig 8.51 and listed in Table 8.14.

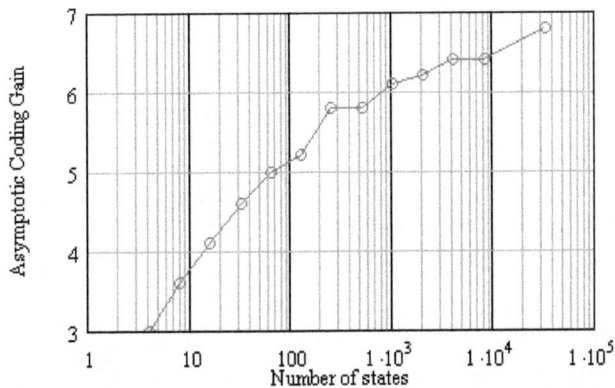

Figure 8.51: Asymptotic coding gain of Ungerboeck 8-PSK codes with respect to uncoded QPSK.

In Fig. 8.51, the increase in the asymptotic coding gain with the number of states is at the expense of complexity, which rapidly increases for a slight

Number of states	4	8	16	32	64	128	256	512	1024	2048	4096	8192	32768
Asymptotic Coding Gain	3	3.6	4.1	4.6	5.0	5.2	5.8	5.8	6.1	6.2	6.4	6.4	6.8

Table 8.14: Asymptotic coding gain of Ungerboeck 8-PSK codes with respect to uncoded QPSK.

increase in the coding gain after approximately 5 dB. Finally, as evident from Fig. 8.49, TCM provides a means to increase the minimum distance between the signal points that are most likely to be confused, *without* increasing the average transmitted power!

8.10.3   Simulation : TCM

⭐ SIMULATION **TCMDecoder:** The DCS shown in Fig. 8.52 is modeled. Simulation results are presented in Fig. 8.53. Notice that the coding gain compared to uncoded BPSK, or equivalently uncoded QPSK, is approximately 2.7 dB for $P_e = 10^{-5}$, which agrees well with the expected asymptotic gain of 3 dB. In Fig. 8.53, the simulation results of the rate 1/2 (7, 5) code punctured to create a rate 2/3 code, using BPSK over an AWGN with 8-level soft-decision Viterbi decoding is also presented. Although both schemes provide a similar coding gain at a high SNR, there is an important difference which is addressed in the next section. Namely, that the punctured rate 2/3 code requires three times the channel bandwidth requirement for the rate 2/3 TCM code.

- **Experiment:** How is the performance effected with the use of Gray coding instead of the current natural mapping scheme within 8-PSK modulator.

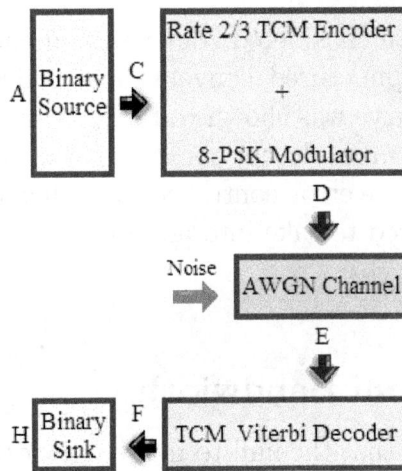

Figure 8.52: Simulated DCS : TCM.

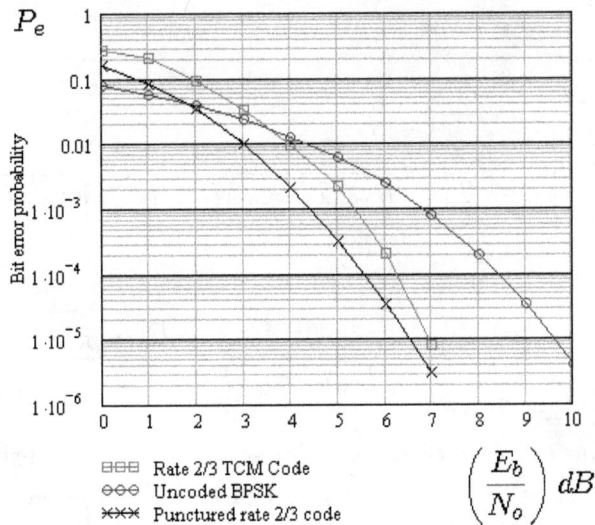

Figure 8.53: Rate 2/3 TCM code performance.

8.11 Bandwidth Efficiency

Up to this point, we have considered convolutional encoding with hard and soft-decision decoding, punctured convolutional codes and TCM. In each case, the performance curve was shown to be P_e versus $\frac{E_b}{N_o}$ in decibels. However, such performance curves do not convey the full picture from which one could select the appropriate error-control coding scheme for a given DCS. For a fair comparison, we need to take into account the bandwidth requirement for each scheme.

8.11.1 Null-to-Null Bandwidth

Recall from section 7.15, that the null-to-null bandwidth B_{MPSK} of an M-ary PSK signal of duration T_s is given by $B_{MPSK} = \frac{2}{T_s}$ for *unfiltered* MPSK i.e. MPSK with rectangular pulses. With pulse shaping via raised cosine filters, the *baseband* Nyquist bandwidth is equal to $\frac{1}{2T_s}$ (section 6.3.1) and therefore in this *bandpass* case, the double-sided Nyquist bandwidth $B_{MPSK}^{(Nyquist)} = \frac{1}{T_s}$. Using error-control coding and with no reduction in the information throughput, recall from section 7.8, equation 7.75, that $T_s = R_{code}T_b \log_2 M$. Therefore, in general,

$$B_{MPSK} = \frac{2}{T_s} = \frac{2}{R_{code}T_b \log_2 M} = \frac{2r_b}{R_{code} \log_2 M} \qquad (8.104)$$

and

$$B_{MPSK}^{(Nyquist)} = \frac{r_b}{R_{code} \log_2 M} = \frac{B_{MPSK}}{2} \qquad (8.105)$$

where r_b is the number of binary digits per second entering the channel encoder given by $r_b = \frac{1}{T_b}$. A comparison of the B_{MPSK} bandwidth requirement for with and without error-control coding is presented in Table 8.15 for the BPSK, QPSK and TCM. The use of a rate 1/2 convolutional code encoder prior to the BPSK modulation scheme doubles the bandwidth requirement i.e. from $2r_b$ to $4r_b$. However, by switching from BPSK to QPSK, the required channel bandwidth is halved. By puncturing a rate 1/2 code to create a rate 2/3 code lowers the channel bandwidth requirement, but the penalty incurred is a reduction in the error-control capability of the rate 2/3 code compared with the original rate 1/2 code. A key feature in Table 8.15 is that a rate 2/3 code with 8-PSK requires the *same* channel bandwidth as uncoded QPSK. Indeed this is the magic of Ungerboeck TCM codes which provide error-control coding without an additional increase in the required channel bandwidth.

- ⭐ SIMULATION **MPSKBandwidth:** A plot of the equation $B_{MPSK} = \frac{2r_b}{R_{code} \log_2 M}$.

8.11.2 Shannon's Channel Capacity Theorem

The channel bandwidth and the transmitted power are the two main precious resources for a communication system. The key question is not which particular modulation scheme and error-control coding is the best for an AWGN channel, but rather what is the theoretical limit on the rate of error-free transmission for a power and bandwidth limited AWGN channel? The answer was given by Shannon in 1949 in the form of the channel capacity theorem that, the channel capacity C (bits/sec) of an AWGN channel of bandwidth B is given by

$$C = B \log_2 \left(1 + \frac{E_b r_b}{N_o B} \right) \qquad (8.106)$$

where r_b is the number of binary digits per second output from a binary source to be transferred over this channel. Error-free communication is only possible if $r_b \leq C$ i.e. a maximum of $B \log_2 \left(1 + \frac{E_b r_b}{N_o B} \right)$ binary digits per second can be transmitted with a probability of error arbitrarily close to zero whereas for $r_b > C$, error-free communication is not possible.

Note that the channel capacity formula finds the maximum possible transmission rate for a given channel **without** specifying the modulation and error control-coding scheme required to achieve this limit! Compare this with Shannon's source coding theorem, that the average number of binary digits per second r_b required to represent a source with an information rate of R bits/sec must be such that $r_b \geq R$. If the capacity of a channel is C bits/sec, then its intuitively satisfying to ensure $r_b \leq C$ for error-free communication.

Manipulating the formula $C = B \log_2 \left(1 + \frac{E_b r_b}{N_o B}\right)$, we find $2^{\frac{C}{B}} = \left(1 + \frac{E_b r_b}{N_o B}\right)$ and hence $\frac{E_b}{N_o} = \frac{2^{\frac{C}{B}} - 1}{\frac{r_b}{B}}$, from which

$$\frac{E_b}{N_o}(dB) = 10 \log_{10} \left(\frac{2^{\frac{C}{B}} - 1}{\frac{r_b}{B}}\right) \tag{8.107}$$

For the ideal case where $r_b = C$,

$$\frac{E_b}{N_o}(dB) = 10 \log_{10} \left(\frac{2^{r_b/B} - 1}{r_b/B}\right) \tag{8.108}$$

A plot of this equation is shown in Fig. 8.54. This curve separates the region of error-free communication where $r_b < C$ and the region for $r_b > C$ where error-free communication is not possible. For a given bandwidth efficiency, the minimum $\frac{E_b}{N_o}(dB)$ required to achieve an arbitrarily low P_e simply corresponds to the $\frac{E_b}{N_o}(dB)$ on the channel capacity curve. For example from Fig. 8.54, the minimum SNR required for a DCS to work at the bandwidth efficiency of 4 bits/sec/Hz is 5.74 dB.

- ⭐ SIMULATION **CapacityCurve**: Plot of Fig. 8.54.

For a channel with a very large bandwidth i.e. $\frac{r_b}{B} \to 0$, the bandwidth efficiency approaches the limiting value of approximately -1.6 dB as shown in Fig. 8.54. This value is referred to as the *Shannon Limit* and is easily

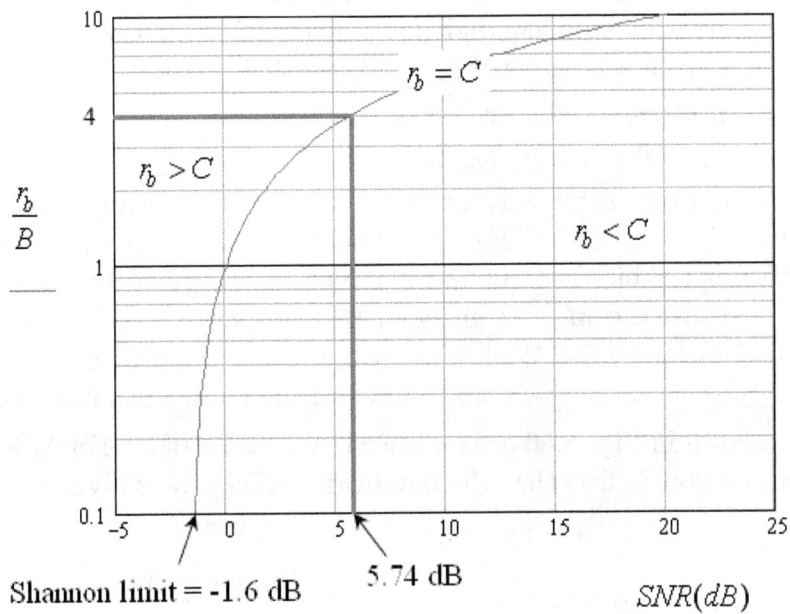

Figure 8.54: Channel capacity curve.

shown (refer to problems section) to be given by $-10 \log_{10} (\log_e 2)$ dB $= -$
1.6 dB. As expected, this result agrees with that from the channel coding
theorem in section 8.1, where as the code rate R_{code} approaches zero (so
that bandwidth efficiency tends to zero), we found the minimum SNR for
essentially error-free communication to approach -1.6 dB.

8.11.3   The Big Picture

A given modulation and error-control coding scheme provides a certain probability of an error in a binary digit for a given SNR. To compare the performance of a given scheme on Fig. 8.54, we shall take $P_e = 10^{-5}$ within a
DCS to represent essentially error-free. In this case, the $\left(\frac{E_b}{N_o}\right)$ dB required
to achieve $P_e = 10^{-5}$ and its corresponding bandwidth efficiency (*with no
pulse shaping*) from Table 8.16 are used to plot the *operating point* for a
given scheme in Fig. 8.55.   For example, the bandwidth efficiency of uncoded BPSK is 0.5 bits/sec/Hz and it requires a SNR of 9.6 dB to ensure
$P_e = 10^{-5}$. Thus, the $M = 2$ uncoded BPSK operating point coordinates
are (9.6, 0.5) in Fig. 8.55. With pulse shaping, the Nyquist bandwidth effi-
ciency $\frac{r_b}{B_{MPSK}^{(Nyquist)}} = R_{code} \log_2 M$ and therefore the bandwidth efficiency of all
the points shown in Fig. 8.55 are increased by a factor of 2. For M-ary FSK,
recall from section 7.15.1, that the bandwidth efficiency is given by

$$\rho_{FSK} = \frac{r_b}{B_{MFSK}} = \frac{r_b 2 T_s}{M} = \frac{2 R_{code} \log_2 M}{M} \qquad (8.109)$$

These bandwidth efficiency points are not shown in Fig. 8.55 to avoid over
complicating this figure. The penalty incurred with the use of a rate 1/2 (7,
5) convolutional code with hard-decision Viterbi decoding is a reduction in
bandwidth efficiency. With the use of soft-decision decoding, this operating
point moves towards the capacity curve by approximately 1.5 dB.   Selected
TCM points are also shown in Fig. 8.55, where the number in brackets is
the number of states. For example, TCM (4) corresponds to the 4 state
code considered in section 8.9. Notice that TCM provides a coding gain
without a reduction in the bandwidth efficiency i.e. unlike traditional error-
coding schemes which reduce the bandwidth efficiency to improve the power
efficiency (by reducing the $\frac{E_b}{N_o}$ required for a given P_e and thus reducing the

average transmitted power of $E_b r_b$ Watts), the bandwidth efficiency of TCM remains the same.

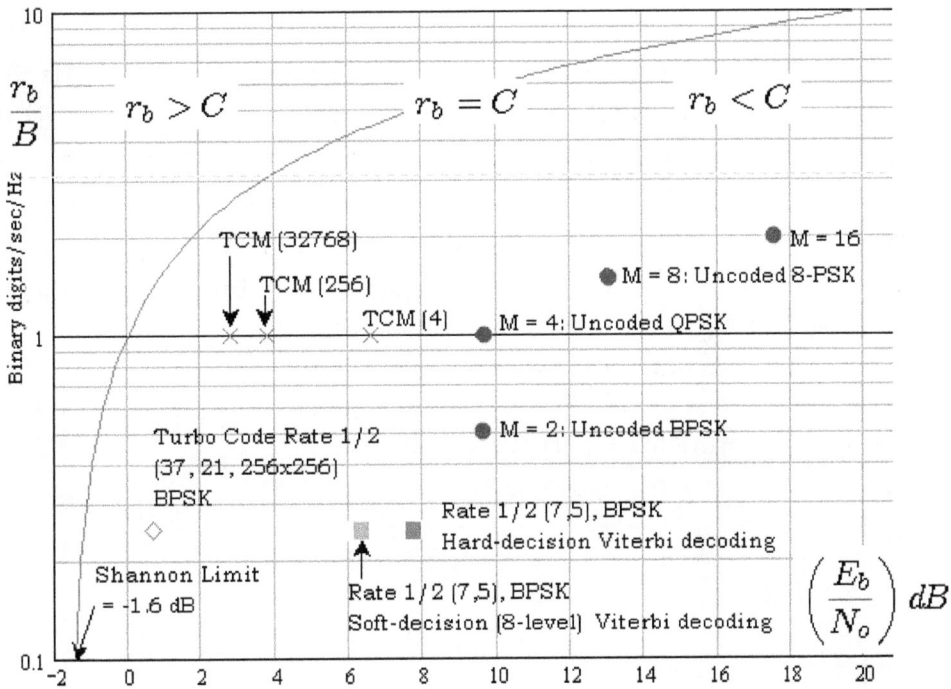

Figure 8.55: Shannon capacity curve for an AWGN channel. All the operating points correspond to $P_e = 10^{-5}$.

The channel coding theorem promises the existence of a rate $1/2$ code encoder and its corresponding decoder, utilized with an appropriate modulation and demodulation scheme with infinite quantization, that will require a minimum channel SNR of 0 dB to provide essentially error-free communication. Since 0 dB corresponds to a bandwidth efficiency of 1 bits/sec/Hz on the capacity curve, we can infer that the modulation bandwidth efficiency can be up to 1 bits/sec/Hz. This implies for example, that using a rate 1/2 code with QPSK, for which the Nyquist bandwidth efficiency is $\frac{r_b}{B_{MPSK}^{(Nyquist)}} =$ $R_{code} \log_2 M = \frac{1}{2} \log_2 4 = 1$ bits/sec/Hz and given that uncoded QPSK requires 9.6 dB to achieve $P_e = 10^{-5}$, we should be able to achieve a coding gain of (9.6 - 0) = 9.6 dB with an appropriate encoder and decoder over an AWGN channel. In the quest to achieve operating points that lie on the capacity curve, complex convolutional codes were used in the past with limited success. For example, a rate 1/4 constraint length $K = 14$ convolutional code used for the Galileo mission [Collins, 1992] achieved $P_e = 10^{-5}$ at $\left(\frac{E_b}{N_o} \right)$ dB = 1.75 dB with a bandwidth efficiency of 0.25 bits/sec/Hz. For many years, researchers tried without success to close the gap even further. The break through came in 1993 with the advent of turbo codes [Berrou and Glavieux, 1993]. For example, the rate 1/2 $(37, 21, 256 * 256)$ turbo code achieves $P_e = 10^{-5}$ at $\left(\frac{E_b}{N_o} \right)$ dB = 0.7 dB with a bandwidth efficiency of 0.5 bits/sec/Hz. Thereafter, Richardson et. al. [Richardson, 2001] presented results using BPSK over an AWGN channel with a probability of a binary digit error of 10^{-6} about 0.13 dB away from Shannon's capacity prediction. Given the importance of this break through, the next chapter is devoted to turbo codes and low density parity check codes.

Modulation Scheme	Bandwidth			
M-ary	B_{MPSK}	Uncoded	Rate 1/2	Rate 2/3
BPSK $M = 2$	$\frac{2r_b}{R_{code}}$	$2r_b$	$4r_b$	$3r_b$
QPSK $M = 4$	$\frac{r_b}{R_{code}}$	r_b	$2r_b$	$\frac{3r_b}{2}$
8-PSK $M = 8$	$\frac{2r_b}{3R_{code}}$	$\frac{2r_b}{3}$	$\frac{4r_b}{3}$	r_b
TCM Uncoded $M = 2^k$	$\frac{2r_b}{k}$			
TCM Coded $M = 2^{k+1}$	$2r_b\left(\frac{k+1}{k}\right)\frac{1}{k+1}$ $= \frac{2r_b}{k}$			

Table 8.15: Channel bandwidth required for no reduction in the information throughput.

Modulation Scheme	Bandwidth efficiency ρ_{MPSK} $= \frac{r_b}{B_{MPSK}} = \frac{R_{code}\log_2 M}{2}$ bits/sec/Hz			
M-ary	$\frac{r_b}{B_{MPSK}}$	uncoded	Rate 1/2	Rate 2/3
BPSK $M = 2$	$\frac{R_{code}}{2}$	$\frac{1}{2}$	$\frac{1}{4}$	$\frac{1}{3}$
QPSK $M = 4$	R_{code}	1	$\frac{1}{2}$	$\frac{2}{3}$
8-PSK $M = 8$	$\frac{3R_{code}}{2}$	$\frac{3}{2}$	$\frac{3}{4}$	1
TCM Uncoded $M = 2^k$	$\frac{k}{2}$			
TCM Coded $M = 2^{k+1}$	$\frac{k}{2}$			

Table 8.16: M-PSK bandwidth efficiency.

8.12   Problems

1. The parity matrix P for a $(7, 4)$ linear block code is given by

$$P = \begin{bmatrix} 1 & 1 & 0 \\ 0 & 1 & 1 \\ 1 & 1 & 1 \\ 1 & 0 & 1 \end{bmatrix}$$

Determine the following:
(a) All the code words of the code.
(b) The error correcting and detecting capability of the code.
(c) The syndrome table for single code digit errors.

2.   The generator matrix may alternatively be developed using $\mathbf{G} = \begin{bmatrix} \mathbf{P} & \mathbf{I}_k \end{bmatrix}$. In this case, the parity-check matrix is given by $\mathbf{H} = \begin{bmatrix} \mathbf{I}_{(n-k)} & \mathbf{P}^T \end{bmatrix}$. For the following two $(7, 4)$ linear block code generator matrices, determine the following:

$$G1 = \begin{bmatrix} 1 & 1 & 0 & 1 & 0 & 0 & 0 \\ 0 & 1 & 1 & 0 & 1 & 0 & 0 \\ 1 & 1 & 1 & 0 & 0 & 1 & 0 \\ 1 & 0 & 1 & 0 & 0 & 0 & 1 \end{bmatrix} \qquad G2 = \begin{bmatrix} 1 & 1 & 1 & 1 & 0 & 0 & 0 \\ 1 & 0 & 1 & 0 & 1 & 0 & 0 \\ 0 & 1 & 1 & 0 & 0 & 1 & 0 \\ 1 & 1 & 0 & 0 & 0 & 0 & 1 \end{bmatrix}$$

(a) All the code words of the code.
(b) The error correcting and detecting capability of the code.
(c) The syndrome table for single code digit errors.
(d) Verify that any two code digit errors cannot be corrected.

3.   A further alterative is to represent the generator matrix by $\mathbf{G} = \begin{bmatrix} \mathbf{I}_k & \mathbf{P}^T \end{bmatrix}$ in which case the parity-check matrix is given by $\mathbf{H} = \begin{bmatrix} \mathbf{P} & \mathbf{I}_{(n-k)} \end{bmatrix}$. Determine the following for the parity matrix P given by

$$P = \begin{bmatrix} 1 & 0 & 1 \\ 1 & 1 & 1 \\ 1 & 1 & 0 \end{bmatrix}$$

(a) All the code words of the code.
(b) The error correcting and detecting capability of the code.
(c) The syndrome table for single code digit errors.

(d) Show that the $\mathbf{H}^T$ rows correspond to the single digit error syndrome.
(e) Decode the received word 1 1 1 0 1 1.

4. From a list of all possible combination of 5 binary digits i.e. 5-bit tuple that begins with 00000 and ends with 111111, select two code words that end with 10 and 01.

(a) Use these two words to create a generator matrix and determine all the code words of this code. Ensure that the sum of any two code words produces a valid code word in that set.
(b) Determine the parity check matrix, code rate, error correcting and detecting capability of this code.

5. (a) Determine the parity-check matrix for the generator matrix shown below and verify your answer.

$$G = \begin{bmatrix} 1 & 0 & 0 & 0 & 1 & 1 & 0 \\ 0 & 1 & 0 & 0 & 0 & 1 & 1 \\ 0 & 0 & 1 & 0 & 1 & 0 & 1 \\ 0 & 0 & 0 & 1 & 1 & 1 & 1 \end{bmatrix}.$$

(b) Determine all the code words of the code for which the parity-check matrix is given by

$$H = \begin{bmatrix} 1 & 0 & 1 & 1 & 0 & 0 \\ 0 & 1 & 1 & 0 & 1 & 0 \\ 1 & 1 & 1 & 0 & 0 & 1 \end{bmatrix}.$$

(c) Determine the generator matrix $\mathbf{G} = \begin{bmatrix} \mathbf{P} & \mathbf{I}_k \end{bmatrix}$ and parity-check matrix corresponding to the parity matrix

$$P^T = \begin{bmatrix} 1 & 1 & 0 \\ 0 & 1 & 1 \\ 1 & 0 & 1 \\ 1 & 1 & 1 \end{bmatrix}.$$

6. (a) Determine the generator and parity check matrix of a $(n, 1)$ repetition code.
(b) Consider the generator matrix $\mathbf{G} = \begin{bmatrix} 1 & 1 & 0 & 1 & 0 \\ 1 & 0 & 1 & 0 & 1 \end{bmatrix}.$

(i) Determine all the code words of this code and the error-correcting capability of this code. (ii) Select any two code words and verify that their modulo-2 addition is also a valid code word. (iii) Determine the parity check matrix and verify that $\mathbf{G}\mathbf{H}^T = 0$.

7. For a $(n, 1)$ repetition code, the parity check matrix H is given by $\begin{bmatrix} P^T & I_{n-1} \end{bmatrix}$ or $\begin{bmatrix} I_{n-1} & P^T \end{bmatrix}$. By evaluating the syndrome for single and double code digit errors, verify that this code can correct up to two errors using either representation of H if $n = 5$.

8. (a) The parity check matrix of a (n, k) Hamming code is given by $\mathbf{H} = \begin{bmatrix} \mathbf{P}^T & \mathbf{I}_{n-k} \end{bmatrix}$ where $n = 2^m - 1$ and $k = 2^m - 1 - m$. Design a hamming code for $m = 4$. Would the generator matrix be of the form $\mathbf{G} = \begin{bmatrix} \mathbf{I}_k & \mathbf{P} \end{bmatrix}$ or $\mathbf{G} = \begin{bmatrix} \mathbf{I}_k & \mathbf{P}^T \end{bmatrix}$ or $\mathbf{G} = \begin{bmatrix} \mathbf{P} & \mathbf{I}_k \end{bmatrix}$?

(b) For a (n, k) hamming code, $n = 2^m - 1$ and $k = 2^m - 1 - m$. The probability of an error in a binary digit is given by $P_e = \alpha - \alpha (1 - \alpha)^{n-1}$, where α is the BSC crossover probability. Plot a graph of P_e versus $\frac{E_b}{N_o}(dB)$ to compare the performance of the $(7, 4)$, $(15, 11)$ and $(31, 26)$ hamming codes with the use of binary phase shift keying (BPSK) over an AWGN channel. Be sure to include the performance of uncoded BPSK.

9. A single parity error detection code is created by adding a single parity digit (1 or 0) to a block of information digits. For even parity, the parity digit is selected to ensure that the modulo-2 summation of all the digits is zero. For example, if the information digits are 100, then the parity digit is 1 to ensure that $1+0+0+1 = 0$. The transmitted code word in this case is 1001.

(a) Determine all the code words for a $(4, 3)$ even-parity error detection code.

(b) What is the probability of N code digit errors, where $N \in \{0, 1, 2, \cdots, n\}$.

(c) If the received word from the channel has even parity, we cannot assume that the received word is a valid code word. Determine the probability of no error detection P_{error} for a given BSC crossover probability and plot P_{error} versus the channel SNR in decibels with the use of binary shift keying (BPSK) over an AWGN channel.

10. The code digits of a (31, 15) block code, that is capable of correcting 3 errors, are transmitted over an AWGN using the BPSK modulation scheme with no reduction in the information throughput.

- Let P_N represent the probability of N errors or less within the received n-bit code word.

- Let P_m represent the probability of an error in the received k-bit word without the use of error-control coding i.e. without using the (n, k) block code and simply transmitting the k-bit word via BPSK.

- Let P_e represent the probability of an error in the decoded k-bit word.

(a) At the channel SNR of 0 dB, plot a graph of P_N versus N for $N = 0, 1, 2, \cdots n$. Plot also a graph of P_N versus $\frac{E_b}{N_o}(dB)$ for $N = 2$. Comment on the graphs.

(b) Plot a graph of P_e versus $\frac{E_b}{N_o}(dB)$. Overlay this graph with a plot of P_m versus $\frac{E_b}{N_o}(dB)$ and estimate the coding gain.

11. For the (6, 3) block code with generator matrix

$$G = \begin{bmatrix} 0 & 1 & 1 & 1 & 0 & 0 \\ 1 & 0 & 1 & 0 & 1 & 0 \\ 1 & 1 & 0 & 0 & 0 & 1 \end{bmatrix}$$

(a) Determine the syndrome for all possible received words in the 6-bit tuple i.e. 000000 to 111111. Verify that all single error patterns correspond to a unique syndrome and that the only unique two error pattern that can be corrected is 0 1 0 0 1 0.

(b) Hence determine an expression for the probability of a decoding error P_e and plot a graph of P_e versus the channel SNR in decibels assuming the use of BPSK over an AWGN channel to transfer the code digits. Overlay the uncoded BPSK curve. Does this code provide a coding gain ?

(c) Repeat part (b), but now do not take into account of the only double error pattern that can be corrected. Verify that the performance curve now corresponds exactly to the expression for a t-error correcting linear block code given by

$$P_e = \sum_{j=t+1}^{n} \binom{n}{j} \alpha^j (1-\alpha)^{n-j}$$

where α is the BSC crossover probability.

12. The generator polynomial for a $(7, 3)$ cyclic code is given by $g(X) = X^4 + X^2 + X + 1$.

(a) Confirm that the generator polynomial is a factor of $(X^n + 1)$.
(b) Determine the code word corresponding to the information digit sequence 100.
(c) Verify your solution to part (b) via $c(X) = a(X)g(X)$.

13. (a) Draw the encoder of type II for the cyclic code of the previous question.
(b) Implement the encoder in Mathcad/MATLAB and verify the solution to the previous question.
(c) Determine all possible code words.
(d) The syndrome table for all single digit errors.
(e) Decode the received word 1011011.

14. Consider a $(15, 7)$ systematic cyclic block code with generator tap connections 100010111.

(a) Implement in Mathcad/MATLAB, the encoder of type I and type II.
(b) Verify that both encoders are equivalent by encoding the input sequence 0001010.
(c) Implement the syndrome former of type I and II corresponding to the encoder of type I and type II, respectively and determine the syndrome for the code word 000101010010110. Add the error vector 100000000000000 to this code word and determine the syndrome for type I and II.
(d) Using the error pattern 000001000000000, verify that if the syndrome former of type II is flushed with zeros until the syndrome is equal to 1 followed by $(n - k - 1)$ zeros, then the number of zeros inserted will correspond to the single code digit error position.
(e) It is possible to correct the burst error pattern 000000010110000 using the syndrome former of type II ?

15. Consider a $(15, 9)$ systematic cyclic block code with generator tap connections 100010111.

(a) Implement in Mathcad/MATLAB, the encoder of type I and type II.

(b) Verify that both encoders are equivalent by encoding the input sequence 000001010.

(c) Implement the syndrome former of type II corresponding to the encoder of type II and determine the syndrome for the code word 000001010100111. Add the error vector 100000000000000 to this code word and determine the syndrome.

(d) Using the error pattern 000001000000000, verify that if the syndrome former of type II is flushed with zeros until the syndrome is equal to 1 followed by $(n - k - 1)$ zeros, then the number of zeros inserted will correspond to the single code digit error position.

(e) It is possible to correct the burst error pattern 000000001110000 using the syndrome former of type II ?

16. The diagram below shows a rate $\frac{1}{2}$ convolutional code encoder.

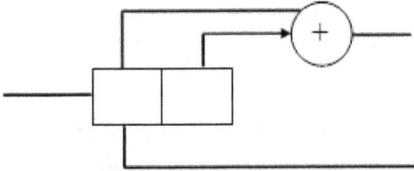

(a)   Write down is generator transfer-function matrix.

(b)   Determine the encoder output produced by the message sequence 10111...1 using the transform-domain approach.

(c)   Determine the state table and draw the corresponding code trellis diagram.

(d)   Assuming the code digits output by the encoder pass through a noiseless channel, use the Viterbi algorithm to decode the message sequence. Be sure to draw a clear diagram of the trellis. Do not assume the encoder started in the all-zero state.

(e)   If the output sequence from the encoder in part (a) enters a channel which corrupts every fourth bit, starting from the first bit that enters the channel, determine the output of the Viterbi decoder. For example is the input to the channel is an all-zero sequence, the output of the channel is 10001000100010001 ...

(f)   Determine the free distance of this code.

17. A rate 2/3 code is described by the following generator matrix:

$$G(D) = \begin{bmatrix} 1+D & 1+D & 1 \\ D & 0 & 1+D \end{bmatrix}$$

(a)   Draw a diagram of the convolutional code encoder and determine its state diagram.

(b)   Determine the free distance of this code.

18. Consider a rate $\frac{1}{2}$, constraint length 3 convolutional code with a generator transfer function matrix $G(D) = \begin{bmatrix} 1+D+D^2 & 1+D^2 \end{bmatrix}$.

(a)   Draw a diagram of the encoder and clearly indicate the shift registers which represent the state of the encoder.

(b) Determine the state table for this code and draw the corresponding code trellis diagram.

(c) If the code digit sequence received by the Viterbi decoder is 11, 00, 00, 10, 11, 00, 00, where 11 is the first code word to be received, determine the following assuming the encoder started in the all-zero state. (i) The decoded information binary sequence (ii) Identify the code digits which were corrupted within the channel.

19. Consider a digital source which outputs one of four possible symbols A, B, C or D. The output of this digital source is input to an encoder which assigns a code word to each symbol as shown in the table below.

SYMBOL	CODE WORD
A	000000
B	101010
C	010101
D	111111

If the received sequence over a binary symmetric channel (BSC) is 111010 and a maximum likelihood decoder is used

(a)   What will be the decoded symbol ?

(b)   What is the probability of receiving the code digit sequence 111010 if the BSC crossover probability is 0.1 and symbol B is transmitted ?

20. Consider a digital communication system (DCS) which consists of a binary information source, a binary symmetric channel (BSC) and an information sink.

(a)   If a channel encoder and decoder are introduced within the DCS, explain what happens to the information throughput.

(b)   What is the solution to this problem ?

21. The diagram shows the state diagram of a rate $\frac{1}{2}$ convolutional code encoder.

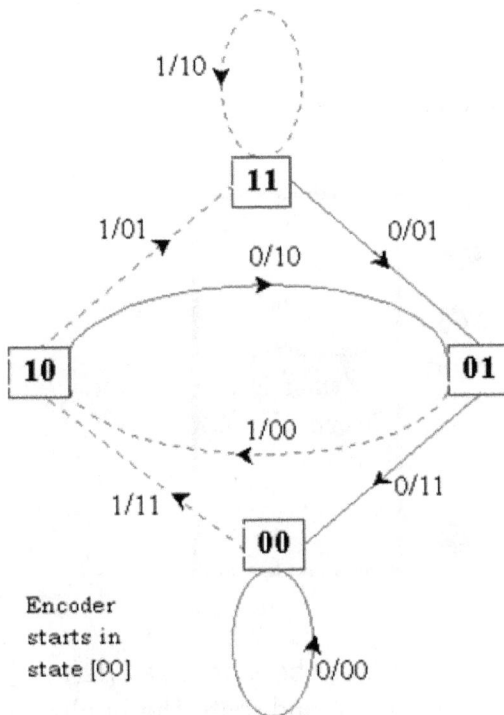

(a) The binary sequence m_1, m_2, m_3 is input to the encoder. Assuming the encoder started in the all zero state and $m_1 = 1, m_2 = 0, m_3 = 1$, determine the output code word sequence. Be sure to flush the encoder.

(b) Feed the output of the encoder into the corresponding trellis in which trellis depth is set to 3. Determine the state metrics and highlight the trace back path. Do not assume the encoder started in the all-zero state.

(c) Feed the output of the encoder into the configuration shown below to determine the output sequence. What do you conclude about this configuration ?

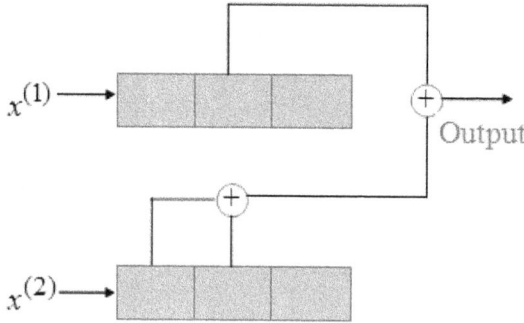

22. Consider the following digital communication system.

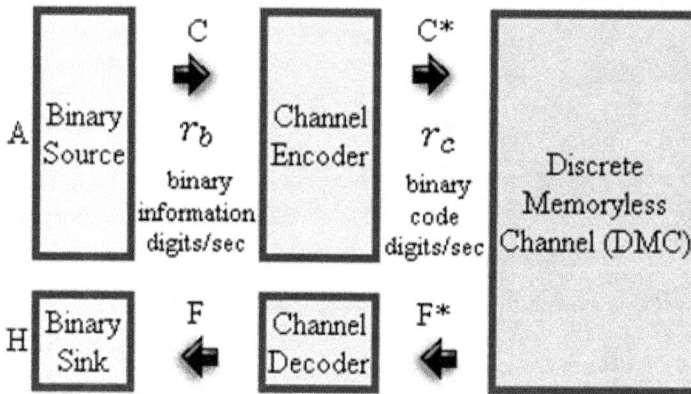

(a) Shannon's channel coding theorem may be stated as $r_b \Omega(p) \leq r_c C$ bits/sec. Explain the meaning of each symbol and state the implication of this theorem. What is the surprising result of the Shannon's coding theorem ?

(b) The *recursive systematic code* (RSC) code trellis shown below is used to transfer a short binary sequence of 4 digits.

The corresponding received code word sequence is [00, 10, 10, 11], where the code word 00 corresponds to the first binary digit entering the RSC code.

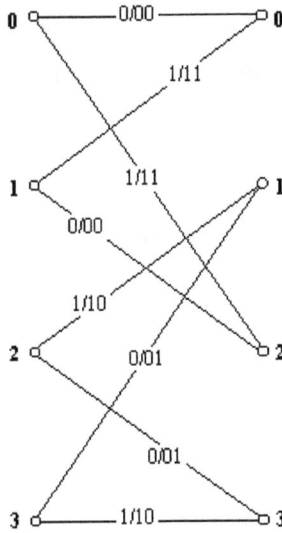

Figure 8.56: Rate 1/2 recursive systematic code trellis.

You may not assume the encoder started in the all-zero state. Decode the binary sequence using the Viterbi algorithm. (c) Identify the code digits that were corrupted within the channel.

23. A rate $\frac{1}{2}$ convolutional code is described by the generator transfer-function matrix $G(D) = \begin{bmatrix} 1 + D + D^2 & 1 + D^2 \end{bmatrix}$. Determine the generator transfer-function matrix $G_p(D)$ of the punctured convolutional code created using the puncturing matrix $P = \begin{bmatrix} 1 & 1 \\ 1 & 0 \end{bmatrix}$ in combination with the rate $\frac{1}{2}$ code.

24. A rate 1/3 convolutional code is described by the generator transfer-function matrix $G(D) = \begin{bmatrix} 1 + D & 1 + D + D^2 & 1 + D^2 \end{bmatrix}$.

(a)   Draw a diagram of the encoder and clearly indicate the shift registers, which represent the state of the encoder.
(b)   Determine the state table for this code.
(c)   Use the corresponding trellis to determine the free distance of this code.

TCM

25. In section 8.10, the asymptotic coding gain was calculated by assuming a value of E_b. Recalculate this coding gain of 3 dB, but now without assuming a value for E_b.

26. A trellis coded modulation scheme uses the following rate 2/3 systematic convolutional code encoder.

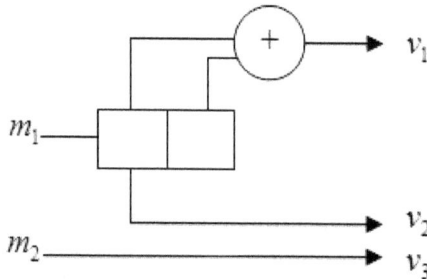

A given code word is transmitted over the channel using 8-PSK, using the mapping scheme shown in the following table.

Codeword	Signal Point
000	0
001	1
011	2
010	3
110	4
111	5
101	6
100	7

(a) Use the above table to show what is meant by Gray encoding.

(b) Calculate the Euclidean distance between each signal point on the 8-PSK signal constellation diagram.

(c) Determine the state table for this convolutional code and hence the corresponding code trellis diagram.

(d) If the input to encoder is the binary sequence 100000, where the first pair of digits to enter the encoder are $m_1 = 1$, $m_2 = 0$, determine the output of the encoder and the corresponding three signal points.

(e) If the first signal point sent through the channel is received as the signal point 2 and the other two are received without error, decode the information sequence.

27. A trellis coded modulation scheme uses the following rate 2/3 systematic convolutional code encoder.

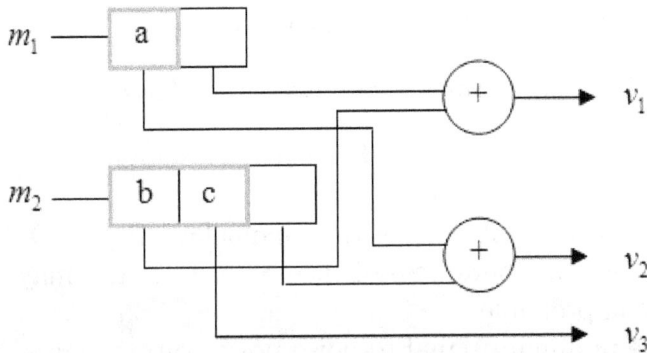

The state of the encoder is the contents of the shift register elements highlighted in the order "abc". (a) Determine the state table for this convolutional code.

(b) Using the code trellis diagram, determine the asymptotic coding gain of this 8 state TCM code.

28. Using uncoded coherent M-ary PSK with Gray coding over an AWGN channel, the probability of an error in a binary digit is found to be 0.0001. If the system bandwidth is 10 kHz and the data rate is 40,000 binary digits per second, what is the operating channel SNR in decibels ?

29. What is minimum $\left(\dfrac{E_b}{N_o}\right)$ dB required to achieve a probability of an error in a binary digit less than 10^{-5} if the data rate required is 50,000 binary digits/sec over a channel bandwidth of 20KHz ?

30. What is the minimum signal power required for reliable communication at a transmission of 56000 binary digits/sec over an AWGN channel of bandwidth 3.3 kHz with a single sided noise power spectral density of 10^{-11} watts/Hz ?

31. If the transmitted signal power is 10^{-5} watts, is it possible to transmit 50,000 binary digits/sec over a channel bandwidth of 10 kHz if the channel noise is 10^{-11} watts/Hz of bandwidth.

32. What is the maximum data rate that can be achieved with essentially error-free communication over a BSC with a crossover probability α of 0.05, if 5000 digits can be transferred over the BSC per second ? Sketch the variation of this maximum data rate as α ranges from 0.01 to 0.5.

33. Consider a telephone channel of bandwidth 3.3 KHz.

(a)   What is the channel capacity if the signal-to-noise power ratio is 25 dB ?

(b)   If the symbols transmitted over this channel are from a DMS of 26 symbols, what is the maximum possible symbol rate ? You may assume that each symbol is equiprobable.

(c)   What is the minimum signal-to-noise power ratio required for the following data rates ? (i) 28800 binary digits/sec (ii) 31200 binary digits/sec and (iii) 56000 binary digits/sec.

34. Consider a telephone 56K modem operating over a channel of bandwidth 3.3 KHz.

(a)   What is the bandwidth efficiency ?

(b)   If the available signal-to noise ratio $\left(\dfrac{E_b}{N_o}\right)$ is 9.6 dB, what the maximum possible data rate over this channel ?

(c)   By how much would $\left(\dfrac{E_b}{N_o}\right)$ have to be increased in order to ensure the modem operates at 56,000 binary digits/sec ?

35. A video camera is able to send one picture frame every 35 seconds. Each frame consists of a million pixels. For a given pixel, there are 256 equally likely brightness levels. If the available signal to noise power ratio is 42 dB, what is the

(a) information conveyed per pixel ?

(b) information conveyed per picture ?

(c) information rate ?

(c) minimum bandwidth required for the transmission of this signal.

36. The channel capacity of a band-limited AWGN channel is under the constraint of signal power $P = E_b r_b$

is given by $C = B \log_2 \left(1 + \dfrac{P}{N_o B} \right)$ bits/s.

(a) Show for $B \to \infty$, the minimum $\frac{E_b}{N_o}$ required for error-free communication is -1.6 dB (Shannon Limit).

(b) Does the channel capacity go to infinity as the channel's bandwidth $B \to \infty$?

Everything should be made as simple as possible, but not simpler.

— *Albert Einstein*

Appendices

As far as the laws of mathematics refer to reality, they are not certain, as far as they are certain, they do not refer to reality.

— *Albert Einstein*

Appendix A

Probability Theory

A.1   Introduction

If a boy hits a bulls-eye with a probability of 0.1, what does this mean ?
It does **not** mean that out of every 10 attempts, he will definitely hit the
bulls-eye once. It means that out of every 10 attempts, he will probably hit
the bulls-eye once. In fact, he may hit the bulls-eye 3 times in the first 10
attempts. Then may miss the target completely in the next 10 attempts.
However, you will find that over say one 1000 attempts, the bulls-eye will be
hit approximately 100 times. Mathematically, we can describe this experiment as follows:

- Let A represent the event that the bulls-eye is hit

- Let $P(A)$ represent the probability of event A

If the experiment is carried out n times and event A occurs n_A times,
then we may estimate $P(A) = \frac{n_A}{n}$ e.g. if the bulls-eye is hit 3 times out of
10 attempts, then we estimate $P(A) = \frac{3}{10} = 0.3$. Of course this estimate will
be more accurate if we increase n to say 10000. In fact the probability $P(A)$
is given by

$$P(A) = \lim_{n \to \infty} \frac{n_A}{n} \tag{A.1}$$

Finally, note that the probability $P(A)$ of any event A is simply a number such that $0 \le P(A) \le 1$.

A.2   Axioms

Probability theory is based on the following three axioms:

- $P(A) \ge 0$

- $P(S) = 1$ where S is the set of all possible outcomes referred to as the *sample space*. For example, in throwing a die, 1, 2, 3, 4, 5 and 6 are the possible outcomes.

- If event A and B are mutually exclusive (i.e. if A occurs, then B cannot occur and vice versa), then the probability of event A or B, denoted by $P(A \cup B)$, is given by $P(A \cup B) = P(A) + P(B)$.

A.3   Independent Events

In a classroom, suppose Mork scratches his head with probability 0.1 and Mindy scratches her head with probability 0.3. They are both unaware of each other. Obviously, the probability that Mork and Mindy will scratch their head at the same time will occur with a very small probability. In fact, its simply given by $(0.1)(0.3) = 0.03$. More formally, the *joint* probability of two statistically independent events A and B, denoted by $P(A \cap B)$, is given by $P(A \cap B) = P(A)P(B)$. Using the axioms introduced in the previous section, it is easily shown that in general,

$$P(A \cup B) = P(A) + P(B) - P(A \cap B) \qquad (A.2)$$

For example, the probability that Mork or Mindy will scratch their head is equal to $0.1 + 0.3 - 0.03 = 0.37$. However, if the scratching of Mork and Mindy is mutually exclusive, then $P(A \cup B) = P(A) + P(B) = 0.1 + 0.3 = 0.4$.

A.4   Conditional Probability

The probability of an event A given that event B has occurred is denoted by $P(A|B)$ i.e. the event to the right of the vertical line "|" has occurred. This *conditional probability* is defined as

$$P(A|B) = \frac{P(A \cap B)}{P(B)} \tag{A.3}$$

Similarly, $P(B|A) = \frac{P(A \cap B)}{P(A)}$. Combining these two equations, we obtain Bayes' rule that

$$P(B|A) = \frac{P(A|B)P(B)}{P(A)} \tag{A.4}$$

Note that for independent events, given that $P(A \cap B) = P(A)P(B)$, the conditional probability $P(A|B) = \frac{P(A \cap B)}{P(B)} = \frac{P(A)P(B)}{P(B)} = P(A)$ which is intuitively satisfying because if event A is independent of event B, then event B does not influence event A i.e. $P(A|B) = P(A)$.

To illustrate the use of conditional probabilities, consider an experiment in which we transmit either a binary digit zero (represented by x_1) or a binary digit one (represented by x_2) through a digital communication system (DCS). Let the output of this DCS be either a binary digit zero (represented by y_1) or one (represented by y_2) as shown in Fig. A.1.

Suppose we transmit and receive 20 binary digits as shown in the table below.

Input   1 0 0 0 1 1 0 0 0 1 0 0 0 1 1 1 1 0 1 0
Output   0 0 1 0 1 1 0 0 0 1 0 1 0 0 1 0 1 0 1 0

From the first row, we can estimate the following probabilities. The probability that x_1 is transmitted $P(x_1) = \frac{11}{20} = 0.55$ i.e. simply count the number of x_1 in this row and divide by 20. Similarly, the probability that x_2 is transmitted $P(x_2) = \frac{9}{20} = 0.45$. Of course given that only either a binary

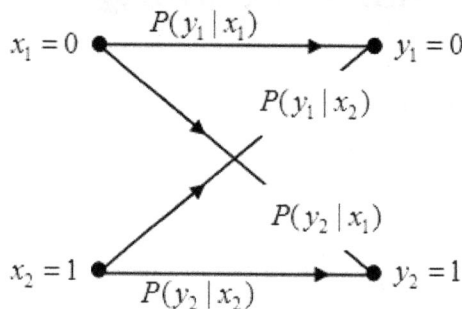

$$x_1 = 0 \xrightarrow{\ P(y_1\,|\,x_1)\ } y_1 = 0$$

$$P(y_1\,|\,x_2)$$

$$P(y_2\,|\,x_1)$$

$$x_2 = 1 \xrightarrow{\ \ } y_2 = 1$$
$$P(y_2\,|\,x_2)$$

Figure A.1: Binary symmetric channel.

digit zero or one can be transmitted, the sum $P(x_1) + P(x_2)$ must be equal to 1 i.e. second axiom. Thus alternatively, $P(x_2) = 1 - P(x_1) = 1 - 0.55 = 0.45$. From the second row, we find that the probability that y_1 is received $P(y_1) = \frac{12}{20} = 0.6$ and the probability that y_2 is received $P(y_2) = \frac{8}{20} = 0.4$ or simply $P(y_2) = 1 - P(y_1) = 0.4$.

To determine the conditional probability $P(y_2|x_1)$, we need to determine the number of times y_2 is received when x_1 is transmitted. There are two such events as highlighted in the table below.

Input   1 0 **0** 0 1 1 0 0 0 1 0 **0** 0 1 1 1 1 0 1 0
Output 0 0 **1** 0 1 1 0 0 0 1 0 **1** 0 0 1 0 1 0 1 0

Given that x_1 was transmitted a total of 11 times, $P(y_2|x_1) = \frac{2}{11}$. If x_1 is transmitted, the output can only be either y_1 or y_2. Thus, $P(y_2|x_1) + P(y_1|x_1) = 1$ and accordingly, $P(y_1|x_1) = 1 - \frac{2}{11} = \frac{9}{11}$. As a quick check, verify for yourself from the second row that y_1 was received a total of 9 times when x_1 was transmitted. Following a similar analysis, we find $P(y_1|x_2) = \frac{3}{9}$ and $P(y_2|x_2) = 1 - P(y_1|x_2) = \frac{6}{9}$. We shall now take a slightly different approach.

Given that y_1 can occur as the result of either x_1 or x_2 being transmitted, the probability that x_1 is transmitted and received as y_1 is equal to $P(y_1|x_1)P(x_1)$ i.e. the joint event of $P(y_1|x_1)$ and $P(x_1)$. Similarly, the probability that x_2 is transmitted and received as y_1 is equal to $P(y_1|x_2)P(x_2)$.

Thus $P(y_1) = P(y_1|x_1)P(x_1) + P(y_1|x_2)P(x_2) = \left(\frac{9}{11}\right)0.55 + \left(\frac{3}{9}\right)0.45 = 0.6$. This confirms our previous result. Similarly, we obtain $P(y_2) = P(y_2|x_1)P(x_1) + P(y_2|x_2)P(x_2) = \left(\frac{2}{11}\right)0.55 + \left(\frac{6}{9}\right)0.45 = 0.4$. Using Bayes' rule, the probability that x_1 was transmitted given that y_1 is received is given by $P(x_1|y_1) = \frac{P(y_1|x_1)P(x_1)}{P(y_1)} = \frac{\left(\frac{9}{11}\right)0.55}{0.6} = 0.75$ and $P(x_2|y_1) = \frac{P(y_1|x_2)P(x_2)}{P(y_1)} = \frac{\left(\frac{3}{9}\right)0.45}{0.6} = 0.25$. This is as expected because having received y_1, it must have been either x_1 or x_2 that was transmitted i.e. $P(x_1|y_1) + P(x_2|y_1) = 1$. Similarly, $P(x_1|y_2) + P(x_2|y_2) = 1$.

From the definition $P(A|B) = \frac{P(A \cap B)}{P(B)}$, the joint event $P(x_1 \cap y_1)$ that x_1 is transmitted and y_1 is received is given by $P(x_1 \cap y_1) = P(x_1|y_1)P(y_1) = (0.75)\,0.6 = 0.45$. The table below highlights these events, from which once again $P(x_1 \cap y_1) = \frac{9}{20} = 0.45$.

Input	1 0 0 0 1 1 0 0 0 1 0 0 0 1 1 1 1 0 1 0
Output	0 0 1 0 1 1 0 0 0 1 0 1 0 0 1 0 1 0 1 0

Similarly, $P(x_2 \cap y_1) = P(x_2|y_1)P(y_1) = (0.25)\,0.6 = 0.15$ or alternatively, $P(x_2 \cap y_1) = \frac{3}{20} = 0.15$.

A.5 Random Variable

The value x of a *random variable* X is a number whose value is determined by a random event e.g. $x = 1$ or 0 if the result of tossing a coin is a head or tail, respectively. If the coin is tossed 10 times, then x may be 1101000101. Note that a capital letter is used to denote the random variable and a lower case letter is used to denote its value.

A function of a random variable is also a random variable. For example, suppose two unbiased dice are thrown. The table below shows the corresponding sample space i.e. all the possible outcomes.

(1,1)	(1,2)	(1,3)	(1,4)	(1,5)	(1,6)
(2,1)	(2,2)	(2,3)	(2,4)	(2,5)	(2,6)
(3,1)	(3,2)	(3,3)	(3,4)	(3,5)	(3,6)
(4,1)	(4,2)	(4,3)	(4,4)	(4,5)	(4,6)
(5,1)	(5,2)	(5,3)	(5,4)	(5,5)	(5,6)
(6,1)	(6,2)	(6,3)	(6,4)	(6,5)	(6,6)

Let X denote the random variable whose value x is equal to the sum of the two numbers thrown. The minimum value of x is 2 and its maximum value is 12. If each dice is unbiased, then the probability a number 1, 2, ... 6 thrown is equal to $\frac{1}{6}$ i.e. each number is equally likely to be thrown. Hence for example, the probability of the joint event (4,3) is then simply $\left(\frac{1}{6}\right)\left(\frac{1}{6}\right) = \frac{1}{36}$. To determine for example the probability $P_X(x = 5)$, we need to know what combinations of throws will add to 5. These are simply (2, 3) or (3, 2) or (1,4) or (4,1). Thus, $P_X(x = 5) = \frac{1}{36} + \frac{1}{36} + \frac{1}{36} + \frac{1}{36} = \frac{4}{36}$. Following the same analysis, the complete list of $P_X(x)$ probabilities are listed in the table below and plotted in Fig. A.2.

x	2	3	4	5	6	7	8	9	10	11	12
$P_X(x)$	$\frac{1}{36}$	$\frac{2}{36}$	$\frac{3}{36}$	$\frac{4}{36}$	$\frac{5}{36}$	$\frac{6}{36}$	$\frac{5}{36}$	$\frac{4}{36}$	$\frac{3}{36}$	$\frac{2}{36}$	$\frac{1}{36}$

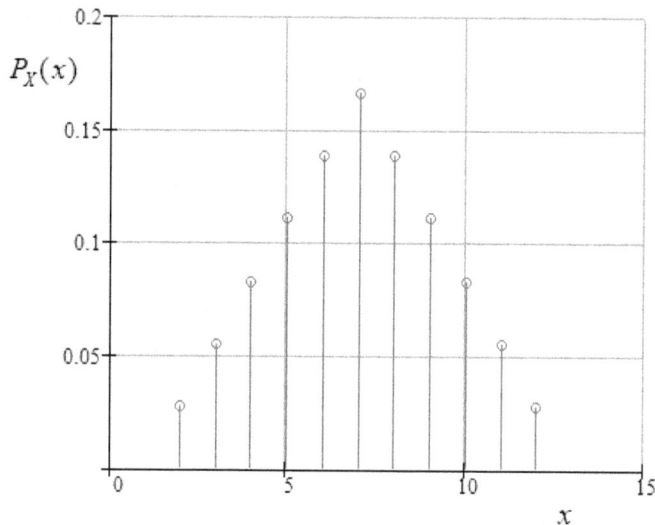

Figure A.2: Probabilities $P_X(x)$.

From Fig. A.2, the most likely value of x is 7 and that its difficult to throw a double-six or a double-one. Remember this the next time you play Ludo!

A.6   Cumulative Distribution Function (CDF)

The *cumulative distribution function* (CDF), denoted by $F_X(x)$, is the probability that the random variable X takes a value less than or equal to x. In the example of throwing two dice, the probability with which $x \leq 5$ is given by

$$F_X(5) = P_X(x \leq 5) = P_X(x = 2) + P_X(x = 3) + P_X(x = 4) + P_X(x = 5)$$
$$= \tfrac{1}{36} + \tfrac{2}{36} + \tfrac{3}{36} + \tfrac{4}{36} = \tfrac{10}{36}.$$ In passing, notice that the probability $P_X(x > 5)$ is simply given by $P_X(x > 5) = 1 - P_X(x \leq 5)$.

A table of all the possible values of $F_X(x)$ are as follows and the corresponding graph is shown in Fig. A.3.

x	2	3	4	5	6	7	8	9	10	11	12
$F_X(x)$	$\frac{1}{36}$	$\frac{3}{36}$	$\frac{6}{36}$	$\frac{10}{36}$	$\frac{15}{36}$	$\frac{21}{36}$	$\frac{26}{36}$	$\frac{30}{36}$	$\frac{33}{36}$	$\frac{35}{36}$	$\frac{36}{36}$

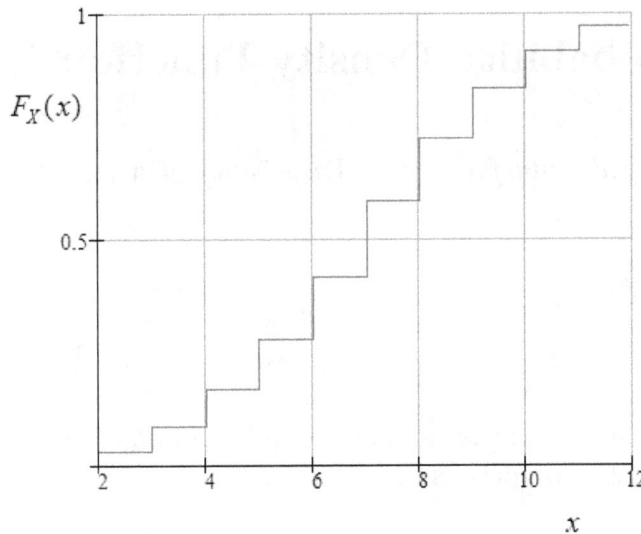

Figure A.3: Cumulative distribution function $F_X(x)$.

- ⭐ SIMULATION **DistFunction:** Implementation of the foregoing analysis. Note that $F_X(12) = 1$.

In general, the properties of $F_X(x)$ are

- $F_X(-\infty) = 0$

- $F_X(\infty) = 1$

- $0 \leq F_X(x) \leq 1$

- $F_X(x_1) \leq F_X(x_2)$ for $x_1 \leq x_2$

- $P(x_1 < X \leq x_2) = F_X(x_2) - F_X(x_1)$

A.7   Probability Density Function (PDF)

The *probability density function* (PDF) $f_X(x)$ of a random variable X is defined by

$$f_X(x) = \frac{dF_X(x)}{dx} \tag{A.5}$$

where X is a *continuous* random variable and not discrete as in the previous section. The properties of $f_X(x)$ are

- $f_X(x) \geq 0$

- The total area under the PDF curve is equal to 1 i.e. $\displaystyle\int_{-\infty}^{\infty} f_X(x)dx = 1$

- The probability that the random variable value is less than or equal to x_2 and greater than x_1 is given by $P(x_1 < X \leq x_2) = F_X(x_2) - F_X(x_1) = \int_{x_1}^{x_2} f_X(x)dx$ i.e. the area under the PDF curve between x_1 and x_2.

To understand the concept of a PDF, consider the *Gaussian* PDF given by

$$f_X(x) = \frac{1}{\sigma\sqrt{2\pi}} \exp\left[-\frac{(x-m)^2}{2\sigma^2}\right] \tag{A.6}$$

A plot of this function is shown in Fig. A.4, in which the mean $m = 4$ and the standard deviation $\sigma = 1$.

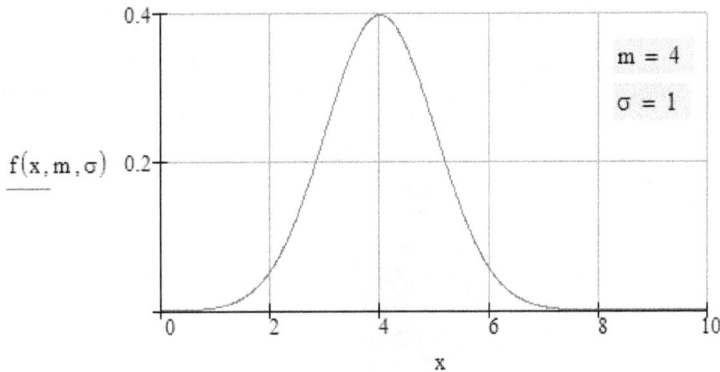

Figure A.4: Gaussian PDF.

- ⭐ SIMULATION **GaussianPDF:** Simulation corresponding to Fig. A.4. Experiment with different values of m and σ. Notice how σ controls the width of this bell-shaped curve.

Do not fall into the classical student-trap of thinking that the height of a PDF curve is a measure of probability! It is a probability *density* function, in which the area under the curve is a measure of probability. Two examples of the area under the curve between x_1 and x_2 are as shown in Figs. A.5 and A.6.

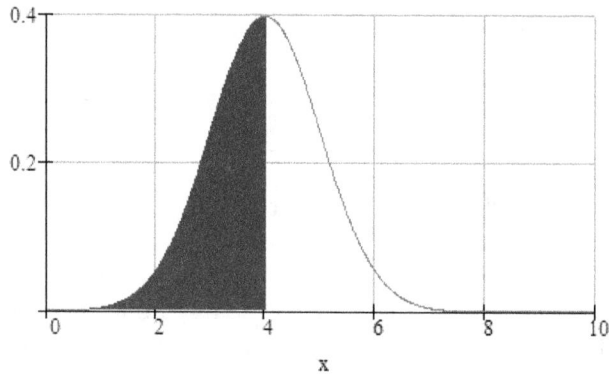

Figure A.5: For $x_1 = 0$ and $x_2 = 4$, $P(x_1 < X \le x_2) = 0.5$.

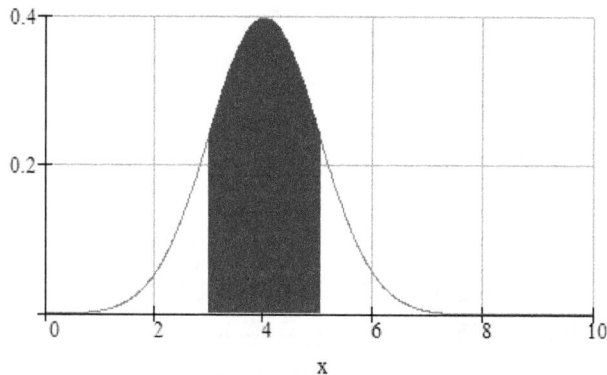

Figure A.6: For $x_1 = 3$ and $x_2 = 5$, $P(x_1 < X \le x_2) = 0.683$.

- ⭐ SIMULATION **GaussianPDF:** For further details on Figs. A.5 and A.6. Experiment with different values of x_1 and x_2 e.g. try $x_1 = -2$ and $x_2 = 10$ and also alter the mean m and standard deviation σ. Have some fun with probability!

An interesting feature is that the integral $\int\limits_{x_1}^{x_2} f_X(x)dx$ can only be solved numerically. A convenient solution is to express it in terms of the *Q-function* defined as

$$Q(u) = \int\limits_{u}^{\infty} \frac{1}{\sqrt{2\pi}} \exp\left(-\frac{z^2}{2}\right) dz \tag{A.7}$$

Notice that the expression $\frac{1}{\sqrt{2\pi}}\exp\left(-\frac{z^2}{2}\right)$ is simply $f_X(x) = \frac{1}{\sigma\sqrt{2\pi}}\exp\left[-\frac{(x-m)^2}{2\sigma^2}\right]$ with $m = 0$ and $\sigma = 1$ i.e. $Q(u)$ is simply the area to the right of the coordinate u, on a unit standard deviation, zero mean Gaussian PDF as shown in Fig. A.7.

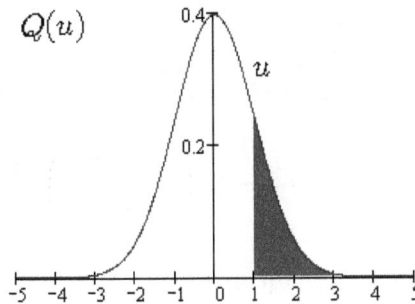

Figure A.7: Zero mean, unit standard deviation Gaussian PDF.

- Refer to Appendix B.1 for the tabulated values of $Q(u)$.

Given that $f_X(x) = \frac{dF_X(x)}{dx}$, where in this case $f_X(x) = \frac{1}{\sigma\sqrt{2\pi}}\exp\left[-\frac{(x-m)^2}{2\sigma^2}\right]$, the cumulative distribution function (CDF) is given by

$$F_X(x) = \int_{-\infty}^{x} f_X(x)dx = \int_{-\infty}^{x} \frac{1}{\sigma\sqrt{2\pi}}\exp\left[-\frac{(x-m)^2}{2\sigma^2}\right]dx \qquad \text{(A.8)}$$

Integrating by substitution, let $z = \frac{(x-m)}{\sigma}$. Then $\frac{dz}{dx} = \frac{1}{\sigma}$ and thus

$$F_X(x) = \int_{-\infty}^{\frac{(x-m)}{\sigma}} \frac{1}{\sqrt{2\pi}}\exp\left[-\frac{z^2}{2}\right]dz \qquad \text{(A.9)}$$

Now $Q\left(\frac{(x-m)}{\sigma}\right) = \int_{\frac{(x-m)}{\sigma}}^{\infty} \frac{1}{\sqrt{2\pi}}\exp\left[-\frac{z^2}{2}\right]dz$ is area under the curve from $\frac{(x-m)}{\sigma}$ to ∞. However, we require $\int_{-\infty}^{\frac{(x-m)}{\sigma}} \frac{1}{\sqrt{2\pi}}\exp\left[-\frac{z^2}{2}\right]dz$ which is the area under the curve from $-\infty$ to $\frac{(x-m)}{\sigma}$. Since the total area under the curve is 1,

$$F_X(x) = 1 - Q\left(\frac{(x-m)}{\sigma}\right) \qquad \text{(A.10)}$$

A plot of this function is shown in Fig. A.8.

- ⭐ SIMULATION **CDF:** Further details in which the PDF from $F_X(x)$ is determined and shown to be the same as $f_X(x) = \frac{1}{\sigma\sqrt{2\pi}}\exp\left[-\frac{(x-m)^2}{2\sigma^2}\right]$.

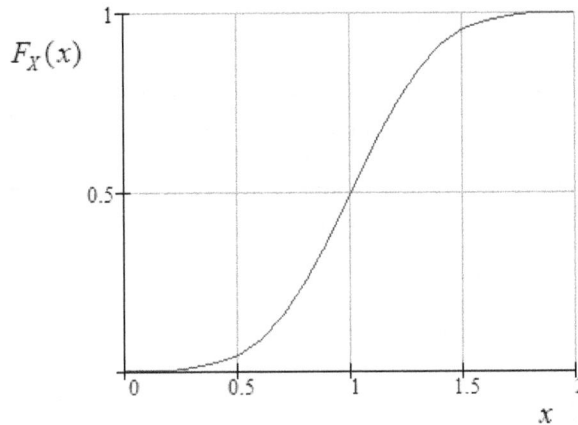

Figure A.8: The cumulative distribution function $F_X(x)$ for a Gaussian PDF.

A.8 Probability Histogram

The probability *histogram* of a random variable X is an estimate of its probability density function developed via simulation, by segmenting the range of X into intervals referred to as *bins*. For example, consider a Gaussian random number generator $GAUSS(m, \sigma)$ with its mean m set to 1 and standard deviation set to $\sigma = 0.3$. A plot of $x_n = GAUSS(1, 0.3)$ where $n = 0, 1, \cdots N - 1$ with $N = 10000$ is shown in Fig. A.9.

A plot of only the first 100 points is shown in Fig. A.10, where the range of x_n (along the vertical axis) is segmented into 8 bins of width 0.25. The number of points that lie within the range 1.25 and 1.5 (shown highlighted in Fig. A.10) divided by the total number of points generated is an estimate of the probability that the random number generator output will lie in the range 1.25 to 1.5, which from Fig. A.10, is $15/100 = 0.15$.

To develop a probability histogram, let

- x_{min} represent the minimum value of x_n

- x_{max} represent the maximum value of x_n

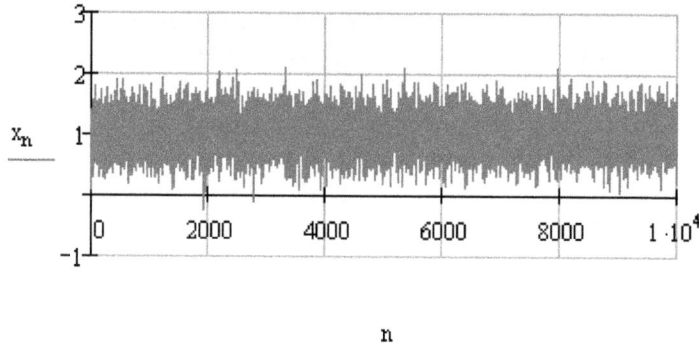

Figure A.9: Gaussian random number generator output.

- $x_{\text{lower}} = floor(x_{\min})$, where the $floor(z)$ function returns the greatest integer $\leq z$. For example, $floor(0.1) = 0$, $floor(1.1) = 1$, $floor(3.6) = 3$.

- $x_{\text{upper}} = ceil(x_{\max})$, where the $ceil(z)$ function returns the smallest integer $\geq z$. For example, $ceil(0.1) = 1$, $ceil(1.1) = 2$, $ceil(3.6) = 4$.

- N_{bin} represent the number of bins in the histogram

- Δ represent the width of a bin $\Delta = \frac{x_{\text{upper}} - x_{\text{lower}}}{N_{\text{bin}}}$

- index $j = 0, 1, \cdots, N_{\text{bin}}$

- horizontal coordinate of the histogram intervals $x_j^{\text{interval}} = x_{\text{lower}} + \Delta j$

- midpoint of the histogram intervals $x_j^{\text{mid}} = x_j^{\text{interval}} + \frac{\Delta}{2}$

- N_j^{mid} represent the number of times x_n falls within the bin corresponding to x_j^{mid} so that $\sum N_j^{\text{mid}} = N$

The probability P_j^{bin} with which x_n falls within a given bin is given by $P_j^{\text{bin}} = \frac{N_j^{\text{mid}}}{N}$ so that $\sum_{j=0}^{N_{bin}-1} P_j^{\text{bin}} = 1$. The average value $\bar{x}$ of x_n is given by

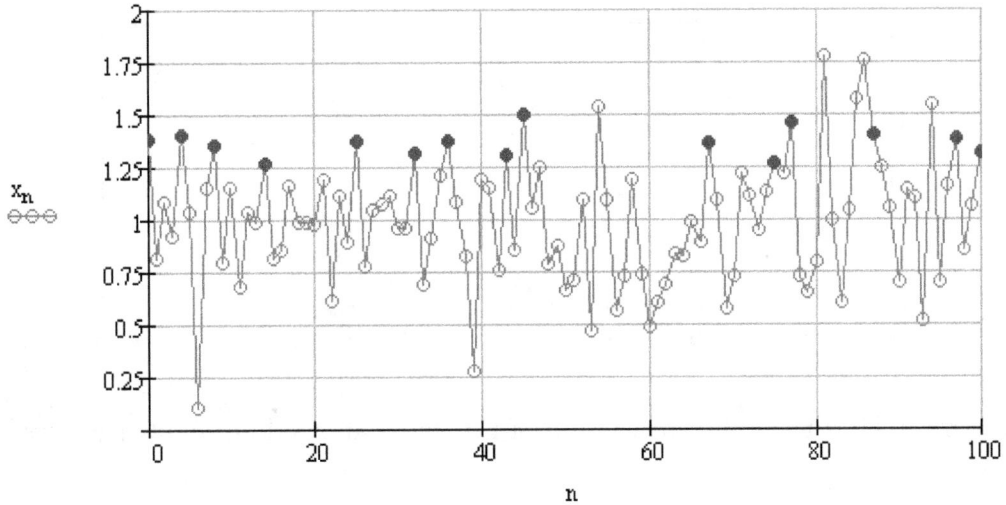

Figure A.10: Highlighted points within the range 1.0 to 1.5.

$$\overline{x} = \frac{1}{N} \sum_{n=0}^{N} x_n = \sum_{j=0}^{N_{bin}-1} P_j^{\text{bin}} \left(x_j^{\text{mid}} \right) \tag{A.11}$$

Figure A.11 shows the probability histogram of x_n for $N_{\text{bin}} = 100$, $\Delta = 0.1$ and $x_{\text{lower}} = -1$. The probabilities P_j^{bin} are shown as solid bars and overlaid on this histogram are the expected values given by

$$P\left(a < X \leq b\right) = \int_{a}^{b} f_X(x) dx \tag{A.12}$$

where $a = x_j^{\text{interval}}$, $b = x_{j+1}^{\text{interval}}$ and $f_X(x) = \frac{1}{\sigma\sqrt{2\pi}} \exp\left[-\frac{(x-m)^2}{2\sigma^2}\right]$ with $m = 1$ and $\sigma = 0.3$. The excellent agreement between the simulation and the theoretical results indicates that the Gaussian random number generator is accurate. There are more stringent tests that one could do to verify the properties of a random number generator.

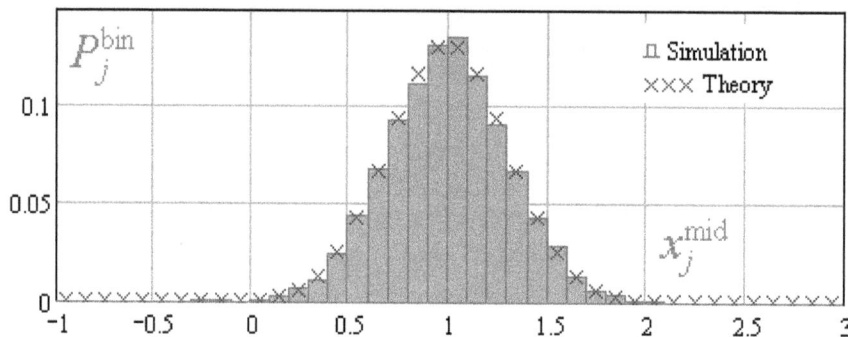

Figure A.11: Probability histogram of the gaussian random number generator.

- ⭐ SIMULATION **Histo:** Simulation corresponding to Fig. A.11.
  Verify for yourself that the total area under the histogram is equal to
  1. Experiment with the values for N_{bin} and N. What do you conclude?

A.9   Binomial Distribution

Consider a binary source that generates a binary digit zero with probability
p.   Then the probability of a binary digit one being generated is simply
$(1 - p)$. The probability that the source will output the sequence 01100 for
example is given by $p(1 - p)(1 - p)pp = p^3 (1 - p)^2$, because we require the
joint event of 01100.   In general, if within a N digit output sequence, there
are n zeros, then the number of binary digit ones is equal to $(N - n)$ and
the probability of this sequence is given by

$$p^n (1 - p)^{(N-n)} \tag{A.13}$$

The table below lists all the possible combinations of 3 binary zeros within
a 5 digit sequence. Of course, the probability of each specific sequence is the
same and is given by $p^3 (1 - p)^2$.

```
1  1  0  0  0
1  0  1  0  0
1  0  0  1  0
1  0  0  0  1
0  1  1  0  0
0  1  0  1  0
0  1  0  0  1
0  0  1  1  0
0  0  1  0  1
0  0  0  1  1
```

The probability $P(n = 3)$ that the source will generate 3 binary zeros within a 5 digit binary sequence in *any position* is given by

$$P(n = 3) = P(11000) + P(10100) + \cdots P(00011) = 10p^3 (1 - p)^2 \quad \text{(A.14)}$$

In general, the number of combinations of n zeros within N binary digits is given by $\begin{pmatrix} N \\ n \end{pmatrix} = \frac{N!}{n!(N-n)!}$, where $N!$ and $n!$ denote N–factorial and n–factorial, respectively. Thus, in general,

$$P(n) = \frac{N!}{n! \, (N - n)!} p^n (1 - p)^{(N-n)} \quad \text{(A.15)}$$

where the probability $P(n)$ of n events within N trials is said to have a *binomial distribution*. It is easily shown that the mean and standard deviation of n are given by Np and $\sqrt{Np\,(1 - p)}$, respectively. For example, the binomial distribution for $N = 5$, $n = 3$ and $p = 0.3$ is shown in Fig. A.12 e.g. $P(3) = \frac{5!}{3!(5-3)!} p^3 (1 - p)^{(5-3)} = \frac{5(4)(3)(2)(1)}{3(2)(1)(2)(1)} p^3 (1 - p)^2 = 10p^3 (1 - p)^2 = 0.132$. The mean and standard deviation of n are given by $Np = 1.5$ and $\sqrt{Np\,(1 - p)} = 1.025$, respectively.

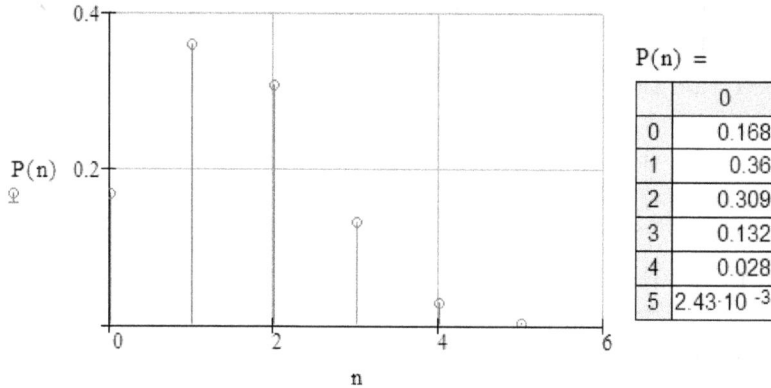

P(n) =	
	0
0	0.168
1	0.36
2	0.309
3	0.132
4	0.028
5	$2.43 \cdot 10^{-3}$

Figure A.12: Binomial distribution.

• ⭐ SIMULATION **Binomial:** Simulation corresponding to Fig. A.12. Increase the value of N to 50 and see what happens as p is increased from a small to a large value. Is the binomial distribution symmetrical about the mean for $p = 0.5$?

A.10   Poisson Distribution

⭐ SIMULATION **Binomial:** Set the value of N to 150, 160, 170, ... and at some point, the software will crash. Fortunately, the binomial distribution can be approximated by the *Poisson distribution* for $N \gg n$ and $p \ll 1$ and is given by

$$P(n) \approx e^{-\alpha} \left(\frac{\alpha^n}{n!} \right) \qquad (A.16)$$

where $\alpha = Np$. The mean and standard deviation of this distribution are given α and $\sqrt{\alpha}$, respectively. For example, the Poisson distribution for $N = 170$, $p = 0.03$ is shown in Fig. A.13 together with the corresponding binomial distribution.

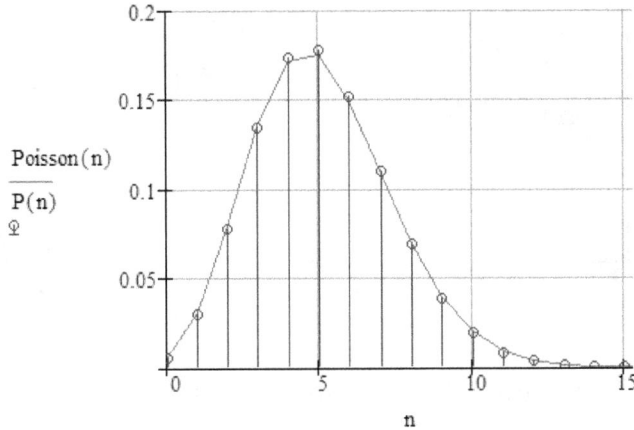

Figure A.13: Poisson distribution.

● ⭐ SIMULATION **Poisson:** Simulation corresponding to Fig. A.13. Try $N = 10000$.

The Poisson cumulative distribution function is given by

$$P(n \leq k) = \sum_{n=0}^{k} e^{-\alpha} \left(\frac{\alpha^n}{n!} \right) \tag{A.17}$$

For example, suppose a million binary digits are transmitted over a channel in which the probability of a binary digit error is 10^{-6}. Figure A.14 shows a plot of $P(n \leq k)$ from which we note that for example, the probability of no more than 4 binary digit errors in any position is 0.9963. Thus, the probability of more than 4 errors $P(n > 4) = 1 - P(n \leq 4) = 0.0037$.

● ⭐ SIMULATION **PoissonCDF:** Simulation code corresponding to Figure A.14.

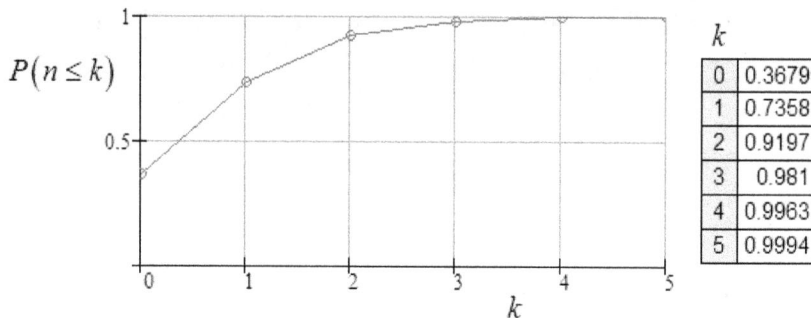

k	
0	0.3679
1	0.7358
2	0.9197
3	0.981
4	0.9963
5	0.9994

Figure A.14: Probability $P(n \leq k)$.

A.11   Expected Value

A.11.1   Discrete Random Variable

The *average* or *mean* value of the numbers 5, 8 and 2 is simply $\frac{(5+8+2)}{3} =$ 5. The inherent assumption made in this calculation is that each of these numbers is equally likely. In general, the mean value m of a *discrete random variable* X, or equivalently its *expected* value, or *expectation* denoted by $E[X]$, is given by

$$E[X] = m = \sum_{x=0}^{N} x P(x) \tag{A.18}$$

where

$$\sum_{x=0}^{N} P(x) = 1 \tag{A.19}$$

and

$$\sigma^2 = E\left[(X-m)^2\right] = \sum_{x=0}^{N} (x-m)^2 P(x) \tag{A.20}$$

If each value of x is equally likely, then $P(x) = \frac{1}{N}$, and $m = \sum_{x=0}^{N} x \frac{1}{N} =$

$\frac{1}{N} \sum_{x=0}^{N} x$ and $\sigma^2 = \frac{1}{N} \sum_{x=0}^{N} (x - m)^2$.

- ⭐ SIMULATION **ExpectedBinomial:** The mean m and standard deviation σ are determined via $E[X]$ and $E\left[(X - m)^2\right]$.

Example : Binary Random Variable

If $P(x = 0) = \alpha$ and $P(x = 1) = 1 - \alpha$, then $m = E[X] = \sum_{x=0}^{N} x P(x) =$

$0(\alpha) + 1(1 - \alpha) = (1 - \alpha)$ and $\sigma^2 = \sum_{x=0}^{N} (x - m)^2 P(x) = (0 - 1 + \alpha)^2 \alpha +$

$(1 - 1 + \alpha)^2 (1 - \alpha) = \alpha (1 - \alpha)$.

A.11.2 Continuous Random Variable

Similarly, the mean, expected value or expectation $E[X]$ of a continuous random variable X is given by

$$m = E[X] = \int_{-\infty}^{\infty} x f_X(x) dx \qquad (A.21)$$

where

$$\int_{-\infty}^{\infty} f_X(x) dx = 1 \qquad (A.22)$$

and the variance σ^2 of X is given by

$$\sigma^2 = E\left[(X - m)^2\right] = \int_{-\infty}^{\infty} (x - m)^2 f_X(x) dx \qquad \text{(A.23)}$$

- ⭐ SIMULATION **ExpectedGaussian:** The mean m and standard deviation σ are determined via $E[X]$ and $E\left[(X - m)^2\right]$.

In general, the *nth moment* of X is defined by

$$E[X^n] = \int_{-\infty}^{\infty} x^n f_X(x) dx \qquad \text{(A.24)}$$

and the *nth central moment* of X is defined by

$$E[(X - m)^n] = \int_{-\infty}^{\infty} (x - m)^n f_X(x) dx \qquad \text{(A.25)}$$

i.e. the mean is the first moment and the variance is the second central moment. The expectation operator $E[.]$ is linear so that for example,

$$\sigma^2 = E\left[(X - m)^2\right] = E\left[X^2 - 2mX + m^2\right]$$

$$= E\left[X^2\right] - 2mE[X] + E\left[m^2\right] = E\left[X^2\right] - 2m^2 + m^2$$

$$= E\left[X^2\right] - m^2 = E\left[X^2\right] - E[X]^2 \qquad \text{(A.26)}$$

For any constant a and a random variable X with variance σ^2, we find that

- $E[aX] = aE[X]$ with variance $(a\sigma)^2$

- $E[a] = a$ with variance 0

- $E[X + a] = E[X] + a$ with variance σ^2

Finally, if a random variable Y is a function of another random variable X, denoted by $Y = g(X)$, then

$$E[Y] = \int_{-\infty}^{\infty} y f_Y(y) dy = \int_{-\infty}^{\infty} g(x) f_X(x) dx \qquad \text{(A.27)}$$

Example : Linear Function of a Gaussian Random Variable

⭐ SIMULATION **GaussianLinearFunct:** The important feature that a linear function of a Gaussian random variable is itself a Gaussian random variable is illustrated. Specifically, if a random variable $Y = aX + b$, where a and b are constants and X is a Gaussian random variable with a mean m and standard deviation σ, then Y is a Gaussian random variable with mean $(am + b)$ and standard deviation $(a\sigma)$.

A.12 Joint Distribution

For two random variables X and Y on the same sample space S, the *joint CDF* $F_{X,Y}(x, y)$ is defined by

$$F_{X,Y}(x, y) = P(X \leq x, Y \leq y) \qquad \text{(A.28)}$$

and the *joint PDF* $f_{X,Y}(x, y)$ is defined by

$$f_{X,Y}(x, y) = \frac{\partial^2 F_{X,Y}(x, y)}{\partial x \partial y} \qquad \text{(A.29)}$$

The properties of $F_{X,Y}(x,y)$ and $f_{X,Y}(x,y)$ are

- $F_X(x) = F_{X,Y}(x,\infty)$ and $F_Y(y) = F_{X,Y}(\infty,y)$

- $f_X(x) = \displaystyle\int_{-\infty}^{\infty} f_{X,Y}(x,y)dy$ and $f_Y(y) = \displaystyle\int_{-\infty}^{\infty} f_{X,Y}(x,y)dx$

- $f_{X,Y}(x,y) \geq 0$

- $\displaystyle\int_{-\infty}^{\infty}\int_{-\infty}^{\infty} f_{X,Y}(x,y)dxdy = 1$

- $F_{X,Y}(x,y) = \displaystyle\int_{-\infty}^{x}\int_{-\infty}^{y} f_{X,Y}(u,v)dudv$ and $F_{X,Y}(\infty,\infty) = 1$

- Random variables X and Y are *statistically independent* if

$$F_{X,Y}(x,y) = F_X(x)F_Y(y) \tag{A.30}$$

or

$$f_{X,Y}(x,y) = f_X(x)f_Y(y) \tag{A.31}$$

The *conditional probability density function* $f_{Y|X}(y|x)$ is defined by

$$f_{Y|X}(y|x) = \begin{cases} \frac{f_{X,Y}(x,y)}{f_X(x)} & if \quad f_Y(y) \neq 0 \\ 0 & otherwise \end{cases} \tag{A.32}$$

in which if X and Y are statistically independent, then

$$f_{Y|X}(y|x) = \frac{f_X(x)f_Y(y)}{f_X(x)} = f_Y(y) \tag{A.33}$$

The expected value of $g(X, Y)$ is given by

$$E\left[g(X,Y)\right] = \int\limits_{-\infty}^{\infty}\int\limits_{-\infty}^{\infty} g(x,y)f_{X,Y}(x,y)dxdy \qquad (A.34)$$

The *joint moment* of X and Y is defined by

$$E\left[X^n Y^k\right] = \int\limits_{-\infty}^{\infty}\int\limits_{-\infty}^{\infty} x^n y^k f_{X,Y}(x,y)dxdy \qquad (A.35)$$

where n and k are any positive integers.   For $n = k = 1$, $E\left[XY\right]$ is referred to as the *correlation* of X and Y and is denoted by

$$R_{XY} = E\left[XY\right] \qquad (A.36)$$

The *covariance* of X and Y, denoted by $C_{X,Y}$, is defined by

$$C_{X,Y} = E\left[XY\right] - E\left[X\right]E\left[Y\right] \qquad (A.37)$$

where the random variables X and Y are said to be *uncorrelated* if their convariance is zero i.e. $C_{X,Y} = 0$ so that

$$E\left[XY\right] = E\left[X\right]E\left[Y\right] \qquad (A.38)$$

The *correlation coefficient* of X and Y, denoted by $\rho_{X,Y}$, is defined by

$$\rho_{X,Y} = \frac{C_{X,Y}}{\sigma_X \sigma_Y} \qquad (A.39)$$

where σ_X and σ_Y are the standard deviations of X and Y, respectively. Note that if X and Y are uncorrelated, then $C_{X,Y} = 0$ and $\rho_{X,Y} = 0$ but this does **not** necessarily imply that X and Y are independent! However, if X and Y are independent, then $C_{X,Y} = 0$. Finally, two random variables are called *orthogonal* if

$$R_{XY} = E[XY] = 0 \tag{A.40}$$

Note that if X and Y are uncorrelated with zero mean, then X and Y cannot be orthogonal.

A.12.1   Cauchy-Schwartz Inequality

Given that $E\left[(X - aY)^2\right] \geq 0$ for any value of a, we can expand this as $E[X^2] - 2aE[XY] + a^2E[Y^2] \geq 0$. To minimize the left-hand of this inequality, we set $a = \frac{E[XY]}{E[Y^2]}$, so that $E[X^2] - \frac{(E[XY])^2}{E[Y^2]} \geq 0$ or equivalently, $(E[XY])^2 \leq E[X^2]E[Y^2]$ which is referred to as the *Cauchy-Schwarz inequality*. Using this inequality, it can be shown that $|\rho_{X,Y}| \leq 1$.

Appendix B

Q Function Table

B.1   Q-Function

The $Q(.)$ function is defined by

$$Q(u) = \int_{u}^{\infty} \frac{1}{\sqrt{2\pi}} \exp\left(-\frac{z^2}{2}\right) dz \qquad \text{(B.1)}$$

For example, $Q(0.52) = 0.3015$, $Q(1.73) = 0.0418$, $Q(2.5) = 6.2097$ x 10^{-3}, $Q(3.19) = 7.1136$ x 10^{-4}, etc.

- ⭐ SIMULATION **QFunctionTable:** How to determine the Q-function table.

The most beautiful thing we can experience is the mysterious. It is the source of all true art and all science. He to whom this emotion is a stranger, who can no longer pause to wonder and stand rapt in awe, is as good as dead: his eyes are closed.

— *Albert Einstein*

Appendix C

References

C.1 References

- Bahl L. R., Cocke J., Jelinek F. and Raviv J. "Optimal decoding of linear codes for minimizing symbol error rate," *IEEE Trans. Info. Theory*, vol. IT-20, pp. 248-287, Mar. 1974.

- Berrou C., Glavieux A. and Thitmajshima P., "Near Shannon Limit Error-Correction Coding and Decoding: Turbo Codes", International Conference on Communications, Geneva, Switzerland, pp. 1064-1090, May 1993.

- Berrou C. and Glavieux A., "Near Optimum Error Correcting Coding and Decoding: Turbo-codes," *IEEE Trans. Commun.*, vol. 44, no. 10, pp. 1261-1271, Oct. 1996.

- Bingham J., "Multicarrier modulation for data transmission: An idea whose time has come", *IEEE Commun. Mag.*, pp. 5-14, May 1990.

- Brown J.L. Jr., "First order sampling of bandpass signals—A new approach.", *IEEE Trans. Information Theory*, IT-26(5), pp. 613–615, 1980.

- Clark G. C. Jr. and Cain J. B., *Error-Correction Coding for Digital Communications*, Plenum Press, 1981.

- Collins O. M., "The Subtleties and Intricacies of Building a Constraint Length 15 Convolutional Decoder", *IEEE Trans. Commun.*, COM 40, pp.1810-1819, 1992.

- Couch L. W., *Digital and Analog Communication Systems*, Sixth Edition, Prentice Hall, 2001.

- Gallager, R. G., "Low density parity check codes", *IRE Trans. Information Theory*, no. 8, pp. 21-28, 1962.

- Gitlin R. D., Hayes J. F. and Weinstein S. B., *Data Communications Principles*, Plenum Press, 1992.

- Haykin S., *Communication Systems*, Fourth Edition, John Wiley, 2001.

- Heller J. A. and Jacobs I. W., "Viterbi Decoding for Satellite and Space Communication", *IEEE Trans. Commun. Technol.*, vol. COM 19, no. 5, pp. 835-848, Oct. 1971.

- Jacobs I. M., "Practical Applications of Coding", *IEEE Trans. Information Theory*, vol. IT 20, pp. 305-310, May 1974.

- Lathi B. P., *Modern Digital and Analog Communication Systems*, Third Edition, Oxford University Press, 1998.

- Lin S. and Costello D. J., Jr., *Error Control Coding: Fundamentals and Applications*, Prentice Hall, 1983.

- Lindsey W. C. and Simon M. K., Telecommunication Systems Engineering, Prentice Hall, 1973.

- Odenwalder J. P., *Error Control Coding Handbook*, Linkabit Corporation, San Diego, Calif., July 1976.

- Odenwalder J. P., "Error Control", *Data Communications, Networks and Systems*, Indianapolis, Howard W. Sams, 1985.

- Pawula, R. F., "A New Formula for MDPSK Symbol Error Probability", *IEEE Commun. Letters*, vol. 2, pp. 271-272, Oct. 1998.

- Peterson W. W. and Weldon, E. J., *Error Correcting Codes*, Second Edition, Cambridge, Mass., MIT Press, 1972.

- Proakis J. G. and Manolakis D. G., *Digital Signal Processing*, Third Edition, Prentice Hall, 1996.

- Qureshi S. U. H., "Adaptive Equalization", *Proc. IEEE*, vol. 73, no. 9, pp. 1340-387, Sept. 1985.

- Richardson T. J., Shokrollahi A. and Urbanke R., "Design of capacity-approaching low-density parity-check codes," IEEE Trans. Inform. Theory, vol. 47, Feb. 2001, pp. 619-637.

- Ryan W. E., "A turbo code tutorial", New Mexico State University, Box 30001 Dept. 3-O, Las Cruces, MN 88003, 1997.

  http://www.ece.arizona.edu/~ryan

- Sklar B., *Digital Communications : Fundamentals and Applications*, Second Edition, Prentice Hall, 2001.

- Sodha J. & Als A., "Shape nature of error-control codes", Signal Processing Journal, Elsevier, vol. 83, pp. 1457-1465, July 2003.

- Ungerboeck G., "Channel Coding with Multilevel/Phase Signals", *IEEE Trans. Information Theory*, vol. IT 28, no. 1, pp. 55-67, Oct. 1971.

- Ungerboeck G., "Trellis-Coded Modulation with Redundant Signal Sets", Parts 1 and 2, *IEEE Communications Magazine*, vol. 25, no. 2, pp. 5-21, Feb. 1987.

- Viterbi A., "Convolutional Codes and Their Performance in Communication Systems", *IEEE Trans. Commun. Technol.*, COM 19, no. 5, pp. 751-772, Oct. 1971.

- Van de Vegte J., *Fundamentals of Digital Signal Processing*, Prentice Hall, 2001.

- Ziemer R. E. and Peterson R. L., *Introduction to Digital Communication*, Second Edition, Prentice Hall, 2001.

My Notes

Appendix D

Mathematical Tables

D.1 Identities

Trigonometric

$\tan \theta = \frac{\sin \theta}{\cos \theta}$

$\cot \theta = \frac{1}{\tan \theta}$

$\sec \theta = \frac{1}{\cos \theta}$

$\operatorname{cosec} \theta = \frac{1}{\sin \theta}$

$\sin(-\theta) = -\sin \theta$

$\cos(-\theta) = \cos \theta$

$\sin(A \pm B) = \sin A \cos B \pm \cos A \sin B$

$\cos(A \pm B) = \cos A \cos B \mp \sin A \sin B$

$\tan(A \pm B) = \frac{\tan A \pm \tan B}{1 \mp \tan A \tan B}$

$\cos A + \cos B = 2 \cos \left(\frac{A+B}{2}\right) \cos \left(\frac{A-B}{2}\right)$

$\cos A - \cos B = -2 \sin \left(\frac{A+B}{2}\right) \sin \left(\frac{A-B}{2}\right)$

$\cos \left(\theta \pm \frac{\pi}{2}\right) = \mp \sin \theta$

$\sin \left(\theta \pm \frac{\pi}{2}\right) = \pm \cos \theta$

$\sin \left(90^{\circ} - \theta\right) = \cos \theta$

$\cos \left(90^{\circ} - \theta\right) = \sin \theta$

$\cos^2 \theta = \frac{1}{2} \left[1 + \cos \left(2\theta\right)\right]$

$\sin^2 \theta = \frac{1}{2} \left[1 - \cos \left(2\theta\right)\right]$

$\cos^3 \theta = \frac{1}{4} \left[3 \cos \theta + \cos \left(3\theta\right)\right]$

$\sin^3 \theta = \frac{1}{4} \left[3 \sin \theta - \sin \left(3\theta\right)\right]$

$\sin^2 \theta + \cos^2 \theta = 1$

$1 + \cot^2 \theta = \operatorname{cosec}^2 \theta$

$\cos^2 \theta - \sin^2 \theta = \cos(2\theta)$

$2\sin\theta\cos\theta = \sin(2\theta)$

$a\cos\theta + b\sin\theta = \sqrt{(a^2+b^2)}\cos(\theta+\alpha)$, where $\alpha = \tan^{-1}\left(\frac{-b}{a}\right)$ and $a \geq 0$

$a\cos\theta + b\sin\theta = -\sqrt{(a^2+b^2)}\cos(\theta+\alpha)$, where $\alpha = \tan^{-1}\left(\frac{-b}{a}\right)$ and $a < 0$

$\sin A \sin B = \frac{1}{2}\left[\cos(A-B) - \cos(A+B)\right]$

$\cos A \cos B = \frac{1}{2}\left[\cos(A-B) + \cos(A+B)\right]$

$\sin A \cos B = \frac{1}{2}\left[\sin(A-B) + \sin(A+B)\right]$

Indicies

$x^0 = 1$

$x^p x^q = x^{p+q}$

$\frac{x^p}{x^q} = x^{p-q}$

$(x^p)^q = x^{pq}$

Logs

$\ln x \equiv \log_e x$

$\log(x^p) = p\log x$

$\log_2 x = \frac{\log_{10}(x)}{\log_{10}(2)}$

$\log A + \log B = \log(AB)$

$\log A - \log B = \log\left(\frac{A}{B}\right)$

Complex

$j = \sqrt{-1}$

$e^{\pm j\theta} = \cos\theta \pm j\sin\theta$

$e^{\pm j\frac{\pi}{2}} = \pm j$

$\cos\theta = \frac{1}{2}\left[e^{j\theta} + e^{-j\theta}\right]$

$\sin\theta = \frac{1}{2j}\left[e^{j\theta} - e^{-j\theta}\right]$

$a + jb = re^{j\theta}$, where $r = \sqrt{(a^2+b^2)}$ and $\theta = \tan^{-1}\left(\frac{b}{a}\right)$

$r\left(e^{j\theta}\right)^n = r^n e^{jn\theta}$

$\left(r_1 e^{j\theta_1}\right)\left(r_2 e^{j\theta_2}\right) = r_1 r_2 e^{j(\theta_1+\theta_2)}$

D.2   Series Expansions

Exponential

$e^x = 1 + x + \frac{x^2}{2!} + \frac{x^3}{3!} + \cdots$

Logarithmic

$$\log_e (1 + x) = x - \frac{x^2}{2} + \frac{x^3}{3} - \cdots \text{ where } (|x| < 1)$$

Binomial

$$(1 + x)^n = 1 + nx + \frac{n(n-1)}{2!} x^2 + \cdots \quad \text{where } |nx| < 1$$

Taylor

$$f(x) = f(a) + (x - a) \left. \frac{df(x)}{dx} \right|_{x=a} + (x - a)^2 \frac{1}{2!} \left. \frac{d^2 f(x)}{d^2 x} \right|_{x=a} + \cdots + (x - a)^n \frac{1}{n!} \left. \frac{d^n f(x)}{d^n x} \right|_{x=a}$$

- ⭐ **SIMULATION Series:** Experiment with the above series expansions and try various functions for the Taylor series.

D.3   Derivatives and Integrals

$$\frac{d(au)}{dx} = a \frac{d(u)}{dx}$$

$$\frac{d(u+v)}{dx} = \frac{d(u)}{dx} + \frac{d(v)}{dx}$$

$$\frac{d(uv)}{dx} = u \frac{d(v)}{dx} + v \frac{d(u)}{dx}$$

$$\frac{d(x^n)}{dx} = nx^{n-1}$$

$$\frac{d(e^u)}{dx} = e^u \frac{d(u)}{dx}$$

$$\frac{d(e^{ax})}{dx} = ae^{ax}$$

$$\frac{d(\sin u)}{dx} = \cos u \frac{d(u)}{dx}$$

$$\frac{d(\cos u)}{dx} = -\sin u \frac{d(u)}{dx}$$

$$\frac{d(\sin ax)}{dx} = a \cos (ax)$$

$$\frac{d(\cos ax)}{dx} = -a \sin (ax)$$

$$\frac{d(\tan \theta)}{dx} = \sec^2 \theta$$

$$\frac{d(\cot \theta)}{dx} = -\operatorname{cosec}^2 \theta$$

$$\frac{d(\sec \theta)}{dx} = \tan \theta \sec \theta$$

$$\frac{d(\operatorname{cosec} \theta)}{dx} = -\cot \theta \operatorname{cosec} \theta$$

$$\frac{d(\ln ax)}{dx} = \frac{d(\ln x)}{dx} = \frac{1}{x}$$

Indefinite Integrals

$$\int au\ dx = a \int u\ dx$$

$$\int (u + v)\ dx = \int u\ dx + \int v\ dx$$

$$\int x^p\ dx = \frac{x^{p+1}}{p+1} \quad (p \neq -1)$$

$$\int \frac{1}{x}\ dx = \ln |x|$$

$$\int u\ dv = uv - \int v\ du$$

$$\int u\ \frac{dv}{dx} dx = uv - \int v\ \frac{du}{dx} dx$$

$$\int \sin(ax)dx = \frac{-1}{a} \cos(ax)$$

$$\int \cos(ax)dx = \frac{1}{a} \sin(ax)$$

$$\int \sin^2(ax)dx = \frac{x}{2} - \frac{\sin(2ax)}{4a}$$

$$\int \cos^2(ax)dx = \frac{x}{2} + \frac{\sin(2ax)}{4a}$$

$$\int x \sin(ax)dx = \frac{\sin(ax) - ax \cos(ax)}{a^2}$$

$$\int x \cos(ax)dx = \frac{\cos(ax) + ax \sin(ax)}{a^2}$$

$$\int x^2 \sin(ax)dx = \frac{2ax \sin(ax) + 2 \cos(ax) - (ax)^2 \cos(ax)}{a^3}$$

$$\int x^2 \cos(ax)dx = \frac{2ax \cos(ax) - 2 \sin(ax) + (ax)^2 \sin(ax)}{a^3}$$

$$\int e^{ax}dx = \frac{e^{ax}}{a}$$

$$\int xe^{ax}dx = \frac{e^{ax}}{a^2}(ax - 1)$$

$$\int x^n e^{ax}dx = \frac{x^n e^{ax}}{a} - \frac{n}{a} \int x^{n-1}e^{ax}dx$$

$$\int xe^{ax^2}dx = \frac{1}{2a}e^{ax^2}$$

$$\int e^{ax} \sin(bx)\, dx = \frac{e^{ax}}{a^2+b^2}\left[a\sin(bx) - b\cos(bx)\right]$$

$$\int e^{ax} \cos(bx)\, dx = \frac{e^{ax}}{a^2+b^2}\left[a\cos(bx) + b\sin(bx)\right]$$

$$\int \frac{1}{(a^2+b^2x^2)}\, dx = \frac{1}{ab}\tan^{-1}\left(\frac{bx}{a}\right)$$

$$\int \frac{x^2}{(a^2+b^2x^2)}\, dx = \frac{x}{b^2} - \frac{a}{b^3}\tan^{-1}\left(\frac{bx}{a}\right)$$

$$\int \frac{x}{(a^2+x^2)}\, dx = \frac{1}{2}\ln\left(x^2 + a^2\right)$$

Definite Integrals

$$\int_0^\infty \frac{x\sin(ax)}{b^2+x^2}\, dx = \frac{\pi}{2}e^{(-ab)} \quad \text{where } a>0 \text{ and } b>0$$

$$\int_0^\infty \frac{\cos(ax)}{b^2+x^2}\, dx = \frac{\pi}{2b}e^{(-ab)} \quad \text{where } a>0 \text{ and } b>0$$

$$\int_0^\infty \operatorname{sinc} dx = \int_0^\infty \operatorname{sinc}^2 dx = \frac{1}{2}$$

$$\int_0^\infty e^{-ax^2}\, dx = \frac{1}{2}\sqrt{\frac{\pi}{a}} \text{ for } a>0$$

$$\int_0^\infty x^2 e^{-ax^2}\, dx = \frac{1}{4a}\sqrt{\frac{\pi}{a}} \text{ for } a>0$$

$$\int_0^\infty x^n e^{-ax}\, dx = \frac{n!}{a^{n+1}}$$

- ⭐ SIMULATION **Integrals:** Experiment with examples.

www.ingramcontent.com/pod-product-compliance
Lightning Source LLC
Chambersburg PA
CBHW061925190326
41458CB00009B/2657

* 9 7 8 0 9 9 2 8 5 1 0 0 2 *